Digital Rock Scour

This book examines the digitalization of rock scour engineering at dams and hydraulic structures. It outlines the current digitalization (technologies, applications, issues) in rock engineering, as well as the digital evolution that has strongly characterized the development of computational methods in state-of-the-art rock scour over recent years. The challenges of rock scour digitalization are also discussed, such as parametric standardization, real-time data acquisition, data analysis and interpretation, quantitative rock mass indices, new ways of thinking and digital twin implementation. Further, it presents the major components and characteristics that are needed to develop an environment that implements rock scour digitally into dam safety procedures and dam risk analyses, such as IT platforms, database availability, topology, physics, computational methods, phase coupling, accessibility, portability, reliability, real-time and ahead-of-time implementations and more.

Features:

- Provides an overview of semi-empirical and physics-based computational methods that have been developed by the engineering community over the last 20 years, which can easily be implemented digitally into cloud-based platforms.
- Offers examples of the next-generation computational environment, combining both real-time computational power and an up-to-date scour database allowing new parametric refinements.
- Includes several case studies of real-life rock scour.
- Presents the latest *Digital Twin* developments, which are novel and new to dam operations.

Digital Rock Scour
Cloud-Based Modelling and Engineering

Erik F. R. Bollaert

CRC Press is an imprint of the
Taylor & Francis Group, an **informa** business

Designed cover image: Kariba Dam plunge pool (Zambia-Zimbabwe); Courtesy of M. Joël Monin, Razel-BEC, France.

First edition published 2024
by CRC Press
2385 NW Executive Center Drive, Suite 320, Boca Raton FL 33431

and by CRC Press
4 Park Square, Milton Park, Abingdon, Oxon, OX14 4RN

CRC Press is an imprint of Taylor & Francis Group, LLC

ISBN: 978-1-032-33399-1 (hbk)
ISBN: 978-1-032-33421-9 (pbk)
ISBN: 978-1-003-31961-0 (ebk)

DOI: 10.1201/9781003319610

Typeset in Times
by codeMantra

This book is dedicated to my parents, who allowed and encouraged me to perform engineering studies, and to my wife Marianne and my two children Loïc and Naya, for their constant and perennial support during my career and especially during the years of writing of this work.

Contents

Foreword

The development of methodologies for the study of all types of rock scour in hydro projects has as its goal the essential guidance and answers that the design engineer requires – to provide procedures, somewhat in the words of Hans Albert Einstein (the younger) in one of his monumental works on sedimentation, "for many, if not all the scour situations in water resources projects". The procedures developed and applied widely by Dr Erik Bollaert comprise one of the many. Notably, however, as presented in this comprehensive work, *Digital Rock Scour: Cloud-based Modelling and Engineering*, there are developments which represent a landmark venture into answers to meet the design challenge.

The behaviour of water has occupied people for millennia with its intrigue, wonder and importance. It has engrossed many with its beauty, immersed many in its study, has been a source of refreshment, demanded intensive investigation of its behaviour and is essential to our existence – able to be managed, utilized and distributed for the good of society. It carries seen and unseen the life that thrives within. It demands appreciation and respect as a life source and as a power source.

Many of the ways that hydraulicians go about their business normally have as their focus the development of procedures that produce the results which are the foundation of the design process for water-related engineering. Einstein observed in a symposium on sedimentation in 1971 that many of the procedures and techniques in its study are "very theoretical, some born entirely in the laboratory, some of them come directly out of the field and some of them may even be introduced as design ideas." Recognizing that it is largely impossible to find complete solutions in theory and in the field alone, he went on, "if we want to design anything we must apply the knowledge from all" of the available sources.

The hydraulic impacts on the natural or design environment bring into play not only the well-established knowledge of fluid mechanics of water, air and air-water behaviour but also the need to know the properties, characteristics and extent of the mediums in which the water flows. The designer needs to know the mutual interaction – the effect of one upon the other. Development of the needed methodologies has engaged the research, observations and skills of the many over many years – and it is an ever-developing pursuit of those who provide what the designer needs to know and have. In 1950 Einstein (again, the son) published what is well known among the many investigators and practitioners in sedimentation engineering – his "bedload function for sediment transportation". His was an attempt to provide a tool which would be "sufficiently general to apply to a large number" of the sediment problems in alluvial channels – "a method which may be used to determine the bed-load function for many but not all types of stream channels". Has that work been superseded? In its entirety and relevance? I think not. As his work from upwards of 70 years ago still endures it is why I have chosen to quote a few of his "gems" of wisdom.

In this book, for which I am privileged to offer this foreword, Dr Bollaert has provided an in-depth discussion of the interaction of water and rock and the characteristics and problems associated with scour. But it is more than that. The book presents the latest rock scour computational methods and the ways they can be digitalized. Most of these methods are used by procedures that the author has studied and worked in for a good many years; the users are invited to subscribe to the platform he presents for their future computations, among others of a similar ilk from other sources. The book presents, what is believed by this writer, a significant step towards digital transformation of rock scour engineering by the creation and presentation of rock scour digital platforms and digital twins.

This book will be of interest to all engineers, companies and institutions that deal with rock scour issues on their hydraulic assets. This has application to dams and spillways, but it is much wider than that. It does, for example, deal with channel scour and bridge scour, and aims at setting the digital scene for rock scour engineering as a stand-alone engineering field, with digitalization of computational methods and their parameters, such that large-scale feedback and experiences may be acquired worldwide. This feedback will in turn allow calibration and standardization of rock scour parameters and data-driven methods.

As one who has worked with and alongside Erik Bollaert on several challenging projects, where scour was a key ingredient of the problem matrix, I am pleased to commend this work to those who have seen the effects, have worked and continue to work where those sometimes-violent characteristics of water flow are faced.

Eric J. Lesleighter
Sydney, Australia

Preface

Many publications and textbooks can be found on the topic of scour and erosion generated by turbulent flows occurring at hydraulic structures. The topic has been of interest to researchers and practising engineers for many years, and many approaches and design methods exist, based mainly on the shear stresses imposed at the water-solid interface. The most recent developments focus on detailed numerical modelling of the interaction between the fluid and the solid (i.e. from the block-sized rock units down to granular soil).

Accordingly, engineers feel *a priori* confident with the available design tools. Nevertheless, when exploring the basic equations and parametric settings of these numerical methods, things are not that simple! Appropriate parametric settings vary from model to model, and no standardization of parameters or models currently exists. It's all up to the engineer to judge; but at least he has some tools to apply to his particular situation.

When embarking on the study of scour and erosion of rock at hydraulic structures in the 1990s, it became clear to the author, even if solutions seemed possible, that it is even more complex than soil engineering, and that available numerical tools to handle the problem were simply non-existent. Moreover, rock scour was governed at that time by empiricism or stream power-based semi-empirical tools. As such, when delving into the study of the importance of fluctuating turbulent pressures on rock scour, there was always the risk that things might become too complex and beyond reasonable and practical solution.

Nevertheless, several early investigations existed already in this field, showing the relevance of the turbulent behaviour and characteristics to the scour of rock; and studying two-phase transient pressures in rock joints was revealed to be a necessary and challenging key to understanding the topic. This realization became a catalyst to the development of more physics-based rock scour prediction methods in the early 2000s.

It has been the author's experience to continually enhance and further develop these methods for more than 20 years, thanks to the confidence and engineering curiosity of many dam owners and practising engineers, for whom I am most grateful. The enhancements comprised the progressive numerical formulation and calibration of these methods, and the regular publication of outcomes of the related scour modelling. However, the engineering community still did not embrace what has existed for instance in soil engineering for a long time – fully customizable models applicable for detailed numerical simulations of rock scour.

To cope with this inadequate status of the rock scour situation and living in a world becoming more digital every second, my endeavours have been to transfer my knowledge and experiences in rock scour so that practitioners and professionals would benefit by means of the present advancements on digital rock scour.

Embracing the worldwide recognition of rock scour computational techniques and methods, the present publication is state-of-the-art – the way those techniques can be digitalized. Furthermore, herein the challenge of digitalization of rock scour engineering is discussed, together with the necessary digital elements needed for a complete digital transformation, such as digital platforms and digital twins.

Paralleling the present work, for the first time a digital platform for rock scour has been generated. Users may subscribe to perform computations in a dedicated and secured space or consult a shared digital database of historical case studies.

Whether the engineer is facing plunge pool scour, channel scour, bridge pier scour or dam abutment scour, this book describes and advances the actual digital application that is needed.

This work aims at setting the digital scene for rock scour engineering as a stand-alone engineering field. As such, readers so engaged are encouraged to participate in the current process of digitalization of rock scour engineering, whether it is to perform numerical simulations, or to generate feedback and new experiences to be shared on the platform.

The genuine intent for the application of the material in this publication is that you the reader and practitioner will derive as much pleasure and satisfaction through your journey as I have experienced over the years in the development of the procedures, description of the occurrences and the applications. Those are applications which can achieve an understanding of the behaviour and the way scour can be quantified to meet the requirements of the design engineer.

Dr Erik Bollaert
Lausanne, Switzerland

Acknowledgements

The author expresses his gratitude to the following professionals who kindly responded to his request to use some of their figures and/or data:

Eric Lesleighter, Lesleighter Consulting, Australia
Frédéric Laugier, EDF-CIH, Le Bourget, France
Matthias Regli, Project Director AFRY Switzerland Ltd.
Guiseppe Pittalis, Studio Ing. G. Pietrangeli Srl, Italy
Nuno Verdelho Trindade, IST, Universidade de Lisboa, Portugal
Biswajit Dasgupta, Southwest Research Institute, USA
Davide Elmo, Prof. UBC Applied Science, Canada
Stéphane Friedli, Responsable exploitation et maintenance, SFMCP, Suisse
Oscar Jiménez, Consultant Senior, DLZ-Carbon Ltd., USA
Mike George, Senior Geological Engineer, BGC Engineering, Canada
Tiziano Crameri, Leiter Planung Bau, Repower AG, Switzerland
Wayne Barnett, Principal consultant, SRK Consulting, Canada
Doug Stead, Prof. Em. Simon Fraser University, Canada
Omar Chang, Owner of XRGeo, Vancouver, Canada
Luca Schenato, Prof., Universita Degli Studi di Padova, Italy
Sithembinkosi Mhlanga, Director, Zambezi River Authority, Lusaka, Zambia
C. Munodawafa, Chief Executive, Zambezi River Authority, Lusaka, Zambia
Tony Wahl, US Bureau of Reclamation, Denver, USA
Jean-Louis Briaud, Distinguished Prof., Texas A&M University, USA
Michael Gardner, Prof., University of Reno, USA
Yii-Wen Pan, Prof., National Chiao Tung University, Taiwan
Dong-Soon Park, Stanford University, USA
James Rhys, Senior Engineer Dam Safety, Seqwater, Australia
Ali Firoozfar, Chief Dam Safety Engineer, Seattle City Light, USA
Alan Chemaly, Specialist Dam Engineer, Zutari Inc., South Africa
Luis G. Castillo, Prof., Universidad Politécnica de Cartagena, Spain
Sarah McComber, Program Manager, Sunwater Ltd., Brisbane, Australia
M. Shepherd, Chief Development Officer, Sunwater Ltd., Brisbane, Australia
Joël Monin, Razel-BEC, France

Dr Erik F. R. Bollaert
Ecublens, Switzerland
September 2023

About the Author

Erik F. R. Bollaert is President of AquaVision Engineering Llc, a company specialized in rock scour engineering and hydraulics of dams and spillways. He holds a MSc in Civil Engineering since 1996 and a PhD degree in hydraulics and rock mechanics from the Swiss Federal Institute of Technology (EPFL) in Lausanne since 2001.

He is an internationally known expert in scour and erosion problems downstream of dams and appurtenant structures and has been involved in many scour problems related to dam spillways worldwide.

He is the developer of the Comprehensive Scour Model, comprising several physics-based computational methods for rock mass fracturing and rock block uplift, and used to design scour mitigation measures at the iconic scour hole of Kariba Dam in Zambia-Zimbabwe.

Dr Bollaert co-organized the "International Workshop on Rock Scour" in Lausanne (EPFL) in 2001. He authored close to 100 scientific papers on the topic and is co-editor of the book *Rock Scour due to Falling High-Velocity Jets*, published in 2002 (Swets & Zeitlinger, The Netherlands). His work has been incorporated into several international guidelines, among others those of the US Society on Dams, the US Federal Highway Administration and the American Society of Civil Engineers.

Dr Bollaert is the developer of the rocsc@r® environment and its related digital platforms allowing advanced comparative scour computations, database storage and digital twin developments for rock scour.

List of Acronyms

The present list of acronyms is valid throughout the whole work, unless otherwise mentioned explicitly in the text.

ASP	Applicable Stream Power per unit area	[kW/m^2]
BEM	Boundary Element Method	[–]
BPM	Bonded Particle Model	[–]
CC	Centre-cracked shaped rock joint	[–]
CFM	Comprehensive Fracture Mechanics Method	[–]
DES	Detached Eddy Simulation	[–]
DI	Dynamic Impulsion Method	[–]
DNS	Direct Numerical Simulation	[–]
DP	Dynamic Pressure Method	[–]
EIM	Erodibility Index Method	[–]
EL	elliptical-shaped rock joint	[–]
FDM	Finite Difference Method	[–]
FEM	Finite Element Method	[–]
GSI	Geological Strength Index	[–]
LES	Large Eddy Simulation	[–]
MDI	Modified Dynamic Impulsion Method	[–]
MQSI	Modified Quasi-Static Impulsion Method	[–]
MULT	Multiplication factor for RMS pressure values	[–]
PDF	Probability Density Function	[–]
QSI	Quasi-Steady Impulsion Method	[–]
RANS	Reynolds Averaged Navier-Stokes	[–]
RMR	Rock Mass Rating index	[–]
RNG	Renormalization Group turbulence model	[–]
RQD	Rock Quality Designate	[%]
RSM	Reynolds Stress Model	[–]
SE	single-edge shaped rock joint	[–]
SST	Shear Stress Transport	[–]
UCS	Uniaxial compressive strength of rock	[MPa]

List of Symbols

The present list of symbols is valid throughout the whole work, unless otherwise mentioned explicitly in the text.

Lower Case

a	joint length for stress intensity computation [m]
a,b,c,d	parameters for RMS pressure rectangular jets [–]
a_1 to a_4	vertical decay coefficients for exponential decay laws [–]
a_{net}	net acceleration given to a block [m/s²]
$b(\varsigma)$	jet width for development length ς [m]
b_j	jet thickness at impact (for rectangular jets) [m]
c	pressure wave celerity [m/s]
c	rock mass cohesion [MPa]
c, c_{add}	viscous damping and added viscous damping [kg/s]
d	rock block size (equivalent cube size) [m]
d_m, d_{90}	grain sizes [m]
e_j	joint width [m]
f	frequency or boundary correction factor [Hz]
f	initial degree of break-up of joint for stress intensity computation [%]
f_{res}	resonance frequency [Hz]
g	gravitational acceleration [m/s²]
$gamma$	coefficient for pressure amplifications in rock joints [–]
h	flow depth [m]
$h_{up/down}$	jet thickness up/down after deflection at water-rock interface [m]
h_2	tailwater depth in riverbed downstream of scour hole [m]
h_{decel}	additional net vertical uplift height block during deceleration phase [m]
$h_{up,net}$	net vertical uplift height block during acceleration phase [m]
$h_{up,tot}$	total net vertical uplift height block [m]
j	number of a joint set [–]
k	turbulent kinetic energy [m²/s²]
k	stiffness of system during block displacement [N/m]
k_s	equivalent sand roughness of interface [m]
$k_{(1...3)}$	parameters celerity-pressure relationship [–]
m	mass of rock block [kg]
m_{am}	added mass of rock block [kg]
m_b	rock mass constant Hoek-Brown criterion [–]
m_r	joint fatigue sensitivity [–]
n	ratio of maximum deviation from mean to RMS coefficient [–]
n_b	critical net uplift displacement factor [–]
$n_{(1...4)}$	radial pressure decay parameters [–]
p	water pressure head [m]
p_m	average water pressure head [m]

$p_{\max}$	maximum instantaneous pressure head [m]
$p_{\min}$	minimum instantaneous pressure head [m]
q	discharge per unit width [m²/s]
r	radial coordinate [m]
r_{ref}	reference distance for radial pressure decay [m]
s	eccentricity between pressure stagnation point and jet impact point [m]
s_b	rock mass constant Hoek-Brown criterion [–]
t	time [s]
u', v'	root-mean-square value of longitudinal/transversal velocity fluctuations [m/s]
t_j	jet width at impact (only for rectangular jets) [m]
x	longitudinal coordinate [m]
x_b	longitudinal rock block sidelength [m]
x_{ult}	longitudinal distance from dam of the ultimate scour depth [m]
y	lateral coordinate [m]
y_b	lateral rock block sidelength [m]
z	vertical coordinate [m]
z_b	vertical rock block height [m]
z_{sc}	scour depth below initial bed level [m]
z_{model}	modelled scour depth below initial bed level [m]
z_{proto}	prototype scour depth below initial bed level [m]

Coefficients

C_p	average dynamic pressure $=(p_{\text{mean}} - Y)/(\phi V_j^2/2g)$ [–]
C'_p	RMS pressure fluctuations $=(\sigma)/(\phi V_j^2/2g)$ [–]
C_p^+	positive deviation from average pressure $=(p_{\max} - p_{\text{mean}})/(\phi V_j^2/2g)$ [–]
C_p^-	negative deviation from average pressure $=(p_{\text{mean}} - p_{\min})/(\phi V_j^2/2g)$ [–]
$C_p^{\max}$	maximum dynamic pressure $=(p_{\max})/(\phi V_j^2/2g)$ [–]
$C_p^{\min}$	minimum dynamic pressure $=(p_{\min})/(\phi V_j^2/2g)$ [–]
C_{sp}	stream power coefficient [–]
C_{BOTTOM}	average dynamic pressure coeff. along bottom of block [–]
C_{DVF}	average dynamic pressure coeff. along downstream vertical face of block [–]
C_I	dynamic impulsion coefficient [–]
C_{SURF}	average dynamic pressure coeff. along surface of block [–]
C_{UVF}	average dynamic pressure coeff. along upstream vertical face of block [–]

Upper Case

B	maximum positive deviation from quasi-steady pressure value [m]
B_i	jet thickness at issuance (rectangular-shaped jets) [m]
B_j	contracted jet thickness at impact in pool (rectangular-shaped jets) [m]
B_{out}	outer jet thickness at impact in pool (rectangular-shaped jets) [m]
C_r	joint fatigue coefficient [–]
C_r	mass strength coefficient for EIM [–]

C_t Courant number [–]
CC centre-cracked shaped rock joint [–]
D_i jet diameter at issuance (circular-shaped jets) [m]
D_j contracted jet diameter at impact in plunge pool (circular-shaped jets) [m]
D_{out} outer jet diameter at impact in plunge pool (circular-shaped jets) [m]
E_m Young's modulus of elasticity [GPa]
F_{buoy} buoyancy forces acting on the block [N]
F_{grav} gravity forces acting on the block [N]
F_{over} dynamic pressure forces acting over the block [N]
F_{under} dynamic pressure forces acting under the block [N]
$F_{\mathrm{up,DI}}$ net dynamic uplift pressure forces acting on the block [N]
F_{sh} shear and interlocking forces acting on the block [N]
$F_{\mathrm{sh},G}$ shear forces along block lateral joint based on weight and buoyancy [N]
$F_{\mathrm{sh},P}$ shear forces along block lateral joint based on differential pressures [N]
$F_{\mathrm{tot(t)}}$ net total external force applied to the block as a function of time [N]
Fr Froude number [–]
G_b immerged weight of block [N]
H velocity head (= $V^2_j/2g$) at issuance [m]
I_{up} net uplift impulsion on rock block [Nm]
J_a joint alteration number [–]
J_f energy slope of flow [m/m]
J_n joint set number [–]
J_r joint wall roughness [–]
J_s relative ground structure number [–]
J_v volumetric joint count [–]
K jet air drag coefficient, or Erodibility Index [–]
K_b block size number [–]
K_I stress intensity at rock joint tip [MPa.($m^{1/2}$)]
K_{Ic} in situ fracture toughness of rock mass [MPa.($m^{1/2}$)]
K_d discontinuity bond shear strength number [–]
L' non-dimensional block length [–]
L_b jet break-up length [m]
L_c jet core length [m]
L_f total length of joint around rock block [m]
H difference between upstream and downstream water level [m]
L_j length of joint, or jet trajectory length in air [m]
M distance from water-rock interface to determine flow velocity [m]
M_s mass strength intact rock [MPa]
N number of joint sets [–]
N number of pressure wave cycles per second [–]
P_j persistency of joint (= fissured length/total possible joint length) [%]
P_{max} maximum dynamic pressure in rock joint [m]
P_{mean} mean dynamic pressure [m]
Q Q-system index [–]
Q_a air discharge [m^3/s]
Q_w water discharge [m^3/s]

Re	Reynolds number [–]
T	Uniaxial tensile strength of rock [MPa]
Tu	initial jet turbulence intensity [%]
T_{up}	time coefficient for pressure pulses on block [–]
$V_{Dtpulse}$	vertical displacement velocity given to a block [m/s]
V_i	average jet velocity at issuance from the dam [m/s]
V_j	average jet velocity at impact in plunge pool [m/s]
V_{air}	minimum air entrainment velocity [m/s]
V_{local}	local jet velocity along water-rock interface [m/s]
$V_{up,net}$	net uplift velocity block following acceleration phase [m/s]
V_z	jet velocity along jet axis through pool [m/s]
$V_{zbottom}$	jet velocity at impact water-rock interface [m/s]
We	Weber number [–]
Y	travel distance of jet through pool [m]
Y_{local}	oblique distance for radial pressure decay [m]
Z_{core}	distance necessary for jet core to diffuse [m]

Greek Case

α_{am}	added mass coefficient [–]
α_b	angle of prismatic rock block with horizontal [°]
α_i	angle of jet with horizontal (at issuance) [°]
α_j	dip angle of joint set [°]
α_j	air concentration of jet at impact [%]
α_p	air concentration at point of jet impact in plunge pool [%]
α_r	air concentration in rock joint [%]
β	volumetric air-to-water ratio [–]
β	non-uniform velocity friction loss coefficient [–]
δ_{out}	outer angle of jet spread [°]
ε	rate of dissipation of turbulent kinetic energy [m^2/s^3]
ϕ	coefficient of non-uniform velocity profile [–]
ϕ	rock mass friction angle [°]
γ	angle for calibration of radial pressure decay [°]
γ_s	particle or rock specific weight [N/m^3]
η	unsteady friction loss coefficient pressures in rock joints [–]
φ	residual friction angle of joint set or of plunge pool side wall [°]
λ	hydraulic friction factor [–]
λ	wavelength rock joint pressures [–]
θ	jet angle with horizontal at impact in plunge pool [°]
ρ_a	density of air [kg/m^3]
ρ_r	density of rock [kg/m^3]
ρ_w	density of water [kg/m^3]
σ	standard deviation of pressure fluctuations (root-mean-square) [m]
τ_w	average wall shear stress of flow [N/m^2]
ω	mean particle fall velocity [m/s]
ω	angular frequency [rad/s]

ς	development length of jet through plunge pool [m]
Δt_{pulse}	time duration of positive pressure pulse on block [sec]
Γ^{+}	amplification coefficient of RMS pressures in rock joints [–]
Ω	reduction factor net uplift force on rock block [–]

1 Introduction to Digitalization in Engineering

INTRODUCTION

Digitalization has become a game changer and strongly influences the way we live, work and relate to one another every day. One of the key issues of the current digital revolution of our society is "digitability", i.e. our ability to fastly and soundly cope with a constantly increasing amount of changes in habits, reactions and communications that are imposed by digitalization. Quoting Charles Darwin, "It is not the strongest of the species that survives, nor the most intelligent that survives. It is the one that is most adaptable to change."

Hence, people get used to continuously adapt themselves to new software, files and applications to download and install, digital payments and receipts, instant messaging and communications, autonomous vehicles and so on. We generally do not bother about the technical background and methodology of all these processes because this is taken care of by professionals in the field. On the contrary, flaws or shortcomings in such digital processes have become unacceptable and generally conduct to the commercial end of the application in question.

Digitalization in engineering does not follow the same rules. Engineering is based on practical experiences and theoretical knowledge and scientific background, generally mixed in a personal cocktail made (and served) by engineers. The cocktail can have different flavours depending on the bartender in question, but the essence remains the same: we do not rely solely on theory and scientific facts and findings, but continuously relate engineering to practical feedback and subjective experiences on similar problems, gained personally or by other engineers in the same field. Also, a number of (non-mathematical) local constraints often influence the choice of the engineering solution, biasing feedback and experience.

Furthermore, for example in rock mechanics and rock engineering, most of the relevant parameters are determined statistically or based upon experience. Few or no sound deterministic approaches currently exist, and parameters and indexes are used based on traditional engineering practice rather than based on their pertinence to the problem in question. One example is use of the Rock Quality Designate (RQD) (Deere and Miller 1967) as a parameter to describe rock mass quality. Recent research has shown the shortcomings of this parameter, but it is still largely used worldwide in about all projects related to rock mechanics and rock engineering.

DOI: 10.1201/9781003319610-1

As a result, it is obvious that, pertaining to adaptability and digitalization, most engineering fields are not yet compliant with Darwin's quote. To succeed in changing the traditional way we conduct engineering projects, one needs a lot of courage, energy, time and new technology able to convince our profession. While the former elements are rather easy to generate, the latter aspect is challenging because it is subject to the availability of relevant data to test and to calibrate parameters, as well as to the willingness of leading experts worldwide. Without the necessary amount of data, even today's popular AI Chatbot software would not be able to convince...

It is within this challenging context that the present work has been developed. It aims at bridging the gap between traditional rock scour engineering and digital rock scour engineering. It points out the need for digitalization and describes the creation of a novel digital framework that allows to digitalize traditional engineering methods and results, as well as to perform future digital calibrations of new engineering parameters that may be of relevance to this field.

It promotes gathering and developing worldwide engineering experiences in an ad hoc digital environment that focuses on creating a database and making our engineering profession adaptable to digitalization, and more generally to any kind of changes, in a sound and reliable manner.

TERMINOLOGY AND ROOTS

TERMINOLOGY

Digitization, *digitalization* and *digital transformation* are commonly used terms and frequently mixed up.

Digitization means to convert something into a digital format, and usually refers to encoding of data and documents. In the simplest of terms, digitization is turning something into bits and bytes, or 1's and 0's. Digitization is basically the process of taking analogue information, such as documents, sounds or photographs, and converting it into a digital format that can be stored and accessed on computers, mobile phones and other digital devices. Straightforward examples include recording the music from a vinyl record into an MP3 file, or capturing an ancient video from a VHS tape onto a DVD.

In business, digitization may involve scanning old documents into PDFs, converting printed photographs into image files or transforming printed reports into meaningful data that can be manipulated and analysed.

The process is being applied by engineers since 50–60 years. The technology used to digitize is constantly changing and accelerates transfer and sharing of data: from the earliest keypunch machines to scanners, emails, cloud sharing, laser scans, remote sensing techniques (Photogrammetry, LiDAR), up to recently developed real-time holograph sharing worldwide.

Digitalization stands for the use of digitized data and digital technologies to change a business model and provide new value-producing opportunities. It represents a process of moving to a digital business, it goes beyond digitization and transforms the way you solve problems as an engineer. As an example, you will digitize a document, but you will digitalize your workflows or engineering processes. Digitalization is subordinate to digitization and would not exist without it. Our society digitalizes through increasing use of digitized forms of communication (emails, etc.). Digitalization advances extremely rapidly in the fields of data collection, interpretation and communication, because easy access to very large datasets becomes possible. In a nutshell, digitization refers to information, while digitalization refers to processes.

The meaning of *digital transformation* is closely tied to digitization and digitalization and has evolved over the years as more and more digital technologies have emerged. Digital transformation is the integration of digital technology into all areas of a business, fundamentally changing how you operate and deliver value to customers. Digital transformation is a broad framework to help businesses respond to emerging technologies, in order to significantly improve performance and presence. Digital transformation shifts the business thinking to include new products, audiences and greater integration that result from digital technologies. It includes all aspects of a business, regardless of the kind of business or how digitalized they are.

At the heart of all these terms, and in accordance with Darwin's principle, is the basic understanding that businesses must adapt and evolve to take advantage of digital technologies in order to remain competitive.

Roots of Digitalization

The roots of digitalization go back to 1679, when Gottfried Wilhelm Leibniz developed the first-ever binary system (Leibniz 1679). He was followed by Samuel Johnson in 1755, who described the binary system in a broader perspective, and by George Boole, who introduced Boolean algebra in 1847.

In 1938, Alec Reeves discovered the PCM (Pulse Code Modulation) technology that enables effective and successful voice calls in the telecommunication industry. This technology was not up for commercial use until the 1950s when the transistor was discovered. A master thesis at MIT in 1940 by Claude Shannon (1940) focused on digital circuits and Boolean algebra to show how those circuits can be used in the telecommunication industry. It forms the basis behind telephone routing digital technology. He introduced the theory that information was a measurable element and defined the basic unit, which would later be called a "bit".

John V. Atanasoff, together with Clifford Berry, introduced and extensively described the first digital calculating machine in 1939. This technology increased the capacity of data that would be stored and processed by digital devices such as computers. In 1943, the first-ever digital voice transmission technology known as the SIGSALY was discovered. This technology was embraced, especially by the troops, to secretly and effectively communicate with fellow allies during World War II.

In 1954 General Electric launched the first UNIVAC 1 computer, the first-ever system to have incorporated a payroll system in its operations in the USA. It significantly increased efficiency in operations and cost-effectiveness. In 1956, IBM announced the release of the first-ever disk storage unit, the IBM 350. The 350 Disk Storage Unit consisted of a magnetic disk memory unit with its access mechanism, the electronic and pneumatic controls for the access mechanism, and a small air compressor. The disk had a maximum storage space of up to 5 megabytes. United Airlines was the first company to purchase the storage disk for its reservation system.

The CCD (charge-coupled device) was officially launched in 1969 by Willard Boyle and George Smith. It turned light into digital information and shaped the future of the development of cameras, telescopes and medical imaging.

The end of the 20th century allowed to strongly accelerate digitization. The 1970s saw the launch of the Pulsar, the first watch in the world to showcase a digital display, and the first digital camera. In 1976, Phillips and Sony developed the compact disc (CD) Offsite Link, an optical disc used to store and playback digital data. It was originally developed to store and playback sound recordings exclusively. CDs can hold up to 700 megabytes. This equates to up to 80 minutes of uncompressed audio.

Finland launched its first 2G network in 1991. In August 1994, one of the first on-line transactions was actually a pizza ordered from Pizza Hut. The first film to be produced using computer systems, Toy Story, was launched in 1995. Digital transmission of television networks took shape in 1998. The beginning of the year 2000 marked the launch of the first digital entertainment and news service. At the same time, digital electronic payments systems surpassed traditional payments for the first time.

In 2004, Google created a standard hub for information, the Google platform. Bitcoin was introduced in 2008 as a digital, decentralized and cashless currency while on-line advertisements surpassed newspaper advertising by a significant margin.

In 2012, e-commerce exploded, with Amazon selling more books on-line than any physical stores. The year 2014 marked the beginning of revenues from streaming songs and movies from various sites. From 2015 to date, digital social platforms such as Facebook, Twitter, Instagram and TikTok have continued to change and digitize our world.

Last but not least, since 2022, artificial intelligence (AI) was discovered by the grand public via Chatbots, i.e. AI programmes that chat with you. They are used to reproduce powerful interactions with users, to aid business processes, to gain information from large groups, as a personal assistant among others. Chatbots are also used by search engines to lag the web and archive new pages for future search.

ChatGPT is a large language model (LLM), i.e. a language model consisting of a neural network with many parameters, founded by the OpenAI team in 2021. It is designed to assist users in generating human-like text based on given input. The model is trained on a massive amount of data, allowing it to generate text that is often difficult to distinguish from text written by a human. It is closely followed by Bard, launched by Google in March 2023.

DAM ENGINEERING DIGITALIZATION

INTRODUCTION

Numerous novel digital technologies have evolved in engineering applications over the last decades. Among them one can cite the Internet of Things (IoT), use of sensors in real time, use of artificial intelligence to handle big data (AI), digital twin (DT) developments and Chatbots starting to evolve.

Nevertheless, use of these emerging technologies in dam engineering has been particularly scarce compared to other engineering fields. A survey conducted by Loughrey et al. (2019) among members of the British Dam Society about the past, present and future adoption of digital technology within the dam industry resulted in the dam engineering industry lagging behind other industries and being resistant to digital advancements. Younger engineers even believed the dam industry is far behind other industries, with the perception that dam engineers are behind the times. Digital technologies that should be used more are monitoring, BIM, data, remote sensing and so on. Especially the need to capture, store and share data digitally was pointed out. Capturing data digitally through terrestrial laser scanning, drones, infrared and thermal imagery, to name a few, would allow better and more accurate monitoring of dams and reservoirs. All in all, without significant big data, no AI will ever be feasible in dam engineering.

As will be shown hereafter, significant efforts are currently being made by the dam industry to catch up, and huge advancements have been noticed over the last few years. The comprehensive literature review performed by Hariri-Ardebili et al. (2023) on AI and digital technologies in dam engineering confirms these ongoing efforts.

Digital Technologies

The present chapter presents digital technologies that have yet been incorporated into dam and/or rock engineering. Some of these technologies relate to the design and construction phases, but most of them are particularly useful during operation and maintenance phases of structures.

Virtual Design and Construction/Building Information Modelling (VDC|BIM)

Virtual Design and Construction (VDC) fundamentally changes the way projects are developed, coordinated and optimized. Owners, designers, equipment suppliers and civil contractors join forces. They develop, coordinate and optimize a project in a collaborative way. This unleashes a vast potential, not possible with a classical design approach. Projects are thereby strongly optimized and thoroughly coordinated using BIM models.

Building Information Modelling (BIM) generates digital models of buildings or any other structures. Each construction element, with all of its individual properties, is included as a data object in a central, multi-dimensional model. As well as the physical elements, this virtual twin includes planned timings and costs, as well as information about the location and other data. All the different specialists involved in the project can access the model so every process in the entire life cycle of a project – from planning to production to management of the finished building – can be designed, simulated and optimized, while problems can be identified early on. This reduces project risks and costs, optimizes timetables and results in better quality planning and execution.

BIM is a 3D model-based process that provides Architecture, Engineering and Construction (AEC) insight and tools for professionals to plan, build, manage buildings and infrastructure more efficiently. It is also useful for generating and managing data such as geometry, spatial relationships, quantities, the properties of components, etc. during the life cycle of a building. BIM also has the ability to transform how a building is designed and built and also to facilitate multi-disciplinary coordination, integration of 3D design, analysis, cost estimation and construction scheduling (McArthur 2015).

According to Azhar (2011), the technology can be used for several purposes including visualization, fabrication/shop drawings, code reviews, cost estimation, sorting of construction, detecting conflicts, disturbances and clashes, forensic analysis and facility management.

Moreover, BIM makes use of different levels of development (models) during the different phases of a project. These include 2D and 3D which are the basic form with limited use such as documentation and visualization, 4D which requires additional time and mostly used to optimize planning through schedule, 5D which comes with additional costs and used to develop a more efficient, cost-effective and sustainable construction through cost/estimation, 6D which provides additional energy analysis and used to regulate overall energy consumption for as-built operation or sustainability, and 7D for the optimization of asset and facility management from the design to operations and maintenance phase.

While BIM has been widely used over the last decade in building infrastructure, its use in dam infrastructure is not so common yet. Also, BIM models are often wrongly considered as digital twin (DT) models. The latter are discussed further on in this chapter.

While BIM models of hydropower schemes and power houses start to become standard during the design and implementation phase, few examples exist today of BIM models of dams.

AFRY Switzerland Ltd. has developed a standard approach of how to apply VDC|BIM in daily engineering work. It has been successfully implemented in several infrastructure and energy projects. Generally, new hydropower projects in Switzerland designed by AFRY are executed with this approach only, making use of a series of sub-models following the IFC standard (Industry Foundation Classes). IFC is a platform-independent, open file format that enables the exchange of BIM content between different software programs, allowing users to share IFC-based IFC models between applications.

A first example of a power house BIM model is shown in Figure 1.1 of the recently commissioned 900 MW Nant de Drance power house cavern in Switzerland. BIM tools were extensively used to support coordination and information sharing among owners, planners and contractors. The project was recognized with the Swiss Arc BIM Award 2017.

Moreover, Figure 1.2 illustrates the renewal of the 100-year-old Schils Hydro Power Plant in the Swiss Alps, where VDC|BIM was consistently implemented, using VDC for the entire coordination and BIM2Field during execution (i.e. BIM2Field is the process of bundling accurate digital data in an information model and using it on site for construction, operation or maintenance). Different civil contractors worked completely model-based, with no drawings needed or produced. Once the

FIGURE 1.1 BIM model of the 52-m-high Nant de Drance power house cavern in Switzerland. Courtesy of AFRY Switzerland Ltd.

coordination had been completed, the detailed models were prepared by AFRY in Zurich and uploaded into the cloud. They were then immediately available for use by the contractors to construct the powerplant. Not a single drawing was printed. One of the advantages is that there is no loss of information between the design office and the construction site. All required information is transmitted digitally throughout and is immediately available to all parties everywhere. With the permanent use of digital models, all information is always up to date, there are no inconsistent drawings or misinterpretations of drawings.

Schils power plant was successfully commissioned in 2021. It demonstrates that VDC|BIM can be successfully applied in hydropower and that the consistent implementation of VDC|BIM results in significant advantages, both for coordination and for execution through BIM2Field.

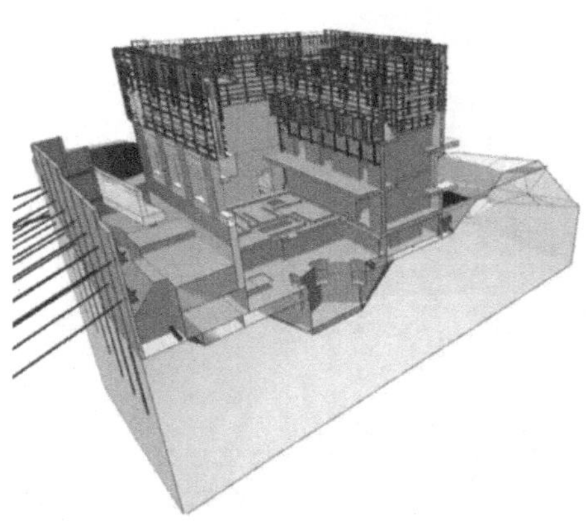
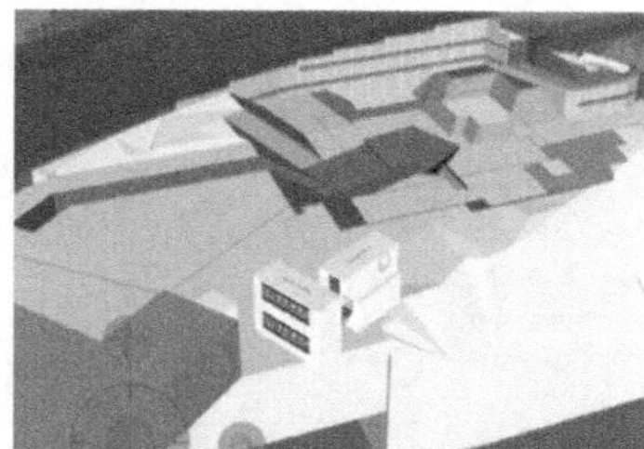

FIGURE 1.2 BIM model of the Schils HPP in the Swiss Alps. Courtesy of AFRY Switzerland Ltd.

Remote Sensing

In recent years, the measurement of dam displacements has benefited from a great improvement of existing technology, which has allowed a higher degree of automation. This has led to data collection with an improved temporal and spatial resolution. Robotic total stations and GNSS (Global Navigation Satellite System) techniques provide efficient solutions for measuring 3D displacements on precise locations on the outer surfaces of dams (Scaioni et al. 2018).

Until the 1950s, the static monitoring of dams was mainly based on geodetic control networks allowing to measure absolute and relative displacements of the structure and the nearby areas. The operations were quite complex and required a team of expert surveyors to work for several days per campaign. Geodetic networks were complemented by sensors able to measure local deformations (e.g. extensometers, inclinometers) and other physical quantities (e.g. piezometers, stress cells). In the 1960s, the introduction of automatic data acquisition and telemetric transmission allowed for the collection of data at a higher rate, up to continuous monitoring. These solutions provided long-term data series that improved the capability of analysing deformation patterns. Over the years, technological developments have continuously increased the precision, degree of automation and data handling capabilities of geodetic technology (Scaioni et al. 2018).

Methods based on Global Navigation Satellite Systems (GNSS) have been widely applied for measurement of dam displacements. Initially, GNSS networks were adopted for periodic measurements of control points, controlled by a set of reference points established in proximal stable areas. In recent years, automatic systems adopting differential GNSS sensors, able to work in continuous mode, have been developed to be integrated into early warning systems for safe dam maintenance. Real-time kinematic (RTK) measurements have also been used to carry out high-accuracy dam deformation monitoring (Scaioni et al. 2018).

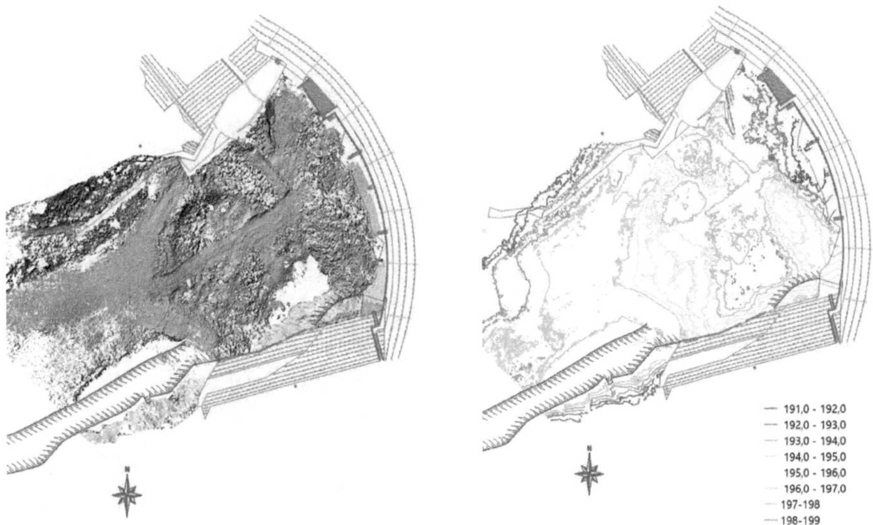

FIGURE 1.3 TLS point cloud and related bathymetry developed for an arch dam in France. Courtesy of EDF SA.

In parallel, during the last decades, dam monitoring has benefited from the development of remote sensing techniques from ground-based and satellite platforms (Scaioni et al. 2018). These have offered unprecedented opportunities for improving the structural analysis, since they extended the monitoring to a large portion of a structure, instead of a limited number of control points.

Among these techniques, three types of ground-based sensors are commonly used: terrestrial laser scanning (TLS), such as illustrated in Figure 1.3 at an arch dam in France (EDF SA), Distributed Fibre Optical Sensor (DFOS) systems such as installed in 3-m-high concrete volumes poured during the heightening of Luzzone Dam in Switzerland (Thévenaz et al. 1999), allowing to monitor the 2D temperature evolution inside the concrete volume, and finally ground-based SAR (GB-SAR, synthetic aperture radar).

DFOS has recently been integrated directly into trenches made in the downstream face of one of the gravity wall dams of Lago Bianco in Switzerland (Crameri et al. 2019). The DFOS were installed on the centre part of the eastern arch in close proximity to the existing monitoring system. The interrogator has been positioned in a box with controlled and constant temperature conditions. After the installation of the cable, the trenches of the grid were filled with a high-strength mortar (usually adopted for dam repair) to avoid relative strain (displacement) between cables and concrete (Figure 1.4).

Satellite-based interferometric SAR (InSAR) uses radar images to measure the surface displacement of the Earth's crust. In case of dam monitoring, InSAR is used to measure the deformation over time. The technique involves the use of two or more radar images of the same area taken at different times. The radar signals bounce off the Earth's surface and are recorded by the satellite. The images are then combined to create an interferogram, which is a map of the phase difference between

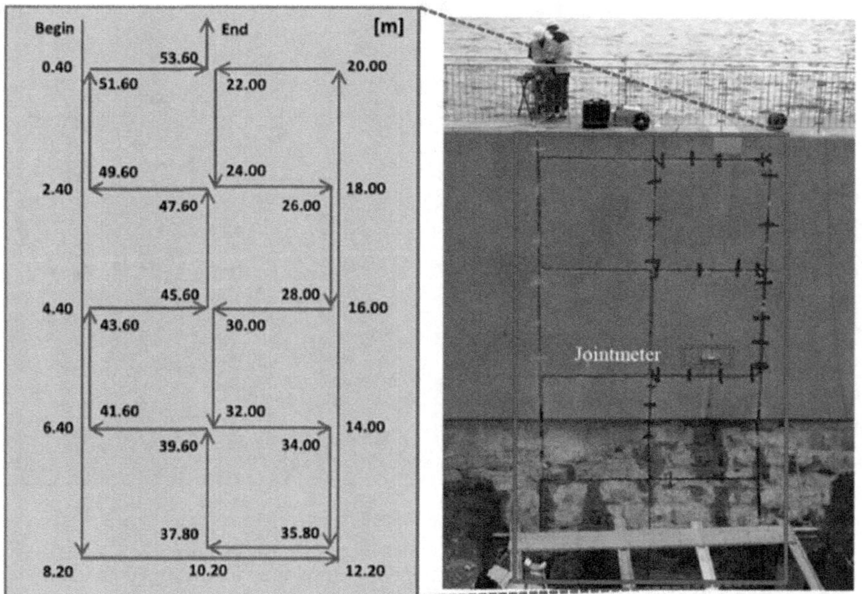

FIGURE 1.4 DFOS cables installed in downstream face of Lago Bianco Dam (Crameri et al. 2019).

the radar signals. The phase difference is related to the surface displacement of the area between the two radar acquisitions. InSAR can cover large areas and provide continuous monitoring, as well as measurements over a long period of time, which may be critical for understanding the long-term behaviour of dams suffering from deterioration (e.g. alkali-silica reaction) (Hariri-Arbelili et al. 2023).

Infrared thermography (IRT) provides a new solution for the inspection of earth rock dam leakage (Zhou et al. 2022). It has the advantages of fast visualization, strong mobility and wide coverage, and allows to detect the leakage areas of earth rock dams and dikes during floods.

Following Opyrchal and Chmielewski (2022), IRT measurements are more and more frequently performed to diagnose the technical condition of hydraulic structures. IRT may be used for detecting areas of increased seepage. They present examples of thermograms of surface damage of concrete surfaces and leakages through hydraulic structures.

Digital Photogrammetry (DP) is less feasible for the measurement of dam deformations but would be possible for the inspection of the conservation state of surface materials. This option is supported by the use of multi-copter drones, which may fly around the dam body and collect high-resolution images. Another possibility offered by image-based measurement techniques is to analyse the 2D surface displacements in specific areas, for example, by means of optical flow/digital correlation techniques or by installing specific targets.

Using photogrammetry software, overlapping photos can be used to generate 3D point clouds and terrain models for analysis and design. Additionally, the photos can be combined and scaled to fit the terrain, creating seamless, spatially accurate aerial

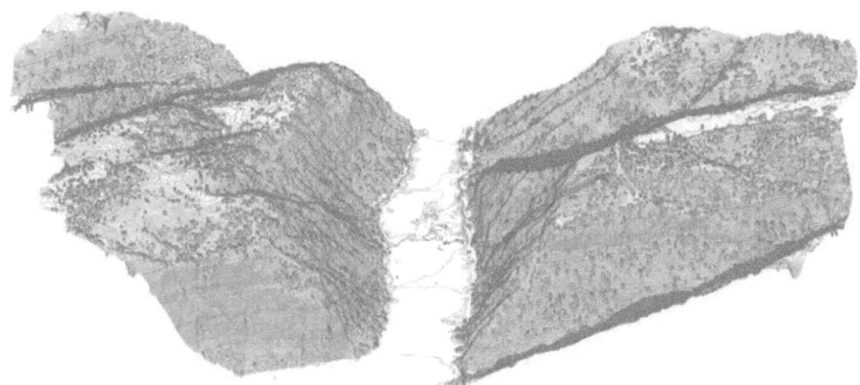

FIGURE 1.5 Axonometric view of Batoka Gorge (Zambia-Zimbabwe) contours from DSM model developed by eBEE drone survey (Pietrangeli et al. 2015).

images for use in CAD and GIS applications. Aerial images can also be overlaid on 3D terrain models, creating a high-resolution 3D environment that can be inspected and analysed virtually. Thermal and multispectral imagery can also be overlaid on the model to assist with analysis and documentation.

Finally, drone-based 3D Digital Surface Models (DSMs) may be very useful for preliminary monitoring and design, such as done by Studio Ing. G. Pietrangeli S.r.l. as shown in Figure 1.5 at Batoka Gorge situated just downstream of Victoria Falls in Zambia-Zimbabwe (Pietrangeli et al. 2015). Nevertheless, where there is tall and/or dense vegetation, the use of drones for topographical surveys must be complemented with ground or LiDAR survey.

Digital Twins and IoT Sensors

A digital twin is a virtual replica of a physical asset that makes use of (IoT) sensors and related engineering models to obtain and maintain a continuous real-time update of the real physical situation. For dams, it may be used for example to monitor structural behaviour and optimize dam operations. To create a digital twin, it is strictly speaking necessary to continuously maintain accurate information in real time about the physical asset.

Typically recorded data are, for example, changes in water level, outflow discharges, weather over the catchment area, structural loads and so on. More details can be found in Zhu et al. (2021) and Conde López et al. (2022).

Such a digital twin may identify potential issues before they occur, such as structural failure, scour potential or downstream flooding. It allows risk-informed decision-making by testing and optimizing in real time different operational strategies. Another application is to monitor and adjust the dam's operation in real time to optimize power generation.

Examples of digital twins of hydraulic structures are Yuansuan and Three Gorges Dam in China, Tarragona Dam in Spain, the Rance tidal power station and the Avignon water lock gate in France, Bluestone Dam (Virginia), New Bullards Dam (California) and Diablo Dam (Seattle) in the USA, and Simjun Dam in South Korea.

The degree to which real-time data is automatically updating these 3D virtual replica is not always clear, however. As such, it is difficult to state whether these examples are real digital twins or just detailed 3D numerical models that can be manually updated whenever needed. For example, the Bluestone, New Bullards and Diablo Dams are called digital twins, but seem to have no automated real-time connection with the physical reality.

Digital twins make use of IoT sensor networks, a network of physical devices and objects embedded with sensors, electronics, software, etc., which enables these objects to connect and exchange data with each other, and with the external environment. IoT systems typically consist of three main components: sensors and actuators that collect data and interact with the physical world; gateways or edge devices that process and filter the data; and a cloud-based platform that stores, analyses and manages the data. IoT devices typically use a variety of communication protocols, such as WiFi, Bluetooth and cellular networks, to connect to the internet and to other devices (Hariri-Ardebili et al. 2023).

Digital twins specifically related to rock scour are discussed more in detail in Chapters 2, 4 and 5 of the present work.

Cloud-Based Virtual and Augmented Reality

Cloud-based virtual reality (VR) and augmented reality (AR) make use of cloud computing technology to deliver VR and AR experiences over the internet. The required computing power and storage are thereby located on servers in remote data centres, and the device being used by the user only needs to have the ability to connect to the internet and receive the content.

In dam engineering projects, VR and AR applications are discussed for example by Spero et al. (2022) to visualize dam failure simulations, Verdelho Trindade et al. (2019; 2020) for inspection and structural health monitoring of dams (Figure 1.6) and Goff et al. (2016) in the field of reservoir operation and maintenance. Following Hariri-Arbedili et al. (2023), VR and AR are used for site surveying and modelling, allowing engineers and designers to explore the site and plan the construction process, for training and simulation of operators on safe and efficient operation of the dam, for inspection and maintenance (i.e. to superimpose for example instructions or repair procedures on top of the real-world view of the dam) and for facilitating remote communication and collaboration among engineers, designers and other stakeholders.

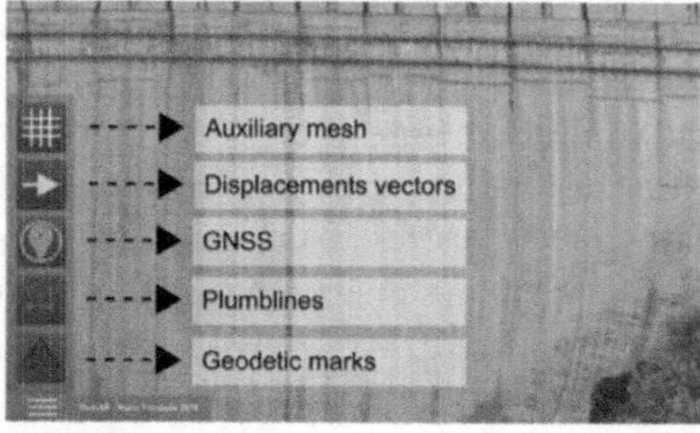

FIGURE 1.6 AR software environment of an arch dam with main menu on the left and network of sensors on the right (Verdelho Trindade et al. 2020).

3D and 4D Printing

3D printing, also known as additive manufacturing, is a process of creating three-dimensional objects by laying down successive layers of material. It can be used to create physical models of structures, such as dams, for testing and analysis (Pacina et al. 2022).

In the industry, 3D concrete printing (3DCP) is increasingly used to construct small- and large-scale concrete structures (Xiao et al. 2021), or any other construction component of a project. Coupling 3D printing with AI can reduce the cost and time required for traditional manufacturing methods and allow for more efficient designs.

Following Hariri-Arbedili et al. (2023), US army engineers used 3D printers to build a 1/240-scale model of the new spillway at Folsom Dam, designed to help reduce the risk of flooding throughout the Sacramento region. Moreover, China plans to build a hydropower dam using AI, construction robots and zero human labour. The 180-m-high Yangqu dam on the Tibetan plateau will be assembled layer by layer, like with 3D printing. A central AI system will be used to manage a massive automated assembly line of unmanned trucks to transport construction materials to the worksite, unmanned bulldozers and pavers (i.e. smart robotics) to transform these materials into distinct dam layers, and rollers equipped with sensors to press each layer. Whenever a layer is constructed, robots will send this information back to the AI system.

4D printing involves printing with multiple smart materials to create structures that can respond to different external stimulations, such as heat, moisture, light, currents, etc. By doing this, the 4D printed structure can change its shape over time (Zhang et al. 2019). Hariri and Abdelili et al. (2023) state that 4D printing could potentially be used in dam construction to create components that can adapt to changing conditions, such as rising water levels. However, the technology is still in its early stages of development.

Smart Robotics

Smart robotics is a field of robotics that involves the integration of advanced technologies, such as AI, ML, computer vision and sensor networks. It allows robots to perform complex tasks with a high degree of autonomy and intelligence.

Smart robotics has multiple applications in dam engineering and construction, such as for example inspection and maintenance. Robots can be used to inspect the dam for signs of surface ageing, like wear, cracks and so on. They can also be used for routine maintenance tasks, such as cleaning and painting, that are difficult or dangerous for humans to perform. Moreover, robots can be used to assist with heavy construction tasks, such as excavating and moving large amounts of earth and rock, or to pour and place concrete and to install steel reinforcement. Finally, robots can be used to monitor the dam and its surrounding area, collecting data on water levels, weather conditions and other factors that are important for the dam's operation and safety.

An extensive review and related references on the topic can be found in Hariri-Abdelili et al. (2023).

Advanced Numerical Modelling

Last but not least, digitalization is also about numerical simulations, which were actually one of the first digitalization technologies developed and widely applied by engineers. Remotely sensed datasets, together with traditional field-based mapping data, are used as input for advanced 2D and 3D numerical modelling analyses. In dam engineering, numerical models are used for example to determine hydraulics and sedimentation of dam reservoirs, design of dams and appurtenant structures, flood inundation mapping following dam break, 3D detailed stress and strain behaviour, soil-structure interaction, sliding stability analysis, drainage and seepage flow analysis, earthquake response analysis and so on.

Industry-available software programs are for example BASEMENT (ETH Zurich), MIKE-11/MIKE-21 (Danish Hydraulic Institute Inc.), TELEMAC-2D (EDF – HR Wallingford – Artelia-Sogréah), CCHE-2D (Computational Hydro-engineering Technology, Inc.) for river and reservoir hydraulics and sedimentation, DIANA (Diana FEA BV), PLAXIS-3D (Bentley Systems Inc.) or 3DEC (Itasca International Inc.) for geotechnic analysis of dams and their foundations, and ANSYS® FLUENT (Ansys Inc.), FLOW-3D® (Flow Science International Inc.) or OpenFOAM® (OpenCFD Ltd.) for detailed turbulent flow modelling at dams and appurtenant structures.

While some of these models allow to couple the dam and reservoir hydraulics to the hydromechanical behaviour of the rock foundation, and other models allow to simulate turbulent flows with a high level of detail, none of these models currently allows fully coupled modelling of rock scour generated by turbulent air-water flows from dam spillways. This is mainly due to the lack of implementation of rock break-up processes and to the difficulty in modelling transient hydraulic water pressures exerted by air-water mixtures inside rock joints.

One of the first attempts of coupled modelling of rock erodibility due to spillway overflows is illustrated in Figure 1.7 (Dasgupta et al. 2011). A 3D turbulent (k–ε) flow model is used in the ANSYS® FLUENT environment, followed by a manual transfer of the interface flow velocities and pressures towards a 2D UDEC (distinct element) geomechanical model of the rock. Multiple consecutive manual transfers of results between both models, including adaptation of the rock bottom, then allow sequentially coupled modelling. The rock mass was modelled using a set of Voronoi

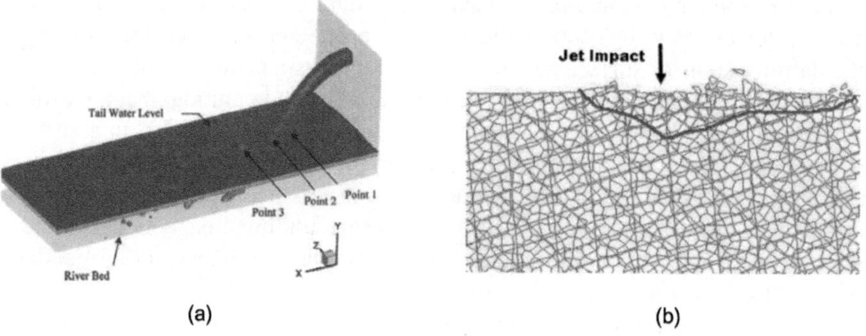

FIGURE 1.7 (a) 3D CFD modelling of spillway flow; (b) geomechanical modelling of the rock mass using polygonal elements (Dasgupta et al. 2011).

polygonal small elements, proposing multiple contact points by normal and shear stiffness springs. Break-up of bounds between the elements was based on tensile strength and shear strength.

Artificial Intelligence (AI) and Machine Learning (ML)

AI refers to the simulation of human intelligence in machines that are programmed to think and learn like humans (Hopgood 2021). The goal of AI is to develop systems that can perform tasks that typically require human intelligence, such as visual perception, speech recognition, decision-making, and language understanding. There are several approaches to creating AI, including rule-based systems, expert systems, machine learning, and natural language processing.

According to Jordan and Mitchell (2015), Machine Learning (ML) is a sub-field of AI that focuses on the development of algorithms and statistical models that enable computers to learn from data and improve their performance on specific tasks over time. It enables machines to learn from experience without being explicitly programmed. With the advent of big data and advances in computing power, ML has become increasingly popular in engineering applications, as it can help engineers make more informed decisions and optimize complex systems more efficiently (Hariri-Ardebili et al. 2023).

A detailed review of AI and ML applications in dam engineering can be found in Hariri-Ardebili et al. (2023). They distinguish between dam safety monitoring and early warning systems, expedited numerical simulations, hydropower generation optimization, hydrological and flood forecasting, water resource management and reservoir operation, dam design optimization, predictive maintenance and finally decision-making and risk assessment.

From these, flood forecasting, improved reservoir operation and risk assessment seem to be the most relevant issues related to rock scour downstream of the dam. Flood forecasting may be optimized by training AI models on historical data and real-time sensor data to predict the likelihood of flooding and help with flood management. Reservoir operation optimization manages the water release from the dam for a given set of constraints. Decision-making and risk analysis in the context of dam safety management are performed by analysing large amounts of data and identifying patterns to define the best course of action in response to potential risks or emergency situations.

None of these systems, however, integrates rock scour potential and risk as a constraint. Nonetheless, the integration of rock scour into risk-informed decision-making scenarios seems obvious for dams where rock scour is an issue.

This absence is merely due to the current lack of both data and appropriate modelling techniques, i.e. no detailed scour measurements are available in real time during flood events, and no appropriate numerical modelling allows to fill up this lack of data in real time.

ROCK ENGINEERING DIGITALIZATION

Introduction

Until a decade ago, development and use of novel digital technologies seemed more advanced and diversified in rock engineering than in dam engineering. One of the reasons was probably significant conservatism in dam engineering on one hand, and the high number of engineering infrastructures where rock investigations are necessary, i.e. buildings, roadways, tunnels, mines, slopes, bridge foundations, etc., on the other hand. However, at the time of writing of the present work, dam engineering is strongly catching up.

Digital Technologies

The present chapter summarizes digital technologies that have been implemented during the last decades into rock engineering. Some of these technologies overlap with dam engineering technologies, and some are more specific to rock engineering and rock mass characterization.

Remote Sensing

Similar to dam engineering, remote sensing represents a significant part of novel digital technologies in rock engineering. Techniques in common with dam engineering are high-resolution photography, LiDAR using drones or helicopters, 3D digital photogrammetry using drones (DP) and terrestrial laser scanning (TLS) and infrared thermography (IRT). Hyperspectral imaging (HIS), however, seems more specific to rock engineering.

In rock engineering, these techniques are used for example to perform varied geological and geotechnical analyses, including discontinuity mapping, rock slope damage characterization, blast damage mapping, lithological mapping, groundwater seepage analysis, rockfall investigation, weathering and alteration mapping.

On the other hand, recently developed remote sensing techniques such as GB-SAR and InSAR do not seem to be fully appropriate yet for rock mass investigations, at least not for investigations where a high precision is requested.

Mixed and Virtual Reality

Onsel et al. (2019) made an overview of the latest digital developments related to Mixed (MR) and Virtual (VR) reality in engineering applications. According to the authors, during the last decade there has been a dramatic increase in the use of MR and VR techniques in mining engineering. The majority of these applications have focused on improving visualization of engineering projects including open pits and surface mine reclamation. But MR/VR can be also used for geotechnical site characterization using the Microsoft® HoloLens (HL) headset. Such a system comprises a short distance 3D scanner that can be used on site for data collection and mapping. It is also capable of visualizing 3D multi-sensor remote sensing data, both in the field and at the office. The headset supports 3D mesh files and bitmap image files. 3D datasets collected using remote sensing techniques (e.g., TLS, DP, SfM) can be used to create mesh files. 2D remote sensing datasets (e.g., HR, IRT, HSI) can be used to create textures that can be overlain onto the 3D meshes.

An example of recently developed applications is rock outcrop mapping by HoloLens applications in the field. These applications allow creating a 3D mesh by scanning, followed by digital tracing of discontinuities and definition of their type, georeferencing and orientation (dip, dip direction), all quasi in real time. Furthermore, joint lengths and roughness can be measured digitally.

Another example is holographic core logging, using a 3D core scanner and allowing to visualize 3D results from within and outside of the core, together with comparison with various 3D spatial datasets, such as acoustic televiewer data (DGI Geoscience 2019) or full-bore formation micro-imager (FMI) data (Schlumberger 2013) (Figure 1.8).

Moreover, field-based investigations and remote sensing can be converted into 3D georeferenced meshes that allow for example to investigate landslides and risk assessment. The same data can then be used to perform 2D and 3D numerical modelling. The results of this modelling can then be integrated into the holographic environment to be compared with each other and be stored in a geodatabase that can be exploited simultaneously at multiple locations. This work has been made by Jesse Mysiorek at Clifton Engineering Group Inc. Calgary, Canada (Jessemysiorek@clifton.ca) and by Doug Stead at Simon Fraser University, Department of Earth Sciences, Burnaby, Canada (doug_stead@sfu.ca).

Both physical and digital worlds can finally be brought together on a multi-platform, mixed reality environment, where engineers and geologists can collect and visualize data in real time. The EasyMineXR platform (Figure 1.9), developed

FIGURE 1.8 Holographic core compared to its real counterpart. Courtesy of SRK Consulting Inc.

FIGURE 1.9 A multi-platform mixed reality environment to collect and visualize rock mass data in real time. Courtesy of SRK Consulting Inc.

by SRK Consulting Inc. in collaboration with Simon Fraser University, enables advanced visualization of subsurface conditions and site interpretation and characterization. The software captures geotechnical data acquired in the field and renders it in 3D, allowing for example digital meetings to discuss the projected features.

Advanced Numerical Modelling

Remotely sensed datasets, together with traditional field-based mapping data, are used as input for advanced 2D or 3D numerical modelling analyses.

Proprietary software can be roughly subdivided into Finite Element Analysis (FEM), such as Rocplane, Slide 3 and RS3 (Rocscience Inc.), Distinct Element Modelling (DEM) such as FLAC3D and 3DEC (Itasca International Inc.), Discontinuous Deformation Analysis (DDA) such as 3D-DDA or DDA-2D methods (by Dr. Shi Genhua), Keyblock Theory such as VisKBT (Engineering Computing Center, UCAS Llc.) and ROCK3D (Geosoft Inc.), and Discrete Fracture Network (DFN) modelling such as DFN.lab (Fractory from Itasca), MoFrac (Mirarco Mining Innovation Inc.) and FracMan (Golder Associates Inc.).

Open-source 3D point cloud processing software is for example CloudCompare (https://www.cloudcompare.net/main.html) or SciPy (https://scipy.org).

Open-source numerical developments specific to the generation and modelling of rock blocks have been made by fractured rock mass generation codes, such as for example UnBlocksgen (Rasmussen 2020) or SparkRocks (Gardner et al. 2017). Furthermore, advancements are made in the field of automatic extraction and recognition of rock discontinuities based on 3D point clouds, involving MATLAB codes and machine learning principles (Riquelme et al. 2014, 2015; Kong et al. 2021; Mammoliti et al., 2022). Finally, automatic extraction from 3D point clouds of rock block volumes and surface areas for kinematic block stability analyses is currently under development (Weidner and George 2023).

As mentioned before in the field of dam engineering, some of these models account for steady or quasi-steady-state flow analysis in joint networks, but do not account for transient dynamic water pressure fluctuations and their propagation and amplification inside rock joints, nor for rock break-up mechanisms such as progressive fracturing by fatigue or block movements. Hence, they are not suited to model rock scour such as generated by spillway flows at dams or turbulent flows at bridges founded on rock.

RIVER ENGINEERING DIGITALIZATION

Introduction

Besides dam and rock engineering, another field of engineering that is regularly exposed to rock scour is river engineering. River engineering englobes both natural rock scour of the riverbed and banks, and rock scour by the presence of structures situated in or near the riverbed, such as bridge piers and abutments, or also river training works.

Scour of bedrock in rivers and near bridges founded in rivers is mainly governed by turbulent flows that generate flow velocities, shear stresses and fluctuating dynamic pressures at the water-rock interface. As such, the hydromechanical interactions and main rock break-up processes are quite similar to the ones encountered at dams and hydraulic structures, i.e. rock joint initiation and propagation, rock block plucking, rock block displacement and finally rock abrasion on the long term.

Digital Technologies

Most of the digitalization processes in river engineering are similar to the ones used in dam and rock engineering, such as remote sensing techniques or advanced numerical modelling. In the following, emphasis is given on some processes that are more specific to river engineering.

Flow Monitoring

Flow monitoring makes use of physical sensors or data loggers to measure essential parameters like flow depth, flow velocity, flow temperature, flow pressure, turbidity, water quality and so on. The data measured by for example Ultrasonic Distance Level (ADV) sensors or by Radar Surface Velocity sensors is acquired locally and in real time. It is generally transmitted automatically and/or continuously using appropriate telemetry to a cloud-based database platform.

Remote sensing in river engineering is currently strongly developing. IoT sensors transmit their data based on 4G, 5G, WAN or satellites. Data can be visualized, analysed and downloaded from the cloud, and real-time monitoring and flood alerts are common practice.

Furthermore, Acoustic Doppler Current Profiling (ADCP) from a moving boat or profiler with GPS allows to perform detailed cross-sectional determination of flow velocity and thus of the river discharge, as well as of the flow depth and of the Suspended Sediment Concentration (SSC) by use of Acoustic Back Scatter (ABS).

Satellite-based river flow monitoring is currently strongly developing. The European Space Agency (ESA) RIDESAT Project (RIver flow monitoring and Discharge Estimation by integrating multiple SATellite data, Contract Number 4000125543/18/I-NB) aimed at developing a new methodology for the joint exploitation of three sensors (altimeter, optical and thermal) for river flow monitoring and discharge estimation. The RIDESAT Project aimed at providing, for the first time, an accurate satellite-based river discharge product for small to large rivers. The project is currently continuing through the STREAMRIDE project, which is an extension of the STREAM (SaTellite-based Runoff Evaluation And Mapping, Contract Number 4000126745/19/I-NB) project and investigates the possibility to improve river discharge estimates by merging the STREAM and RIDESAT projects (http://hydrology.irpi.cnr.it/projects/streamride). An application of satellite data to the Mississippi river basin is described in Camici et al. (2022).

Riverbed Monitoring

Besides the ADCP technology, riverbeds can be monitored using Multi-Beam Echo Sounding (MBES), which emits a multidirectional radial beam of sound waves from a boat to obtain information within a fan-shaped swath. The timing and direction of the returning sound waves provide detailed information on the depth of water and the shape of the river channel, lake bottom or any underwater features of interest.

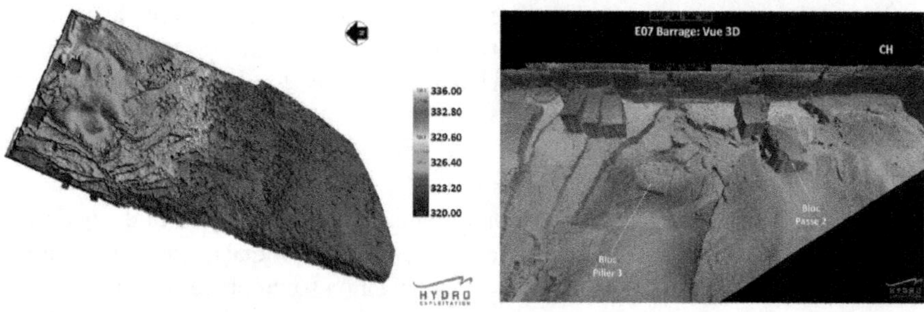

FIGURE 1.10 Plan view and detailed 3D view of rock riverbed downstream of Chancy-Pougny Dam obtained by MBES. Hydro-Exploitation SA 2019, courtesy of SFMCP SA.

The method is widely used to track scour formation of riverbeds, for example, at bridge piers and abutments or downstream of dams. An example of MBES is shown in Figure 1.10 illustrating a plan view and a detailed 3D view of the bedrock downstream of Chancy-Pougny Dam on the Swiss-French border (Hydro-Exploitation SA 2019, courtesy of SFMCP SA).

Terrestrial laser scanning (TLS), a fast and highly accurate remote laser scanning providing 3D point cloud data, allows even more detailed mapping of the riverbed in cases where most of the bed is sub-aerially exposed at low flows. TLS offers the opportunity to collect high-precision and high-accuracy data over large spatial extents. When coupled with a suitable bathymetric mapping technique, high-resolution digital elevation models (DEMs) can be developed for both dry and wet areas.

On the other hand, airborne laser scanning (LiDAR) from an Unmanned Aerial Vehicle (UAV or drone) or from a plane or helicopter can provide detailed information on both the emerged (topographic) and submerged (bathymetric) terrain, provided that sensors with two wavelengths are being used (near-infrared for topography, blue-green laser for bathymetry) and that the water is clear and of minimum 1.5 m of depth. A water penetration depth of up to 50 m for a point density of 40,000 points per second can be obtained with the latest high-performance airborne sensors and systems, or a point density of 140,000 points per second for shallow water surveys down to 25 m of depth (https://leica-geosystems.com/products/airborne-systems/bathymetric-lidar-sensors).

Similar laser scans can also be obtained from underwater LiDAR, based on sensors operated by a diver or carried by a towed body, a remotely operated vehicle (ROV) or an autonomous underwater vehicle (AUV). This particular field of laser scans is strongly developing, mainly in the fields of ocean engineering and oil and gas industry, as well as for military operations (demining for example). Current research focuses on deeper water penetration as well as penetration of trouble waters (mud, etc.), together with increased accuracy and precision.

Latest technological developments make use of satellites to deploy multispectral and multi-angular spaceborne LiDAR sensors (Conaway et al. 2019). The NASA Applied Sciences Program is helping establish an operational river gauging system in Alaska that uses satellites to measure river levels and flow. Scientists from the US Geological Survey, NASA and the University of Ohio are collaborating with

Alaska agencies that depend on river monitoring data, including the Alaska Dept. of Transportation, Alaska Dept. of Fish & Game, US Fish & Wildlife Service and National Weather Service (https://swot.jpl.nasa.gov).

The primary goal is to provide objective, accurate and timely assessments of river water levels and flows in remote areas. Satellite laser altimeters, such as on the ICESat-2 mission, use lasers operating within the green wavelength to measure water surface altitudes and water body depths.

Within the framework of the European Union's Horizon 2020 Research and Innovation Programme, the HYdro-POwer-Suite (HYPOS, https://hypos-project.eu/) project is another example of satellite-based river monitoring. It uses multi-sensor satellite-based measurements to determine water quality and sediment-related parameters, such as suspended sediment concentration, harmful algae blooms, temperature and evaporation rates, etc. This on-line tool for the hydropower industry combines Earth Observation (EO) technologies with hydrologic modelling, and can be complemented by local on-site information. It offers business-oriented tools for environmental and economic investment planning to the hydropower industry.

Bridge Scour Monitoring

Riverbed scour near bridge piers and abutments is caused by turbulent flow acting at the water-rock interface. Scour potential significantly increases during times of high discharges and can compromise the safety of the structure.

Bridge scour monitoring, and scour monitoring at other structures, can identify and prevent disasters before they strike. As scour alters the elevation of the riverbed at a pier, a monitoring system can indicate when a bridge becomes structurally deficient and dangerous. While monitoring does not make a structure less susceptible to scour, a real-time alert can be sent out if a bridge becomes scour critical and must be closed.

A real-time monitoring system can observe scour progression and indicate when and what actions must be taken to maintain structural integrity. These monitoring systems can be used to detect the need for immediate mitigation measures during a flood event or simply to establish a schedule for structural countermeasures. Following the HEC-23 circular of the US Federal Highway Administration (FHWA 2009), different scour monitoring instruments have been extensively tested and are available for practice, such as a sonar-based depth sensor, a magnetic sliding collar or float-out devices. Based on Prendergast and Gavin (2014), other monitoring instruments measuring scour depths are electromagnetic pulse or radar devices or electrical conductivity devices, among others.

All these instruments measure the change in riverbed elevation. Sonar is the most widely used due to the ease of installation and the wider range of applicable site conditions. In addition, sonar instruments provide continuous bed level data, while magnetic sliding collars and float-out devices only indicate when scour has reached a certain depth.

Furthermore, for bridges founded on rock, sonar systems are easy to implement, as they do not need to be installed or buried in the riverbed. Sliding collar solutions are comprised of a rod and a ring driven into the riverbed. As the bottom erodes, the collar slides down the rod and is detected by magnetic triggers. This solution is very susceptible to debris and not compatible with rocky foundations. Float-out devices are buried at varying depths in a potential scour area. When scour occurs, removing the sediment covering the device, the instrument will float to the surface. A wireless signal is transmitted to a nearby data logger to indicate its release. Like for magnetic collars, such devices are not compatible with rocky foundations.

The measured data are generally sent to an integrated telemetry system. A data logger can support multiple sonar and water level sensors and log scour data from each at pre-defined intervals. With telemetry, whether radio, cellular or satellite, the system can then securely transmit the scour data on-line in real time for viewing from any computer.

Integrated data logging and telemetry stations are customizable based on bridge scour monitoring needs. The integrated system, whether mounted on a bridge, pier or along the riverbank, can connect to and support multiple sensors, even when they are mounted on different piers. All equipment is powered via a central battery and recharging solar panel system. With multiple telemetry options to choose from, continuous real-time data are available from any computer. This ensures that the scour conditions are constantly monitored, and any control or countermeasures can be implemented immediately if the scour becomes critical.

Examples and more detailed descriptions of monitoring instruments can be found for example at https://www.nexsens.com or http://www.fondriest.com. An extensive review of bridge scour monitoring techniques can be found in Zheng (2013) or Prendergast and Gavin (2014).

River Dike Monitoring

Since the first field test application in France in the 1990s, Distributed Fibre Optical Sensor (DFOS) Systems for leakage detection are more frequently installed in river dikes, over significant lengths. Furthermore, based on the German national research programme, "Risk Management of Extreme Flood Events", or RIMAX (https://www.hydroc.de/rimax-guide-risk-management-in-extreme-flood-events), smart monitoring systems have been developed allowing to detect incipient effects, such as strain and temperature differences, of failure of hydraulic engineering structures. DFOS has thereby been integrated into geosynthetics used in the stabilization of dams and dikes. More information on geo-hydrological applications can be found in Barrias et al. (2016) and Schenato (2017) (Figure 1.11).

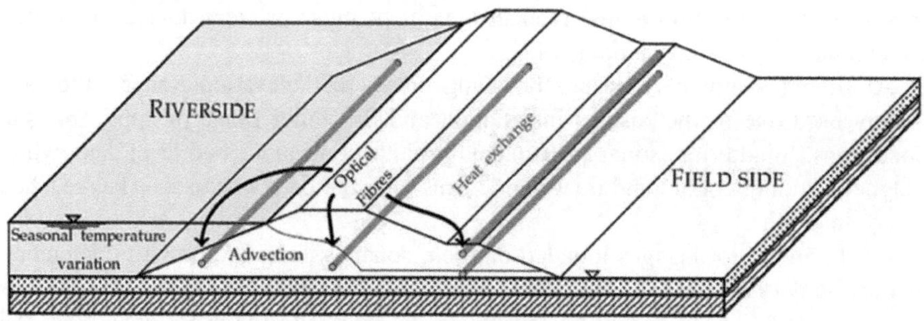

FIGURE 1.11 The principle of monitoring for seepage detection by distributed temperature sensor (DTS) (Schenato 2017).

Advanced Numerical Modelling

Over the past decades, river engineering has benefited from a wide range of numerical models allowing to model turbulent flows, sediment transport and scour of granular foundations in rivers in two and three dimensions, but the hydromechanical interaction with a rocky foundation has received few or no attention.

A similar situation holds for bridges founded on rock, for which a lot of attention has been paid by researchers to the 3D turbulent flow structures (vortices) and their scour potential at piers and abutments founded on granular soils, but very few developments exist pertaining to pier or abutment scour in rocky foundations.

Niemann et al. (2017) propose a risk-based screening strategy for bridges potentially affected by rock scour. This screening strategy includes three tiers: Tier I shows evidence of long-term channel stability and requires no further action in assessing rock scour; Tier II requires consideration of abrasion as the only mode of scour; and Tier III requires consideration of quarrying/plucking as the primary or significant secondary mode of scour.

Current state-of-the-art is summarized in the NCHRP Report 717 by Keaton et al. (2012). During this research project, different modes of rock scour at bridges have been investigated, among which are abrasion, wearing, block quarrying and plucking, dissolution and cavitation.

Within the framework of the NCHRP research, the Comprehensive Scour Model (Bollaert 2004) was applied to quarrying and plucking of regularly shaped rock blocks near bridge pier foundations. Numerical modelling has been performed of various hydrodynamic conditions appropriate for natural channels where bridge foundations might be located, resulting in design curves of critical velocity and ultimate scour depths at bridge piers (Bollaert 2010; Keaton et al. 2012) (Figure 1.12).

While this numerical modelling has allowed developing scour and velocity threshold charts for rock block plucking at bridge piers, it does not provide a numerical framework for coupled hydromechanical modelling of flow and rock.

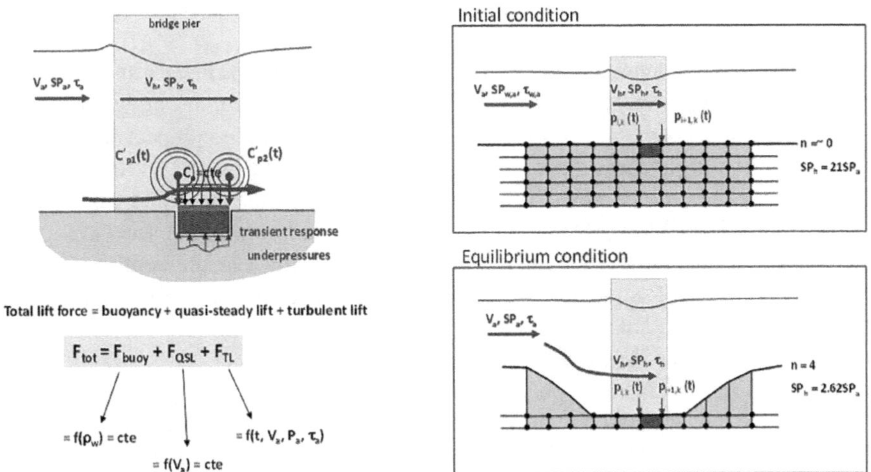

FIGURE 1.12 Main parameters of numerical modelling of single rock block plucking at bridge piers.

COASTAL ENGINEERING DIGITALIZATION

INTRODUCTION

Rock scour in coastal areas deals with different physical phenomena. The sudden impact of air-water waves on natural rock cliffs or blockwork coastal structures generates localized distinct pressure pulses that may be propagated through the joints towards the interior of the cliff or structure. These pulses may then further propagate existing cracks and generate dynamic pressures behind blocks that may expulse these blocks towards the sea. An example is provided in Chapter 3 at the Pointe du Hoc cliffs in French Normandy (Briaud 2008).

DIGITAL TECHNOLOGIES

Most of the digitalization processes in coastal engineering are similar to the ones described before, such as remote sensing techniques or numerical modelling. In the following, emphasis is given on those processes that are specific for coastal engineering.

Wave Monitoring

As described in Rossi et al. (2022), wave monitoring generally makes use of buoys. Buoys, whether moored or drifting, have the ability to communicate, in a programmed way, in real time via satellite telecommunication systems, transmitting acquired observations to collection centres. A buoy can be used to yield high-quality directional wave spectra by installing on it a downward-looking acoustic Doppler current profiler (ADCP), used to derive two-dimensional spectra from wave orbital velocities, or a global positioning system (GPS), to measure the motion of the drifter at different frequencies. Wave drifters are generally developed based on GPS receivers and deployed in the open sea. Directional wave spectra (DWS) drifters are a new generation of GPS-based tracking devices, able to compute directional Fourier coefficients used to derive wave parameters such as significant wave height, swell direction and directional spread.

A recent development in buoy sensors is to derive wave parameters and the directional wave spectrum using the buoy GPS output velocities, representing the velocities of water particles. GPS buoys, which are typically more compact in size, can facilitate three-dimensional movement with an accuracy of 1–2 cm up to wave periods of 100 s, and an accuracy of between 0.5% and 1% of the measured value.

Satellite Micro-Wave (MW) remote sensing allows monitoring of sea waves especially for regions that are not easily accessible for in-situ measurements, as well as Synthetic Aperture Radar (SAR). The Significant Wave Height (SWH) is measured by a radar altimeter that transmits pulses into the sea surface to accurately estimate the distance between the satellite and the sea.

Furthermore, wave parameters can also be monitored by coastal HF (High-frequency) radars. These are land-based remote sensing instruments that have attained great popularity in the last few decades. The reason for their wide distribution lies in the fact that they are able to provide synoptic data (i.e. simultaneous over a relatively large area), repeated in time at unprecedentedly high spatial and temporal resolutions.

More detailed information on these measurement technologies can be found in Rossi et al. (2022).

Seabed/Shoreline Monitoring

Shoreline change is one of the most common natural processes that prevail at coastal areas. The most important aspect is to identify the location and change over time of shorelines. This requires frequent monitoring of the shoreline. Shoreline monitoring is done with conventional surveying techniques such as Total Stations, GNSS, etc., or can be performed by more advanced techniques such as satellite imageries over time or ground-based high-resolution digital photogrammetry, which may provide higher spatial and temporal resolutions.

Liu et al. (2019) developed a seabed real-time sensing system for in-situ long-term multi-parameter observation applications. As illustrated in Figure 1.13, it consists of a seabed acoustic-based observation system, a sea surface relay transmission buoy and a remote monitoring system. The system communication link is implemented by underwater acoustic communication and satellite communication.

Furthermore, Tritonia Scientific Ltd., an underwater 3D imaging firm located in Oban (UK), is partnering with researchers at The Lyell Centre of the Heriot-Watt University in Edinburgh (UK), seafood producer Mowi (Rosyth, UK) and the Sustainable Aquaculture Innovation Centre (SAIC, University of Stirling, UK) to create 3D digital models for seafood producers to map and monitor complex marine environments and habitats.

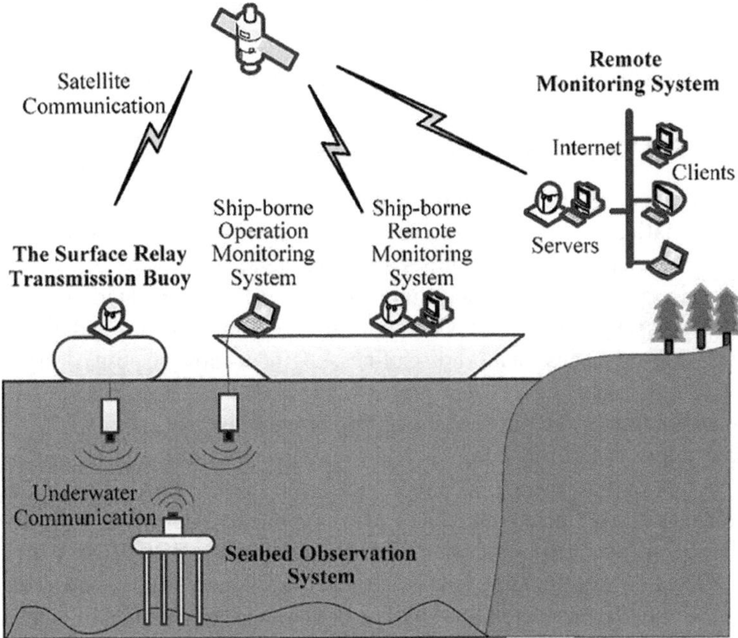

FIGURE 1.13 System structure of the seabed real-time sensing system for in-situ long-term multi-parameter observation applications (SRSS/ILMO) developed by Liu et al. (2019).

By using remotely operated vehicles (ROVs), a 3D digital twin of the seabed is generated. The future 3D model will be able to show various physical characteristics and be used alongside advanced comparison software to monitor changes on the seabed. By regular and/or systematic repetition of campaigns, quasi-real-time comparisons would allow a digital twin to be set up.

Furthermore, photogrammetric techniques are used to accurately verify the seabed, for example, to estimate the levels of coral bleaching and coral mortality (Sheppard et al. 2017).

Advanced Numerical Modelling

Numerous numerical models exist for simulation of wave generation, propagation, transformation and break-up, as well as for related shoreline erosion, in 1D, 2D or 3D.

Proprietary examples are the SWAN, Unibest, Pharos, Sobek, Delft3D, XBeach and XBeach-VEG software packages (Deltares models, https://www.deltares.nl), the MIKE21-SW, MIKE21-BW, MIKE3 and LITPACK software packages (DHI models, https://www.dhigroup.com) or also FLOW-3D (Flow Science Inc., https://www.flow3d.com).

Open-source models are for example SWASH (Tu Delft, https://swash.sourceforge.io), ARTEMIS (Open Telemac-Mascaret Consortium at http://www.opentelemac.org) and OpenFOAM (https://www.openfoam.com).

Nevertheless, in the field of waves impacting rock cliffs or blockwork coastal structures, numerical modelling procedures and techniques are scarce. Wave impact phenomena at rock cliffs and cracked coastal structures have been studied analytically and experimentally over the last decades (Wolters et al. 2004; Müller et al. 2003), but to the author's opinion no numerical model allowing to simulate progressive rock break-up by wave impact is currently available for practice.

FUTURE TRENDS IN DIGITALIZATION

This introductory chapter has shown that the process of digitalization and even digital transformation is currently ongoing in all engineering disciplines that are related to rock scour, even if some disciplines are more advanced than others.

All in all, the process of (data) digitization is well implemented, and digitalization of engineering models has benefited from huge improvements and developments over the last few decades. In most disciplines, engineers now dispose of an increasing number of numerical models that may provide additional details and complement traditional analytical and physical modelling procedures.

Nevertheless, the ultimate digital goal to attain a complete digital transformation of the engineering business is still to be performed for most disciplines. Some particular fields already started this process, for example hydraulic machines design and surveillance, or power house design and execution using VDC|BIM, where digital twins and related workflows have been strongly developed over the last decade.

Future trends in most disciplines will be of similar nature and be oriented towards further digitalization of distinct models, processes and workflows, in order to fulfil the final goal of a complete digital transformation of the discipline.

A key element in reaching the goal of complete digital transformation is undoubtedly the development of Digital Twins (DT), a digital replicate of the physical reality of an infrastructure, offering continuous and complete physical-to-digital connectivity using data sampled by sensor networks and generally transmitted through the Internet of Things (IoT). More details on digital twin developments in the field of dam engineering and rock scour can be found in Chapters 4 and 5.

Furthermore, once a DT is created, artificial intelligence (AI) and machine learning (ML) algorithms will continuously drive the digital twin to simulate future possible states and scenarios of the real system, allowing predictive or prescriptive maintenance, or to become more pertinent and accurate in digital replication and business decisions, using real-world datasets, all within the relative safety of a virtual world.

For example, AI has already been implemented in smart operation and maintenance systems of power houses, allowing to perform automated remote inspection by robots and advanced transmission technologies. This allows condition-based maintenance, with suggestions for start-up and shut-down sequences of machines, and lifetime predictions based on neural network modelling. As such, DTs and AI allow optimizing decision-making and enhancing outcomes through the use of data and subsequent simulation models, in the context of day-to-day operations or during extreme events, allowing the engineering assets to be tested and exposing potential problems on assets before they arise in the real world.

With a strongly growing interest in AI in all fields of our society, implementation of DTs and AI in rock scour engineering is only a question of time and of availability of the necessary and adequate technology for continuous and reliable real-time data acquisition. This digital challenge in the field of rock scour engineering is discussed in the next chapter.

REFERENCES

Azhar, S., "Building Information Modeling (BIM): Trends, Benefits, Risks, and Challenges for the AEC Industry", *Leadership & Management in Engineering*, 11, 241–252, 2011. https://doi.org/10.1061/(ASCE)LM.1943-5630.0000127

Barrias, A., Casas, J.R. and Villalba, S., "A Review of Distributed Optical Fiber Sensors for Civil Engineering Applications", *Sensors*, 16, 748, 2016. https://doi.org/10.3390/s16050748

Bollaert, E.F.R., "A Comprehensive Model to Evaluate Scour Formation in Plunge Pools", *International Journal of Hydropower & Dams*, 1, 94–101, 2004.

Bollaert, E.F.R., "Numerical Modeling of Scour at Bridge Foundations on Rock", in *Proceedings of the 5th Intl. Scour and Erosion Conference*, San Francisco, VA, 2010.

Briaud, J.-L., "Case Histories in Soil and Rock Erosion: Woodrow Wilson Bridge, Brazos River Meander, Normandy Cliffs, and New Orleans Levees", in *The 9th Ralph B. Peck Lecture, Journal of Geotechnical and Geoenvironmental Engineering*, Vol. 134, No. 10, ASCE, Reston, VA, USA, 2008.

Camici, S., Giuliani, G., Brocca, L., Massari, C., Tarpanelli, A., Farahani, H. H., Sneeuw, N., Restano, M. and Benveniste, J., "Synergy between Satellite Observations of Soil Moisture and Water Storage Anomalies for Global Runoff Estimation", *Geoscientific Model Development Discussion*, 15, 18, 6935–6956, 2022. https://doi.org/10.5194/gmd-15-6935-2022

Conaway, J., Eggleston, J.R., Legleiter, C.J., Jones, J.W., Kinzel, P.J. and Fulton, J.W., "Remote Sensing of River Flow in Alaska-New Technology to Improve Safety and Expand Coverage of USGS Streamgaging", *U.S. Geological Survey Fact Sheet*, 2, 4, 2019. https://doi.org/10.3133/fs20193024

Conde López, E.R., Toledo, M.Á. and Salete Casino, E., "The Centroid Method for the Calibration of a Sectorized Digital Twin of an Arch Dam", *Water*, 14, 19, 3051, 2022.

Crameri, T., Höttges, A. and Rabaiotti, C., "Distributed Fibre Optics Monitoring of the Lago Bianco Dam in Switzerland", in *Proceedings SMAR 5th conference on Smart Monitoring, Assessment and Rehabilitation of Civil Structures*, 2019.

Dasgupta, B., Basu, D., Das, K. and Green, R., "Development of Computational Methodology to Assess Erosion Damage in Dam Spillways", in *USSD 31st Conference*, San Diego, CA, April 11–15, 2011.

Deere, D.U. and Miller, D.W., "The Rock Quality Designation (RQD) Index in Practice, Classification Systems for Engineering Purposes", in *ASTM STP, American Society for Testing and Materials*, Philadelphia, PA, pp. 91–101, 1967. https://doi.org/10.1520/STP48465S

DGI GeoScience, Televiewer: Acquisition Services, Retrieved from: https://www.dgigeoscience.com/en/acquisition-services/, 2019.

FHWA, HEC-23, Report NHI-09-111, "Bridge Scour and Stream Instability Countermeasures Experience, Selection, and Design Guidance Third Edition, Volume 1", *US Department of Transportation - FHWA*, 2009.

Gardner, M., Kolg, J. and Sitar, N., "Parallel and Scalable Block System Generation", *Computers and Geotechnics*, 89, 168–178, 2017. https://doi.org/10.1016/j.compgeo.2017.05.001

Goff, C.A., Atyeo, M.S., Gimeno, O. and Wetton, M.N., "Dealing with Data: Innovation in Monitoring and Operation and Maintenance of Dams", *Dams and Reservoirs*, 26, 1, 5–12, 2016.

Hariri-Ardebili, A., Mahdavi, G., Nuss, L.K. and Lall, U., "The Role of Artificial Intelligence and Digital Technologies in Dam Engineering: Narrative Review and Outlook", *Engineering Applications of Artificial Intelligence*, 126, part A, 106813, 2023. https://doi.org/10.1016/j.engappai.2023.106813

Hopgood, A.A., *Intelligent systems for engineers and scientists: a practical guide to artificial intelligence*, CRC Press, Boca Raton, FL, 2021.

Hydro Exploitation SA, "SFMCP - Barrage de Chancy-Pougny, RELEVÉ BATHYMÉTRIQUE ETAT-E07", Internal Report, 2019.

Jordan, M.I. and Mitchell, T.M., "Machine Learning: Trends, Perspectives, and Prospects", *Science*, 349, 6245, 255–260, 2015.

Keaton, J., Mishra, S.K. and Clopper, P.E., *NCHRP report 717: scour at bridge foundations on rock,* Transportation Research Board, Washington, DC, 2012.

Kong, D., Wu, F., Saroglou, C., Sha, P. and Li, B., "In-situ Block Characterization of Jointed Rock Exposures Based on a 3D Point Cloud Model", *Remote Sensing*, 13, 2540, 2021. https://doi.org/10.3390/rs13132540

Leibniz, G.W., "De progressione dyadic", in E. Hochstetter & H.-J. Greve (eds), *Archives Leibniz de la Niedersächsische Landesbibliothek Hannover, fac-similé publié dans.*, Siemens Aktiengesellschaft, Berlin, 1679.

Liu, L., Liao, Z., Chen, C., Chen, J., Niu, J., Jia, Y., Guo, X., Chen, Z., Deng, L., Xu, H. and Liu, T., "A Seabed Real-Time Sensing System for In-Situ Long-Term Multi-Parameter Observation Applications", *Sensors*, 19, 5, 1255, 2019. https://doi.org/10.3390/s19051255

Loughrey S., McHugh, R. and MacDonald, K., "Future Dam Engineers in a Digital World", *Dams and Reservoirs*, 29, 4, 164–167, 2019. https://doi.org/10.1680/jdare.19.00060

Mammoliti, E., Di Stefano, F., Fronzi, D., Mancini, A., Maliverni, S. and Tazioli, A., "A Machine Learning Approach to Extract Rock Mass Discontinuity Orientation and Spacing from Laser Scanner Point Clouds", *Remote Sensing*, 14, 10, Art. 10, 2022. https://doi.org/10.3390/rs14102365

McArthur, J.J., "A Building Information Management (BIM) Framework and Supporting Case Study for Existing Building Operations, Maintenance and Sustainability", *Procedia Engineering,* 118, 1104–1111, 2015. https://doi.org/10.1016/j.proeng.2015.08.450

Müller, G., Wolters, G. and Cooker, M.J., "Characteristics of Pressure Pulses Propagating Through Water-Filled Cracks", *Coastal Engineering*, 49, 1, 83–98, 2003. https://doi.org/10.1016/S0378-3839(03)00048-6

Niemann, W.L., Wait, I.W. and Keaton, J.R., "A Proposed Risk-Based Screening Strategy for Bridges Potentially Affected by Rock Scour", *Environmental and Engineering Geoscience*, 23, 3, 221–241, 2017. https://doi.org/10.2113/gseegeosci.23.3.221

Onsel, I.E., Chang, O., Mysiorek, J., Donati, D., Stead, D., Barnett, W. and Zorzi, L., "Applications of Mixed and Virtual Reality Techniques in site Characterization", in *26th Vancouver Geotechnical Society Symposium,* 2019.

Opyrchal, L. and Chmielewski, R., "The Application of Infrared Thermography in the Diagnostics of Hydraulic Structures", *Dams and Reservoirs*, 33, 8, 1–17, 2022. https://doi.org/10.1680/jdare.22.00087

Pacina, J., Cajthaml, J., Kratochvílová, D., Popelka, J., Dvořák, V. and Janata, T., "Pre-Dam Valley Reconstruction Based on Archival Spatial Data Sources: Methods, Accuracy, and 3D Printing Possibilities", *Transactions in GIS*, 26, 1, 385–420, 2022.

Pietrangeli, G., Pittalis, G., Millesi, V. and Cifres, R., "Innovative Applications of Drone Technology in the Engineering of Large Dams", in *Proceedings of the International Conference on Hydropower and Dams*, Bordeaux, 2015.

Prendergast, L.J. and Gavin, K., "A Review of Bridge Scour Monitoring Techniques", *Journal of Rock Mechanics and Geotechnical Engineering*, 6, 138–149, 2014. https://doi.org/10.1016/j.jrmge.2014.01.007

Rasmussen, L.L., "UnBlocksgen: A Python Library for 3D Rock Mass Generation and Analysis", *SoftwareX*, 12, 100577, 2020. https://doi.org/10.1016/j.softx.2020.100577

Riquelme, A.J., Abellan, A. and Tomas, R., "Discontinuity Spacing Analysis in Rock Masses Using 3D Point Clouds", *Engineering Geology*, 195, 185–195, 2015. https://doi.org/10.1016/j.enggeo.2015.06.009

Riquelme, A.J., Abellan, A., Tomas, R. and Jaboyedoff, M. "A New Approach for Semi-Automatic Rock Mass Joints Recognition from 3D Point Clouds", *Computers and Geosciences*, 68, 38–52, 2014. https://doi.org/10/j.cageo.2014.03.014

Rossi, G.B., Cannata, A., Iengo, A., Migliaccio, M., Nardone, G., Piscopo, V. and Zambianchi, E., "Measurement of Sea Waves", *Sensors*, 22, 1, 78, 2022. https://doi.org/10.3390/s22010078

Scaioni, M., Marsella, M., Crosetto, M., Tornatore, V. and Wang J., "Geodetic and Remote-Sensing Sensors for Dam Deformation Monitoring", *Sensors*, 18, 3682, 2018.

Schenato, L., "A Review of Distributed Fibre Optic Sensors for Geo-Hydrological Applications", *Applied Science*, 7, 9, 896, 2017. https://doi.org/10.3390/app7090896

Schlumberger, FMI Fullbore Formation Microimager, https://www.slb.com/services/characterization/geology/wireline/fullbore_formation_microimager.aspx, 2013.

Shannon, C.E., "A symbolic analysis of relay and switching circuits", Thesis (M.S.) Massachusetts Institute of Technology, Dept. of Electrical Engineering, 1940.

Sheppard, C., Sheppard, A., Mogg, A., Bayley, D., Dempsey, A.C., Roche, R., Turner, J. and Purkis, S., "Coral Bleaching and Mortality in the Chagos Archipelago", *Atoll Research Bulletin*, 2017, 613, 1–26, 2017. https://doi.org/10.5479/si.0077-5630.613

Spero, H., Vazquez-Lopez, I., Miller, K., Joshaghani, R., Cutchin, S. and Enterkine, J., "Drones, Virtual Reality, and Modeling: Communicating Catastrophic Dam Failure", *International Journal of Digital Earth*, 15, 1, 585–605, 2022.

Thévenaz, L, Prince, R., Facchini, M. and Dardel, B., "Monitoring of Large Structure Using Distributed Brillouin Fibre Sensing", in *Proceedings of SPIE - The International Society for Optical Engineering,* 1999.

Verdelho Trindade, N.V., Ferreira, A. and Oliveira, S., "DamAR: Augmented Reality in Dam Safety Control", *International Journal of Hydropower Dams*, 26, 5, 56–62, 2019.

Verdelho Trindade, N.V., Ferreira, A. and Oliveira, S., "Extended Reality in the Safety Control of Dams", in *4th International Dam World Conference*, LNEC, Lisbon, Portugal, pp. 1–19, 2020.

Weidner, L. and George, M.F., "Generating Realistic Blocky Rock Masses from Point Clouds for Spillway Scour Assessments Using Semi-Automated Python Tools", in *4th Meeting of European Working Group on Overflow and Overtopping Erosion,* Lyon, 2023.

Wolters, G., Müller, G., Bullock, G., Obhrai, C., Peregrine, H. and Bredmose, H., "Field and Large Scale Model Tests of Wave Impact Pressure Propagation into Cracks", in *Proceedings of the 29th International Conference, National Civil Engineering Laboratory*, Lisbon, Portugal, 19–24 September 2004, https://doi.org/10.1142/5808

Xiao, J., Ji, G., Zhang, Y., Ma, G., Mechtcherine, V., Pan, J., Wang, L., Ding, T., Duan, Z. and Du, S., "Large-Scale 3D Printing Concrete Technology: Current Status and Future Opportunities", *Cement and Concrete Composites*, 122, 104115, 2021.

Zhang, Z., Demir, K.G. and Gu, G.X., "Developments in 4D-Printing: A Review on Current Smart Materials, Technologies, and Applications", *International Journal of Smart and Nano Materials*, 10, 3, 205–224, 2019.

Zheng, W., "Instrumentation and Computational Modeling for Evaluation of Bridge Substructures across Waterways", *FHWA report MS-DOT-RD-13-229, Department of Civil and Environmental Engineering Jackson State University,* 2013.

Zhou, R., Su, H. and Wen, Z., "Experimental Study on Leakage Detection of Grassed Earth Dam by Passive Infrared Thermography", *NDT & E International*, 126, 102583, 2022. https://doi.org/10.1016/j.ndteint.2021.102583

Zhu, X., Bao, T., Yeoh, J.K., Jia, N. and Li, H., "Enhancing Dam Safety Evaluation Using Dam Digital Twins", *Structure and Infrastructure Engineering*, 1–17, 904–920, 2021.

2 Rock Scour and Its Digital Challenge

INTRODUCTION

Terminology

The term "scour" etymologically means "to cleanse by hard rubbing" and probably has links with the Middle Dutch scuren, schuren "to polish, to clean" and with the Old French "securer". Both these old terms probably come from the Late Latin "excurare" (Medieval Latin scurare) or "clean off", literally "take good care of" (https://www.etymonline.com).

In current hydraulic terms, scour stands for the erosion of granular material or rock occurring over a region of limited extent of the stream bed or banks due to local flow conditions caused by the presence of hydraulic structures. As such, scour needs three components: flowing water, bed material and a hydraulic structure.

When speaking about the scour depth of rock at a hydraulic structure, however, different definitions may exist in literature. One of the most widely used terms is the ultimate or equilibrium scour depth, defined as a sort of end-condition of deepest scour that may occur at a hydraulic structure, generally related to a well-defined design flood event (hydrograph).

In reality, however, scour continuously progresses during the lifetime of the structure, by a large number of consecutive (minor and major) flood events that progressively fracture the rock mass and break up the rock blocks into smaller pieces (i.e. ball-milling). This actually means that all these flood events and their time durations should be accounted for to determine the ultimate scour at the end of the lifetime of the structure, and not just the design flood event. Nevertheless, in the remaining of this work, the computed ultimate scour depths are simplified and related to pre-defined unique design flood events.

Scour Mechanisms

A rock mass may break up by different mechanisms, depending on its physical, chemical and geomechanical characteristics, as well as on the intensity and duration of the flow action. A brief overview is given here, further information can be found in Hancock et al. (1998) and Keaton et al. (2012).

Rock Joint Fracturing

In the short term, i.e. during flood event durations of typically several hours, days or weeks, fracturing of the rock mass into distinct blocks is most often the primary break-up mechanism, especially in rock-containing joints and/or bedding planes.

DOI: 10.1201/9781003319610-2

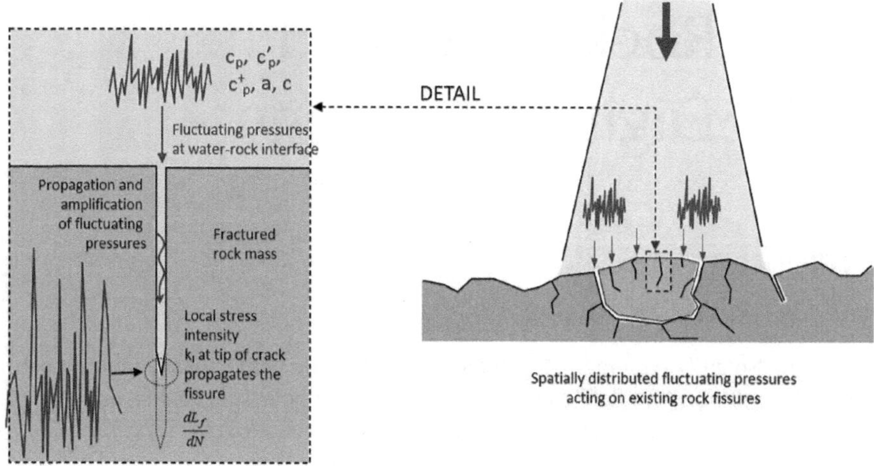

FIGURE 2.1 Progressive or instantaneous fracturing of an existing rock joint.

The dynamic water pressures generated by the impact of a turbulent flow onto a rock joint generate cyclic hydraulic jacking of the joints by pressure wave propagation and potentially amplification (Bollaert and Schleiss, 2005; Bollaert, 2021). As such, the corresponding rock mass stresses occurring at the tip of a joint may induce further fracturing (Figure 2.1). This process distinguishes between brittle (or instantaneous) joint propagation and time-dependent joint propagation. The former happens for stresses higher than the fracture toughness of the rock. In the opposite case, joints may be propagated by fatigue, depending on the frequency and amplitude of pressure load cycles.

Progressive or fatigue fracturing of rock joints is directly responsible for time dependency of rock scour formation during flood events. This break-up mechanism involves time scales that are generally on the same order of magnitude than time scales of spillway duration curves on dams or flood duration curves in rivers, i.e. from a few hours to a few thousands of hours.

Rock Block Plucking

Plucking of rock blocks may occur after a period of preconditioning of the rock by means of progressive or instantaneous fracturing of its joints. During this period, blocks progressively become completely detached from the bed along their bounding discontinuities. The joint surfaces generally are weathered, wedged apart and weakened by turbulent flows.

Once the blocks are detached, the impacting forces of turbulent flows may potentially generate a series of block movements expressed by translations, rotations or combinations of both, depending on the 3D orientation of the discontinuities bounding the block. The movement that is generally most pertinent to scour is block lifting generated by instantaneous pressure differences around the block upper and lower faces. From the scour modelling point of view, most engineering models do not consider time evolution of scour but make use of a scour threshold, under the form of an index, a velocity, an impulsion or a pressure difference, to express block lifting.

Some models do also consider other block failure modes. The more failure modes are considered, the more complex becomes the modelling and its digitalization process.

Rock Mass Abrasion

Rock break-up by abrasion mainly occurs by impact forces of sediment-laden flows, or ball-milling of rock or concrete material in stilling basins or plunge pools of hydraulic structures.

The former phenomenon is characterized by small grain sizes and may occur over a very wide range of time scales. It typically forms sculpted forms, potholes and pitted surfaces at the bedrock surface. For weak sedimentary rock, for example, significant abrasion may happen within a few hours, while for more resistant rock, any significant abrasion might need years or even much more to be notified.

Ball-milling of large pieces of rock or concrete, in contrast, generally occurs during a single flood event on a spillway of a dam. Most often, large downstream eroded pieces of rock and/or eroded concrete pieces are recirculating due to turbulent flows inside the stilling basin or plunge pool, generating significant damage to the concrete or bedrock, and accentuating in this way removal of other blocks, triggering additional scour.

The susceptibility of rock to abrasion is controlled primarily by density, hardness and fracture mechanics properties of both the rock and the impacting grains. The abrasion erosion rate is mainly controlled by bedrock susceptibility and density, mass concentration of the sediments and flow velocity (Hancock et al. 1998).

Figure 2.2 illustrates abrasion scour to the concrete apron of Paradise Dam in Queensland, Australia, following the 2013 flood event. Strong recirculating flows containing rock pieces and debris have caused extensive damage to the concrete of the stilling basin, together with significant scour of the downstream bedrock, and resulted in the destruction of part of the end wall of the apron (Lesleighter et al. 2016).

FIGURE 2.2 Abrasion erosion of concrete apron at Paradise Dam following the 2013 flood event (Lesleighter et al. 2016).

Accurate modelling of abrasion processes of bedrock at these structures is currently out of the realm of engineering and research. A comparative approach has been developed by Keaton et al. (2012), based on the definition of a geotechnical scour number (GSN) determined by modified Slake Durability Tests on rock samples.

Rock Mass Weathering

Weathering of rock may occur by physical processes, i.e. wetting-drying cycles, freezing-thawing cycles, plant-root growth and so on, or also by chemical processes, such as the effects of hydrolysis, carbonic acid and organic acids. Weathering may rapidly degrade the conditions of a rock mass and, as such, strongly influence its scour susceptibility.

Physical or mechanical weathering enhances fracturing or disintegration. An example is frost weathering, in which liquid water in rock cracks freezes and splits the rock. Temperature changes cause rock to expand (with heat) and contract (with cold). Over time, the rock weakens and crumbles. Another type of mechanical weathering occurs when clay or other materials near rock absorb water. Clay, more porous than rock, can swell with water, weathering the surrounding, harder rock. Salt also works to weather rock. Saltwater in cracks evaporates and salt crystals that are left behind grow and break the rock apart. Finally, plants and animals can be agents of mechanical weathering.

Chemical weathering softens, weakens and alters the rock material, by changing its molecular structure. For instance, carbon dioxide from the air or soil may combine with water. This produces carbonic acid that can dissolve rock. Carbonic acid is especially effective at dissolving limestone, creating karst formations containing sinkholes and caves.

Other types of chemical weathering are iron-rich rocks that generate rust by oxidation, or hydration, which is a form of chemical weathering in which the chemical bonds of the mineral are changed as it interacts with water. This occurs as anhydrite reacts with groundwater, transforming it into gypsum. Another familiar form of chemical weathering is hydrolysis, where a new solution is formed as chemicals in rock interact with water. In many rocks, for example, sodium minerals interact with water to form a saltwater solution. Living or once-living organisms can also be agents of chemical weathering. The decaying remains of plants and some fungi form carbonic acid, which can weaken and dissolve rock. Some bacteria can weather rock in order to access nutrients such as magnesium or potassium.

Similar to dissolution, from the scour modelling point of view, weathering processes are not modelled as such. Weathering is implemented indirectly, however, based on adequate preliminary geologic investigations and modification of rock resistance parameters such as joint shear strength or UCS strength, or initial degree of fracturing of the rock mass.

Rock Mass Cavitation

Cavitation may occur at hydraulic structures where the threshold conditions of velocity (and pressure) are attained, generating shock waves by implosion of vapour bubbles that may weaken and pit the concrete or underlying rock, such as for example in spillway tunnels or along sloped chutes of high-head dams. For example, cavitation-induced erosion of sandstone bedrock produced 10-m-deep cavities in 1983 in the 12.5 m diameter Glen Canyon Dam spillway tunnels that were discharging

900 m^3/s. Cavitation is generally triggered at flow separations induced by joints, bedding planes or other surface irregularities.

In contrast with chute spillways and flow diversion tunnels in dam engineering, cavitation-induced damage or erosion of bedrock in natural channels has not been documented widely. Research into the threshold slope, velocity and water depth for cavitation in rivers and channels has shown that it is not a significant erosion process in rivers and naturally flowing channels.

From the scour modelling point of view, cavitation is most often not considered to be a predominant scour mechanism. Today's hydraulic structures where the phenomenon might arise are adequately aerated to prevent it from happening, as are plunge pools that receive impinging turbulent flows.

Furthermore, in case of cavitation erosion, this will be immediately accompanied by one or more other rock break-up mechanisms, such as plucking or fracturing, making it very difficult to point out the relevance of cavitation to the final scour formation.

Rock Mass Dissolution

Highly soluble rocks, such as gypsum and salt, are generally not used as foundations for engineering structures. Lowly soluble rocks, such as limestone and dolostone, may be used as foundations but should be checked on their ability to generate caves and subsurface voids and sinkholes within the rock mass. These areas may contain debris or rubble following collapse of voids, and may be pertinent to scour potential.

Besides the incorporation of these voids, dissolution of rock as a process is generally not modelled during rock scour computations, but rather implemented indirectly based on adequate preliminary geologic investigations and modification of rock resistance parameters such as joint shear strength or UCS strength, or initial degree of fracturing of the rock mass.

Synthesis

A synthesis of the main rock break-up mechanisms that are relevant for scour at hydraulic structures is illustrated in Figures 2.3 and 2.4.

Figure 2.3 presents these mechanisms as subsequently occurring processes, from an initially intact rock mass towards a rock mass that is completely disintegrated into numerous small fragments, ready to be transported towards downstream as bedload.

Weathering processes generally occur at the surface of the rock mass during the whole process of progressive destruction.

Next, initially (quasi-) intact rock starts to be fractured along its weakest planes, i.e. local faults, joints, etc. This fracturing phase may take more or less time, depending on the vulnerability and the initial degree of fracturing of the rock. For example, sedimentary rocks will generally fracture much more quickly than igneous rocks.

Next, once distinct rock blocks are detached from their mass, plucking of these elements may start. For rather massive, intact rocks, the initial water-rock interface may be rather smooth. In such a case, block lifting will be generated primarily by (instantaneous) turbulent eddies and related turbulent lift forces.

For a rougher and more angular water-rock interface, block protrusion may generate additional quasi-steady lift and drag forces that enhance block ejection.

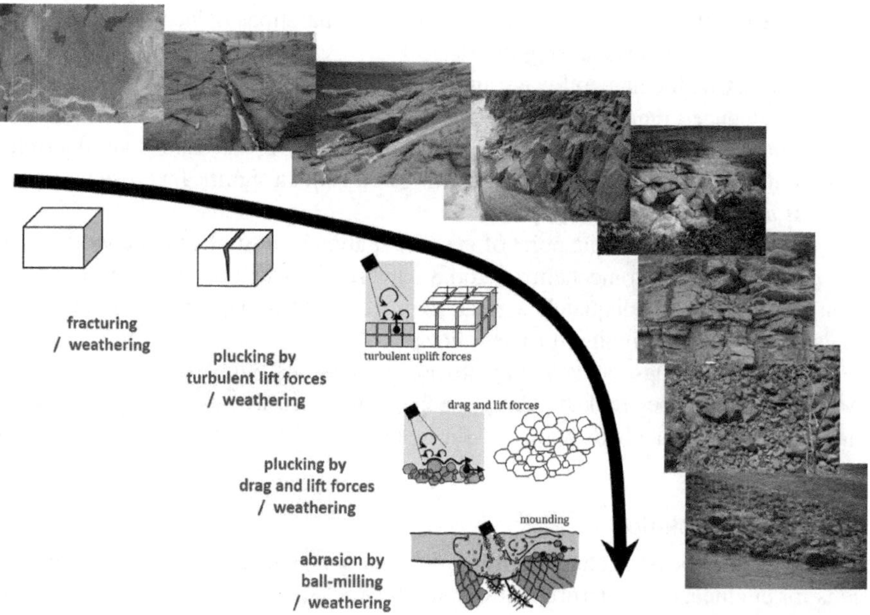

FIGURE 2.3 Schematic of consecutive break-up mechanisms of a rock mass.

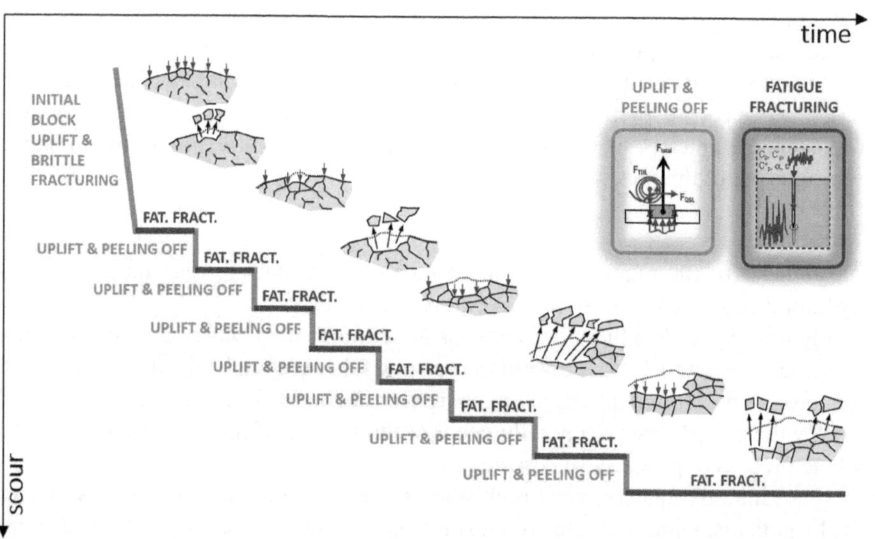

FIGURE 2.4 Sequence of break-up mechanisms of a rock mass at hydraulic structures.

During these lifting processes, blocks will start recirculating inside the turbulent flow and generate abrasion by ball-milling, i.e. blocks will progressively break up further into smaller pieces, until they can be transported towards downstream.

Figure 2.4 presents the sequential occurrence of both fracturing and plucking as it is most often encountered at hydraulic structures.

At first, depending on the initial degree of fracturing of the rock mass, lifting of already detached blocks and instantaneous fracturing of vulnerable joints, followed by lifting, will occur. Lifting thereby distinguishes between vertical uplift or ejection, based on quasi-steady and/or turbulent lift and drag forces, and peeling off, triggered by quasi-steady flows oriented quasi-parallel to the water-rock interface (Bollaert, 2012).

Second, time-consuming fatigue fracturing will progressively detach new blocks ready to be lifted. Simultaneously, abrasion of the rock interface may occur by recirculating blocks. This may further enhance break-up of rock joints, split already detached recirculating pieces and detach new rock elements.

The scour process takes an end when no further blocks can be detached anymore by fracturing, and/or when no newly detached blocks can be expelled anymore, despite their detached nature and despite ongoing abrasion and ball-milling.

Climate Change Relevance

Within the current context of climate change and more frequently occurring scour-generating flood events, rock scour risks or problems are increasing not only on many hydraulic infrastructures, such as dams, weirs, stilling basins, unlined channels, diversion tunnels, but also at bridges (piers, abutments) over waterways.

The phenomenon may result in serious damage to the infrastructure in question, and may even lead to its destruction. A recent example is the massive scour that occurred during a major flood event in 2017 at both the service and emergency spillways of Oroville Dam (US). This event has resulted in increased awareness and preparedness, and has triggered increased interest in this field. Nevertheless, major challenges remain to be tackled in rock scour engineering, as will be pointed out later in this chapter.

ENGINEERING DISCIPLINES

Dams and Appurtenant Structures

This field of engineering englobes a large and diverse number of hydraulic structures, such as dams, spillways, stilling basins, diversion tunnels, weirs, bottom outlets, etc.

The total number of dams suffering from rock scour during their operational lifetime is not known precisely but, based on literature and feedback from practice, it may be roughly estimated on the order of several hundreds to thousands worldwide. This number continuously increases due to climate change, generating extreme flood events more frequently and thus also increasing spillway functioning or risk for dam overtopping.

For some historic cases, scour risks or problems were such that they initiated extensive modelling and search for mitigation measures within the scientific or engineering community. As such, a body of experimental evidence, together with analytical and numerical techniques, is available today to assess the potential for rock scour downstream of dams. An overview is provided in Chapter 3, pointing out a severe lack of digitalization of these techniques.

When discussing rock scour at dams, distinction can be made between different types of structures and turbulent flows. Among others, the following situations of rock scour are encountered:

- plunge pool scour by free surface (crest) overflowing jets,
- plunge pool scour by ski-jump jets,
- plunge pool scour by pressurized jets,
- unlined channel scour downstream of weirs/spillways/stilling basins,
- bedrock scour following failure of concrete lining of spillway chute/stilling basin,
- dam abutment scour by overflowing jets and/or resulting high-velocity flows,
- tunnel scour by free surface or pressurized flow transfer.

For all these situations, scour may occur both along the bottom and the sidewalls of the plunge pools and channels, or along the entire tunnel cross-section. Sidewall scour formation may sometimes be initiated by return currents and secondary flows generated by an impacting jet in a confined space.

These situations are briefly discussed below, together with some explanatory examples.

Plunge Pool Scour by Free Surface Overflowing Jets

This type of scour may easily threaten dam safety, because the jet trajectories and related impact of the turbulent flows are situated close to the dam toe. As such, free surface overflows without protection measures are only used in case of good quality rock. Otherwise, scour mitigation measures are needed, such as jet splitters, rock bolting, a tailpond dam ensuring a minimum tailwater level or even an anchored concrete lining. Based on feedback from practice and literature, this type of scour is relatively rare compared to the other types of scour.

Laouzas Dam is a 50-m-high arch dam in the south-west of France, constructed in 1965 and owned and operated by EDF SA. The dam has three gated spillway bays near its crest, generating free surface flow conditions at full gate openings and lowly pressurized flow conditions at partial gate openings. The positioning of the gates and the shape of the concrete profile are designed such that jet trajectories easily detach from the invert. The downstream bedrock is a migmatite with a high UCS strength, but suffers from locally significant fracturing with depth.

As illustrated in Figure 2.5, scour of up to 4–5 m depth has formed in the unlined bedrock downstream, mainly along the right-hand side of the thalweg, and this already for a low outflow discharge of 62.5 m^3/s ($q \sim 5–10\,m^2/s$). Scour mitigation measures implemented by EDF SA in 2007 are rock bolting of the most heavily fractured areas.

Plunge Pool Scour by Ski-Jump Jets

This type of scour occurs frequently in unlined (and sometimes pre-excavated) plunge pools, but is most often not critical to dam safety, because ski-jump spillways generally allow the turbulent flows to partly dissipate their energy during their fall (by dispersion and aeration) and to impact the bedrock at a safe distance from the dam and its appurtenant structures.

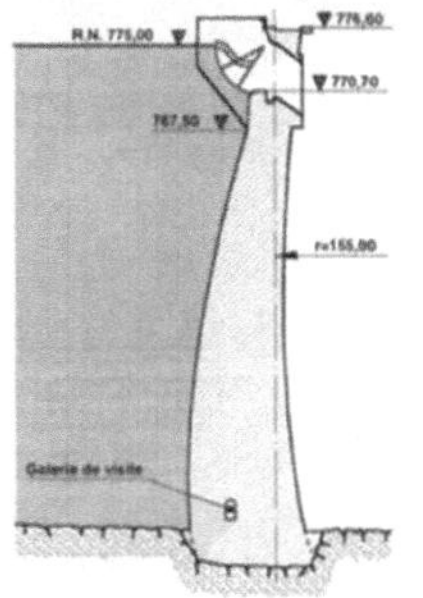

FIGURE 2.5 Turbulent flows and scour formation at Laouzas Dam. Courtesy of EDF SA.

FIGURE 2.6 Scour damage following the 2017 flood event at Chucás Dam, Costa Rica. Courtesy of M. O. Jimenez.

Chucás Dam is a 63-m-high concrete gravity dam on the Tarcola river in Costa Rica. The dam experienced its 100-year flood of record in 2017, i.e. only one year after dam commissioning. A 25-m-deep scour hole was formed in the andesitic lava, displacing 15,000 m^3 of rock towards downstream, and damaging the left bank of the plunge pool, fortunately without threatening the penstock (Figure 2.6).

Plunge Pool Scour by Pressurized Jets

This type of scour may also threaten dam safety, but this depends on the jet trajectory and location of impact of the turbulent flows. For low-head pressurized jets approaching free surface overflowing conditions, the location of impact will be close to the dam and may potentially threaten dam safety. For high-head pressurized jets, however, the location of impact will be further downstream, which reduces the risk for the dam.

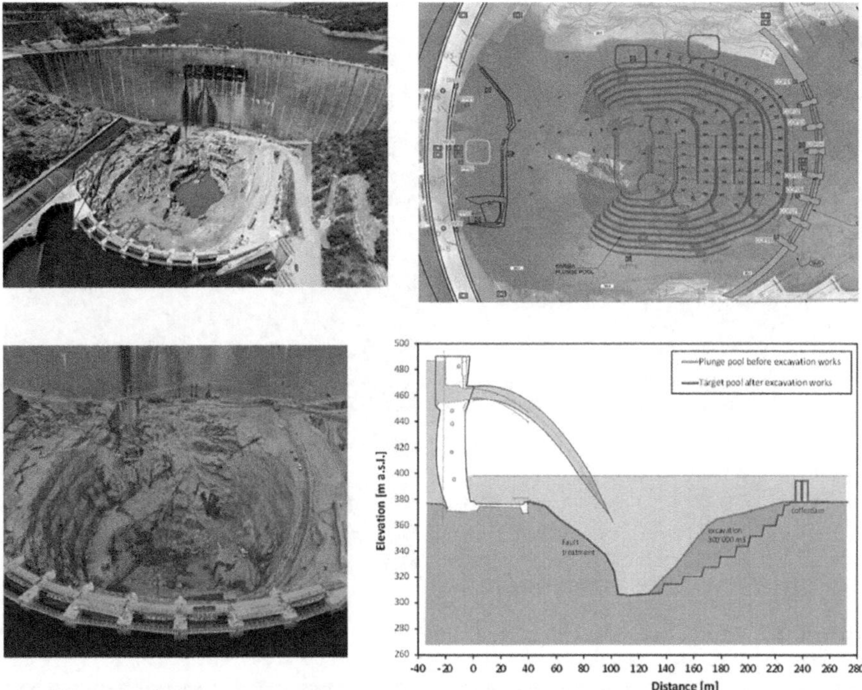

FIGURE 2.7 Scour mitigation measures implemented at Kariba Dam. Courtesy of M. C. Munodawafa, Zambezi River Authority.

The most iconic case is the more than 80-m-deep scour hole downstream of the 128-m-high Kariba Dam on the border between Zambia and Zimbabwe, where rock scour mitigation measures are being implemented by excavation of about 300,000 m^3 of rock towards downstream (Figure 2.7), to deviate the flow turbulence generated by the impacting jets away from the dam foundation. Furthermore, adequate treatment of a geological fault along the upstream rock face of the plunge pool is performed. These works are performed under (quasi) dry conditions created by the presence of a temporary cofferdam between the plunge pool and the powerhouse.

Unlined Channel Scour Downstream of Weirs/Spillways/Stilling Basins

River reaches or man-made channels situated immediately downstream of weirs, spillways or stilling basins are most of the time unlined. Under certain circumstances, however, the residual turbulence and intensity of the outflows from these hydraulic structures may be able to generate scour in the downstream bedrock. This type of scour may potentially threaten structural safety, especially in case of scour regression towards the structure.

Paradise Dam, a 37-m-high RCC gravity dam located in Queensland, Australia, experienced a 180-year return period flood event in 2013, which generated a quasi-15-m-deep local scour hole downstream of the left side of the spillway concrete apron and a quasi-5-m-deep local scour hole downstream and undercutting the

FIGURE 2.8 Scour damage following 2011 and 2013 flood events at Paradise Dam (up) and Boondooma Dam (down), Australia. Courtesy of Sunwater Ltd.

right-hand side of the primary spillway concrete apron (Figure 2.8). Provisional remediation measures were implemented by backfill concrete steps and anchor bars, partially filling the scour hole and minimizing the risk for apron undermining, and by a repair and strengthening of the concrete apron itself. Nevertheless, general dam safety concerns finally resulted in a 5.8 m temporal lowering of the crest of the primary spillway in 2020–2021.

In December 2021, the Queensland Government announced that Paradise Dam will be returned to its original height, as part of significant safety improvement works. This decision follows detailed technical investigations, which show it is possible to safely re-raise, strengthen and stabilize the dam wall. Paradise Dam Improvement Project (PDIP) works will include buttressing with mass concrete across the primary and secondary dam spillways and left abutment wall, new training walls, extending the existing downstream apron, and finally raising and replacing sections of the secondary spillway.

Finally, in January 2024, Sunwater Ltd. announced that a new dam will be constructed just downstream of the existing dam, because it was found that no amount of improvement work would allow fixing the latter.

Another example is the concrete ogee crest spillway of Boondooma Dam in Queensland, Australia, which experienced the same flood events and suffered from severe scour of the unlined spillway channel and sidewalls and of the plunge pool downstream of an erosion control structure. Spillway repair works have been done in 2017–2018 under the form of chute wall stabilization, concrete capping of local dykes, defensive rock anchoring, strengthening of the ogee crest with corrosion protected anchoring and finally construction of a secondary control structure upstream of the existing one.

FIGURE 2.9 Local bedrock scour and newly formed channel during the 2017 flood event at Oroville Dam Service Spillway.

Bedrock Scour Following Failure of Concrete Lining

Oroville Dam is a 235-m-high earthfill embankment dam in Northern California, US. The dam has been constructed from 1961 to 1967 and is owned and operated by the California Department of Water Resources (DWR). In February 2017, massive scour damage occurred to both the service and emergency spillways while releasing flood waters. Concerns for the stability of the spillway crest structures resulted in the temporary evacuation of 188,000 residents downstream of the dam.

Figure 2.9 shows the bedrock scour that occurred following local failure of the concrete lining of the service spillway. The flow not only created a local scour hole underneath the destroyed area of the lining, but was subsequently deviated into a naturally created channel in the bedrock adjacent to the damaged spillway.

Both service and emergency spillways have been fully reconstructed following the latest engineering standards since 2019, including many sensors producing automated data that is collected daily to check the structural performance.

Dam Abutment Scour by Overflowing Jets and/or Resulting High-Velocity Flows

This type of scour is rarely described in literature, despite its pertinence in practice for quite some dams. It is mostly related to unwanted dam crest overtopping, generating free surface jets impacting the abutments and resulting in high-velocity flows that may disperse towards downstream or progressively accumulate as bed-parallel flows close to the dam toe. Nevertheless, it might also occur by regular functioning of a free surface spillway directly situated above the abutment. Abutment scour is often critical to dam safety.

A well-known example is Gibson Dam, a 61-m-high concrete arch dam in Montana, US, that was overtopped in June 1964 during an extreme storm event. Fortunately, the abutment rock mass was of good quality and resisted quite well to the generated high-velocity flows, and thus only minor damage occurred (Frizell 2006).

In 1981, overtopping protection was constructed over portions of both abutments. Due to the fractured nature, the right abutment rock mass was protected by a concrete cap (Figure 2.10a). The concrete cap was reinforced with wire mesh and anchored with fully grouted rock bolts to resist potential hydrodynamic pressures. The rock mass along the left abutment was considered not erodible, except for a couple of weaker areas that were grouted in the more competent rock, together with bars

FIGURE 2.10 Upper graphs: Overtopping protection measures at Gibson Dam, US (Anderson et al. 1998, USBR): (a) right bank, (b) left bank. Lower graph: Overtopping protection measures at Arieskraal II Dam: (c) before measures, (d) after measures, (e) close-up of overburden rock erosion at dam toe. Courtesy of A. Chemaly.

anchored 3.3 m into the rock (Figure 2.10b). Finally, eight splitter piers were installed on the top of the dam, spaced about 30 m apart radially along the length, to provide aeration of the overtopping flows (Anderson et al. 1998; Frizell 2006).

A second example is Arieskraal II Dam, a 29-m-high double curvature concrete arch dam constructed in 1969 on the Palmiet River, some 14 km south of Grabouw in the Cape province of South Africa. The rock is a locally folded and fractured sandstone interbedded with shale.

Similar to Gibson Dam, a concrete lining was installed in 1998–1999 along the right bank to protect the rock from overflowing jets. This lining was completed by a downstream sidewall (A. Chemaly, personal communication 2023).

Bridge Piers and Abutments

Overview

Besides dams, bridges constructed over waterways may also face rock scour at their piers or abutments (Figure 2.11). The phenomenon of rock scour at bridge foundations has only recently been put under the spotlights of scientists and engineers, mainly by research conducted by the California Department of Transportation (CDOT) in the 1990s (Annandale and Smith 2001) and by the Transportation Research Board of the US Federal Highway Administration (FHWA) (i.e. NCHRP 717 Report "Scour at Bridge Foundations on Rock", Keaton et al. 2012). The latter report describes the different mechanisms of rock scour at bridge piers, as well as associated computational methods.

A 2014 report in the *Journal of Mechanics and Geotechnical Engineering* mentions that scour of bridge foundations is the number one cause of bridge collapse in the USA, and 600 bridges have failed due to scour problems between 1970 and 2000 (Briaud et al. 1999), causing major operating disruption and financial losses. State Department of Transportation Services (DOTs) in the USA have identified more than 26,000 bridges over waterways to be scour critical. From these, generally only a few percent is founded on rock, resulting nevertheless in an estimate of several hundreds of bridges nationwide.

Bridges founded on rock in Kentucky (US) were rated for scour condition by Hopkins and Beckham (1999) using a risk-based rating system that included the proximity of scour to the footing, the depths of holes caused by construction or scour, the distance that a hole might extend under a footing and the traffic exposure in terms of average annual daily traffic.

Froehlich et al. (1999) report that 366 Kentucky bridges have at least one pier founded on rock. Of these, the scour hazard is high for 8.5%, moderate for 12.1% and low for 79.4%. Hopkins and Beckham (1999) reported that scour is controlled by rock properties, including rock type, spacing of discontinuities, abrasion resistance and weathering rate.

Bridges in West Virginia are supported on rock at an average depth of about 6 m. Based on field review of scour modes in West Virginia, block plucking and quarrying is the most significant form of scour. Currently, dislodgment scour is estimated empirically due to the complexity of the hydraulics problem. When thinly bedded

FIGURE 2.11 View upstream at headcut across Montezuma Creek (Utah, US), a short distance upstream of the bridge. Courtesy of Jeff Keaton.

rock is exposed to high flow velocities, as much as almost 2 m of scour over the life of the bridge has been observed in West Virginia (source: https://transportation.wv.gov/highways/mcst/geotech/Pages/Geotechnical-Design/Rock-Scour.aspx).

Another example is the bridge crossing Montezuma Creek in Utah about 275 miles southeast of Salt Lake City (Figure 2.11). The bridge is founded on fluvial sandstone with claystone interbeds. By 2003, headcutting of the sandstone riverbed had migrated upstream from the bridge, attaining a height of 2.4 m in 2008 and exposing friable claystone under hard sandstone. Concrete retaining walls were constructed in 2004 to protect exposed claystone interbeds under the bridge foundations from further erosion (Keaton et al. 2012).

The Oregon Department of Transportation (Oregon, US) has determined that several bridges are supported by footings on potentially erodible rock material (Dickenson and Baillie 1999). They acknowledged that erosion of weak and jointed rock can lead to ongoing expensive maintenance problems at bridge footings and require frequent monitoring, channel modification and footing protection.

Engineering Tools

Very few engineering tools are currently available in the field of rock scour at bridges over waterways. Based on research conducted by the Colorado Department of Transportation (Smith 1994), Annandale and Smith (2001) applied the Erodibility Index Method (EIM) to bridge piers by proposing an equation relating the stream power in the scour hole near the bridge to the upstream stream power. The residual stream power in the scour hole reduces with increasing scour until equilibrium is obtained.

A conceptual stream power model has been proposed by Costa and O'Connor (1995), accounting for the total flood duration in calculating the total stream power on the extent of channel degradation, and thus focusing on the long-term exposure of rock to flow and not just the short-term exposure during peaks of flood events. Focusing on abrasion rather than block plucking, this model has been used by

Dickenson and Baillie (1999) to develop resistance of rock to abrasion by means of abrasion tests and an abrasion number, based on cumulative or integrated stream power.

Keaton et al. (2012) propose a method to compute bridge pier scour by abrasion, based on cumulative stream power and a modified slake durability test (Keaton and Mishra 2010) that allows determining a Geotechnical Scour Number (GSN).

Based on detailed transient numerical modelling made by Bollaert (2010), scour threshold charts and critical velocity charts for rock block plucking around bridge foundations have been developed and are available as downloadable appendices of the NCHRP 717 Report at www.trb.org. The charts account for the channel slope, the block protrusion and the block height respectively.

An overview of current rock scour calculation field practices in the US has been discussed by Liu et al. (2020). A state of the practice survey was conducted, including a literature review covering the concepts and understanding of scour susceptibility, material sampling, erodibility testing techniques, bridge scour calculation methods and other factors that can affect the scour analysis in scour-susceptible rock. Guidelines for practice are proposed.

Coastal Areas

Rock scour may occur in coastal areas containing natural cliffs. Similar to coastal blockwork structures and seawalls, these cliffs may contain cracks. The propagation of wave impact-induced pressures into such cracks is the reason for observed damages.

Due to transient effects into a confined space, a pressure difference arises between the front face and the interior of the cliff or the structure. This causes progressive cracking of existing joints and finally seaward removal of individual blocks. The mechanism was verified by model scale experiments and numerical simulation of impact pressure propagation, i.e. Müller et al. (2003), Wolters and Müller (2004), Wolters (2004).

As discussed in Briaud (2008), an example is the Pointe du Hoc cliffs in French Normandy, a World War II historical site. These cliffs are vertical walls which are about 25 m high and mainly consist of fractured limestone and sandstone. Since 1944, the cliffs have retreated by about 10 m, threatening an ancient observation post situated on top.

The bottom of the cliffs is attacked by waves especially during large winter storms. These waves have been recorded to reach 6 m and are superposed to the tides which can fluctuate by as much as 7 m. Inspection of the bottom of the cliffs showed the presence of caverns some of which were 3 m high and 3 m deep (Briaud 2008).

Observations indicated that large masses of cliffs had collapsed and that the failure plane was vertical, generating flat-shaped blocks of 1 m thick by 4 m long. Retreat of the cliffs is generated by the removal of these blocks by storm waves at the bottom, until the depth of the caverns becomes too large for the rock mass to sustain the weight of the overhang. Since the cliff line has lost 10 m in 60 years, such massive collapse would occur about every 25 years.

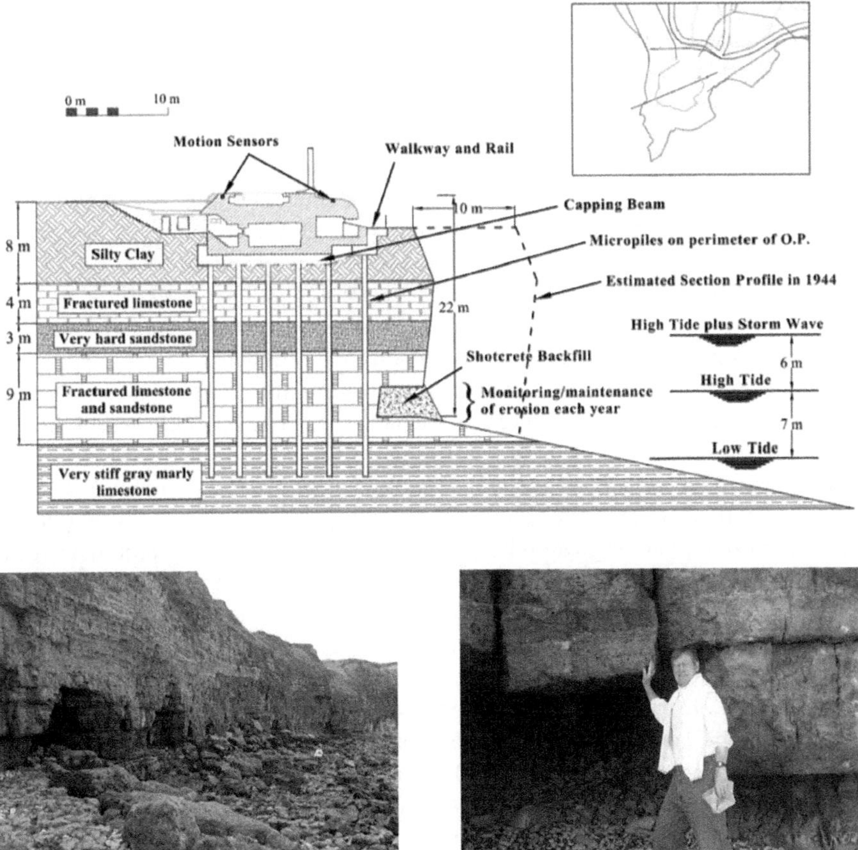

FIGURE 2.12 Rock scour of cliffs at Pointe du Hoc in French Normandy. Courtesy of Prof. J.-L. Briaud.

The remediation measures are shown in Figure 2.12 and consist of backfilling the caverns with grout under pressure to support their roof, to place the observation post on micropiles and to monitor to provide maintenance of the cliff base as necessary (Briaud 2008).

CHALLENGES IN ROCK SCOUR ENGINEERING

Compared to other fields of engineering, rock scour as an engineering discipline has received relatively few interests and attention in the past among engineers and researchers. Computational methods and concepts are available to the practitioner, but remain analytical or empirical-based, without any digitalized framework and with quasi no feedback from experience. While this current lack of advanced engineering tools may sound surprising at first sight, several explanations exist.

One reason may be the relative complexity of the phenomenon, given the interaction between two-phase air-water turbulent flows and scour of multi-layered fractured rock. In fact, rock scour engineering is transboundary and situated somewhere in between hydraulic, civil, geomechanical and geotechnical engineering. As such, a background in these fields is needed to fully understand and numerically reproduce the different break-up processes. As a result, while plenty of engineers and researchers work in hydraulic, civil or geomechanical engineering, those working in rock scour represent a kind of microcosmos in the world of infrastructure engineering.

Second, it has for a long time been considered that scour of rock can be simply resolved using the laboratory physical modelling that is generally performed preliminary to the realization of a major hydraulic engineering structure, such as a large dam. Making mostly use of granular soils and sediment transport approaches for mobile bed testing, or some weakly cemented mixture in the more elaborated version, the deepest scour measured on the laboratory has generally been taken for granted by designers.

Progressive feedback from practice, however, has shown that prototype scour may be significantly different from laboratory estimates. This has progressively initiated a paradigm change among hydraulic engineers, who finally had to admit that a rocky foundation obeys to laws described by rock mechanics and rock engineering rather than by small-scale laboratory hydraulics and sediment transport engineering.

Furthermore, in addition to this paradigm, the basic physical principles of break-up of a rock that is being scoured by flowing water do not match with the basic physical principles that are traditionally studied in rock mechanics and rock engineering. The former principles deal with fracture propagation by hydraulic jacking and with rock block plucking by dynamic pressure fluctuations and impulsions and quasi-steady lift and drag forces. The latter principles primarily focus on cohesion, effective shear strength, joint frictional forces, sliding resistance and so on.

Finally, advancements in engineering need data and feedback from experience. Traditional conservatism in the field of dam engineering (bound by confidentiality rules regarding potential problems on structures that form the base of hydropower generation in a competitive electricity market) makes it difficult to obtain relevant and complete data of observed or computed rock scour. For example, from the scarce list of scientific articles that have been published over the last half century, very few are those that allow retrieving a complete set of data to set up a case study.

As a result, it should be no big surprise that digital transformation in the field of rock scour engineering is only at its earliest beginnings. Besides the abovementioned issues, the remaining of this chapter points out the major technical issues and shortcomings of the digital transformation of rock scour engineering, based on a diagnosis of dam and rock engineering fields, and subdividing the digital challenge into digitization, digitalization and digital transformation phases.

ROCK SCOUR DIGITIZATION

Digitization of engineering data is a constantly ongoing process. Huge advancements have been made over the last decades in collection, cleaning and filtering of large amounts of novel digitally acquired data, and the term *big data* now starts to be used in engineering.

Nevertheless, the question of how to really profit from these data still remains open. Quasi no progress is noticed in standardization and sound interpretation of acquired digital data, i.e. ways are lacking to get relevant engineering parameters out of these data. An example is the development of new rock mass classification systems, which logically need a large number of precise values to be analysed by engineers following general standards. Following Yang et al. (2021), the current lack of both availability and standardization of geomechanical data acquired in the field prevents the development of new rock mass classification systems.

Furthermore, due to the advanced age of most of our dams and bridges, i.e. a lot of civil engineering infrastructure has been constructed following World War II, the original data used during the construction phase, or made available shortly after, is often only available through old and bad-quality copies and scans of original paperwork. Also, georeferencing of these data is based on ancient coordinate systems that are not used anymore today. Finally, the data is often incomplete and may have been obtained through old-fashioned imprecise or subjective measurement techniques.

As a result, data analysis and interpretation may quickly become a hazardous task, especially when historical data has to be completed by much more recent digital data obtained through up-to-date measurement techniques.

An example is provided in Figure 2.13, illustrating bathymetric and topographic data used for a rock scour analysis at an arch dam in the south-west of France (source: EDF-CIH, France). Digging into EDF's archives has allowed discovering ancient plunge pool scour measurements performed in the 1950s using primitive techniques (i.e. the water depth in the plunge pool is measured by a graduated stick manipulated from a boat, following a more or less fixed grid of positions of the boat through the pool).

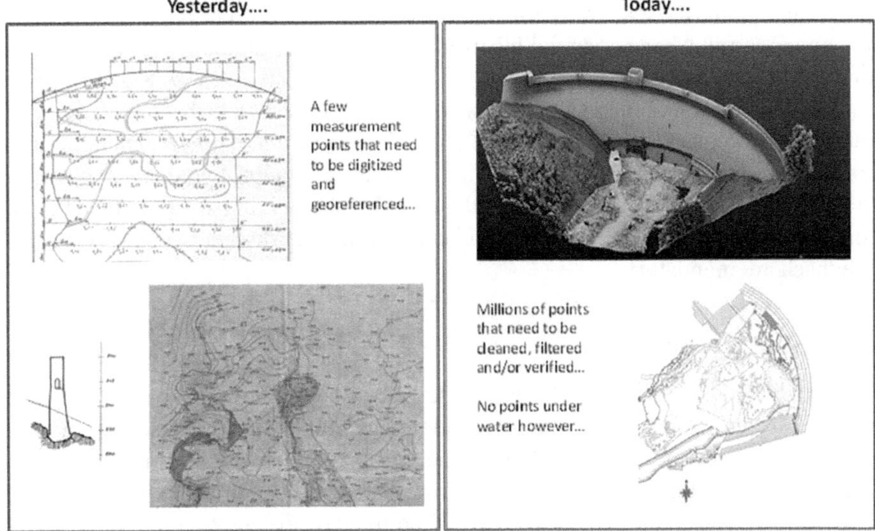

FIGURE 2.13 Comparison of ancient and recent bathymetric data available at an arch dam. Courtesy of EDF-CIH.

This information has to be compared with point clouds from TLS data recently obtained at ultra-high precision and following the latest French coordinate system. Moreover, after plunge pool drawdown, residually submerged areas may subsist along the water-rock interface. These prevent TLS from providing estimates of the bedrock stratum at those locations that are essential to estimate scour development with time.

Accurate mixing of ancient and recent data is essential to the calibration of rock scour numerical models, but clearly suffers from imprecisions and incompatibility issues.

ROCK SCOUR DIGITALIZATION

Numerical Modelling

As pointed out earlier in this chapter, the last decades have offered substantial and continuous progress in numerical modelling in all fields of engineering, and particularly in dam and rock engineering. For most of the physical processes of interest in hydraulics or rock mechanics, a fair number of commercial or open-source user-friendly software programs have become available to the practising engineer.

Nevertheless, most of these models are strongly limited in implementing the interactive physical processes between hydraulics and rock mechanics. As rock scour is an interdisciplinary field, involving aeration issues, 3D hydrodynamics and flow turbulence and complex 3D rock mechanics, there's actually still a lack of understanding of basic physical processes involved, and especially a lack of implementation of those physical processes into computational methods. Knowing that these interactions are essential to a sound and accurate understanding and reproduction of rock scour formation, there's still a fair amount of progress to be made.

Moreover, most professionals are modelling autonomously, applying their own models for specific cases, and monopolizing specific parametric settings and gained user experience at the end of the process. Whenever case studies are published or presented publicly, the source data is not made available for reasons of confidentiality or competitivity, and essential feedback regarding the parametric settings is often non-existent or incomplete.

As a result, most of the recent rock scour engineering projects that make use of the latest technologies and tools unfortunately do not allow our engineering community to advance. These projects rather function as isolated attempts to redefine the current state-of-the-art. We should not forget that, for the engineering art to be recognized, and for engineering models to be validated, worldwide dissemination, review and feedback are mandatory.

Rock Mass Classification

According to Yang et al. (2021), and as described before, digitization of rock engineering suffers from lack of standardization and sound statistical analysis of large databases created by a process of quantification of basically qualitative assessments. In other words, the principles of geotechnical data collection in the field, some of which are half a century old, have not been modified by the current digital revolution. We continue to subjectively interpret and analyse collected data, even if these data are acquired following the latest technologies. The data is being digitized, but the related workflow is not being digitalized.

One of the main reasons is the crucial role that empirical rock mass classification systems play in rock engineering, since their origin in the 1950s and 1960s (Elmo and Stead 2020). These systems allow a qualitative assessment of the rock mass quality, based on basic geologic descriptions and fieldwork, and allow then to transform this into quantitative data to be used by engineers. They may be considered today as worldwide industry standards.

The rock mass classification systems mostly used in rock mechanics and rock engineering are the Rock Mass Rating (RMR) system (Bieniawski 1976, 1989), the Q-system (Barton et al. 1974) and the Geological Strength Index (GSI) by Hoek (1994), Hoek et al. (1995) and Hoek and Brown (2019). Another system is the Kirsten (1982) excavatability index, which has profoundly inspired subsequent developments of rock erodibility approaches in the field of rock scour engineering, such as done by van Schalkwijk (1994), Moore et al. (1994), Annandale (1995), Wibowo et al. (2005) and Pells (2016).

Most of these classification systems strongly rely on the Rock Quality Designation (RQD; Deere et al. 1967), a widely used subjective parameter expressing an initial assessment of the rock mass quality. RQD has been widely discussed and debated since its origin, notably regarding its sensitivity to borehole orientation, its inability to cope with initial fractures, or its scale dependency and the 10 cm core lengths adopted as industry standard (Pells et al. 2017; Yang et al. 2021).

Elmo and Stead (2020) describe in more detail an empirical-digital paradox that rises out of this contradiction between digital input data and qualitative analysis. Despite the overwhelming amount of digital data acquired using the latest techniques, geologists and engineers still rely on old-fashioned subjective and qualitative parameters to describe and interpret the rock mass quality. These biased parameters are then used as quantitative input into next-gen digital modelling (Figure 2.14).

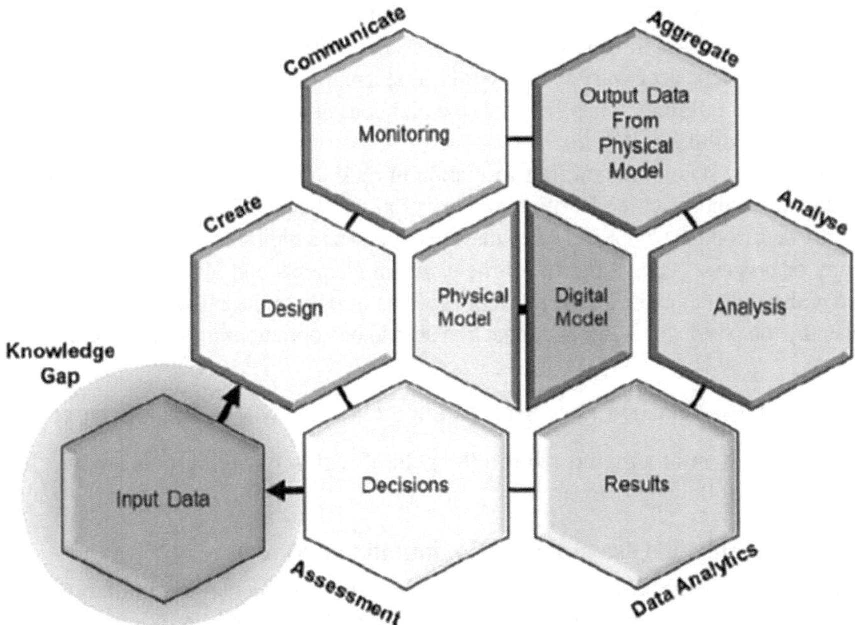

FIGURE 2.14 Empirical-digital paradox (Elmo and Stead 2020).

More details on the geology database at the origin of these classification systems can be found in Elmo and Stead (2020), as well as guidelines on the generation of new rock mass indicators and parameters.

The latter aspect would need accessible, detailed and standardized databases of different types of rock masses, to be used and compared with the original data at the base of the current classification systems. This will allow objective digital databases that should be shared.

As a conclusion, the process of digitalization of the main workflows in the field of rock engineering, and thus also in rock scour engineering, is far from being adopted today. Some attempts are nevertheless being made towards digitalization and even digital transformation, for example, by implementation of Discrete Fracture Network (DFN) modelling.

ROCK SCOUR DIGITAL TRANSFORMATION

INTRODUCTION

Digital transformation can be defined as the implementation of new, fast and frequently changing digital technology to solve processes and workflows. Digital transformation involves the generation and application of new technologies, but this is not sufficient for a successful transformation.

The most important aspect of digital transformation is not really the technology itself, but the transformation of culture and practices of the professionals in the field. Undergoing digital transformation requires professionals to change their approach to virtually everything they do. Truly embracing digital transformation requires that engineers and researchers challenge the status quo and learn to experiment and implement changes frequently, to adopt the digital mindset.

Digital transformation is different for every entity, and this makes it tricky to describe the steps necessary for a digital transformation. Digital transformations can be defined by talking in terms of what the end goal of a digital transformation is: the creation of a digital platform.

The utmost state of digital transformation in rock scour requests implementation of digital twins and machine learning and artificial intelligence. Digital twins have been shortly described in Chapter 1. They may be defined as a digital replication of a physical entity or process, accounting for all the essential elements and allowing to accurately and realistically simulate all the relevant aspects throughout the entity/process lifecycle, whereby the physical-to-digital connection should be continuous and automatized.

WHAT ARE THE CHALLENGES IN SETTING UP A DIGITAL TWIN IN ROCK SCOUR?

Successful set-up of a digital twin in the field of rock scour represents a huge challenge for the following reasons.

Lack of Reliable and Accurate 3D Computational Models

Such models are necessary for both hydraulic and geomechanics simulations. No numerical models allow to accurately simulate the formation of rock scour in a 3D space, by accounting for the coupling between the air, the hydraulics and the rock mechanics, and by reliably simulating the different break-up mechanisms of the rock mass.

Lack of Accurate Geomechanical Data

Detailed geomechanical data is very difficult to acquire (and to manage), especially with depth in the rock mass. Stochastically determined 3D rock mass models, such as DFN, may represent the rock mass in a probabilistic manner, but are not virtual replicates of the real rock mass.

Lack of Automatic Connection between the Physical and the Virtual World

While real-time or even ahead-of-time follow-up is feasible for the hydraulics of a spillway, for example, by hydrological forecasting and real-time monitoring of reservoir levels and in- and outflows, things become much more challenging when monitoring rock that is submerged during flood events. Depending on the infrastructure in question, we most often do not have the capability to closely follow-up in real time what is happening at the interface between the water and the rock during a flood event.

This is because of the lack of visual contact and the huge difficulty to monitor the rock mass from inside by IoT sensors. For example, high-velocity turbulent flows such as jets plunging into a pool make real-time remote sensing of the water-rock interface quasi-impossible by currently available monitoring techniques, because of difficulties in access, fixing of sensors and signal transfer through highly aerated and turbulent flows.

Real-time bathymetric monitoring of rock is feasible for example around bridge piers and abutments, by fixing the submerged sensors on the bridge structure and by measuring through the water. Nevertheless, only local scour values are generally obtained instead of a complete bathymetry.

Extremely Widespread Time and Spatial Scales Should Be Covered

Time scales range from a few milliseconds for transient pressure fluctuations to tens of years of digital twin waiting time before a flood event is occurring on the infrastructure. Spatial scales range from millimetres for 3D rock joints to hundreds of metres of water-rock interface, and tens of metres of rock mass depth, to virtually replicate.

Lack of Big Data

The twin should allow continuous and automated recalibration and refinement of the implemented modelling processes, based on experienced flood events and using ML and AI techniques. Few or even no events may occur during the lifetime of the structure, making it very difficult to apply artificial intelligence and machine learning for model enhancements.

As a summary, setting up a complete digital twin for rock scour seems not feasible with currently available technology, and cornerstone elements that are absolutely needed prior to any digital twin development are still in their infancy phase, such as 3D digital scour modelling, or sound understanding of how a rock mass may break up during a flood event. Nevertheless, it is still worthwhile considering to develop (incomplete) rock scour digital twins and the reader should not be discouraged. This is explained hereafter.

Why Should a Digital Twin in Rock Scour Be Considered?

Despite the multiple constraints and challenges, use of digital twins in rock scour engineering should be developed for the following reasons:

- digital twins are the only reliable and efficient alternatives to Structural Health Monitoring (SHM), which only offers pre-defined periodic sampling of structures,
- digital twins allow to perform predictive (i.e. pro-active) and prescriptive (i.e. pro-active with action plan) maintenance on infrastructures,
- partial (or incomplete) digital twins, omitting automatic real-time updates of the rock mass during flood events and making use of simplified models, may still be useful when using correctly calibrated/validated simulation models, i.e. the type and amount of useful information that can be obtained in real time or even ahead-of-time is still significant and promising for sound risk assessment pertaining to rock scour,
- digital twins of dams and bridges are currently under development for other issues than rock scour, as such the inclusion of rock scour into these twins may be imagined,
- part of these digital twins under development will use hydraulic and/or rock mass data that can also be used for the rock scour digital twin,
- once a rock scour digital twin is set up, relatively minor efforts are needed to maintain the twin connected with the physical reality during the lifetime of the infrastructure, as flood events do not occur very frequently,
- implementation of digital twins forces infrastructure owners and stakeholders to perform a complete digital transformation of their business and workflows, in order to optimize the lifetime of their assets.

Machine Learning (ML) and Artificial Intelligence (AI) Principles in Rock Scour

As discussed in Chapter 1, ML and AI principles are increasingly used to solve dam and rock engineering problems, based on the immensely growing amount of digital data from remote sensing. In rock scour engineering, ML and AI a priori cannot be used when such digital data of the rock mass is not available.

Nevertheless, ML and AI are not only applicable to digital data obtained from remote sensing but also to classical approaches such as geomechanical indices for rock mass characterization (RMS, Q, GSI). In fact, the process allows to continuously correct and increase the precision of these indices during the lifetime of an infrastructure. Whenever automatized high-precision digital data from the rock mass is not readily available to the engineer, which unfortunately is often the case, ML and AI offer a platform for getting most out of the biased classical approach.

Within this context, a digital twin does not necessarily need to be connected with automatized high-precision real-time available data, but can also be imagined to be connected with new data acquired using less digital techniques and involving a not fully automatized workflow.

As an example, a rock scour digital twin may rely on a manually governed update of the bathymetry of a plunge pool scour hole following a flood event, using traditional sensing techniques. The main issue is not the way the bathymetry has been obtained, but rather the quality of the workflow that asks for an immediate update of the bathymetry after each flood event on the dam. The availability of such an updated bathymetry will then allow the digital twin to trigger a process of recalibration of the relevant parameters of the rock scour modelling process, by potential use of ML and AI algorithms. At the end, a sound validation of the rock scour model is obtained.

Care should be taken in determining machine algorithms during these learning processes, and the amount and quality of underlying data should be as precise and objective as possible.

Digital Outlook

As a conclusion, rock scour engineering is a relatively young field of engineering, with strong roots into empiricism and personal experiences, and with a process of digital transformation that is currently still in its infancy phase.

To benefit from scientific and technological advancements, rock scour engineering is strongly dependent on other engineering fields that have their own digitalization progress and issues (river and reservoir hydraulics, civil engineering, rock mechanics and engineering, geotechnical engineering, dam engineering), and each of these should digitally advance, for rock scour to advance and become fully digitalized.

A logical first step towards digital transformation of rock scour engineering is undoubtedly to develop and make available digitalized computational models implementing and coupling the necessary physics and relying on digital data. These models should run in (quasi-) real time and allow the engineer to obtain reliable and sufficiently precise information on rock scour at infrastructures. They should be easily updatable and refinable.

In line with the current lack of precise and objective geomechanical indices defining the quality of a rock mass, these digitalized models should allow significant progress in this area based on data-driven intelligence, i.e. by allowing systematic refinement of parameters based on a large number of cases and data. In this regard, future models should be able to benefit from any digital data on rock scour that are worldwide available.

In parallel, digital platforms should be created with the aim to offer an on-line environment that is easily accessible and updatable. These platforms should house all the elements of the lifecycle of the digital counterpart of the real asset, such as its current geometrical state and structural health, its health history and related data, the evolution of essential parameters, diagnostic and prognosis of near-future potential states. These platforms should also house the numerical engines needed for diagnoses and prognoses, allowing to learn from experiences and to project future potential states of the physical asset.

The aforementioned digital computational models represent the necessary engines of future rock scour digital twins and platforms, while the latter guarantee the body framework. Both elements are strongly interrelated and mandatory to construct a digital twin.

Hence, as a first step towards rock scour digital twins at hydraulic structures, there is a need to develop an adequate digital environment with the following main requirements:

- generation of computational models with reliable and sufficiently precise implementation of the physics, phase coupling and rock break-up mechanisms, on sufficiently detailed replica of the physical reality,
- preservation of adequate computational times and efforts, such that real-time and ahead-of-time usage becomes feasible, i.e. such that we can get the answers when we need them,
- instant, universal accessibility and portability of the digital environment, i.e. easy personalized and secured access and usage, uniqueness of the results,
- easy and fast connection and compatibility with external digital systems (i.e. models, platforms, sensors, twins, etc.),
- possibility to run as an independent digital twin for rock scour, making use of the available digital engines and digital data transmitted by external sources,
- possibility to implement model recalibration and/or refinement following experienced flood events, for example, based on machine learning techniques or reliability analyses.

The remainder of this work describes an interrelated set of digital advancements performed with the aim to fulfil all these digital requests, and to generate the best possible digital twin one can imagine within the framework of currently available technology in rock scour engineering. These advancements cover digital platforms (Chapter 4), digital twins (Chapter 5), digital engines (Chapter 6) and finally digital applications (Chapter 7).

REFERENCES

Anderson, C., Mohorovic, C., Mogck, L., Cohen, B. and Scott, G., "Concrete Dams Case Histories of Failures and Nonfailures with Back Calculations", *Report DSO-98-05, U.S. Department of Interior Bureau of Reclamation,* Dam Safety Office, Denver, US, 1998.

Annandale, G.W., "Erodibility", *Journal of Hydraulic Research*, 33, 471–494, 1995.

Annandale, G.W., and Smith, S.P., "Calculation of Bridge Pier Scour Using the Erodibility Index Method", *Technical Report CDOT-DTD-R-2000-9*, California Department of Transportation, US, 2001.

Barton, N., Lien, R. and Lunde, J., "Engineering Classification of Rock Masses for the Design of Tunnel Support", *Rock Mechanics*, 6, 189–236, 1974.

Bieniawski, Z.T., "Rock Mass Classification in Rock Engineering", in Bieniawski, Z.T., Ed., Symposium Proceedings of Exploration for Rock Engineering, 1, 97-106, 1976.

Bieniawski, Z.T., *Engineering rock mass classification*, Wiley, New York, 1989.

Bollaert, E.F.R., "Rock scour at hydraulic structures: a practical engineering approach", *Geo*-Strata, 2010.

Bollaert, E.F.R., "Wall Jet Rock Scour in Plunge Pools: A Quasi-3D Prediction Model", *International Journal on Hydropower & Dams*, 19, 4, 70–76, 2012.

Bollaert, E.F.R., "The Rocsc@r Cloud: An Innovative Digital Platform to Compute and Record Rock Scour", *International Journal on Hydropower & Dams*, 28, 5, 60–70, 2021.

Bollaert, E.F.R. and Schleiss, A.J., "Physically Based Model for Evaluation of Rock Scour due to High-Velocity Jet Impact", *Journal of Hydraulic Engineering*, 131, 3, 153–165, 2005. https://doi.org/10.1061/(ASCE)0733-9429(2005)131:3(153)

Briaud, J.-L., "Case Histories in Soil and Rock Erosion: Woodrow Wilson Bridge, Brazos River Meander, Normandy Cliffs, and New Orleans Levees", in *The 9th Ralph B. Peck Lecture, Journal of Geotechnical and Geoenvironmental Engineering*, Vol 134 No. 10, ASCE, Reston, VA, 2008.

Briaud, J.-L., Ting, F.C.K., Chen, H.C., Gudavalli, R., Perugu, S. and Wei, G., "SRICOS: Prediction of Scour Rate in Cohesive Soils at Bridge Piers," *Journal of Geotechnical Engineering*, 125, 4, 237–246, 1999.

Costa, J.E. and O'Connor, J.E., "Geomorphically effective floods", in J.E. Costa, A.J., Miller K.W., Potter and P.R. Wilcock, (eds.), *Natural and Anthropogenic Influences in Fluvial Geomorphology-Wolman*, American Geophysical Union Geophysical Monograph, Washington, DC, vol. 89, pp. 45–56, 1995.

Deere, D.U., Hendron, A.J., Patton, F.D. and Cording, E.J., "Design of surface and near surface construction in rock", in C. Fairhurst (ed.), *Proceedings of the 8th U.S. Symposium on Rock Mechanics-Failure and Breakage of Rock*, American Institute of Mining, Metallurgical and Petroleum Engineers, Inc., New York, pp. 237–302, 1967.

Dickenson, S.E. and Baillie, M.W., "Predicting Scour in Weak Rock of the Oregon Coast Range", *Unpublished Research Report*, Department of Civil, Construction, and Environmental Engineering, Oregon State University, Corvallis, OR, Final Report SPR 382, Oregon Department of Transportation and Report No. FHWA-OR-RD-00-04, 1999.

Elmo, D. and Stead, D., "Disrupting rock engineering concepts: is there such a thing as a rock mass digital twin and are machines capable of learning rock mechanics?", in P.M. Dight (ed.), *Slope Stability 2020: Proceedings of the 2020 International Symposium on Slope Stability in Open Pit Mining and Civil Engineering*, Australian Centre for Geomechanics, Perth, pp. 565–576, https://doi.org/10.36487/ACG_repo/2025_34, 2020.

Frizell, K., "Hydraulic Investigations of the Erosion Potential of Flows Overtopping Gibson Dam Sun River Project, Montana Great Plains Region", *Hydraulic Laboratory Report HL-2006-02 US Bureau of Reclamation*, Denver, US, 2006.

Froehlich, D.C., Hopkins, T.C. and Beckham, T.L., "Preliminary assessment of local scour potential at bridge piers founded on rock", in E.V. Richardson and P.F. Lagasse (eds.), *Stream Stability and Scour at Highway Bridges: Water Resources Engineering Division*, ASCE, Reston, VA, pp. 976–980, 1999.

Hancock, G.S., Anderson, R.S. and Whipple, K.X., "Beyond power: bedrock incision process and form", in K.J. Tinkler and E.E. Wohl (eds.), *Rivers Over Rock: Fluvial Processes in Bedrock Channels: American Geophysical Union, Geophysical Monograph*, vol. 107, pp. 35–60, 1998.

Hoek, E., "Strength of Rock and Rock Masses", *ISRM News Journal*, 2, 2, 4e16, 1994.

Hoek, E. and Brown, E.T., "The Hoek-Brown Failure Criterion and GSI - 2018 Edition", *Journal of Rock Mechanics and Geotechnical Engineering*, 11, 3, 445–463, 2019. https://doi.org/10.1016/j.jrmge.2018.08.001

Hoek, E., Kaiser, P.K. and Bawden, W.F., *Support of underground excavations in hard rock*, A.A. Balkema, Rotterdam/Brookfield, 1995.

Hopkins, T.C. and Beckham, T.L., "Correlation of Rock Quality Designation and Rock Scour Around Bridge Piers and Abutments Founded on Rock", *Kentucky Transportation Center*, College of Engineering, University of Kentucky, Report No. KTC-99-57, 1999.

Keaton, J.R. and Mishra, S.K., "Modified Slake Durability Test for Erodible Rock Material", in Burns, S. E., Bhatia, S. K., Avila, C. M. C., and Hunt, B. E., eds., *Scour and Erosion: Proceedings of the Fifth International Conference on Scour and Erosion*, November 7–10, 2010, San Francisco, American Society of Civil Engineers Geotechnical Special Publication No. 210, 743–748, 2010.

Keaton, J.R., Mishra, S.K. and Clopper, P.E., "NCHRP Report 717 : Scour at Bridge Foundations on Rock", *Transportation Research Board*, Washington, DC, 2012.

Kirsten, H.A.D., "A Classification System for Excavation in Natural Materials", *The Civil Engineer in South Africa*, 24, 7, 292–308, 1982.

Lesleighter, E.J., Bollaert, E.F.R., McPherson, B.L., Scriven, D.C., "Spillway Rock Scour Analysis - Composite of Physical & Numerical Modelling", Paradise Dam, Australia, International Symposium on Hydraulic Structures, June, Portland, USA, 2016.

Liu, Z.L., Sheefa, D.E., Vitton, S. and Barkdoll, B., "Improved Calculation of Scour Potential in Cohesive Soils and Scour-Susceptible Rock", Michigan Department of Transportation, Report OR 19-132, Michigan, US, 2020.

Moore, J., Temple, D. and Kirsten, H., "Headcut Advance Threshold in Earth Spillways", *Bulletin of the Association of Engineering Geologists*, 31, 2, 1994.

Müller, G., Wolters, G. and Cooker, M., "Characteristics of Pressure Pulses Propagating through Water Filled Cracks", *Coastal Engineering*, 49, 1–2, 83–98, 2003.

Pells, S., "Erosion of rock in spillways", *PhD Dissertation*, School of Civil and Environmental Engineering Faculty of Engineering, University of New South Wales, 2016.

Pells, P.J., Bieniawski, Z.T., Hencher, S.R. and Pells, S.E., "Rock Quality Designation (RQD): Time to Rest in Peace", *Canadian Geotechnical Journal*, 54, 6, 825–834, 2017. https://doi.org/10.1139/cgj-2016-0012

Smith, S.P., "Preliminary Procedure to Predict Bridge Scour in Bedrock", *Colorado Department of Transportation*, US, 1994.

van Schalkwijk, A., "Minutes - Erosion of Rock in Unlined Spillways", *ICOLD*, Q.71 R.37, 1056–1062, 1994.

van Schalkwijk, A., Jordaan, J. and Dooge, N., "Die erodeerbaarheid van verskillende rotsformasies onder varierende vloeitoestande," Tech. Rep. WNK Verslag No. 302/1/95, *verslag aan die waternavorsingskommissie deur die Departement of Geologie*, Universiteit van Pretoria, South Africa, 1994a.

Veronese, A., "Erosion de fond en aval d'une décharge", *IAHR Meeting for Hydraulic Works*, Berlin, 1937.

Wibowo, J., Villanueva, E., Temple, D. and Yule, D., "Earth and rock surface spillway erosion risk assessment", *U.S. Symposium on Rock Mechanics*, Paper No. 05-813. Alexandria, VA: American Rock Mechanics Association, 2005.

Wolters, G., "Characteristics of wave impact induced pressure pulse propagation in cracks of coastal structures", *PhD Thesis*, Queen's University Belfast, 2004.

Wolters, G. and Müller, G., "The Propagation of Wave Impact Induced Pressures into Cracks and Fissures", *Coastal Chalk Cliff Instability, Engineering Geology Special Publications*, Geological Society, London, 20, 121–131, 2004.

Yang, B., Mitelman, A., Elmo, D. and Stead, D., "Why the Future of Rock Mass Classification Systems Requires Revisiting Their Empirical Past", *Quarterly Journal of Engineering Geology and Hydrogeology*, 55, 1, qjegh2021-039, 2021. https://doi.org/10.1144/qjegh2021-039

3 Rock Scour Computational Methods

INTRODUCTION

The discipline of rock scour engineering is still very young, but yet a large number of computational methods exist to describe short- or medium-term scour formation, i.e. scour that forms during time frames that can be associated with the duration of one or more flood events in rivers and reservoirs. A recent review of such methods can be found for example in Kashtiban et al. (2021).

The present chapter provides a brief overview of these methods from the viewpoint of digitalization. To do this, distinction has been made between the three phases that govern rock scour formation: the air (gas phase), the water (liquid phase) and finally the rock (solid phase). The computational methods have been checked on the implementation of each of the phases, and on the degree of digitalization of this implementation.

A visual comparison of the computational methods is expressed by the digital knowledge cube illustrated in Figure 3.1. This cube represents the three phases along its three axes. By advancing along the axes from the 0-origin, the degree of digitalization progressively increases. One can notice for example the rock mass axis, starting with empiricism and rock mass indices on the left-hand side, close to the 0-origin, and stretching towards numerical modelling in the centre, to finally reach phase coupling and digital twin integration of the numerical modelling on the right-hand side. Similar subdivisions are performed along the other two axes, representing both the air and water phases.

As such, the ultimate digital state is situated in the upper right corner of the digital cube and represents a full digital twin and artificial intelligence integrated modelling that allows sound and automated asset management of hydraulic structures in real time and ahead-of-time. The needs and constraints of such a full digital twin were described more in detail in Chapter 2.

Each computational method may be attributed a positioning inside this 3D space, depending on the degree of digitalization of the method pertaining to each of the three phases of rock scour. For example, methods that are purely empirical or analytical based will be positioned close to the 0-origin, inside the grey area dedicated to analytical methods in Figure 3.1. In the same way, 1D, 2D and 3D numerical-based methods may be positioned inside grey areas that extend much further inside the cube. Furthermore, for numerical methods that implement coupling between the different phases, such as for example aeration of turbulent flow or coupling between turbulent flow and scour formation, the corresponding space allocated inside the digital cube extends progressively towards the upper right corner of the cube. This upper right corner represents the full digital twin described in Chapter 4.

DOI: 10.1201/9781003319610-3

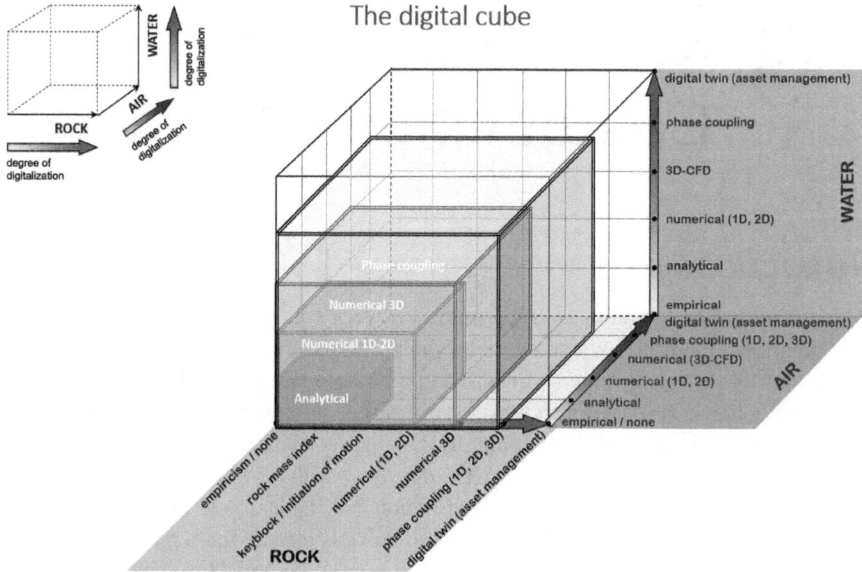

FIGURE 3.1 The digital cube: a 3D three-phase digital space to represent and compare rock scour computational methods.

In the following, the rock scour computational methods are subdivided into groups of equal degree of digitalization. The basic principles of each group are outlined, together with a brief overview of the most relevant methods involved. These methods mostly deal with rock scour at hydraulic structures such as dams, but may sometimes also apply to rock scour in rivers, i.e. at bridge piers or abutments.

GROUP I: EMPIRICAL FORMULAE AND LABORATORY PHYSICAL MODELLING

The first group of computational methods concerns purely empirical-based methods (i.e. empirical equations) and scour assessment by laboratory physical modelling. The degree of digitalization is considered very low to non-existent and, as such, these methods are positioned close to the 0-origin of the digital cube (Figure 3.2).

A fair number of empirical equations for rock scour at hydraulic structures have been developed since the 1930s. For further information, the reader is referred to summarizing referential work by Whittaker and Schleiss (1984) or Castillo and Carrillo (2017).

For laboratory physical modelling, only the most recent and/or most pertinent modelling is briefly presented. Other referenced physical model studies can be found for example in Montgomery (1984), Reinius (1986), Otto (1989) or Dubinski and Wohl (2013).

A first observation to make is that this group of methods has been intensively growing over the last decade, while in fact it is considered the least digitalized component in the digital cube! Excellent research work has been performed and is

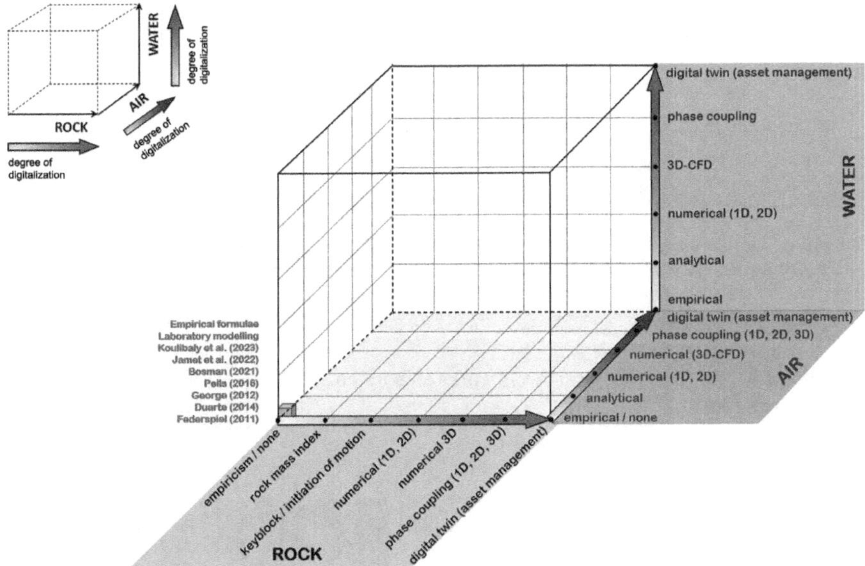

FIGURE 3.2 Group I of computational methods: empirical approaches and laboratory physical modelling.

actually ongoing in a number of laboratories worldwide, focusing on detection and measurement of the physics and break-up mechanics of rock at hydraulic structures.

Furthermore, while most of the empirical formulae have been developed decades ago, new approaches are still being developed, such as the recent work by Dr A. Bosman at Stellenbosch University in Cape Town, South Africa, proposing an empirical-numerical approach for scour assessment based on detailed regression analysis of laboratory physical modelling (Bosman 2021; Basson and Bosman 2020). As such, old-fashioned laboratory modelling seems to resist current digitalization tendencies!

Bollaert (2002), Manso (2006), Federspiel (2011) and Duarte (2014)

Bollaert (2002), Manso (2006), Federspiel (2011) and Duarte (2014) made use of the physical model developed by Dr E. Bollaert at the Laboratory of Hydraulic Constructions of the Swiss Federal Institute of Technology in Lausanne, Switzerland (Bollaert 2002; Bollaert and Schleiss 2005). This laboratory installation simulates vertical jet impingement into an artificially created circular-shaped plunge pool at near-prototype jet velocities of up to 32 m/s. Jet diameters involved are of 57–72 mm, and the plunge pool has a water depth of max. 1 m, for a diameter of 3 m (Figure 3.3).

Bollaert (2002) initiated this sequence of research studies on rock scour by developing the installation and by performing detailed high-frequency measurements of dynamic pressure signals inside 1D and 2D rock joints that were artificially created (Figure 3.4a). This work allowed for the first time to detect and numerically describe the generation and propagation of transient pressure waves responsible for progressive break-up of rock joints by fatigue. This research allowed detecting transient

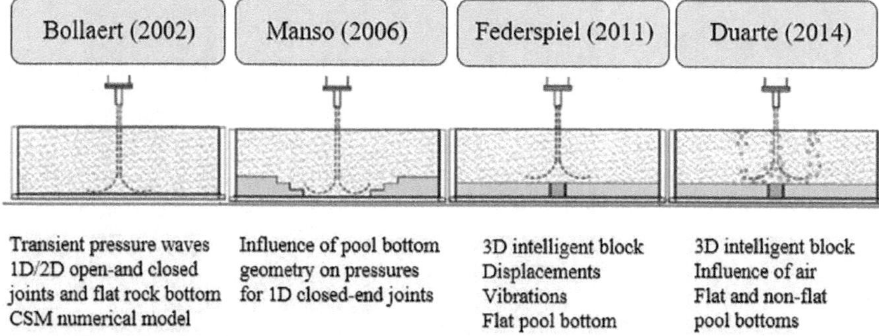

FIGURE 3.3 Series of doctoral thesis works elaborated at EPFL from 1998 to 2014. Adapted from Duarte (2014).

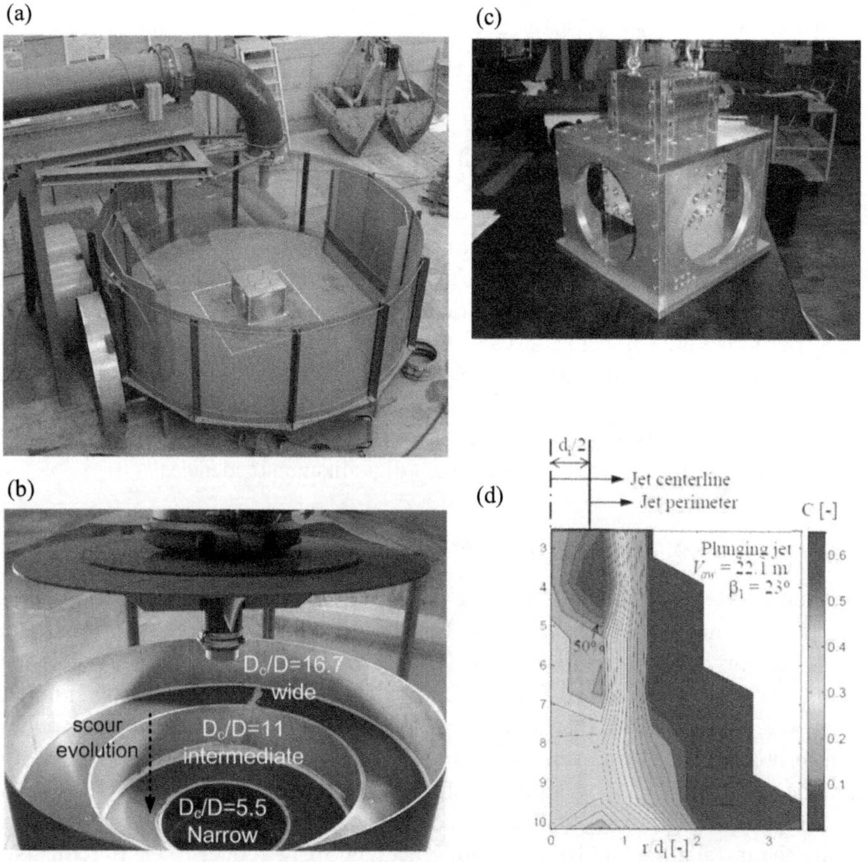

FIGURE 3.4 (a) Experimental facility at near-prototype jet velocities as developed by Bollaert (2002); (b) narrow-shaped circular geometry of plunge pool bottoms as investigated by Manso (2006); (c) intelligent 3D cubic block made of stainless steel (Federspiel 2011); (d) air concentration of an artificially aerated plunging high-velocity jet (Duarte 2014).

amplification of pressure waves, as well as two-phase damping effects. It resulted in the development of the Comprehensive Scour Model (CSM; Bollaert 2004; Bollaert 2012), a numerical model for rock scour computations based on several rock break-up methods, such as dynamic rock block uplift, peeling off of rock blocks by wall jets, and finally rock joint fracturing by fatigue or brittle fracturing.

Manso (2006) extended this work by investigating the influence of more realistically shaped plunge pool bottoms on the generation of dynamic pressures and transient pressure waves inside rock joints. Especially the influence of narrow-shaped circular scour formation was studied (Figure 3.4b).

Federspiel (2011) transformed this installation by installing on the pool bottom a steel-made intelligent 3D cubic rock block, equipped with pressure sensors, accelerometers and displacement sensors (Figure 3.4c). The cubic block had a side length of 20 cm, for a circumferential joint of 1 mm with the surrounding pool bottom. The measurements allowed relating the block movements and accelerations to the instantaneous net pressure differences and dynamic impulsions given to the block.

Finally, Duarte (2014) studied the air entrainment on the installation in more detail (Figure 3.4d). He made detailed measurements of air concentration profiles and air bubble velocities both at jet impact and inside the plunge pool, close to the water-rock interface. The measurements allowed to quantify the influence of air on the dynamic water pressures at the water-rock interface and on the displacements and vibrations of an embedded 3D rock block. Furthermore, the combined influences of pool bottom confinement and jet aeration on the pressures around a block inserted in a plunge pool bottom were assessed experimentally using near-prototype jet velocities.

This line of doctoral works developed by Prof. Emeritus Dr A. Schleiss, using near-prototype scaled physical modelling, undoubtedly allowed detecting new phenomena and procured significant novel insight into the physics responsible for rock break-up.

George (2012)

Innovating research has been performed by Dr M. George (George 2015; George et al. 2015; George and Sitar 2016) at the University of California at Berkeley, USA, focusing on the rigid body behaviour of 3D tetrahedral small-scale blocks.

A programme of experimental and analytical studies was carried out to investigate the influence of geologic structures on the erodibility of blocky rocks. Until this study, very little data existed regarding hydraulic loads on 3D blocks or regarding the mechanics of erodibility of 3D blocks beyond simple cubes or prisms. Both experimental results and observations from physical model experiments have allowed to determine 3D block erodibility.

The physical modelling was performed at the University of California's Richmond Field Station (RFS). Tests were conducted in a 28 m long × 0.85 m wide × 0.91 m deep flume with an overall grade of 1%. A wooden ramp was constructed, containing in its downstream section a rotatable block mould that housed a removable tetrahedral "rock" block made of concrete. The block protrusion into the flow could be adapted, as well as its orientation with respect to the flow direction.

Block displacements were monitored using proximity sensors, while multiple pressure sensors were used to measure hydraulic loads applied to block faces. By applying key block theory (Goodman and Shi 1985), block stability and removability

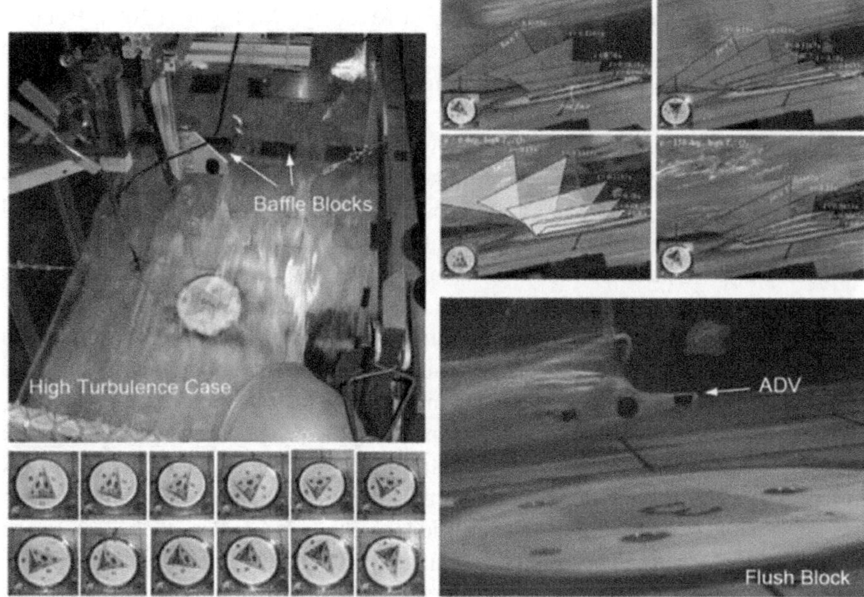

FIGURE 3.5 Uplift of protruding tetrahedral blocks on small-scale flume (George 2012).

are thereby represented using a Limit-equilibrium Stereonet Projection (George and Sitar 2016). The results of the modelling have finally been used to develop a block theory framework for analysis of block erodibility, based on the major potential kinematic failure modes of 3D blocks, such as lifting, sliding and rotation (Figure 3.5).

George (2015) tested the so measured block erodibility thresholds on a real-life situation by instrumenting two real-life blocks in the unlined channel downstream of Spaulding Dam in California. Two flood events allowed to closely follow the block displacements as a function of time and to represent the path of resultant force vectors on a Stereonet projection. The so obtained real-life critical threshold velocity for rock block movement was found in good agreement with the one predicted by theory.

Pells (2016) and Douglas et al. (2018)

The research performed at the University of New South Wales (UNSW) in Sydney, Australia (Pells 2016; Pells et al. 2016, Douglas et al. 2018), was funded as part of an Australian Research Council – Linkage Project LP110100389. It involved both field studies from dams in USA, Australia and South Africa with significant erosion, and extensive small-scale laboratory modelling focusing on potential pressure variations that can be induced in rock joints from parallel spillway flow over protruding prismatic blocks.

The small-scale laboratory measurements performed by Pells (2016) describe lift and drag forces on blocks that are protruding into the flow environment. These studies allowed defining a new geomechanical index that corresponds to a modification of the Geological Strength Index (GSI) developed by Hoek (1994) and Hoek et al. (1995), by accounting for increased erosion vulnerability of rock blocks depending on unfavourable orientations of the joints compared to the flow. The dip of the joints covers thereby 180°, for different joint set spacing ratios, in line with the original relative ground structure number J_s used in numerous existing indices for rock erodibility.

The eGSI index developed by Pells (2016) thus corresponds to a combination of two existing rock mass indices, the GSI and part of Kirsten's index (Kirsten 1982), based upon small-scale flume studies. It represents a gradual description of rock erodibility, and not a threshold.

Second, erosion in 33 unlined spillways in rock has been studied for dams in Australia, South Africa and the USA. Geological factors which influence erosion, such as the orientation, persistence, spacing and nature of rock defects, have been identified using available data and site inspections. The presence of kinematically viable blocks which can be detached and the persistence of the basal defect for these blocks are the most important factors.

A rock mass characterization index, the 'Rock Mass Erodibility Index' (RMEI), which considers spillway flow conditions and erosion mechanisms, has been developed by Douglas et al. (2018). It can be used as a guide to spillway erosion and, when coupled with stream power for spillway flows, provides a method for preliminary assessments of likely amounts of spillway erosion (Figure 3.6).

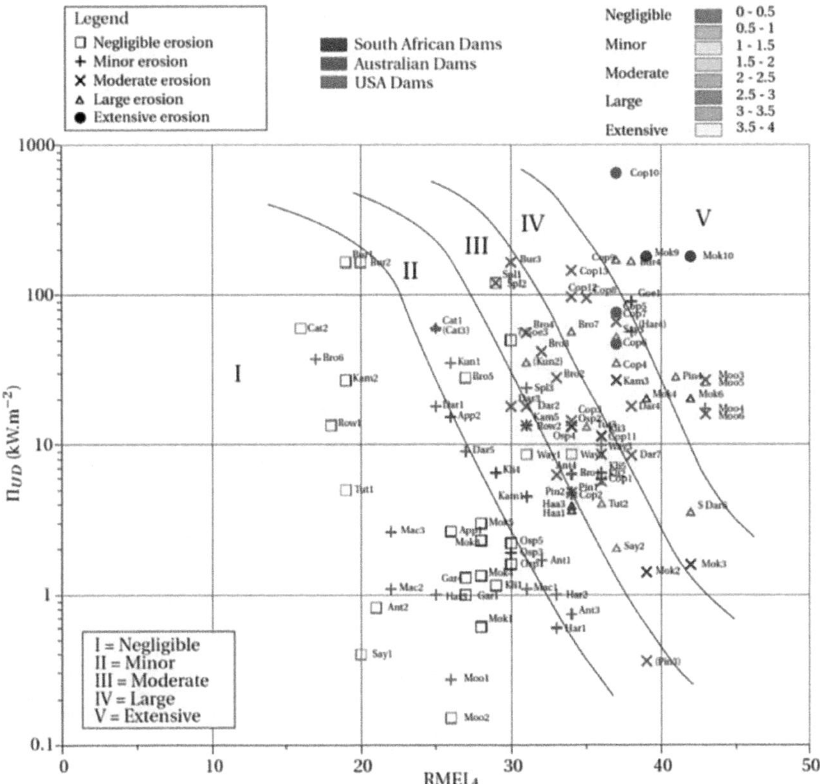

FIGURE 3.6 Contours of erosion class – Unit stream power dissipation versus RMEI index (Pells 2016).

Calitz (2015) and Bosman (2021)

Medium- and large-scale laboratory experiments of scour hole development in broken-up rock, generated by rectangular-shaped oblique jets impinging in a plunge pool, have been performed at the hydraulics laboratory of the Civil Engineering Department at Stellenbosch University, South Africa (Calitz 2015; Bosman and Basson 2020; Bosman 2021).

Calitz (2015) performed 1/40 scaled physical model experiments of rectangular jets obliquely impinging in a pool made of PVC blocks to represent broken-up rock. The blocks had a side length of only 25 mm. His master's thesis work focused on a description of the generated jets and a comparison between the measured scour depth and well-known computational approaches. The study was complemented by 2D CFD computations (Calitz 2015).

Second, Bosman (2021) performed her PhD work on a larger-scale version (1/20) of the same test facility. The physical model is illustrated in section and plan view in Figure 3.7 and consists of a zero-sloped 0.25-m-wide rectangular canal at issuance, replicating an uncontrolled dam spillway. The canal could be adjusted to three different fixed heights above the movable rock bed: 3, 4 and 5 m (prototype heights for 1:20 model scale: 60, 80 and 100 m respectively). The tailwater depth at the plunge pool was adjustable between 0.5 and 1 m (prototype: 10 and 20 m respectively). The plunging jets generated unit discharges between 0.020 and 0.045 $m^3/s/m$ (prototype: 35–80 $m^3/s/m$).

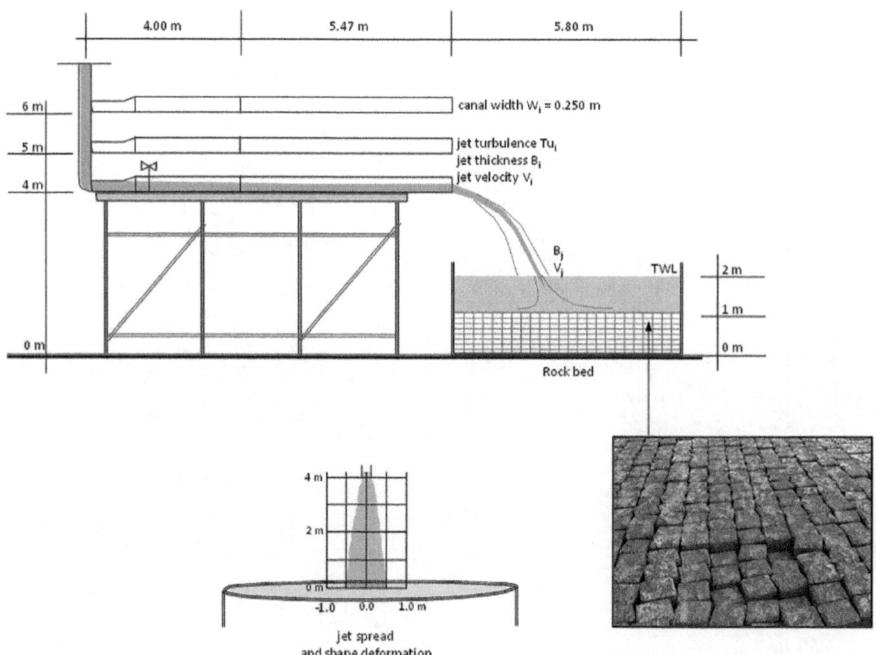

FIGURE 3.7 Experimental set-up of plunging jets impinging on broken-up rock bed and illustrative example of cobblestones such as used for the tests. Adapted from Bosman (2021).

The broken-up rock bed was modelled by using tightly hand-packed concrete paver blocks (cobblestones), generating a uniform three-dimensional open-ended horizontal and vertical rock joint network. The two rock sizes tested were rectangular concrete pavers with x, y, z dimensions of $0.1 \times 0.1 \times 0.05$ m and $0.1 \times 0.1 \times 0.075$ m (prototype: $2 \times 2 \times 1$ m and $2 \times 2 \times 1.5$ m respectively). The paver block densities used were 2,355.4 kg/m^3 (for 1-m-high blocks) and 2,388.1 kg/m^3 (for 1.5-m-high blocks) respectively (Bosman 2021).

Bosman (2021) subsequently developed an empirical equation for equilibrium rock scour depth, scour hole length, width and volume, based on an extended regression analysis and dimensional analysis. The regression equation was finally used to perform manual iterations of 3D CFD modelling of the scour that formed during the laboratory tests.

Kashtiban et al. (2021) and Koulibaly et al. (2022)

Laboratory modelling of rock block detachment processes is performed at the Université de Québec at Chicoutimi, by Kashtiban et al. (2021), within the framework of research work on "Large Scale Experimental Simulations of Scour Mechanisms of Unlined Spillways". This research work is funded by the 2020–2025 NSERC-Collaborative R&D Programme and by the 2020–2023 MITACS Accelerate Programme.

The work focuses on the main parameters of influence as determined by Boumaiza et al. (2019), i.e. the rock block volume, the rock block's shape and orientation relative to flow direction, the nature of the eroding surface (i.e. protrusion of the block) and the degree of opening and the shear strength of the rock joints. Parameters that are relevant to the fracturing process of rock blocks, such as the UCS strength or the initial degree of fracturing of the rock mass, are not considered in this study.

Within the same research group, Koulibaly et al. (2021, 2023) present a laboratory-scale physical model to determine the effects of rock mass parameters on erosion. This model is designed to determine individual and interactive effects of several hydraulic and rock mass parameters on erosion, including joint opening, block size, joint shear strength and the nature of potentially erodible surfaces, as well as water pressure (static and dynamic), variations of flow rate and velocity, and channel roughness.

Jamet et al. (2022)

Laboratory modelling is performed on a rectangular experimental facility built at the Laboratório Nacional de Engenharia Civil (LNEC, Lisbon) with a length of 4 m and a width of 2.65 m. The facility can produce circular water jets with a maximum velocity at issuance V of 18 m/s. The jet nozzle diameter D is 0.072 m, and the water pool depth, Y, can vary from 0.3 to 0.9 m, which corresponds to a pool depth to diameter ratio Y/D ranging from 4.2 to 12.5 (Figure 3.8).

A series of high-frequency pressure transducers are installed on a metallic plate at the pool bottom, allowing to record dynamic pressures and their spectral content at frequencies up to 2,400 Hz. The sensors are placed at different radial distances

FIGURE 3.8 Laboratory modelling of plunging jet impingement on flat pool bottom: UP: general view and side view of facility; DOWN: general view and side view of jet impingement for a jet velocity of 18 m/s and a pool depth of 0.50 m. Adapted from Jamet et al. (2022).

r from the jet centreline (or stagnation point), to investigate the radial decrease of mean and fluctuating pressure coefficients, and focus especially on sudden changes in pressure inside or outside of the jet diameter footprint. Furthermore, the influence of the jet issuance velocity on the dynamic pressure coefficients has been pointed out. Figure 3.8 illustrates the facility (up) and the jet impingement for a jet velocity of 18 m/s and a pool depth of 0.50 m (down) (Jamet et al. 2022).

GROUP II: SEMI-EMPIRICAL METHODS

The second group of computational methods concerns semi-empirical methods that are index-based. A geomechanical index represents the vulnerability of the rock mass to scour, and the hydraulic action is represented by means of stream power. The onset of scour is simply determined based on a large dataset that allows to formulate a mathematical threshold. No analytical equations are implemented to determine this threshold.

This group of methods entirely relies on historical in-situ scour observations and some laboratory physical model tests. Without implementation into a dedicated numerical framework, they must be manually applied on a block-by-block basis. Hence, their degree of digitalization is considered low and, as such, these methods are positioned in the lower left part of the digital cube (Figure 3.9).

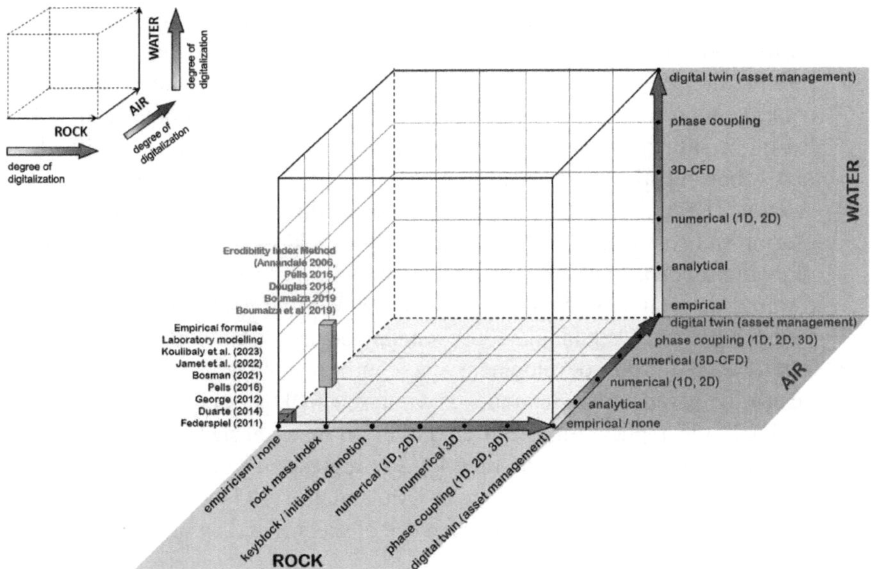

FIGURE 3.9 Group II of computational methods: semi-empirical methods (i.e. index-based).

Index-based methods use unitary stream power expenditure or dissipation rate (i.e. rate of energy dissipation of the flow) to express the flow hydraulics, and compare this with a geomechanical index that characterizes the rock mass quality and vulnerability against scour. The origin of most of these methods lies in the research work performed since the 1980s at the University of Pretoria by Prof. Dr. van Schalkwyk, making use of a database of over 30 cases of erosion in unlined spillway channels in South Africa. This research work has used the Kirsten geomechanical index that was originally developed to assess the excavatability or rippability of rock during excavations. Other methods make use of alternative well-known geomechanical indices, such as the GSI (Geological Strength Index) developed by Hoek et al. (1995).

Some of these methods make use of a well-defined scour criterion (threshold), others rather propose quantitative or qualitative scour categories or classes. Examples of erodibility index-based methods are:

- *van Schalkwyk, Jordaan and Dooge (1994b):* analysis of 18 case studies and categorization of erosion into "non", "some" and "excessive", based on unitary stream power dissipation.
- *van Schalkwijk (1994):* update and completion of previously defined erosion categories and thresholds by inclusion of data by Moore et al. (1994).
- *van Schalkwijk et al. (1994b):* same case studies but use of multiple erosion categories, from "negligible" (<0.2 m) and "minor" (0.2–0.5 m) erosion to "moderate" (0.5–2.0 m) and "large" (>2.0 m) erosion, instead of a threshold.

- *Moore, Temple and Kirsten (1994):* use of Kirsten index as "headcut erodibility" index to compute regressive scour of rock in unlined channels.
- *Annandale (1995, 2006):* use of Kirsten index as "erodibility" index to compute scour of rock and earthen materials for different geometrical and hydraulic situations, based on the datasets of van Schalkwijk et al. (1994b and Moore et al. (1994), supplemented by the case of Bartlett Dam in Arizona, USA. Determination of a universal threshold criterion between erosion and no erosion. Validation of this threshold by large-scale field tests of jet impact on a series of lightweight concrete blocks under a 45° dip angle towards downstream (Annandale et al. 1998). Applicable to bridge piers founded on rock (Annandale 2009; Annandale and Smith 2001).
- *Kirsten et al. (2000):* development of a scour threshold between erosion and no erosion based on Kirsten index and datasets by Dooge (1993) and Moore (1991). Large applicability from sand to intact material such as steel.
- *Wibowo et al. (2005):* by multiple logistic regression analysis for spillway erosion, they used the dataset from Annandale (1995) and proposed an equation for the probability of occurrence of scour based on scour threshold by Annandale (1995).
- *Pells (2016):* qualitative index-based scour assessment by use of Hoek's GSI rock quality index, completed with a qualitatively defined discontinuity orientation adjustment (E_{doa}) factor inspired from the Kirsten (1982) relative ground structure number. Definition of "eGSI" index to express the qualitative vulnerability of rock in unlined channels. Manual adaptation of qualitative erosion categories as defined by van Schalkwijk et al. (1994a) instead of use of a threshold. Database based on South African, Australian and US dams.
- *Pells (2016) and Douglas (2018):* development of the RMEI classification index, including the likelihood of block detachment and the weighted importance of kinematically viable mechanism for detachment, the nature of the potentially eroding surface, the nature of the joints, the joint spacing and the block shape. Use of manually defined qualitative erosion categories instead of use of a threshold. Database based on South African, Australian and US dams.
- *Boumaiza et al. (2019):* development of a new index making use of Pell's qualitative discontinuity orientation adjustment factor, the block volume instead of the block size number (based on Palmström 1995) and the nature of the potentially eroding surface (NPES) as defined by Pells (2016), which may be simply considered as the protrusion of the rock block compared to its surroundings and the degree of opening of the joints. Omission of the UCS strength of the rock.

The methods that incorporate a clearly defined scour threshold are illustrated in Figure 3.10, while the van Schalkwijk (1994) method that proposes interpretative erosion categories is visualized in Figure 3.11. Finally, Pells (2016) reinterpreted the erosion categories as defined by van Schalkwijk et al. (1994a) as a function of the eGSI index instead of the generally used Kirsten index, and Douglas (2018) refined erosion categories developed by Pells (2016) as a function of the RMEI index. These reinterpretations have been performed qualitatively and have no clear scour threshold.

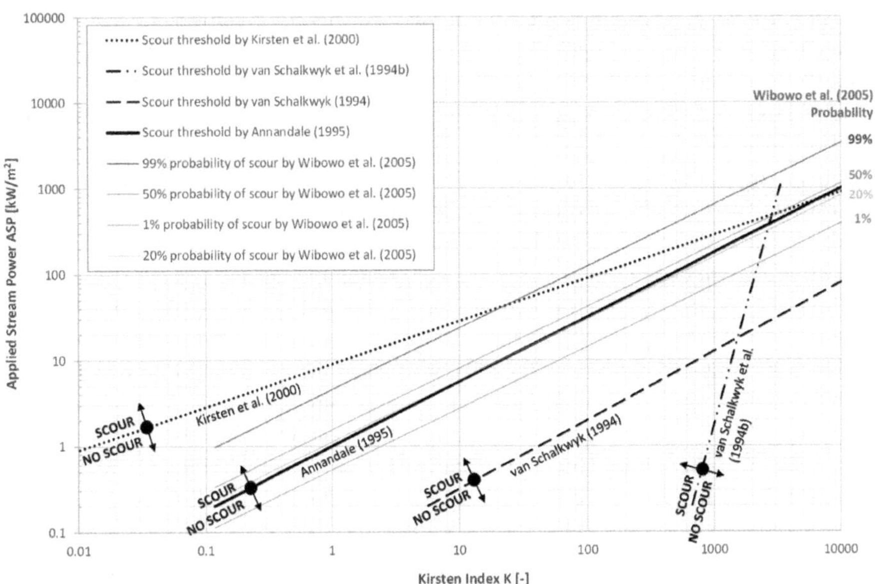

FIGURE 3.10 Comparison of scour thresholds as determined by index-based methods.

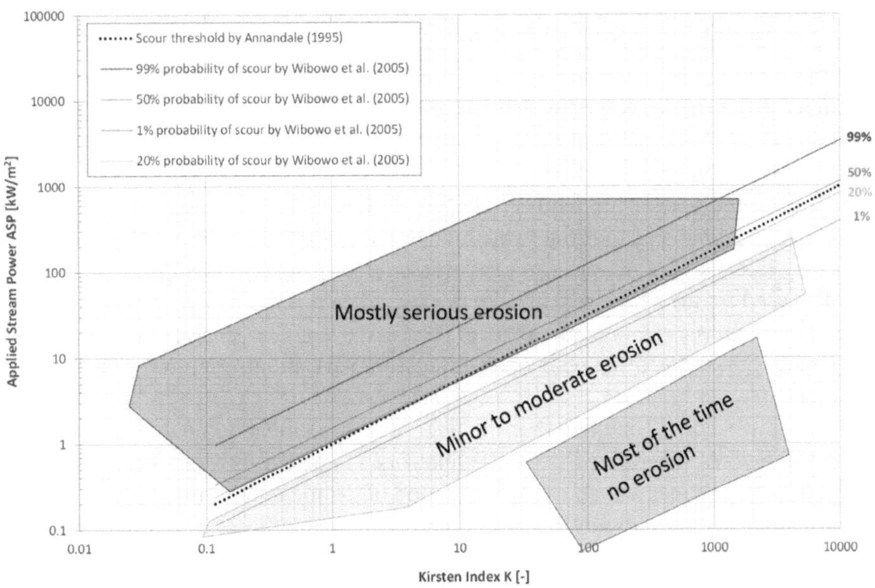

FIGURE 3.11 Erosion categories as defined by van Schalkwyk et al. (1994a) compared to erosion probability thresholds as defined by Wibowo et al. (2005) based on Annandale (1995).

Significant scatter may be observed between the different methods. A detailed comparison of efficiency of the different index-based methods can be found in Boumaiza (2019), based on a set of 86 cases from 23 dam sites. In general, none of the presented methods is superior to the other methods in all of the erosion categories.

The success of these methods is not very surprising given the historic importance and relevance of geomechanical indices in the field of rock mechanics and rock engineering. While these methods are rather straightforward to apply, they rely nevertheless on a subjective and simplified representation of both the rock mass and the flow hydraulics.

Furthermore, some of them make use of qualitative thresholds that subdivide the results into interpretable erosion categories (such as "some" or "excessive" erosion, or also "minor", "moderate" or "large" erosion, each one corresponding to difference depths of erosion depending on the qualified person in question).

Also, index-based methods do not provide insight into time evolution of scour and do not consider the physics of rock break-up processes. As such, they remain very limited in describing and explaining the phenomena and their potential future evolution.

It is interesting to notice that the latest developments in this field (Pells 2016; Douglas 2018; Boumaiza et al. 2019) clearly focus on replacing classical rock mass parameters by other parameters that better represent the vulnerability of rock blocks to detachment. Examples are the NPES parameter that accounts for rock block protrusion and degree of opening of surrounding joints, the relative shape of the blocks and the E_{doa} parameter that attempts to attribute more weight to the orientation of the joints and the shape of the blocks compared to the flow.

As will be seen later in this work, most of these parameters are accounted for since quite some time by physics-based methods that focus on the break-up processes of fractured rock. In this sense, the current evolution towards use of physically more relevant parameters in index-based methods is encouraging.

Finally, despite their relative ease of application, and despite specific research attempts (Rock 2015), no engineering tool is currently available allowing a generalized numerical application of these thresholds for different turbulent flow situations and for an in-situ rock mass area containing a large diversity and number of rock blocks and lithologies with depth.

Moreover, input data must be gathered based on field investigations that are generally very interpretative and subject to large degree of scatter. For instance, most of the data used for the abovementioned index-based methods do not dispose of detailed topographic surveys from before or after the flood event that generated the scour, and the stream power has often been determined in a very simple manner (uniform flow conditions).

Hence, it becomes obvious that the degree of digitalization of these methods is currently low. A generalized digital framework is needed for practising engineers, together with a refinement and standardization of the too qualitative procedures used to compute both the hydraulics and the geomechanical indices.

GROUP III: ANALYTICAL METHODS

The third group of computational methods concerns analytical methods that express initiation of motion of rock blocks (i.e. kinematic block analysis) or brittle or progressive fracturing of rock joints (i.e. linear elastic fracture mechanics applications). Similar to the semi-empirical methods, without implementation into a dedicated

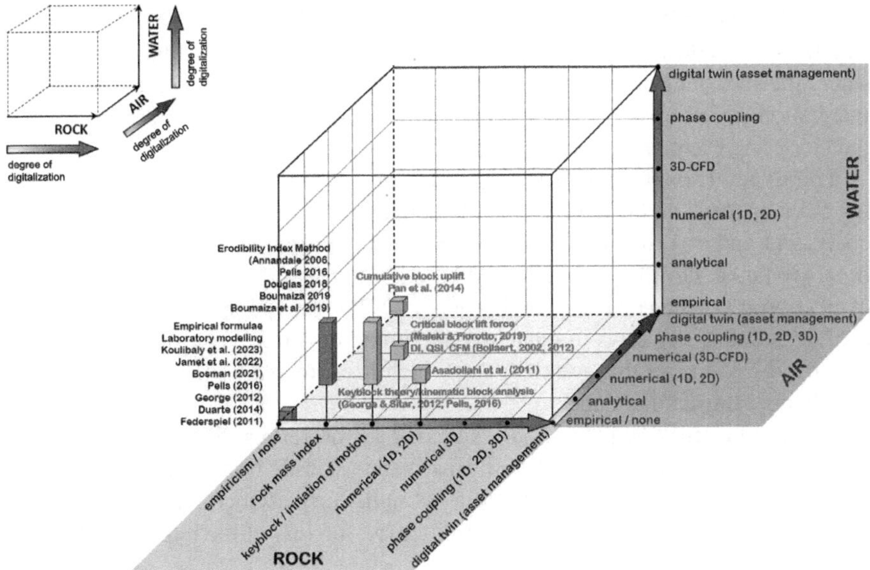

FIGURE 3.12 Group III of computational methods: analytical methods (i.e. initiation of motion-based or fracture mechanics-based).

numerical framework, these methods have to be manually applied on a block-by-block basis. Hence, their degree of digitalization is considered low and, as such, these methods are positioned in the lower left part of the digital cube (Figure 3.12).

This group of methods may partially rely on laboratory physical model tests, but combines these tests with well-defined analytical equations expressing rock mass resistance to different break-up mechanisms, such as rock block plucking and rock joint fracturing.

Initiation of Motion-Based Methods

First, analytical methods based on initiation of motion of rock blocks allow to define the main forces acting on single rock blocks, as well as the related initiation of block displacements. As such, they are considered much more relevant to describe the physical processes of rock scour by block displacements. Nevertheless, most of these methods only describe the start of block movement and do not allow describing the movement of the block with time.

Currently, their application in practical engineering is mostly cumbersome, and must be performed manually and on a block-by-block basis. Automated numerical application to a 2D or 3D rock mass containing a large number of distinct rock blocks and lithologies is only available through the CSM (Bollaert 2004; Bollaert 2012), which is a proprietary non-commercial software. Hence, despite the relative complexity of some of these methods in describing multiple potential block movements, their degree of digitalization may be considered low.

Dynamic Impulsion (DI) Method (Bollaert 2002; Bollaert and Schleiss 2005)

The DI method developed by Bollaert (2002) and Bollaert and Schleiss (2005) considers the maximum dynamic impulsion given to a rock block computed by time integration of the pressure forces under and over the block, the immerged weight of the block and eventual shear and interlocking forces.

This dynamic impulsion physically represents the product of a net uplift force during a time period, and is mathematically defined making use of a net uplift pressure coefficient and of a time coefficient. Both the net uplift pressure coefficient and the time coefficient have been determined based on large-scale laboratory experiments of near-prototype vertically plunging jets and 2D rock joints (Bollaert 2002). An analytical equation allows to relate the maximum dynamic impulsion to the degree of diffusion of the jet through the pool, which is expressed by the ratio of jet travel length to jet impact diameter.

Failure of a block by plucking is expressed by the vertical displacement it undergoes due to the dynamic impulsion. This is obtained by transformation of the net uplift velocity given to the block into a net uplift displacement. The ratio of block displacement to block height then determines the stability of the block, expressed by a critical uplift ratio.

Based on experiences and calibrations performed with the DI method since its development, for practice, a critical uplift ratio ~0.10 is recommended for prototype situations at hydraulic structures.

The validity of this plucking method is limited to the turbulent shear layer of the plunging jet that is diffusing through the plunge pool. It can be relatively easily applied by use of a spreadsheet. A specific application to rock block quarrying and plucking at bridge piers has been developed within the framework of the NCHRP Research Program and is available as a set of design curves for critical velocity and ultimate scour depth (Keaton et al. 2012; Bollaert 2010).

The DI method developed by Bollaert (2002) offers potential for digitalization and is discussed in more detail in Chapter 6.

Quasi-Steady Impulsion (QSI) Method (Bollaert 2012)

The QSI method developed by Bollaert (2012) computes the scour potential along the water-rock interface that is situated outside of the turbulent shear layer of the impacting jet. Subsequent to the quasi-2D diffusion of jets through the plunge pool, the water-rock interface strongly deflects the jets towards up- and downstream. This quasi-steady flow parallel to the bottom may then be potentially deviated by protruding rock blocks, generating quasi-steady net uplift forces on these blocks.

This allows defining the shape of the scour hole and to assess the risk for regressive scour towards the dam toe. Analytical equations express the axial jet velocity decay through the water depth of the pool as well as the discharge distribution between upstream and downstream. This allows defining the velocities and dimensions of the wall jets parallel to the water-rock interface.

In case of rock blocks at the water-rock interface that protrude into the flow, lift and drag forces are generated. These forces are mostly of quasi-steady character (Bollaert and Hofland 2004), but may also contain a fluctuating part. The lift forces are defined by an uplift coefficient expressing the pressure as a function of the kinetic energy of the quasi-parallel flow. Based on Reinius (1986), net uplift coefficients are between 0 and 0.5, depending on the degree of protrusion and the shape of the blocks. For practice, coefficients of 0.1–0.2 are considered plausible for low to very low block protrusions, while coefficients of 0.3–0.5 correspond to moderate to significant block protrusions, i.e. for rough and irregular pool bottoms. The latter values are considered most plausible for a real water-rock interface.

Finally, based on block shape and dimensions, block stability under quasi-steady flow impact is computed by defining the net uplift force during jet impact, by accounting for the submerged weight of the block as stabilizing force. The stability of the block is expressed by a simple force balance equation, defining a mathematical threshold, and no block movements are computed.

The QSI method developed by Bollaert (2012) offers potential for digitalization and is discussed in more detail in Chapter 6.

Dynamic Pressure (DP) Method (Maleki and Fiorotto 2019)

Maleki and Fiorotto (2019) compute the uplift forces on rock blocks based on small-scale laboratory measurements of mean and fluctuating dynamic pressures generated by a jet impinging onto a pool bottom, for different Y/D_j ratios, and by analytical determination of the jet characteristics in the fully developed flow region.

By accounting for pressure correlation functions in both the streamwise and transversal directions, they express the standard deviation of the net fluctuating uplift forces on a block as a function of the maximum standard deviation of the fluctuating pressure field on the block, by means of an Ω coefficient. Furthermore, they make use of the ratio "n" of the maximum net uplift pressure coefficient on a block ($C^+_p - C^-_p$) to the RMS coefficient of the fluctuating pressures C'_p on the block. The authors finally multiply both Ω and n values with the RMS coefficient of pressure fluctuations at the block upper face to determine the max. net uplift pressure (force) on a block generated by dynamic pressures over and under the block. The approach has been developed for prismatic and triangular-shaped 2D and 3D blocks.

Their approach is restricted to the determination of the max. possible net uplift force on a block and to the related necessary min. thickness for a block to remain stable, in which stability is preserved whenever the submerged weight counteracts the net uplift force. A dynamic analysis of the movement of the block is not considered.

The method developed by Maleki and Fiorotto (2019) offers potential for digitalization and is discussed in more detail in Chapter 6.

Block Theory Rock Erodibility (BTRE) Method (George 2015; George et al. 2022)

The research work performed by M. George (George 2015; George et al. 2015; George and Sitar 2016) at the University of California at Berkeley, USA, allowed determining hydraulic loads on 3D tetrahedral small-scale blocks. Both results and observations from physical model experiments have allowed to determine 3D block erodibility. By applying key block theory (Goodman and Shi 1985), block stability and removability are represented using a Limit-equilibrium Stereonet Projection (George and Sitar 2016).

The BTRE method allows to express the pseudo-static equilibrium of a single block, based on the vector of the resultant of the main forces involved. The hydraulic forces acting on the blocks are computed based on analytical estimates of time-averaged and fluctuating dynamic pressure coefficients. These are determined as a function of the protrusion of the block, of the block surroundings and of the angle between the flow velocities and the upstream block face. Corresponding flow velocities in the immediate vicinity of the block can be defined by hydraulic (CFD) modelling. A simple force balance is checked for each of the potential failure modes (i.e. face detachment, sliding, etc.). Values of the potential block displacements are not predicted by the model, only a steady-state threshold of equilibrium is expressed.

The method originates from key block theory developed for local stability of unlined tunnels and, as such, mainly focuses on one or more specific blocks, called key blocks, that might trigger erosion or instability of the structural ensemble. The method is only applicable to rock blocks that are clearly exposed at the surface of the water-rock interface, i.e. for which the different block faces and orientations may be determined with precision and confidence. Hence, it is mainly useful to detect scour potential at the surface of the rock mass and to gain insight into relevant scour processes at a site, but does not allow computing a scour hole.

A more generalized application to rock masses in unlined flow channels or plunge pools would need to redefine the key blocks and their state of equilibrium for all potential failure modes, as well as the detailed flow hydraulics and analytical (subjective) pressure coefficients, after scour of a first series of key blocks at the surface of the rock mass. This, however, would need detailed knowledge on the 3D shape of the underlying blocks and their faces and orientations. Research is ongoing on digitalization of the 3D rock mass surface discontinuities and surface rock blocks based on terrestrial LiDAR scanning (Weidner and George 2023), but the techniques being tested are currently out of realm of engineering practice.

Hence, potential automatization to numerically perform these computations with depth for a real-life situation containing many blocks seems a priori feasible, but would need a significant amount of detailed data from the rock mass, not only at the rock surface but also with depth.

BSS (Kieffer and Goodman 2012)

The Block Scour Spectrum (BSS) is a three-dimensional analytical method for determining the erosive resistance of rock blocks subjected to an impinging jet. The BSS aims to determine the erosive resistance of rock blocks, which depends on the rock mass and the orientation of the resultant force or loading applied to the rock block. Both gravity and hydrodynamic forces are considered in the BSS method.

The BSS comprises kinematic, stability and spectrum analysis modules. The method can prioritize the rock blocks according to their scour resistance and thus identify potential locations where scour initiation is likely to occur, thus allowing the implementation of efficient scour control measures to be implemented. The BSS method assumes that block removal occurs along existing joints by separation or translational sliding (Kieffer and Goodman 2012).

The method seems a priori suitable for digitalization.

BS3D (Asadollahi and Tonon 2010; Asadollahi et al. 2011)

Asadollahi et al. (2011) developed a method expressing stability of a rock block based on the extreme force during half of the natural joint frequency. Block stability in three dimensions (BS3D) is an incremental-iterative algorithm and code that analyses all failure modes of rock blocks due to water forces and pressure loadings (Asadollahi et al. 2011).

The method makes use of the max. possible net uplift force on a block. This extreme uplift force is applied during the max. duration of an extreme pulse. By considering pure vertical translational failure, the method assumes that the block moves with the initial velocity for a time duration of half of the natural period of the characteristic length of the joint. A threshold similar to the DI method is applied, i.e. the ratio of block displacement to block height, to determine block stability.

The method is based on the assumption that an extreme uplift pressure on the block is systematically followed by an equivalent downwards-oriented extreme pressure (overpressure) that pushes the body back into its surrounding mass during the second half of the natural period of the joint.

Calibration revealed that the wave celerity had to be adjusted to values as low as 25–40 m/s to match the design criterion, which are extremely low values. Also, the assumptions of the method seem inconsistent with the large-scale block tests by Bollaert (2002) and Federspiel (2011), who show that block movements are mainly governed by low-frequency turbulent pressures and not by short-lived extreme values. Asadollahi et al. (2011) developed their method by reproducing some of the Federspiel (2011) tests for 3D blocks, but their computational results were erroneously compared with the sum of the measured uplift height and the initial block position, i.e. a 0.85 mm fixed gap of the facility, and not with the net measured uplift height. As such, the physical background of this method as well as its relevance for practice seem questionable.

Finally, while this method has made use of analytical modelling using the BS3D model, and has potential for digitalization, no generic numerical model is available for practising engineers allowing to apply this method to a 3D rock mass.

Cumulative Plucking Model (Pan et al. 2014)

Pan et al. (2014) developed a theoretical framework for cumulative vertical rock block plucking based on simplified, sinusoidal pressure signals acting over and under the blocks. The net uplift pressure is spatially averaged along the lower face of the block, and is mathematically defined by the RMS value of the pressure fluctuations at the block surface, multiplied by a unitary sinusoidal pattern.

Their approach accounts for the in-situ horizontal stress field and determines the frictional forces generated along the lateral faces of the blocks, based on the pressures transmitted by the block surface, and on the local in-situ stress field.

The model assumes that frictional forces in lateral joints may be neglected during upwards-oriented block movements (i.e. large underpressures), because the rock mass is not in compression anymore. Frictional forces are mobilized, on the contrary, during downwards-oriented block movements (i.e. large overpressures), but are limited to the sum of the effective downwards-oriented force. For a low frictional resistance coefficient, the friction is not able to counteract the downward forces and the block may move downwards. For a high frictional resistance coefficient, however, friction may counteract forces and the block gets stuck in its uplifted position. As such, the block may become detached from its matrix by considering an accumulation of small upwards movements.

Pan et al. (2014) applied their theory on a case study of the scour hole downstream of Yi-Xing Dam on the Da-Han River, Taiwan. For this, they related the RMS pressure coefficients to the Y/D_j ratio as defined by Castillo (1989) for rectangular jets. The mathematical developments made for their case study, however, do not seem to match with the Castillo (1989) relationship, and make it elaborate to use this method for practical engineering applications. Pan et al. (2014) only considered one single cycle of uplift to determine the ultimate scour depth, while their approach is based on cumulative uplifts.

Finally, as all the computations are performed analytically, the current degree of digitalness of this method is considered low, but digitalization might be envisaged.

Transient Uplift (TU) Method (Bollaert 2010; Keaton et al. 2012)

Rock block quarrying and plucking has been modelled by Bollaert (2010) within the framework of a National Cooperative Highway Research Program of the Federal Highway Administration in the USA (i.e. NCHRP Project 24-29, FHWA, Keaton et al. 2012). The objective of this project was to develop a methodology for estimating the time-rate of scour and the design scour depth of a bridge foundation on rock.

The developed model is a modification of the CSM model (Bollaert 2002) and is applicable to plucking of regularly shaped rock blocks in channel flows and flows around bridge piers (Bollaert 2010). Various hydrodynamic conditions appropriate for natural channels where bridge foundations might be located have been tested numerically.

The two-phase transient numerical model simulates the time evolution of quasi-steady and turbulent forces around a single, completely detached rock block and expresses the potential vertical movements of the block as a function of the flow turbulence and the stream power in the scour hole that forms around a bridge pier. The hydraulic action on layers of identical rock blocks is automatically adapted numerically during scour hole formation by assuming that the rock block layer is removed once a single block is entrained into the flow.

The model is 1D because a single rock block is considered on a layer-by-layer approach. The model may be considered to have limited two-dimensional characteristics because fluctuating pressures are considered to act in joints on both sides of the single rock block. Rock block entrainment is defined in this numerical model as block lifting equal to 20% of the block dimension perpendicular to the channel bed. Both the ultimate scour depth and the scour threshold flow velocity are determined as a function of (1) the shape, dimensions, and protrusion of the rock blocks, (2) the average upstream river bed slope and (3) the dip angle of rock joints.

The hydraulic parameters used in the model consist of the approach of stream power and factors for adjusting the stream power based on pier shape and angle of attack (HEC-18). Fluctuating pressures caused by pier-induced turbulence were modelled as a simple sine curve with a frequency of 10 Hz and the assumption that the fluctuating pressure peak acts simultaneously in joints on both sides of a rock block.

Model results have been published as ultimate scour depths and velocity threshold values for differently sized and shaped rock blocks and channel flows and slopes. Scour depths have been normalized to the pier diameter (Keaton et al. 2012). Design curves for ultimate scour depth and critical scour velocity can be freely downloaded from *https://www.trb.org/Publications/Blurbs/167222.aspx* as Excel spreadsheets.

Shear Stress-Based Uplift (Coleman et al. 2003; Lamb et al. 2015; Melville et al. 2006)

Rock block plucking has been modelled and studied by several researchers based on shear forces induced by turbulent flow over a rocky channel bottom. The Shields-based critical shear stress is the indicator used for threshold motion of particles on riverbeds.

Lamb et al. (2015) developed theoretical equations based on a classical force balance, but cast in terms of a critical Shields stress, and by expressing the lift force on a block as a linear function of both drag and shear forces. Their values show good agreement with those determined experimentally by others for the case of vertical prismatic block entrainment (Coleman et al. 2003; Melville et al. 2006) and of block sliding (Carling et al. 2002; Dubinski and Wohl 2013).

Results indicate a strong dependence of block incipient motion on the relative protrusion height (P) of the block with respect to the horizontal block dimension (L). When the ratio of $P/L < \sim 0.5$, block removal becomes exponentially more difficult for rough channel beds. For smooth channel beds, this ratio is much lower as channel velocities are higher closer to the bed. According to Lamb et al. (2015), there is no influence of the vertical block dimension (H) when compared to L except for the case of toppling. Intuitively, as the ratio H/L increases, toppling becomes the most dominant failure mode.

Finally, the effect of block wall stresses is negligible when block sides are relatively smooth, such as is the case for most idealized laboratory experiments. This was noted by both Dubinski (2009) and Coleman et al. (2003). As wall stresses increase (associated with greater block wall roughness, more typical of field/prototype settings) the critical block threshold also increases.

Digitalization of these empirical-analytical-based methods seems a priori challenging.

Fracture Mechanics Methods

Analytical methods based on linear elastic fracture mechanics (LEFM) theory applied to rock joints allow to describe progressive, time-dependent failure of a pre-fractured rock mass by turbulent flow impact. These methods are mathematically more complex but have the advantage to describe the time evolution of scour formation.

Comprehensive Fracture Mechanics (CFM) Method (Bollaert 2002; Bollaert and Schleiss 2005)

This method represents a simplified version of Linear Elastic Fracture Mechanics (LEFM) theory to mathematically describe brittle or progressive fracturing of rock joints due to cyclic dynamic pressures generated by turbulent flows. The fractured rock mass is assumed to be a perfectly linear elastic, homogeneous and isotropic material. The method is applicable to any type of solid material that can break up by fatigue failure of bonds, such as rock, concrete, ceramic, etc.

Hydrodynamic pressures inside rock joints are described by a stress intensity factor, representing the amplitude of the rock mass stresses generated by the water pressures at the tip of the joint. The corresponding resistance of the rock mass against joint propagation is expressed by its fracture toughness (Figure 3.13).

Joint propagation distinguishes between brittle (or instantaneous) propagation and time-dependent propagation. The former happens for a stress intensity that is equal to or higher than the fracture toughness. The latter is occurring when the maximum water pressure results in a stress intensity that is inferior to the material's resistance. Joints may then be propagated by fatigue. Failure by fatigue depends on the frequency and the amplitude of the load cycles. The fracture mechanics implementation of the hydrodynamic loading consists of a transformation of water pressures in joints into stresses in the rock.

These stresses are mathematically defined based on the maximum dynamic water pressure inside the joint, the length of the joint to fracture and a boundary correction factor depending on the shape and the initial degree of fissuring (i.e. persistency) of the joint, as well as on the lateral support of the surrounding rock. A parabolic-shaped pressure distribution is assumed inside the joints.

The fracture toughness is related to the mineralogical type of rock and to the tensile strength or the unconfined compressive strength of the rock mass. Furthermore, corrections are made to account for the effects of the loading rate and the in-situ stress field of the rock mass. The analytical equation expressing the corrected in-situ fracture toughness is based on a linear regression of available literature data.

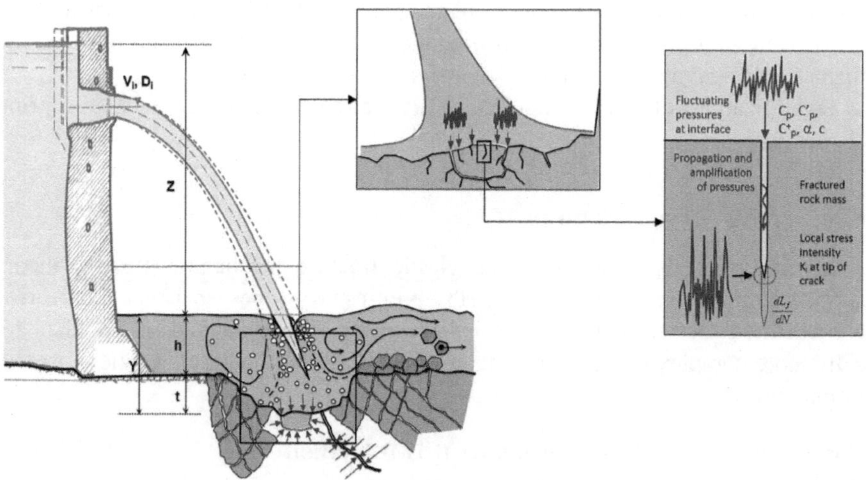

FIGURE 3.13 Application of Linear Elastic Fracture Mechanics to propagation of fractures in rock joints downstream of hydraulic structures.

Instantaneous joint propagation will occur if the stress intensity exceeds the fracture toughness. Otherwise, joint propagation needs a certain time to happen. This is expressed by a material-dependent Paris-type equation allowing to describe fatigue growth of fractures with time. Fracture growth depends on the number of water pressure cycles per second, i.e. on the wave celerity and the resonance frequency of the joint. For practice, high-velocity turbulent flows at hydraulic structures generate wave celerities between 100 (high air content) and 200 (low air content) m/s.

To implement time-dependent joint propagation into a comprehensive engineering model, the vulnerability of rock to fatigue must be known. Relevant fatigue-law parameters have been derived from available laboratory data on the sensitivity of rock to quasi-steady break-up by water pressures in joints (Atkinson 1987). Furthermore, based on 20 years of practical experience and feedback based on in-situ scour holes, Bollaert (2021) defined a range of parametric values to be used for both soft or easily erodible rock and hard or lowly erodible rock. A refinement of these parametric settings, based on a new database of case studies, can be found in Chapter 6.

As a stand-alone analytical method, the CFM can be easily applied on a single rock block by use of a spreadsheet. Its application has been fully automatized by implementing it into a digital environment, generating 2D rock scour with depth for 2D turbulent flows and multi-layered 2D rock masses (see Chapters 6 and 7).

Scour Model (SM) (Maleki and Fiorotto 2022)

Maleki and Fiorotto (2022) developed a stochastically based analytical method to compute rock break-up by brittle or progressive fracturing of rock joints, accounting for the detailed dynamic pressure distribution along the rock joint length and for the fracture width and roughness influence on fluid resistance. This allows to define the probability of occurrence of both brittle and fatigue failure of rock joints.

The authors were inspired by the original developments made in this field by Bollaert (2002, 2004, 2006, 2012) and Bollaert and Schleiss (2005), implemented over the last two decades in the CSM model, to propose a more complex analytical framework of the basic physics behind rock block plucking and rock joint fracturing.

The stress intensity at the tip of the joints was approached probabilistically by a Gaussian distribution, and found dependent on the joint length, the joint roughness, the air volume fraction and the joint thickness. Moreover, the influence of the fluid resistance for very long rock joints was found to strongly reduce the stress intensity values at the crack tip. This, however, was obtained for joint lengths of several tens of metres, i.e. far outside of the typical values for real-life fractured rock masses, for which uninterrupted joint spacings are typically on the order of a few decimetres to max. a few metres.

Furthermore, the influence of the pressure distribution inside the joints was quantified based on a transient wave model without air release and resolution effects during pressure changes. As such, the highly non-linear effects of the instantaneous spatially distributed free air content in the joints, such as modelled by Bollaert (2002) and considered responsible for pressure amplifications in joints, are not accounted for.

The more detailed analysis of the stress intensity factor and its parameters of influence offers interesting additional insight and thus has theoretical merit, but unfortunately also needs considerable (detailed) additional field data to be relevant, and generates additional parameters and complexities to the (already complex) theoretical framework.

The method is a priori digitalizable, but the need for very detailed data will make it particularly challenging for an engineer to apply. Acquiring such large sets of very detailed data on jointed rock masses, together with detailed feedback from a large set of scour cases on these rock masses, will be needed to point out the real-life relevance of all these parameters, as well as to determine plausible parametric values to be used by the engineer.

GROUP IV: CFD MODELLING WITH SINGLE WATER-ROCK COUPLING

A fourth group of models represents CFD modelling of turbulent flows in an unlined channel, plunge pool or stilling basin. Such modelling makes use of open-source or proprietary software and mainly focuses on a detailed 3D description of pressures and flow velocities at the water-rock interface.

Most often, CFD modelling is used as direct hydraulic input to an empirical, analytical or numerical method for rock scour assessment. No automated sequential coupling, during which the computations are performed and updated on a step-by-step basis between the CFD model and the rock scour model, is generally considered.

Today's available CFD packages allow to model turbulent flows with a fair degree of reality and confidence, and most of the available software allows to include air entrainment and detailed geometrical features in a more and more reliable manner.

Nevertheless, it should be kept in mind that a complete numerical reproduction of the entire turbulent spectrum of pressure and velocity fluctuations (turbulent eddies), such as might theoretically be obtained by Direct Numerical Simulation (DNS) techniques, is still out of the realm of practical engineering. Only fluid flows with small Reynolds numbers can be considered by DNS modelling, because the grid must become increasingly finer with increasing Reynolds number and smaller eddy scales. Computational time and efforts remain a huge challenge for all CFD models, most turbulence models asking for very small grid sizes and time steps to generate reliable results.

Researchers and engineers working in rock scour generally use CFD to determine the flow parameters that are needed to apply an empirical, semi-empirical or analytical rock scour computational method, such as a regression equation, an index-based method, an initiation of motion block stability analysis or one of the rock break-up methods implemented in the physics-based Comprehensive Scour Model.

As such, the hydraulics part becomes strongly digitalized today, while the rock mechanics part is generally much less digitalized by rock scour engineers. Also, a sequential coupling between the flow and the rock mass is rarely performed, and if done, it is generally performed manually (Figure 3.14).

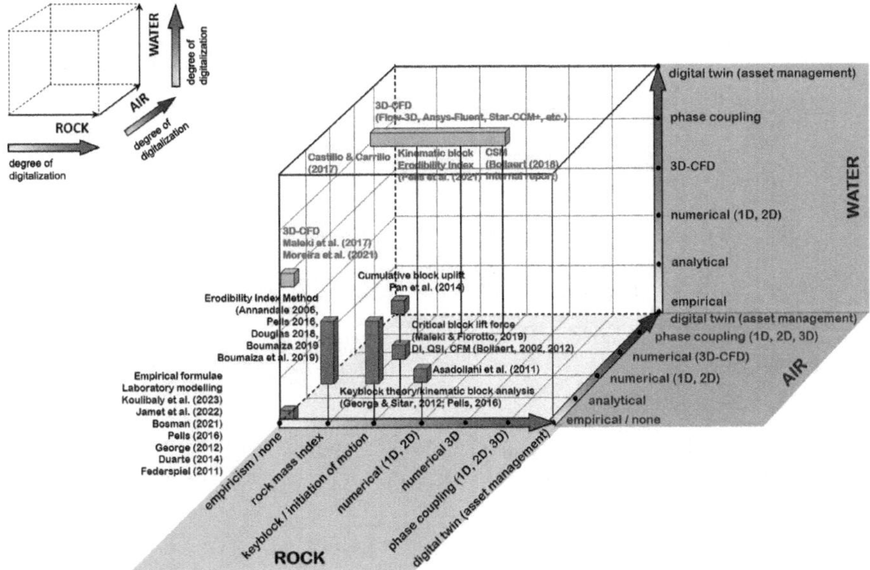

FIGURE 3.14 Group IV of computational methods: CFD modelling of turbulent flows.

CFD Modelling

This first subgroup represents CFD modelling of 3D turbulent flows for spillways, unlined channels and plunge pools without involvement of a rock scour model or method, i.e. for a fixed (solid) bottom. Some examples are briefly presented, implementing the following modelling techniques:

- *Smooth Particle Hydrodynamics (SPH) modelling*: Moreira et al. (2021 and Moreira (2021) performed SPH (meshfree, Lagrangian particle) modelling of a vertically impinging jet onto a flat pool bottom, as well as of a flow deviation gallery with flip bucket jets into a downstream plunge pool (Caniçada Dam, Portugal) (Figure 3.15). This meshfree modelling technique is currently strongly developing in the field of dam spillways.
- *RANS and LES modelling (FLOW-3D®)*: A practical feedback on the capabilities and challenges of 3D modelling of air-water flows at hydraulic structures has been carried out by Oukid et al. (2015). Aspects like aeration, shock waves, ski-jump hydraulics and hydraulic jumps are discussed. An example of 3D CFD modelling is illustrated in Figure 3.16 for the ski-jump tunnel spillways of Koman Dam in Albania (De Cesare et al. 2011).

CFD-CSM Modelling (Bollaert 2018)

Bollaert (2018, unpublished report) combined a 3D-CFD hydraulic model with a 2D rock mass model to determine scour formation in an unlined channel downstream of a

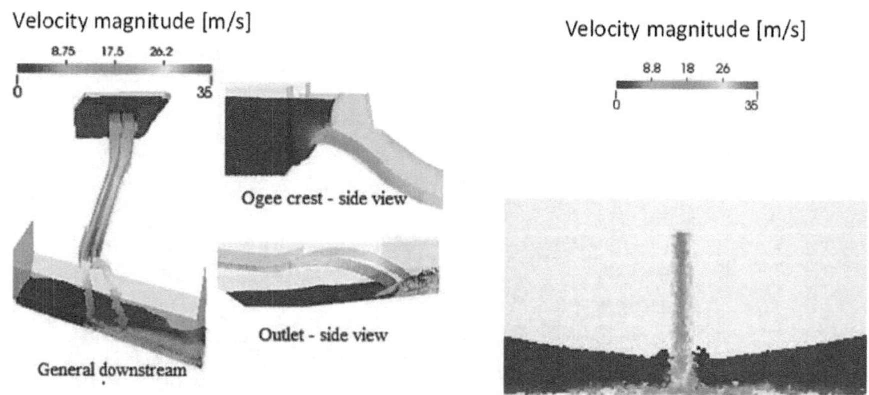

FIGURE 3.15 SPH modelling of turbulent flows in spillways: LEFT: flip bucket jets of Canaçada Dam; RIGHT: vertical high-velocity jet impinging onto a flat pool bottom. Adapted from Moreira (2021).

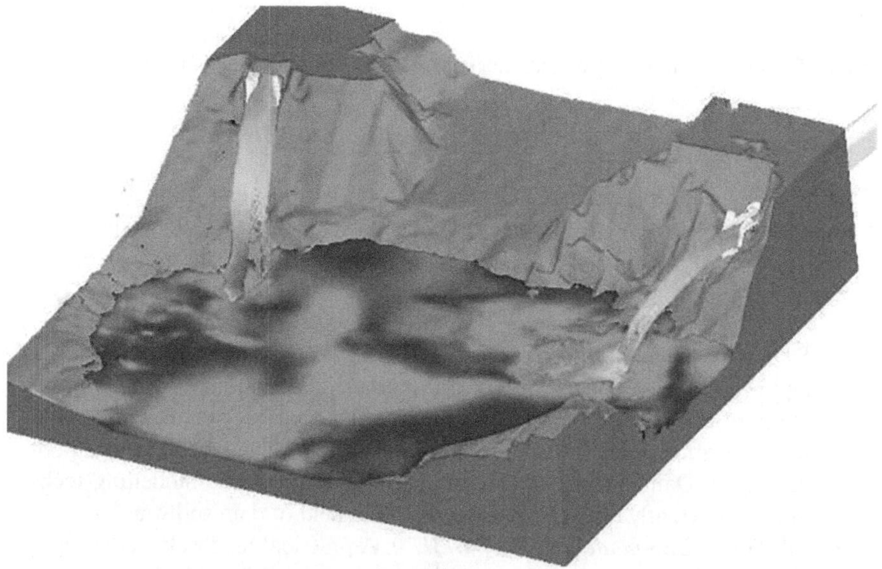

FIGURE 3.16 3D CFD modelling of the ski-jump spillway outflows at Koman Dam in Albania (De Cesare et al. 2011). Copyright© Société Hydrotechnique de France, reprinted by permission of Taylor & Francis Ltd, http://www.tandfonline.com on behalf of Société Hydrotechnique de France.

spillway concrete apron of a dam in the USA. FLOW-3D® was used to compute the flow velocities, turbulent kinetic energies and average dynamic pressures at the water-rock interface for different discharge scenarios and accounting for aeration.

Pressure probes were used to record the CFD computed dynamic pressures at short intervals and to estimate the root-mean-square values of the pressure fluctuations.

Despite the difficulties to generate pressure fluctuations using the Reynolds Averaged Navier-Stokes (RANS) equations, FLOW-3D® was found able to detect areas of higher turbulence and flow perturbations, allowing to make reasonable estimates of the RMS values. Comparison with literature values of RMS pressures in hydraulic jumps (Toso and Bowers 1988; Narayanan 1978) showed that the FLOW-3D® computed RMS values were relatively close, i.e. within the range of 2%–8%, pointing out the plausibility of the approach.

These CFD-generated hydraulic parameters have then been used as input to the 2D Comprehensive Scour Model (CSM) (Bollaert and Schleiss 2005; Bollaert 2012), to compute scour formation in a multi-layered 2D rock mass based on instant or progressive fracturing of rock joints and based on subsequent potential for rock block plucking, detachment and displacement towards downstream.

Finally, sound model calibration for historical flood events has allowed to predict future scour potential for the design flood.

The CSM reproduces the most essential phases of rock scour, i.e. the fracturing phase and the subsequent block detachment phase, together with their respective relevance to scour. The most relevant of these phases to scour depends on the initial degree of fracturing of the rock, on the main characteristics of single rock blocks and on the duration of the flow hydrograph. More details on the different rock break-up modules are provided in Chapter 6.

The key issue of the CSM is its relatively comprehensive character, despite the complex physical interactions between the different phases involved. The physics are mathematically simplified and described in a manner that a practising engineer can understand and handle them, while maintaining a degree of pertinence and precision guaranteeing results that are relevant for practice.

The described methodology allows a (single) coupling between the 3D CFD hydraulics and the 2D geomechanics and break-up mechanisms of the rock mass. Moreover, in contrast with the other modelling examples presented in this Group III, automated procedures have been developed in Visual Basic® for this coupling, and thus a certain degree of digitalization is available.

Nevertheless, only one iteration of coupling has been performed to determine scour formation based on the CFD hydraulic results, and no iterative feedback was given to the CFD model to update the flow hydraulics. The methodology is, however, easy to digitalize and to couple sequentially, and is of particular interest to the practising engineer. This is further described in Chapter 7 (fluid-solid coupling between CFD and CSM).

CFD – Index-Based Modelling (Pells et al. 2022)

Pells et al. (2022) have performed a qualitative hydro-geotechnical rock scour assessment at a dam spillway in French Guiana by relating the unitary stream power determined by FLOW-3D® modelling to the scour vulnerability of the rock mass expressed by the eGSI index (Pells 2016).

Unitary stream power was thereby determined in different ways: by integration with depth of the turbulent dissipation computed by the flow model; by combining density, mass flow and energy gradient; and finally by combining shear stress and flow velocity.

Moreover, for chosen key blocks, a simplified 2D kinematic block analysis was performed to verify block stability under average dynamic water pressures and shear forces exerted by the flow on the block in question. The pressures and forces on a block are thereby determined based on engineering judgement and interpretation, based on a qualitative analysis of the average dynamic pressure and shear forces computed by the CFD at the block upper boundary.

The approach developed by Pells et al. (2022) is thus of qualitative nature and only provides tendencies or proxies for scour potential. Comparison of the outcomes with in-situ observations may nevertheless provide insight into the involved processes.

As such, despite the CFD modelling of the flow, the degree of digitalization of this approach is considered low, because of the multiple manual and subjective assessments to be made relative to the main parameters involved during the process.

When using a threshold-based geomechanical index, digitalization would become easier to implement.

CFD – Kinematic Block Modelling (Hurst et al. 2021; George et al. 2022)

Hurst et al. (2021) have performed a prismatic block stability analysis based on analytical equations for block sliding and block toppling, whereby the pressure forces and shear (frictional) forces on the blocks are derived from CFD (k–ε) computations for fully developed 2D stationary subcritical channel flow. Block vertical uplift has not been considered, however. Their developments are based on the work by Lamb et al. (2015).

For a prismatic block that is situated at the base of a downstream facing step in a riverbed, the authors account for the negative water pressure arising just downstream of this block, as well as for potential block protrusion into the main flow, the channel bed slope and finally the location of the pivot point for toppling, to express a factor of safety against sliding and toppling. This is finally expressed as a function of the shape factor of the block, i.e. the height compared to the side length.

Second, George et al. (2022) have performed a kinematic block analysis (using the aforementioned BTRE method) on a series of surface-exposed key blocks of different shapes on an unlined spillway channel downstream of a dam in northern California (USA). The detailed dimensions, shapes and orientations of the blocks and their multiple faces have thereby been defined based on a 3D point cloud obtained through LiDAR measurements.

Furthermore, the local flow velocities and their orientation compared to the block faces were determined by detailed FLOW-3D® modelling. In contrast with Pells et al. (2022), dynamic pressure coefficients were determined analytically by the authors, based on the local block geometrical situation compared to its surroundings (i.e. block protrusion, slope, etc.) and comparing this situation with laboratory tests performed by George (2012).

By adopting a simplified pressure pattern around the block, i.e. high pressures underneath the block and lowest possible pressure at the block surface, a force balance allows to determine the vector of the resultant force on each of the blocks, for each of the potential failure modes (i.e. face detachment or plucking, sliding, toppling), and to check block stability.

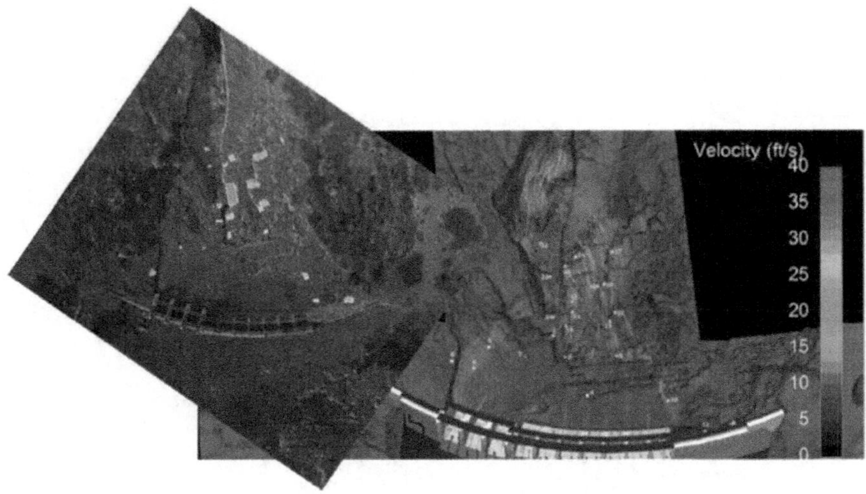

FIGURE 3.17 Digitized 3D block locations and CFD modelling of local flow velocities (George et al. 2022).

The method allows determining whether rock blocks that are exposed at the surface of the water-rock interface will remain stable or not under given (simplified) flow conditions. However, it does not allow determining scour formation with depth, based on consecutive failure of a series of key blocks, and should rather be considered as a sort of proxy for onset of scour formation. Also, likewise Pells et al. (2022), the real pressure pattern around the blocks is not computed, only a simplified interpretation is provided based on pressure coefficients determined at the block surface.

Finally, George et al. (2015) compared their findings with the EIM method applied by determining the stream power at the water-rock interface, for the same blocks (Figure 3.17).

GROUP V: SEQUENTIALLY COUPLED WATER-ROCK MODELLING

The fifth group represents computational methods proposing automated sequential coupling between a hydraulic model and a rock mass scour model (in 2D or 3D). The computations are performed on a step-by-step basis, and both the flow hydraulics and the rock mass scour formation are updated at each step. In other words, this group contains the highest degree of digitalness among currently available rock scour computational methods.

It should be emphasized here that the numerical modelling is not necessarily 3D CFD modelling. The current group of methods specifically focuses on the sequential coupling between water and rock, and not on the complexity of the numerical modelling (Figure 3.18).

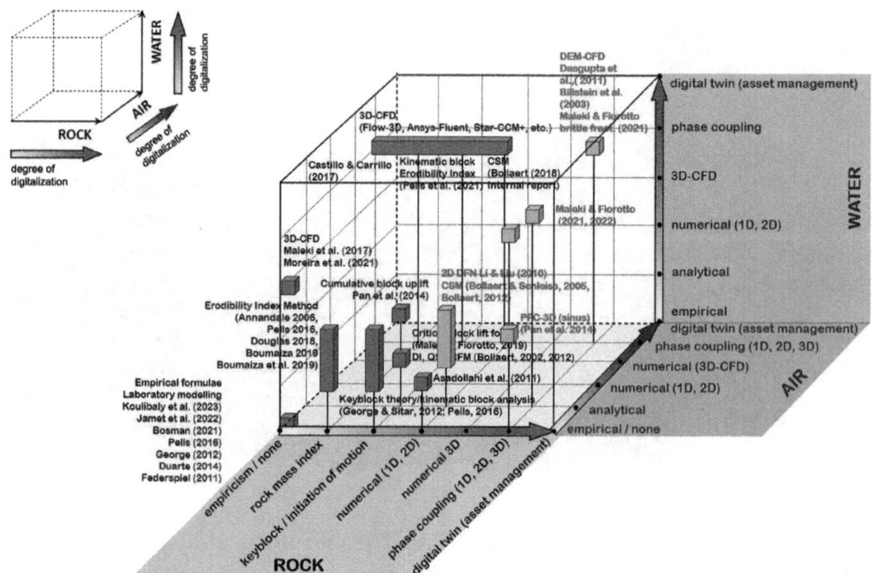

FIGURE 3.18 Group V of computational methods: sequentially coupled water-rock modelling.

CSM Water-Rock Coupling (Bollaert and Schleiss 2005; Bollaert 2012)

The Comprehensive Scour Model (CSM) was initiated by Bollaert (2002) and Bollaert and Schleiss (2005) and further developed numerically by Bollaert (2004; 2006; 2012; 2021). The model proposes an automated sequential coupling between a 2D hydraulic model of plunging air-water jets and a 2D geomechanics model of the rock mass with depth. For other turbulent flows, such as channel flows or hydraulic jumps, the hydraulic model must be fed with dynamic pressures and flow velocities taken from CFD models or from laboratory physical models. The coupling can still be performed and partially automated, but becomes more demanding on input data to be provided.

Second, the rock mass model implements different break-up mechanisms of fractured rock, i.e. plucking of single rock blocks by dynamic impulsions on the blocks (DI method), detachment and downstream displacement of rock blocks by quasi-steady impulsions on the blocks (QSI method) and finally instantaneous or progressive fracturing of the rock mass by frequency-dependent pressure pulsations propagating through the joints around the blocks (CFM method). The basics of these break-up methods are discussed in Chapter 6.

Even though these break-up methods are empirical-analytical based and are also categorized under initiation of motion-based methods (i.e. Group III), the CSM offers a degree of digitalization and coupling between the water and the rock that justifies its place in the current group of sequentially coupled models. This proprietary software is written in Visual Basic® and allows fully automated sequential coupling for plunging jets, accounting for user-defined inflow hydrographs, and computing 2D scour formation in multi-layered rock as a function of time. It has been replaced since 2021 by the rocsc@r® software, written in C# and cloud-based, as described more in detail in Chapters 6 and 7.

Figure 3.19 illustrates an application of the CSM to a scour study performed for the primary spillway of Folsom Dam, California (Bollaert et al. 2006). The left-hand side of the figure illustrates the spatial distribution of dynamic pressures generated by the outflows, while the right-hand side shows the corresponding scour potential in the rock mass underneath the concrete lining of the spillway, based on the assumption that the lining fails. A full 3D assessment of scour potential is obtained as a function of time duration of flood events.

Another example is illustrated in Figure 3.20 for the scour potential in the unlined plunge pool downstream of a French arch dam computed for a 1,000-year flood return period.

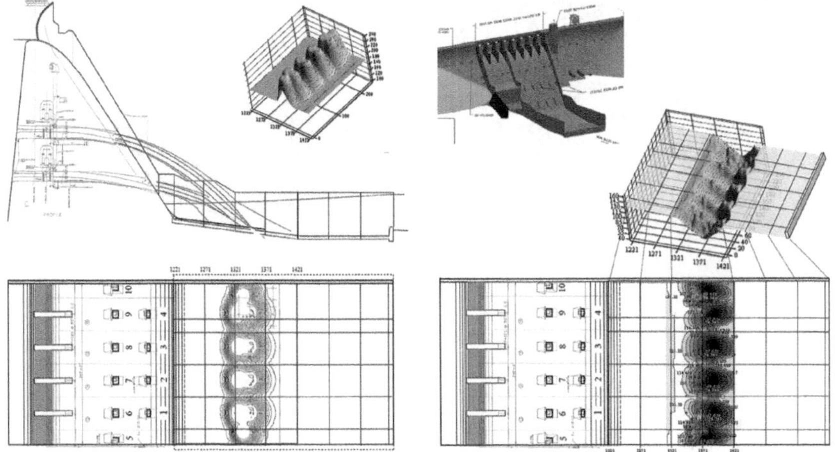

FIGURE 3.19 Quasi-3D CSM water-rock coupling (Bollaert et al. 2006): LEFT: dynamic pressures of outflows; RIGHT: 3D scour potential in rock mass underneath concrete lining.

FIGURE 3.20 Quasi-3D CSM water-rock coupling (Bollaert 2012): 3D perspective view of scour potential for a 1,000-year flood event on a French arch dam. Courtesy of EDF SA.

Finally, Figure 3.21 illustrates a 2D numerical reconstitution of the scour hole (Figure 3.21b) that was formed in the unlined bedrock just downstream of the concrete apron of Paradise Dam in Australia, following a flood event in 2013. The hydraulic parameters (dynamic pressures and flow velocities) have been defined for different scour situations based on a 3D physical model built at Manly Hydraulics Laboratory in Sydney. Pictures illustrate the physical model and the scour observed downstream of the apron.

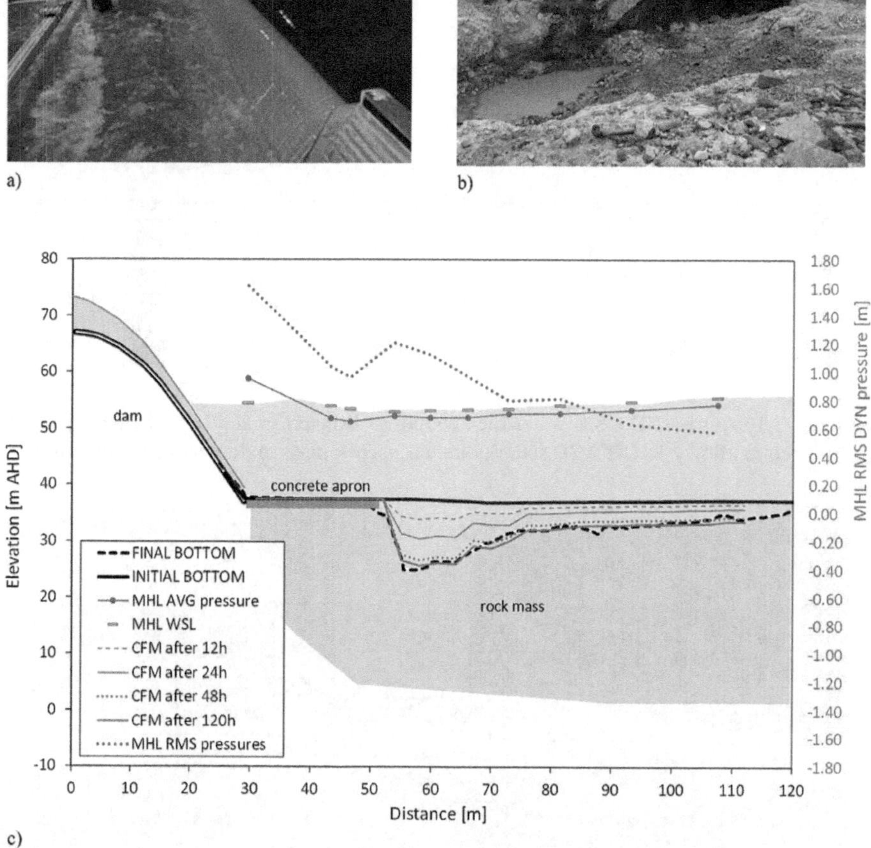

FIGURE 3.21 Fluid-solid coupling of rock scour formation during the 2013 flood event at Paradise Dam, Australia, using a 3D physical model and the CSM numerical model. Courtesy of Sunwater Ltd.

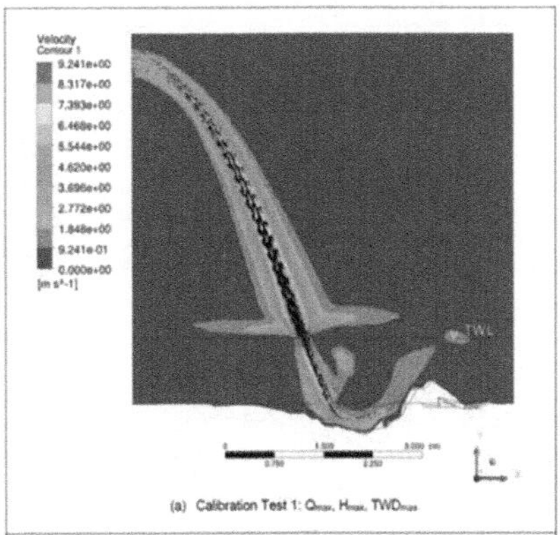

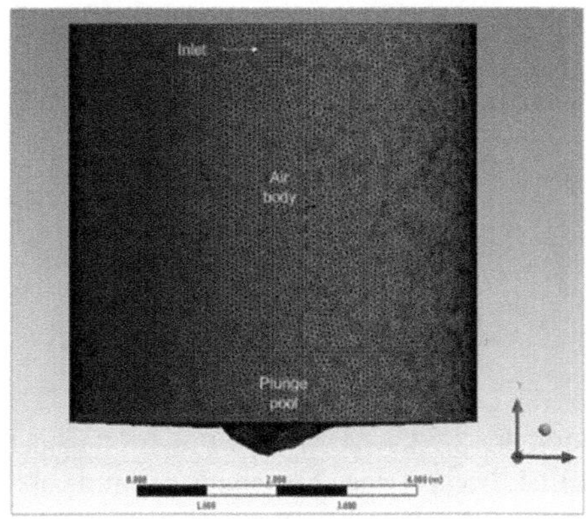

FIGURE 3.22 3D ANSYS-FLUENT® modelling of plunging jet flows by Bosman (2021): LEFT: flow velocities along centreline of jet; RIGHT: meshing along centreline of jet.

CFD – Regression Coupling (Bosman 2021)

Bosman (2021) applied a 3D ANSYS-FLUENT® modelling of her laboratory-generated jets, to relate the computed pressures and velocities to a regression equation that defines scour. The results were compared with the scour holes as measured during the physical model tests with cobblestones.

The computations were performed using a Shear Stress Transport (SST) k–ω turbulence model. The model domain was discretized into a hybrid mesh containing both hexahedron and tetrahedron meshes, using an average mesh size of 0.06 m and a max. mesh size of 0.10 m. The surface roughness parameter for the 0.1 × 0.1 × 0.05 m rock blocks was determined to be approximately 0.2 m, corresponding to a Chézy roughness value of 0.2 m for "hand placed pitching" according to Rooseboom and Van Vuuren (2013). The surface roughness parameter was determined to be 0.4 m for the 0.1 × 0.1 × 0.075 m rock blocks (Figure 3.22).

A time step size of 0.001 s was specified with a maximum of 10 iterations per time step. The time step size was obtained after several trial runs which were found adequate to achieve solution stability and convergence. The number of time steps used in the analyses was 30,000, which equates to 30 s (physical model time). Data was collected at 1-s intervals.

While the jet impact velocity was quite well reproduced numerically, both the jet thickness and jet width were underestimated by 15%–20% for all computations. Also, the dynamic pressures at jet impact were way too high, up to 200%–300% of the laboratory-measured ones. The disparity between the simulated jet footprint at impingement with the plunge pool water surface and the observed measurements was suspected by Bosman to be due to the rate at which air is entrained into the jet. This rate seemed strongly under-predicted by the numerical model. Use of a finer mesh or increase of the initial turbulence from 4% to 9% did not improve the results however.

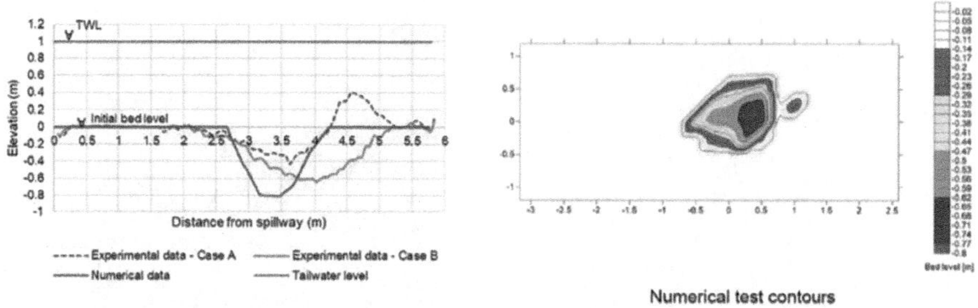

FIGURE 3.23 3D ANSYS-FLUENT modelling of plunging jet flows by Bosman (2021): LEFT: Comparison of longitudinal profile of computed scour hole with laboratory-measured scour hole. RIGHT: Contours of computed scour hole depths.

The computational methodology developed by Bosman (2021) uses the CFD computed dynamic pressures and flow velocities as input to a regression equation to manually perform post-processing and define ultimate scour formation. The so computed scour hole is then used as new bottom for a subsequent CFD computation to update pressures and velocities. Iterations are manually driven and repeated until quasi no scour change is detected anymore between them.

A comparison between computed and laboratory-measured scour holes is made in Figure 3.23 for a tailwater depth of 20 m, a jet discharge of 80 m³/s, a jet issuance velocity of 12 m/s and an upstream head of 100 m. Prototype rock block dimensions were of 2 × 2 m side length for a height of 1 m. The computed scour hole is deeper and significantly more compact than the laboratory scour hole. The reason for this discrepancy most probably lies in the numerical jet being too compact and not enough aerated compared to the large-scale, highly broken-up laboratory jet.

The approach developed by Bosman (2021) is of interest because, despite the uncoupled (manual) CFD post-processing for scour hole determination, the procedure would be relatively easy to digitalize. Nevertheless, one cannot speak about a real sequential coupling, because of the absence of step-wise (block-wise) scour computations like it occurs in reality (i.e. at each step the ultimate scour depth and not the really occurring scour depth is computed by the regression equation).

CFD – Sediment Transport Coupling (Castillo and Carrillo 2016)

Castillo and Carrillo (2014, 2015, 2016) performed a 3D CFD analysis of scour potential downstream of the Paute-Cardenillo Dam, located in Ecuador, which is a double curvature arch dam with a maximum height of 135 m to the foundations.

FLOW-3D® software was firstly used to determine the mean dynamic pressure field at the point of jet impact and to compare the results with analytical equations. The authors performed a sensitivity analysis of the turbulence models, the air entrainment, the grid size and the type of solver, for both free surface overflow jets and intermediate outlet pressurized jet flows.

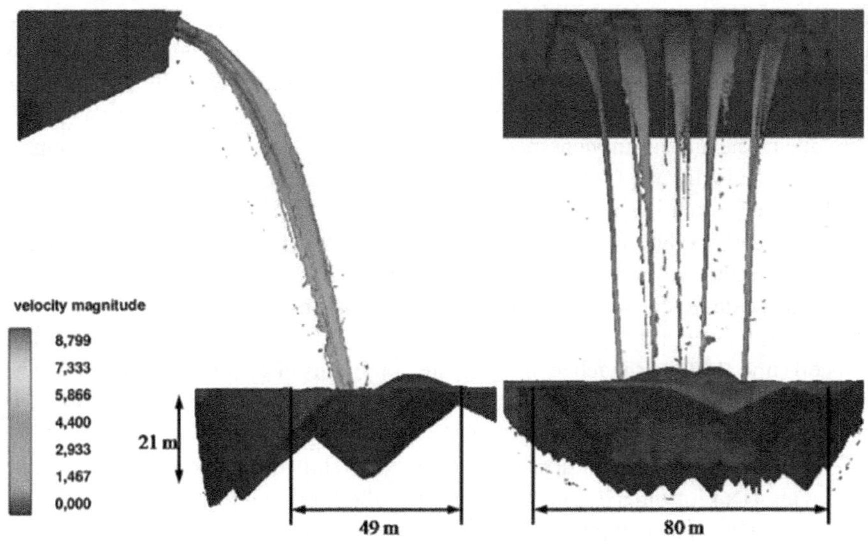

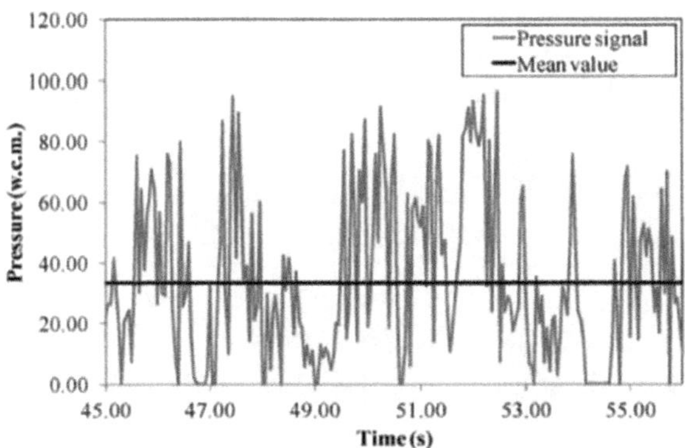

FIGURE 3.24 3D CFD modelling of free surface overflow jets and scour potential at Paute-Cardenillo Dam. UP: Perspective view, DOWN: Pressure fluctuations by impact of free surface jets and water cushion of 2 m. Adapted from Castillo and Carrillo (2016).

Figure 3.24 presents a perspective overview as well as the dynamic pressure signal and fluctuations as computed by FLOW-3D® at the point of impact of the free surface overflow jet into a 2-m-deep water cushion.

For these free surface overflows, the authors found anomalous results when a one-fluid free surface method is used in combination with a RNG k–ε turbulence model. However, when the RNG k–ε model is used with the Split Lagrangian free surface method, then the results are more like the expected values. The most accurate results were obtained by using a mesh size of 0.2 m, and large eddy simulation (LES) turbulence model. In the solver options, the stability and convergence method was used and the free surface was solved with the Split Lagrangian method. Following the authors, the rather (too) high mean dynamic pressures computed at jet impact were due to the air entrainment model not resolving the two-phase flow in an appropriate way (Castillo and Carrillo 2014, 2015).

The scour formation was computed in FLOW-3D® using the implemented sediment transport module (based on suspended load or bed load equations), which is strictly speaking only valid for grain sizes of up to 35 mm. As such, a 1–50 Froude similitude scale model was simulated by the authors instead of the prototype scale, and the results were transformed to prototype scale. The model had more than 4 million cells and the scour reached steady-state conditions after 60 s of simulation. The required computational time was about 26 days on an Intel® Core™ i7-2600 CPU @3.40 GHz processor and 24 GB RAM (Castillo and Carrillo 2016).

The CFD computed scour depth was 21 m, i.e. slightly deeper than the mean value of the empirical formulae (17 m) and similar to the semi-empirical erodibility index method (~ 20 m). The computed scour depth remained within an alluvial layer on top of weathered rock. Scaling effects were not discussed by the authors.

Next, the authors numerically modelled the pressurized jet flows issuing from the half-height outlets of the dam. Figure 3.25 illustrates the CFD model, for which a meshing of 1 m (prototype value) could be used because of the greater dimensions of the jets compared to the overflow jets. The computed scour depth of 34 m was in good agreement with the empirical formulae and slightly less than the erodibility index method.

FIGURE 3.25 Flow velocity in the air and in the pre-excavated stilling basin (prototype scale, units in m/s): Spatial view of the half-height outlet (Castillo and Carrillo 2016).

As a summary, the approach proposed by Castillo and Carrillo (2016) makes use of the direct coupling offered within the FLOW-3D® environment between 3D hydraulics and related 3D sediment transport and deposition. Hence, the approach is fully automated and has a high degree of digitalization regarding the numerical computations.

The basic equations behind these scour computations, nevertheless, have been initially developed for granular soils and not for fractured rock. As such, the used coupling is not applicable to fractured rock and careful engineering judgement is needed when using this approach.

LBM-DEM Coupling (Gardner and Sitar 2019; Gardner 2023)

Gardner and Sitar (2019) and Gardner (2023) model the interaction between turbulent flow and rock blocks by a direct modelling of both the solid and fluid phases, i.e. hydraulic plucking of individual rock blocks is modelled using the Discrete Element Method (DEM), while flow turbulence is modelled using the Lattice Boltzmann Method (LBM) with integration of a Large Eddy Simulation (LES) model for capturing subgrid scale turbulent eddies and dynamic effects between flow and particles (Gardner 2023). The LBM mesh is entirely independent of the DEM discretization, making it possible to refine the LBM mesh such that transient and varied fluid pressures acting on the rock surfaces are directly modelled. This provides the capability to investigate the effect of water pressure inside the fractured rock mass, along potential sliding planes (Gardner and Sitar 2019; Gardner 2023).

The DEM implementation is able to capture detailed 3D rock mass kinematics by considering the detailed 3D shape and orientation of each individual block. The contact detection algorithm used for the DEM computations is able to consider three-dimensional polyhedral blocks. The coupling between DEM and the weakly compressible LBM is achieved through application of a volume-fraction approach that is able to consider three-dimensional polyhedral blocks interacting with fluid.

Gardner (2023) implemented the DEM-LBM-LES methods together in a customized version of the open-source software waLBerla (Bauer et al. 2021), which is a massively parallel computational fluid dynamics framework with the ability to couple fluid simulations with rigid body dynamics solvers. By adding additional functionalities to waLBerla, such as coupling of the polyhedral DEM model with the fluid solver among others, the parallel computing capability already available through this framework can be leveraged on high-performance computing (HPC) systems, which ultimately makes this class of simulation possible. A case considering a backwards-facing step illustrates how sliding and toppling failure modes are captured automatically by the proposed modelling approach (Figure 3.26).

Although this numerical approach is able to model the governing mechanics and kinematics in dynamic fluid-solid interaction, there are still limitations to the scale at which it can be applied. Even with the extensive parallelization and distributed computing capabilities, the computational demands for this type of numerical simulation are such that it is not reasonable to perform simulations at project scale.

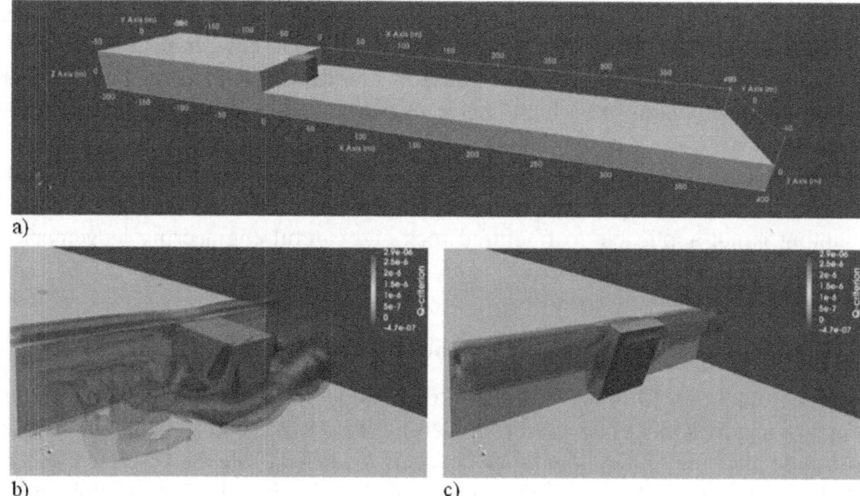

FIGURE 3.26 Different failure modes during hydraulic plucking of rock block by DEM-LBM-LES coupled modelling at backwards-facing step: (a) Simulation domain showing block located at backwards step, (b) sliding failure mode; and (c) toppling failure mode. Which failure mode is dominant depends on the rock particle geometry. Adapted from Gardner (2023).

For example, the simulations of the backwards-facing step with 48 million cells needed a simulation runtime on the order of 9 h and was executed on 144 cores with approximately 384 GB of memory (Gardner 2023). Thus, access to supercomputing resources is absolutely necessary for applying this analysis technique and simulating an entire dam spillway currently is not computationally feasible.

As a summary, this innovating and promising approach offers a very strong degree of digitalization, but has the disadvantage to be still too computationally intensive for engineering purposes.

CFD-DEM Coupling

Several researchers performed a mechanical coupling between the flow and the rock blocks by the Distinct Element Method (DEM), i.e. based on shear stress or tensile bonds between multiple blocks of different shapes.

Billstein et al. (2003) used the UDEC 2D software (Itasca International Inc.) to model the rock behaviour in a spillway channel downstream of Midskog Dam in Sweden. The model is based on the distinct element method that simulates the response of discontinuous media (such as a jointed rock mass) to static or dynamic loading. The medium is represented as an assemblage of discrete blocks. The discontinuities are treated as boundary conditions; large displacements along discontinuities and rotations of blocks are allowed.

Fluid flows through the discontinuities of the blocks. A fully coupled mechanical-hydraulic analysis is performed in which fracture conductivity is dependent on mechanical deformation of the joint aperture. Flow is idealized as laminar viscous flow between parallel plates.

Erosion or lateral movement of blocks on the surface of the channel occurs when horizontal forces acting on a block exceed the shear resistance of joints defining the block. The horizontal forces acting on a block include the surface shear (i.e., drag) force resulting from water passing over the upper surface of a block, and the horizontal water in sub-vertical joints acting to push the block. The shear resistance depends on the shear strength of bounding discontinuities and the surface area on which shear acts, based on Mohr-Coulomb's Law.

Output results are of qualitative nature and generated as displacement plots, scaled at 24 h of real flow. Blocks with large horizontal displacements can thereby be interpreted to correspond to eroded blocks.

Moreover, Dasgupta et al. (2011, 2012) also used the UDEC 2D software to perform a hydraulic-mechanical coupling between spillway flows and downstream rock mass. CFD (ANSYS® FLUENT) is used to calculate the unsteady turbulent flow field. Time-dependent pressure and shear stress on the spillway surface provided by the CFD were used in a geomechanical discontinuum model to determine the dynamic response of the rock mass. No transient wave propagation or amplification of surface pressures inside rock joints is modelled however. The standard k–ε solver with enhanced wall function is used in the present study. A two-fluid multiphase flow model is used to solve the governing equations. The Volume of Fluid (VOF) (Hirt and Nichols 1981) approach is used to track the interphase between two fluids.

The geomechanical model explicitly simulates block removal and brittle failure of rock. The dynamic response of the rock is simulated using UDEC. The joints can fail either in tension or shear, resulting in a joint opening (separation) or slip, allowing displacement and rotation of the blocks, which results in block removal or ejection. Brittle failure of intact blocks (between joints) is modelled by a Voronoi tessellation, i.e. the assembly of relatively small polygonal blocks formed by randomly distributed and oriented fractures with relatively short edges. Fractures are initiated as a function of the strength and forces in the rock and propagate along the boundaries of the Voronoi blocks. Where a network of fractions separates the Voronoi blocks, brittle (tensile) failure occurs, such as shown in Figure 1.6 of Chapter 1.

The modelling approach is presented by the authors as an iterative process between CFD and geomechanical analyses. Applicability of their methodology is nevertheless only shown through examples where only the first iterative loop is completed, because the sequential coupling should be performed manually.

Finally, Pan et al. (2014) developed a theoretical framework to determine the response of a single rock block to plucking. They performed a direct coupling between sinusoidal pressure signals at the water-rock interface and a 3D representation of the rock mass by means of an ensemble of balls that are glued together through tensile strength bonds with a certain tensile strength. The PFC^{3D} (Itasca International Inc.) software used applies a discrete element method (DEM) to model the inter-granular reaction and the displacement of a collection of granular systems (Figure 3.27).

Internal damage in a rock gradually develops by successive bonding failure between adjacent particles. In PFC^{3D}, a bundle of inseparable balls (particles) can be glued together to form a "clump". A set of clumps can model the rock blocks in a jointed rock mass (Mas Ivars et al. 2011). A smooth joint is placed between any pair of neighbouring clumps to model the mechanical behaviour of the discontinuity between a pair of rock blocks. Even if a weak plane with rock bridges is not persistent

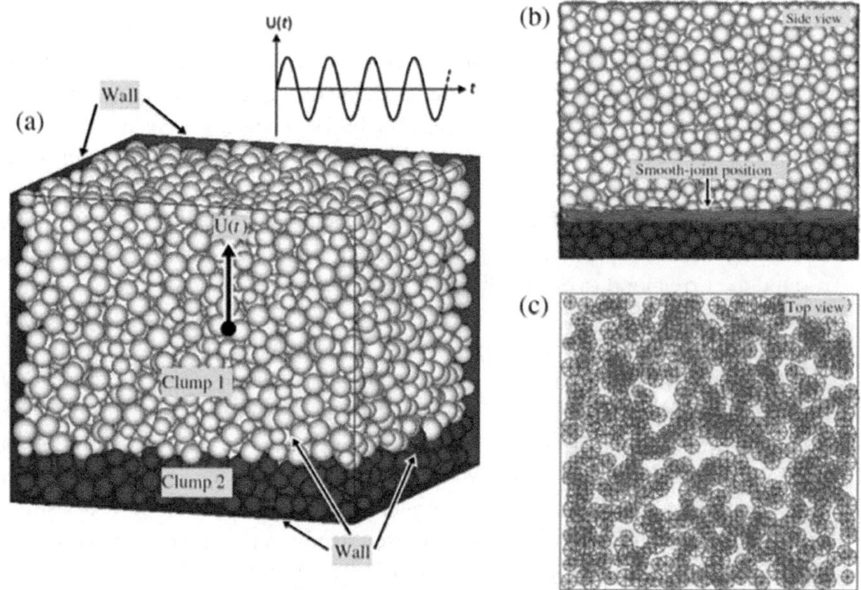

FIGURE 3.27 DEM modelling in PFC^{3D} of progressive damage to a rock block represented by a clump of glued balls, containing bonds with tensile strength: (a) the model of particle assemblage and the applied loading; (b) the side view of the model; (c) the contacts on the smooth joint (Pan et al. 2014).

at the beginning, a progressive loss of rock bridges may occur, along with a cumulative bond failure when repeated loadings are large enough. This will eventually lead to a total loss. Once all the rock bridges have been broken, the weak plane becomes fully persistent.

The performed modelling is purely qualitative and aims at demonstrating the importance of repeated pressure fluctuations to progressively damage (fracture) the rock mass. Pan et al. (2014) in parallel developed a theoretical framework for cumulative rock block plucking generated by sinusoidal pressure fluctuations. This framework is described in more detail in the subchapter on initiation of motion methods.

As a summary on hydraulic-mechanical coupling using DEM models, the often-used UDEC model is primarily based on shear strength or pure tensile strength break-up of bonds between rock blocks, without any fracture mechanics. While the coupling of DEM models with CFD models is a priori very encouraging and sounds digitalized, the coupling still must be performed manually. This clearly affects its degree of digitalization, as well as its engineering applicability.

DFN Water-Rock Coupling (Li and Liu 2010)

This subgroup of computational methods makes use of Discrete Fracture Network (DFN) models to represent the rock mass geometry with depth. DFN models allow to statistically generate 2D and 3D rock masses and their discontinuities, based on

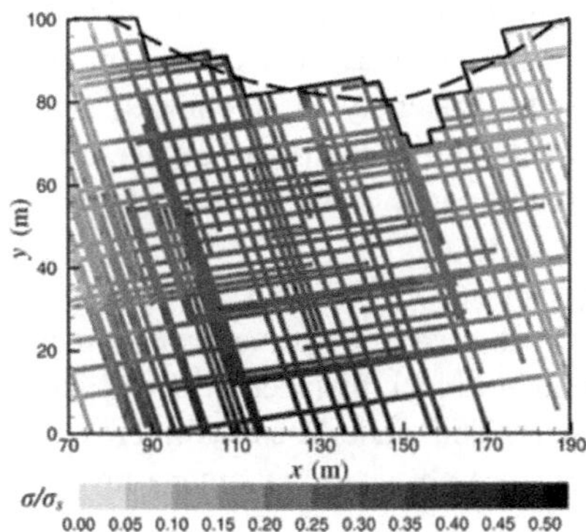

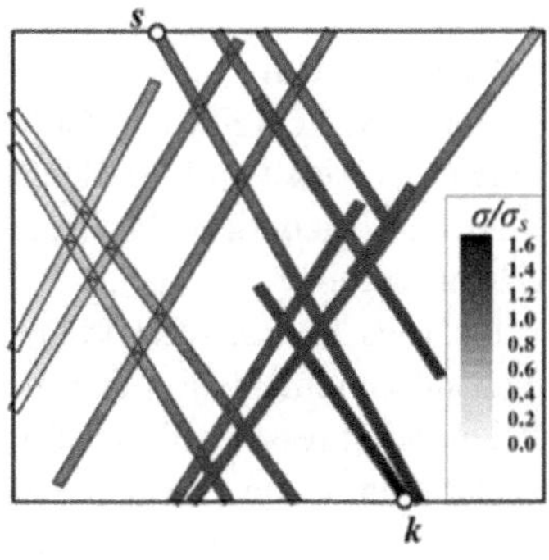

FIGURE 3.28 DFN-based scour modelling at Xi Luo Du HPP by Li and Liu (2010): LEFT: 2D DFN framework; RIGHT: spatial distribution of RMS amplification in joint network.

a stochastic generation of joints statistically representative for the in-situ situation. Different joint sets and their characteristics can be implemented by the user.

Applying the Monte-Carlo probabilistic method, a 2D DFN was developed by Li and Liu (2010) using a FORTRAN® program. Also, the authors performed 1D transient flow modelling to generate pressure wave propagation within the joints between the blocks.

Based on an empirical equation for RMS pressure fluctuations at the rock bottom due to jet impact, they stochastically generated a fluctuating pressure signal by use of a series of 40 sine and cosine functions with different frequencies and amplitudes. This signal was then locally applied at the entrance of one of the joints along the rock surface. Li and Liu (2010) found remarkable amplification of pressure fluctuations inside joints, especially at closed-end joints.

A kinematic stability analysis of the blocks located at the surface of the pool bottom then allows determining the blocks that will be ejected. Following manual adaptation of the new pool bottom, the next sequence of fluctuating pressure wave propagations is then computed inside the adapted fracture network, and so on until no blocks are unstable anymore.

The authors applied their framework on the arch dam of the Xi Luo Du hydroelectric power plant located on the Jing Sha River in China and defined the detailed shape and extension of the potential scour hole (Figure 3.28).

SYNTHESIS AND OUTLOOK

This chapter provides a brief overview of existing rock scour computational methods and approaches from the viewpoint of digitalization and usability/applicability for practising engineers.

A first striking observation to make is the strong influence of physical model tests on newly developed approaches over the past decade. Since the large-scale tests performed at the Swiss Federal Institute of Technology in Lausanne, under supervision of Prof. Dr. Schleiss and Dr. Bollaert, with several PhDs generated between

1998 and 2015, novel test installations have been developed and extensively used in Australia (University of NSW, Dr. Pells and Dr. Douglas), USA (University of California at Berkeley, Dr. George, Colorado State University, Dr. Dubinski), South Africa (University of Stellenbosch, Dr. Bosman) and recently at University de Québec à Chicoutimi (Dr. Boumaiza and currently ongoing with J. Kashtiban and A. Koulibaly).

Test runs have thereby progressed from simple visual observations and subsequent scour measurements to much more detailed fully digitized real-time measurements of dynamic pressure, flow velocity, air entrainment, block displacement and block acceleration, using adequate electronic equipment. This underlines the interest and need for deeper insight into the complex three-phase physical phenomena involved in rock scour at hydraulic structures.

Furthermore, over the past two decades, a tendency is observed towards increasing use of numerical modelling techniques. While rock scour at hydraulic structures has been historically solved by empirical or semi-empirical approaches, mainly based on dimensional analysis and laboratory model tests, a current novel trend is clearly towards integration of numerical models into the solution process. Specific rock scour numerical methods being unavailable until very recently, mainly CFD (open-source or commercial) or DFN models have been solicited.

While this trend is a priori useful and evident, some flaws are inherent to it. First, the used models do not solve all the phases and their interactions. Generally, only the water or the rock phase is tackled numerically and the remaining phases, interactions and deductions must be performed analytically by the engineer, based on empirical or semi-empirical approaches. This multiplies the diversity of methods and makes their standardization cumbersome. Second, a fair number of these numerical models are not straightforward to apply and have not been developed for rock scour at hydraulic structures. Hence, specific use and adaptations of standard equations and techniques must be made to make these models suitable for rock scour. This, again, makes their application more complex.

As a matter of fact, it appears that, with increasing digitalization, the diversity increases and the usability and applicability of the computational methods become more and more questionable. While empirical and analytical methods are relatively easy to be understood and to be applied by any engineer, their digitized counterparts give us a hard job.

As such, the current trend of digitalization certainly favours development of a handful of specialists in the field, but is counteracting worldwide dissemination and standardization of modelling techniques and relevant parameters, while exactly the opposite would be expected.

The reason for this paradox may be explained by the lack of synergy and communication between engineers and between researchers working in this field, as well as by the lack of easily accessible useful data.

Lack of synergy and communication finds its origin, among others, in individualization of research interests (i.e. each researcher prefers developing their own method rather than building further upon an existing method). Lack of data originates from data protection and ownership proper to the very competitive energy sector, and from the lack of a standardized digital platform allowing exchange of modelling techniques, results and relevant parameters. Pertaining to this last aspect, one can state it's the snake that bites its own tail…no data without digitalization and no digitalization without data…

As a result, it may be observed that, from the quite large amount of computational methods developed, unfortunately very few are being used on a worldwide basis by engineers.

Without being too restrictive, most of the rock scour issues today are solved by use of Group I empirical/physical approaches, or by application of one of the index-based methods of Group II, Annandale's (1995) erodibility index method being the most popular one because of its ease of understanding and application.

In parallel, only specialized engineers or researchers tackle rock scour by use of one of the Group IV or Group V methods, i.e. by coupling hydraulics to rock mechanics. While the number of approaches and techniques is quickly multiplying for these groups, only the professionals that developed these models are using them... Hence, one should not speak about models but rather about prototypes.

As a summary, the currently ongoing process of digital transformation of the rock scour engineering field is very individual and not trending towards any standardization of digitalization. For engineers to apply any of the Group IV and V methods, digitalization has to succeed in offering a standardized worldwide available digital platform for any aspects related to rock scour engineering (i.e. modelling techniques, results, relevant parameters) as well as an adequate database of case studies and experiences to work with.

The remaining of this book aims at setting up the first steps towards such a digital standardization, and focuses on the digital path to be taken by all parties involved in rock scour at hydraulic structures, i.e. infrastructure owners and investors, engineers and researchers.

Let's go digital!

REFERENCES

Annandale, G.W., "Erodibility", *Journal of Hydraulic Research*, 33, 471–494, 1995.

Annandale, G.W., *Scour technology: mechanics and engineering practice*, McGraw-Hill Professional, New York, 1 ed., 2006.

Annandale, G.W., "Processes and Proxies Defining Maximum and Time Rate of Scour of Rock around Bridge Piers", in *Contemporary Topics in In Situ Testing, Analysis, and Reliability of Foundations*, pp. 514–521, 2009.

Annandale, G.W. and Smith, S.P., "Calculation of Bridge Pier Scour Using the Erodibility Index Method", *Technical Report CDOT-DTD-R-2000-9*, California Department of Transportation, US, 2001.

Annandale, G.W., Wittler, R.J., Ruff, J.F. and Lewis, T., "Prototype Validation of Erodibility Index for Scour in Fractured Rock Media", in *ASCE Conference on Water Resources*, Memphis, TN, USA, 1998.

Asadollahi, P. and Tonon, F., "Stability of Rock Blocks Subjected to High-Velocity Water Jet Impact", in *44th US Rock Mechanics Symposium and 5th US-Canada Rock Mechanics Symposium*, American Rock Mechanics Association, Salt Lake City, 2010.

Asadollahi, P., Tonon, F., Federspiel, M.P.E.A. and Schleiss, A.J., "Prediction of Rock Block Stability and Scour Depth in Plunge Pools," *Journal of Hydraulic Research*, 49, 6, 750–756, 2011. https://doi.org/10.1080/00221686.2011.618055

Atkinson, B.K., *Fracture mechanics of rock*, Academic Press Inc., London, 1987.

Bauer, M., et al., "Walberla: A Block-Structured High-Performance Framework for Multiphysics Simulations", *Computers & Mathematics with Applications*, 81, Jan, 478–501, 2021. https://doi.org/10.1016/j.camwa.2020.01.007

Billstein, M., Carlsson, A., Söder, P.-E. and Lorig, L, "Midskog Gets Physical and Numerical", *International Water Power & Dam Construction*, 55, 12, 22–26, 2003.

Bollaert, E.F.R., "Transient water pressures in joints and formation of rock scour due to high-velocity jet impact", *PhD Thesis*, LCH-EPFL, Lausanne, Switzerland, 2002.

Bollaert, E.F.R., "A Comprehensive Model to Evaluate Scour Formation in Plunge Pools", *International Journal of Hydropower & Dams*, 2004, 1, 94–101, 2004.

Bollaert, E.F.R. and Mason, P.J., "A Physically based Model for Scour Prediction at Srisailam Dam", *International Journal of Hydropower & Dams*, 2006, 4, 96–103, 2006.

Bollaert, E.F.R., "Rock scour at hydraulic structures: a practical engineering approach", *Geo-Strata*, 2010.

Bollaert, E.F.R., "Wall Jet Rock Scour in Plunge Pools: A Quasi-3D Prediction Model", *International Journal on Hydropower & Dams*, 131, 4, 153–165, 2012.

Bollaert, E.F.R., "Scour potential at spillway", *Internal Report*, 2018.

Bollaert, E.F.R., "The Rocsc@r Cloud: An Innovative Digital Platform to Compute and Record Rock Scour", *International Journal on Hydropower & Dams*, 28, 5, 60–70, 2021.

Bollaert, E.F.R. and Hofland, B., "The Influence of Flow Turbulence on Particle Movement Due to Jet Impingement", in *2nd Scour and Erosion Conference*, Singapore, 2004.

Bollaert, E.F.R. and Schleiss, A.J., "Physically Based Model for Evaluation of Rock Scour due to High-Velocity Jet Impact", *Journal of Hydraulic Engineering*, 131, 3, 153–165, 2005. https://doi.org/10.1061/(ASCE)0733-9429(2005)131:3(153)

Bollaert, E.F.R. Vrchoticky, B. and Falvey, H.T., "Extreme Scour Prediction at High Head Concrete Dam and Stilling Basin (United States)", in *Proceedings of the 3rd International Conference on Scour and Erosion,* Amsterdam, The Netherlands, 2006.

Bosman, A., "High dam scour hole geometry prediction for fully developed jets plunging into shallow pools on bedrock", *PhD Dissertation*, Civil Engineering Faculty, Stellenbosch University, Cape Town, South Africa, 2021.

Bosman, A. and Basson, G., "Physical Model Study of Bedrock Scour Downstream of Dams Due to Spillway Plunging Jets", *Journal of the South African Institution of Civil Engineering*, 62, 3, 36–52, 2020.

Boumaiza, L., "Rock mass parameters governing the hydraulic erodibility of rock in unlined spillways", *PhD Dissertation*, Université du Québec à Chicoutimi, Canada, 2019.

Boumaiza, L., Saeidi, A. and Quirion, M., "A Method to Determine Relevant Geomechanical Parameters for Evaluating the Hydraulic Erodibility of Rock", *Journal of Rock Mechanics and Geotechnical Engineering*, 11, 5, 1004–1018, 2019. https://doi.org/10.1016/j.jrmge.2019.04.002

Calitz, J.A., "Investigation of air concentration and pressures of a stepped spillway equipped with a crest pier", *Master's Thesis*, Stellenbosch, South Africa, Stellenbosch University, 2015.

Carling, P.A., Hoffmann, M., Blatter, A.A. and Dittrich, A., "Drag of emergent and submerged rectangular obstacles in turbulent flow above bedrock surface", in E.F.R. Bollaert and A.J. Schleiss (eds.), *Rock Scour Due to Falling High-Velocity Jets*, Lisse, Swets and Zeitlinger, pp. 83–94, 2002.

Castillo, L.G., "Metodología experimental y numérica para el diseño de los cuencos de disipación de energía. Aplicación aldisipación de energía. Aplicación al vertido en presas bóveda", *PhD Thesis*, Departamento de Ingeniería Hidráulica, Marítima y Medio Ambiental, Universidad Politécnica de Cataluña, Spain, [in Spanish], 1989.

Castillo, L.G. and Carrillo, J.M., "Scour Estimation of the Paute-Cardenillo Dam", in *Proceedings of the International Perspectives on Water Resources & the Environment*, Quito, Ecuador, 2014.

Castillo, L.G. and Carrillo, J.M., "Characterization of the Dynamics Actions and Scour Estimation Downstream of the Dam", in *Dam Protections against Overtopping and Accidental Leakage -1st International Seminar on Dam Protections Against Overtopping and Accidental Leakage*, pp. 231–243. CRC Press, Madrid, Spain, 2015.

Castillo, L.G. and Carrillo, J.M., "Scour, Velocities and Pressures Evaluations Produced by Spillway and Outlets of Dam", *Water*, 8, 68, 2016. https://doi.org/10.3390/w8030068

Castillo, L.G. and Carrillo, J.M., "Comparison of Methods to Estimate the Scour Downstream of a Ski Jump", *International Journal of Multiphase Flow*, 92, 171–180, 2017.

Coleman, S.E., Melville, B.W. and Gore, L., "Fluvial Entrainment of Protruding Fractured Rock", *Journal of Hydraulic Engineering*, 129, 11, 872–884, 2003.

Dasgupta, B., Basu, D., Das, K. and Green, R., "Development of Computational Methodology to Assess Erosion Damage in Dam Spillways", in *USSD 31st Conference*, San Diego, CA, April 11–15, 2011.

Dasgupta, B., Das, K., Basu, D., Green, R. and McGinnis, R., "Computational Approach to Predict Rock Erosion in Unlined Spillways", in *24th ICOLD Congress*, 94, Kyoto, Japan, 2012.

De Cesare, G., Daneshvari, M., Federspiel, M., Malquarti, M., Chauvins, G.E. and Schleiss, A.J., "Étude sur Modele physique et Numerique des evacuateurs de crue et des fosses d'erosion du barrage de Koman en Albanie", *La Houille Blanche*, 3, 48–55, 2011. https://doi.org/10.1051/lhb/2011032

Dooge, N., "Die hidrouliese erodeerbaarheid van rotmassas in onbelynde oorlope met spesiale verwysing na die rol van naatvulmateriaal", *M.Sc. Thesis*, University of Pretoria, South Africa, 1993.

Douglas, K., Pells, S., Fell, R. and Peirson W., "The Influence of Geological Conditions on Erosion of Unlined Spillways in Rock", *Quarterly Journal of Engineering Geology and Hydrogeology*, 51, 219–228, 2018. https://doi.org/10.1144/qjegh2017-087

Duarte, R.X.M., "Influence of air entrainment on rock scour development and block stability in plunge pools", *PhD Thesis*, LCH-EPFL, 2014.

Dubinski, I.M., "Physical modelling of jointed bedrock erosion by block quarrying", *Ph.D. Dissertation*, Colorado State University, USA, 2009.

Dubinski, I. and Wohl, E., "Relationships Between Block Quarrying, Bed Shear Stress, and Stream Power: A Physical Model of Block Quarrying of a Jointed Bedrock Channel", *Geomorphology*, 180–181, 66–81, 2013.

Federspiel, M.P.E.A., "Response of an embedded block impacted by high-velocity jets", *PhD Thesis*, LCH-EPFL, 2011.

Gardner, M., "Toward a Complete Kinematic Description of Hydraulic Plucking of Fractured Rock", *Journal of Hydraulic Engineering*, 149, 7, 2023. https://orcid.org/0000-0002-5238-630X

Gardner, M. and Sitar, N., "Modeling of Dynamic Rock-Fluid Interaction Using Coupled 3-D Discrete Element and Lattice Boltzmann Methods", *Rock Mechanics and Rock Engineering*, 52, 12, 5161–5180, 2019. https://doi.org/10.1007/s00603-019-01857-x

George, M.F., *Ph.D. Dissertation*. University of California, Berkeley, 2015.

George, M.F. and Sitar, N., "Mechanics of 3D Rock Block Erodibility", in *Proceeding of the 50th U.S. Rock Mechanics Symposium (ARMA)*, 2016.

George, M.F., Sitar, N. and Sklar, L.S., "Experimental Evaluation of Rock Erosion in Spillway Channels", in *Proceeding of the 49th U.S. Rock Mechanics Symposium (ARMA)*, San Francisco, CA, 2015.

George, M.F., Christiansen, C., Rickel, A., Israel, B. and Annandale, G.W., "Erodibility Evaluation of an Unlined Rock Spillway: Comparison Between the Erodibility Index Method and a New Method Based on Block Theory", in *Proceedings of the 4th International Seminar on Dam Protections against Overtopping*, Madrid (Spain), 2022. https://doi.org/10.26077/56a7-33b7

Goodman, R.E. and Shi, G. *Block theory and its application to rock engineering*, Prentice Hall, Englewood Cliffs, NJ, 1985.

Hirt, C.W. and Nichols, B.D., "Volume of Fluid (VOF) Method for the Dynamics of Free Boundaries. *Journal of Computational Physics*, 39, 201–225, 1981. https://doi.org/10.1016/0021-9991(81)90145-5

Hoek, E., "Strength of rock and rock masses", ISRM News Journal, 2(2), 4-16, 1994.

Hoek, E., Kaiser, P.K. and Bawden, W.F., *Support of underground excavations in hard rock*, A.A. Balkema, Rotterdam/Brookfield, 1995.

Hurst, A., Anderson, R.S. and Crimaldi, J.P., "Toward Entrainment Thresholds in Fluvial Plucking", *Journal of Geophysical Research Earth Surface*, 126, 5, e2020JF005944, 2021.

Jamet, G., Muralha, A., Melo. J.F., Manso, P.A. and De Cesar, G., "Plunging Circular Jets: Experimental Characterization of Dynamic Pressures Near the Stagnation Zone", *Water*, 14, 173, 2022. https://doi.org/10.3390/w14020173

Kashtiban, Y.J., Saeidi, A., Farinas, M.-I. and Quirion, M., "A Review on Existing Methods to Assess Hydraulic Erodibility Downstream of Dam Spillways", *Water*, 13, 22, 3205, 2021. https://doi.org/10.3390/w13223205

Keaton, J.R., Mishra, S.K. and Clopper, P.E., "NCHRP Report 717 : Scour at Bridge Foundations on Rock", *Transportation Research Board*, Washington, DC, 2012.

Kieffer, D.S. and Goodman, R.E., "Assessing Scour Potential of Unlined Rock Spillways with the BLOCK Scour Spectrum", *Geomechanics and Tunnelling*, 5, 5, 527–536, 2012. https://doi.org/10.1002/geot.201200039

Kirsten, H.A.D., "A Classification System for Excavation in Natural Materials," *The Civil engineer in South Africa*, 24, 7, 1982.

Kirsten, H.A.D., Moore, J.S., Kirsten L.H. and Temple, D.M., "Erodibility Criterion for Auxiliary Spillways of Dams", *International Journal of Sediment Research*, 15, 93–107, 2000.

Koulibaly, A., Saeidi, S., Rouleau, A. and Quirion, M., "Identification of Hydraulic Parameters Influencing the Hydraulic Erodibility of Spillway Flow Channels", *Water*, 13, 21, 2950, 2021. https://doi.org/10.3390/w13212950

Koulibaly, A.S., Saeidi, A., Rouleau, A. et al. "A Reduced-Scale Physical Model of a Spillway to Evaluate the Hydraulic Erodibility of a Fractured Rock Mass", *Rock Mech Rock Eng*, 56, 933–951, 2023. https://doi.org/10.1007/s00603-022-03101-5

Lamb, M.P., Finnegan, N.J., Scheingross, J.S. and Sklar, L.S., "New Insights into The Mechanics of Fluvial Bedrock Erosion Through Flume Experiments and Theory", *Geomorphology*, 244, 33–35, 2015. https://doi.org/10.1016/j.geomorph.2015.03.003

Li, A. and Liu, P., "Mechanism of Rock-Bed Scour due to Impinging Jet", *Journal of Hydraulic Research*, 48, 1, 14–22, 2010. https://doi.org/10.1080/00221680903565879

Maleki, S. and Fiorotto, V., "Blocks Stability in Plunge Pools under Turbulent Rectangular Jets", *Journal of Hydraulic Engineering*, 145, 4, 2019. https://doi.org/10.1061/(ASCE)HY.1943-7900.0001573

Maleki, S. and Fiorotto, V., "Rock Scour Model for Unlined Plunge Pools and Stilling Basins", *Rock Mechanics and Rock Engineering*, 55, 4159–4181, 2022. https://doi.org/10.1007/s00603-022-02808-9

Manso, P.A., "The influence of pool geometry and induced flow patterns on rock scour by high-velocity plunging jets", *PhD Thesis*, LCH-EPFL, 2006.

Mas Ivars, D., Pierce, M.E., Darcel, C., Reyes-Montes, J., Potyondy, D.O., Paul Young, R. and Cundall, P.A., "The Synthetic Rock Mass Approach for Jointed Rock Mass Modelling", *International Journal of Rock Mechanics and Mining. Sciences*, 48, 2, 219–244, 2011.

Melville, B.W., van Ballegooy, R. and van Ballegooy, S., "Flow-Induced Failure of Cable-Tied Blocks", *Journal of Hydraulic Engineering*, 132, 3, 324–327, 2006.

Montgomery, R.A., *Investigations into rock erosion by high velocity water flows*, Hydraulics Laboratory, Stockholm, 1984.

Moore, J., *The characterization of rock for hydraulic erodibility*, Water Resources Publications, Littleton, CO, 1991.

Moore, J., Temple, D. and Kirsten, H., "Headcut Advance Threshold in Earth Spillways", *Bulletin of the Association of Engineering Geologists*, 31:2, 1994.

Moreira, A.M.B., "Numerical modelling of spillways and energy dissipators using the smoothed particle hydrodynamics method", *PhD Dissertation*, University of Porto, 2021.

Moreira, A.M.B., Manso, P.A., Violeau, D. and Taveira-Pinto, F., "Single-Phase SPH Modelling of Plunge Pool Dynamic Pressures at a Near-Prototype Scale", *Journal of Hydraulic Research*, 59, 6, 888–902, 2021. https://doi.org/10.1080/00221686.2020.1862320

Narayanan, R., "Pressure Fluctuations Beneath Submerged Jump", *Journal of the Hydraulics Division*, 104, 9, 1331–1342, 1978. https://doi.org/10.1061/JYCEAJ.0005066

Oukid, Y., Daux, C. and Libaud, V., "3D CFD Modelling of Spillways: Practical Feedback on Capabilities and Challenges", *Hydropower & Dams*, 22, 67–73, 2015.

Otto, B., "Scour potential of highly stressed sheet-jointed rocks under obliquely impinging plane jets", *PhD thesis*, James Cook University of North Queensland, Townsville, 1989.

Palmström, A., "Characterizing the Strength of Rock Masses for Use in Design of Underground Structures", in *International Conference of Design and Construction of Underground Structures*, New Delhi, 1995.

Pan, Y.-W., Li, K.-W. and Liao, J.-L., "Mechanics and Response of a Surface Rock Block Subjected to Pressure Fluctuations: A Plucking Model and Its Application", *Engineering Geology*, 171, 1–10, 2014.

Pells, S., "Erosion of rock in spillways", *PhD Dissertation*, School of Civil and Environmental Engineering Faculty of Engineering, University of New South Wales, 2016.

Pells, S., Douglas, K., Pells, P.J., Fell, N. and Peirson, W.L., "Rock Mass Erodibility", *Journal of Hydraulic Engineering*, 143, 06016031, 2016. https://doi.org/10.1061/(ASCE)HY.1943-7900.0001243

Pells, S., Al-Qassab, F., Weir, F., Faivre, A., Leturcq, T. and Blancher, B., "Hydro-geotechnical Rock Scour Assessment at Petit-Saut Dam", in *4th International Seminar on Dam Protections against Overtopping Madrid (Spain)*, November 30th–December 2nd, 2022.

Reinius, E., "Rock Erosion", *International Journal on Water Power and Dam Construction*, 38, 6, 43–48, 1986.

Rock, A., "A semi-empirical assessment of plunge pool scour: two-dimensional application of Annandale's erodibility index method on four dams in British Columbia, Canada", in M.Sc. Thesis submitted to the Faculty and the Board of Trustees of the Colorado School of Mines, Colorado, USA, 2015.

Rooseboom, A. and Van Vuuren, S., "*Drainage Manual. 6th Edition*", Chapter 5 - Surface drainage. The South African National Roads Agency SOC Ltd Drainage Manual, South Africa, 2013.

Toso, J.W. and Bowers, C.E., "Extreme Pressures in Hydraulic Jump Stilling Basins", *Journal of Hydraulic Engineering*, 114, 8, 829–843, 1988. https://doi.org/10.1061/(ASCE)0733-9429(1988)114:8(29)

van Schalkwijk, A., "Minutes - Erosion of Rock in Unlined Spillways", *ICOLD*, Q.71 R.37, 1056–1062, 1994.

van Schalkwijk, A., Jordaan, J. and Dooge, N., "Die erodeerbaarheid van verskillende rotsformasies onder varierende vloeitoestande", *Tech. Rep. WNK Verslag No. 302/1/95, verslag aan die waternavorsingskommissie deur die Departement of Geologie*, Universiteit van Pretoria, South Africa, 1994a.

van Schalkwijk, A., Jordaan, J. and Dooge, N., "Erosion of Rock in Unlined Spillways," *International Commission on Large Dams*, Paris. Q.71–E.37, 555–571, 1994b.

Weidner, L. and George, M., "Generating realistic blocky rock masses from point clouds for spillway scour assessments using semi-automated Python tools", *4th Meeting of European Working Group on Overflow and Overtopping Erosion*, Lyon - July 5-7, 2023 (unpublished)

Whittaker, J.G. and Schleiss, A.J., *Scour related to energy dissipators for high head structures*, VAW Mitteilungen, Zurich, N°73, 1984.

Wibowo, J., Villanueva, E., Temple, D. and Yule, D., "Earth and Rock Surface Spillway Erosion Risk Assessment", in *U.S. Symposium on Rock Mechanics*, Paper No. 05-813. American Rock Mechanics Association, Alexandria, VA, 2005.

4 Digital Rock Scour Platforms

DEFINITION AND KEY FEATURES

A digital platform may be defined based on different viewpoints (Asadullah et al. 2018). Technical viewpoints focus on the technical elements and processes that interact to form a digital platform. For example, Spagnoletti et al. (2015) define a digital platform as "a building block that provides an essential function to a technological system and serves as a foundation upon which complementary products, technologies, or services can be developed".

Other studies have defined digital platforms based on a non-technical viewpoint and present platforms as a commercial network or market that enables transactions in the form of business-to-business or business-to-customer exchanges. Digital platforms are described by Koh and Fichmann (2014) as "two-sided network…that facilitate interactions between distinct but interdependent groups of users, such as buyers and suppliers". The focus in this view is on the interactions between the different groups that join a platform.

According to Stephen Watts (https://www.bmc.com/blogs/digital-platforms), a digital platform can be thought of as a place for exchange of information, goods or services to occur between producers and consumers as well as the community that interacts with the said platform. He states that it is imperative to understand that the community itself is an essential piece of the digital platform. Without that community, the digital platform has very little inherent value. In his point of view, a successful digital platform performs two key functions:

- Facilitating exchange of goods, services or information
- Leveraging the involved community to provide enhanced value to everyone within the related ecosystem

The key features of digital platforms can be summarized as follows (https://www.bmc.com/blogs/digital-platforms):

- Ease of use and immediate appeal for users
- Trustworthiness and security (clear terms and conditions are necessary as well as privacy protection and assurances for intellectual property and data ownership)
- Connectivity through the use of APIs that allow third parties to extend the capabilities of the platform
- Facilitation of exchanges between users

DOI: 10.1201/9781003319610-4

- Providing value to the community and as a function of the size of the community (the bigger the community, the more value the platform can provide to all parties involved)
- Ability to scale without causing performance degradation

As such, application of digital platforms to the field of rock scour engineering should focus on providing scour computational services and exchange of scour feedback and data. Furthermore, use of such a platform by professionals should allow leveraging added value to the community, for example, by means of parametric standardizations, model validations, feedback from experiences and so on.

TYPES OF DIGITAL PLATFORMS

In the following, digital platforms are subdivided as a function of their business model, their ownership, their IT network and communication, their user intervention or also their "live-ability", i.e. the degree to which they communicate with the physical reality, within the framework of an IoT system network. Potential applicability to future rock scour digital platforms is discussed.

Based on Business Model

Digital platforms have different forms depending on the business model they employ and the specific purposes they serve. Examples of successful digital platforms are:

- *Social media platforms*: allow users to have conversations, share information and create web content (TikTok, Instagram, WhatsApp, Facebook, LinkedIn, Twitter (now X), …)
- *Knowledge platforms*: centralized hub for sharing information, guidelines, policies, instructions, process documentation and more (Quora, Reddit, etc.)
- *Content and media sharing platforms*: allow users and brands to share audio-visual content including images, videos, music and live streams (YouTube, Spotify, etc.)
- *Service-oriented platforms*: Platform-as-a-Service (PaaS) product composed of people, processes and tools; allows to develop, iterate and operate digital services (Uber, Airbnb, Booking, etc.)

Rock scour digital platforms would fall within the group of service-oriented platforms. The provided services are software-based and/or data-based. The corresponding community that interacts with such engineering platforms can be two-fold:

- Users that share common applications under a given subscription, i.e. group of people from the same company (i.e. that owns the hydraulic structures being described and studied), with secured and dedicated access to the software platform, or a group of engineers and managers from the same engineering office.
- Worldwide user group that shares and interacts on a publicly accessible part of the digital platform, for example, a freely accessible database of case studies, a universal user forum, etc.

Rock scour digital platforms may be set up such that both types of community can be present and interact with each other.

Based on Ownership Structure

Proprietary Platforms

These platforms reserve rights only to authorized individuals and do not allow any other user intervention into the functioning and developments of their software and hardware. Examples are Microsoft platforms.

Open-Source Platforms

Open-source platforms allow access to the source code to any other users or developers. Open-source software aims to allow equal access to anyone and everyone and aims to ensure that software users are free to run software, study the software, modify the software and share such modifications. Examples are Linux platforms or Firefox web browser.

Based on IT Framework

Locally Hosted (Private) Platforms

These platforms operate on locally hosted private IT networks, such as the server of a company or institute, and are only accessible by controlled users/groups of the network in question, locally or remotely. These platforms typically use local versions (licence copies) of applications installed on the server or on PCs, subject to a licence for dedicated use of the software and eventual regular updates and maintenance.

External access to such servers is generally only feasible for users of the company in question. As such, these platforms are not designed to grant access, dedicate space and deliver engineering services upon request of potential external users.

Cloud-Based Platforms

These platforms operate on a cloud-based centralized server, generally hosted by an independent third-party provider such as a datacentre. Applications only run on the cloud, and not on local IT networks or PCs. Hence, no software copy is installed on the local PC, omitting the need for regular updates and maintenance following a traditional licence contract.

External access is available for any user, by logging in by means of a private email/identification and related password. Users benefit from a dedicated secured access to the platform, easily accessible from any IT device worldwide. The services provided on the dedicated space are unique.

From the engineering point of view, such an IT structure has its advantages and defaults. As stated before, universal feedback and access to user and project experiences would allow leveraging added value to an engineering community. This could be imagined for example by the creation of a universally accessible database of rock scour use cases and related parameters. Moreover, standardization of relevant parameters, or even creation of new parameters, might be imagined by the engineering community. Furthermore, within a dedicated (company-based) cloud space, it allows everybody involved on a project to discuss on a single and unique version of information (i.e. computational results, input data, etc.), in a centralized manner.

On the other hand, anyone in that company with access to this information may potentially alter it, requesting rigorous application of rules regarding user interventions on the platform, or number of users with access.

Rock scour digital platforms would typically need cloud-based operations, implementing both company-based private access to confidential projects and data, as well as worldwide free access to a public database for leveraging added value to the community.

Based on User Intervention

Passive Digital Service Platform

A passive digital service platform offers data to its users following a well-defined and organized structure, but does not allow any tool for on-line analysis or reorganization of such data. An example of a passive digital platform is typically a weather application, allowing users to consult current weather conditions and weather forecasts at different locations, without any potential further use of this data. The data represents the added value of the platform, but the user only consults without further intervention.

An engineering example would be to set up a database of available case studies of rock scour, containing the most relevant parameters for each case study. The user might benefit from this information, but could not re-engineer the information on the database platform.

Active Digital Service Platform

An active digital service platform offers services to its users following a well-defined and organized structure. These services allow users to actively perform on-line analyses, generate on-line results, or buy on-line goods or services, based on pre-installed software and data. Used terminology is SaaS or PaaS platforms, standing for "Software-as-a-Service" or "Platform-as-a-Service".

An example of an active digital service platform is typically a platform for personalized travel services (driver with vehicle, time duration, trajectory and so on), such as, for example, Uber or Lyft, or also a platform for personalized sport, fitness and mobility (running, cycling, etc.), such as, for example, Strava or Stryd, allowing to generate mobility data and personal statistics in (quasi) real time.

An engineering example would be a digital software platform allowing to perform rock scour computations based on user-generated input data. The added value is generated by the computations made on the platform, following active user intervention.

Based on Data Connectivity

Low Connectivity Platform

A lowly connected digital platform offers data and/or services to its users with low or no direct relationship with a real-time physical or digital situation or value. In other words, the data and/or services depend on information that is not related to or does (quasi) not change with the currently ongoing physical world.

An example of such a platform is an application allowing to scan and obtain nutritional values of foods. Whether the scan is performed today, tomorrow or next week, the same product scan will provide the same information to the user. Another example is an application allowing to access your bank account platform, based on rigorous screening of password and/or additional safety features installed.

An engineering example would be a software platform for rock scour that allows the user to manually update certain data whenever these become available, and to recompute certain scenarios. One might typically think about yearly bathymetric updates of scour holes, or occasional updates because of design flood hydrograph modifications. Because of the low frequency of the updates, manual processing may be plausible.

High-Connectivity Platform

A highly connected digital platform offers data and/or services to its users by maintaining a rather busy direct relationship with a physical or digital situation or value. The connectivity is frequent; however, it is not available or exploited by the users on a continuous basis. In other words, the data and/or services depend on information that is frequently updated by the platform to the ongoing physical or digital world. Such updates are generally performed by using IoT sensors.

An example of such a platform is on-line banking applications. Once accessed through a secured static application, such platforms allow the user to make payments, check bank accounts and salary payments, and so on, but generally no real-time continuous follow-up is performed by the user, because the updates on such a platform are typically sequential.

An engineering example would be a software platform for rock scour that allows manual or automatic updates of data whenever these become available. Examples of data to be frequently updated might be the plunge pool bathymetry, reservoir water levels, flood spillage discharge values and so on, generating the need to frequently update the related scour computations. Typical frequency of such updates would be after every flood event. Depending on the frequency of the updates, automatization of the update process may be plausible, transforming the digital platform into a kind of digital twin making use of IoT sensors or derived values that may function on a continuous basis but are only relevant from time to time (i.e. during flood spillages).

Live Platform

A live digital platform offers data and/or services to its users with a direct and continuous relationship with a physical or digital situation or value. The connectivity changes continuously or very rapidly, and is available or exploited by the users on a real-time basis. In other words, the data and/or services depend on information that changes rapidly and continuously with the currently ongoing physical or digital world.

Examples are the aforementioned platforms for personalized travel services, or for personalized sport, fitness and mobility, allowing to generate and exploit mobility data in (quasi) real time. Other examples are platforms informing on real-time traffic, on weather via real-time radar data or on real-time available space in urban parking hubs.

An engineering example would be a software platform for rock scour that functions as a real digital twin, i.e. that allows automatic updates of continuously changing data, whenever these become available. Examples of data to be continuously updated might be the plunge pool bathymetry, reservoir water levels, flood spillage discharge values and so on, generating the need to continuously update the related scour computations.

KEY FEATURES OF A ROCK SCOUR DIGITAL PLATFORM

This subsection presents the main features judged mandatory to any digital platform for rock scour engineering purposes. These features have been determined based on the needs and challenges of the rock scour engineering discipline outlined in Chapter 2 and based upon the main characteristics and types of digital platforms available today. They are described in more detail hereafter (see Figure 4.1).

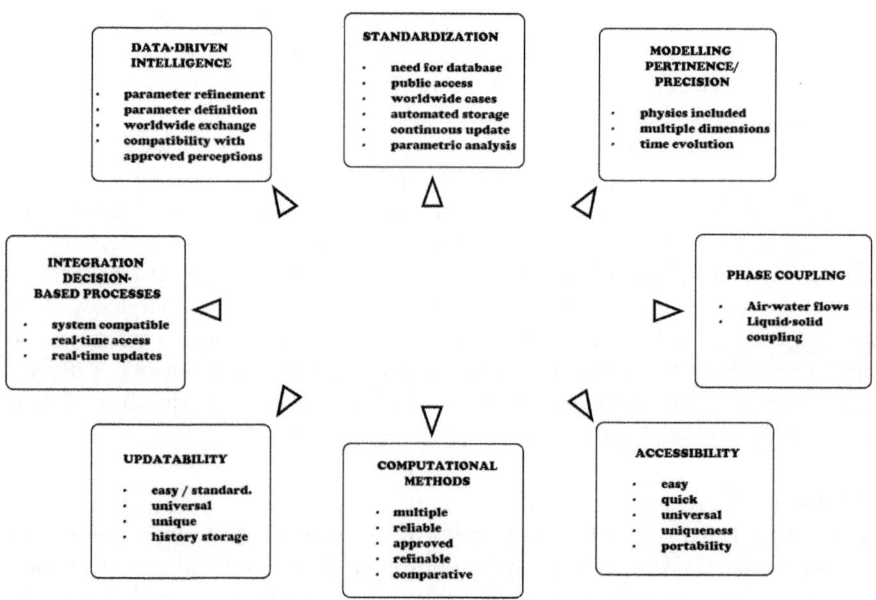

FIGURE 4.1 Summary of key features needed for a digital scour platform.

Modelling Pertinence and Precision

The digital platform should offer computational models and methods that are sufficiently precise and pertinent, such that the obtained results constitute significant added value to the users. For example, multiple geometrical model dimensions are mandatory, i.e. 2D at minimum, and even 3D whenever possible. Rock mass fracture networks and rock block shapes and dimensions should be as realistic as possible.

Also, all relevant physics should be included, such as air entrainment and turbulent water pressure fluctuations, as well as the most relevant rock break-up mechanisms such as rock mass fracturing, rock block plucking, rock block peeling off and so on. Finally, the time evolution of the phenomenon should be implemented. Rock scour should become predictable during future or incoming flood events, depending on the intensity and the time duration of estimated flood hydrographs.

Phase Coupling

Rock scour is a three-phase phenomenon, involving air (the gas phase), water (the liquid phase) and rock (the solid phase). These phases interact with each other during scour formation and, as such, a digital platform should offer computational models and methods that implement phase coupling.

A summary of the most relevant phase couplings is illustrated in Figure 4.2. The first important coupling concerns the air and the water. Air entrainment is essential

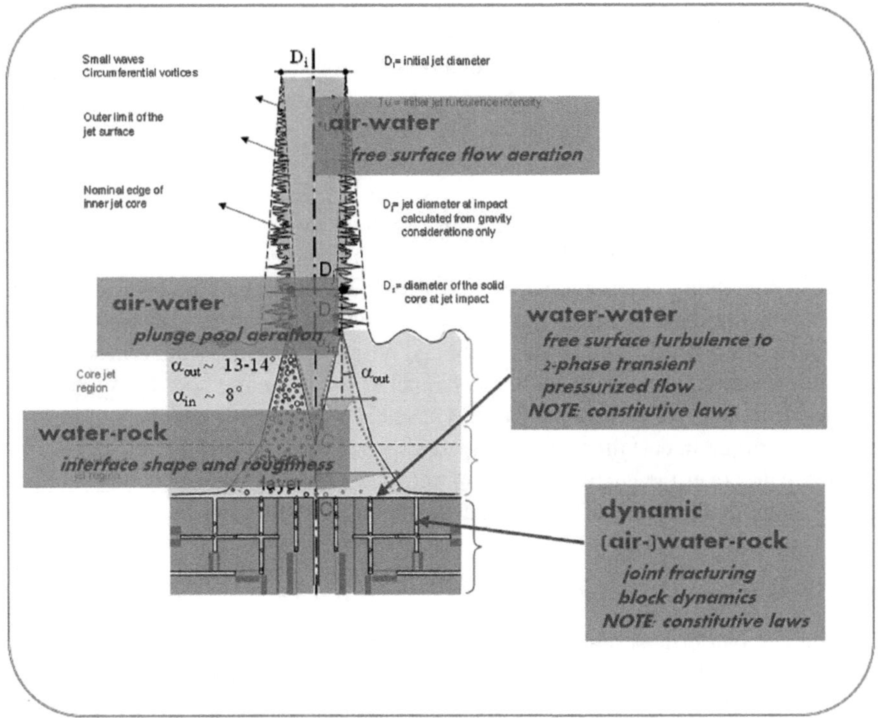

FIGURE 4.2 Summary of phase couplings relevant to rock scour.

to the development and the characteristics of turbulent flows, especially in case of turbulent jets. Second, air entrainment in the body of water of a plunge pool, a river, a channel or a stilling basin directly influences the dynamic pressure forces exerted by the water at the interface with the rock mass. Third, air entrainment and dissolution in the water, inside rock joints, directly governs the speed of propagation of pressure waves relevant to the fracturing process of the rock mass and to the dynamic uplift of formed rock blocks.

A second relevant coupling is the water-rock coupling at the bedrock. The forces exerted by the air-water mixture on the rock mass are highly influenced by the shape, the protrusion and the roughness of the rocky bottom. These characteristics govern the formation of lift and drag forces on rock blocks, as well as the spectral content and intensity of turbulent pressure fluctuations that act at the water-rock interface. These pressures enter the joints of the rock mass and may stimulate the joints to resonating conditions, finally resulting in progressive or brittle fracturing of the rock mass. They may also generate short-lived pressure differentials that provide a destabilizing force and impulsion on rock blocks, which may be expelled from their matrix.

A third coupling is the water-water coupling acting at the bedrock. While this coupling seems a priori evident, it is one of the less understood and modelled couplings in rock scour. At the water-rock interface, the water transforms from a free surface aerated turbulent flow, containing differently sized eddies and kinetic turbulent energy, into an ensemble of two-phase non-linear pressurized waves propagating at a much higher speed through the underlying rock joints. The spectral energy of the free surface turbulent flow, together with the joint geometry and varying air content of the water, determines the intensity and duration of high pressures that install underneath rock blocks and that may temporarily move a block, following the theory of rigid body dynamics.

Computational Methods

The digital platform should offer a toolbox of reliable and worldwide approved computational models and methods.

Most rock scour computational methods have been developed for a particular type of rock or hydraulic structure, or for a particular type of rock mass. As such, some of these methods are mostly used and calibrated for plunging jets, while other methods are more suited and applied to unlined channel flows. Some methods are universally applicable to any type of rock, other ones have been specifically developed for a certain type of rock in certain parts of the world.

Hence, to cover the currently existing knowledge and experience at best, a diversity of methods should be offered to the user.

Accessibility and Updatability

The digital platform should be easily accessible and updatable, offering unique results that can be rendered and discussed quasi-instantaneously and universally, for example, from several offices worldwide, within the same company. Rapid and secure storage and archiving of results should be implemented.

Security

The digital platform should benefit from sound cyber-protection, such that users are and feel safe with their data storage and trust the platform. High-level data encryption and adequate storage redundancy should complement clear guidelines on intellectual property of the stored data as well as data ownership.

Data-Driven Intelligence and Parametric Standardization

The digital platform should allow and trigger large-scale worldwide usage and applications, as well as large-scale generation of related experiences and feedback. This will allow sound determination and standardization of the main parameters of interest of the different computational methods, as well as of their range of values to be used for different possible real-life situations. In other words, by data-driven intelligence, models might be universally validated rather than just calibrated on a particular case study.

Moreover, based on progressively increasing user feedback and experiences, parametric refinement and/or redefinitions may become feasible, as well as the inclusion of newly defined parameters. Especially the link between parameters and the subjective personal experiences and perceptions regarding the rock mass quality and description are of interest.

Connectivity

Connectivity with third-party services is essential to digital platforms. This is generally performed by means of API connections (Application Programming Interface), which allow to extend the capabilities of a platform.

The digital scour platform should be compatible with future digital twin environments of hydraulic structures, such that the services provided by the platform may be implemented into these twins. Furthermore, the platform should be compatible with SCADA (Supervisory Control And Data Acquisition) systems generally used for acquisition, storage and regulation of data related to dams and hydropower stations.

This will allow real-time and ahead-of-time assessment of rock scour potential during flood events. For example, by automated generation and transmission of reservoir level and spillway outflows from the SCADA towards the digital scour platform, the latter may compute in real time the ongoing scour formation in the downstream rock mass and send back the results to the twin environment.

This ahead-of-time scour prediction may also be envisaged during real-time flood events, whenever the near-future outflow hydrograph may be predicted and tested on the scour platform. Under such circumstances, the digital scour platform may even be triggered to check and compare different gate opening scenarios of spillways ahead-of-time, in order to detect the scenario with the lesser risk of scour formation in the downstream rock mass.

In other words, when connected with a digital twin or SCADA, the digital scour platform is able to benefit from real-time data and thus to participate to real-time risk-informed decision-making processes, and allows dam owners or operators to implement rock scour into their decisions regarding dam safety during flood events.

These aspects are discussed further on in this chapter, as well as in Chapter 5.

THE ROCSC@R® DIGITAL ENVIRONMENT

INTRODUCTION

One answer to the rock scour engineering challenges and needs for digital transformation is provided by the rocsc@r digital environment. This proprietary cloud-based environment contains two digital platforms offering services and data in the field of rock scour engineering (Figure 4.3):

- **X_pl@re** is a subscription-based digital software platform for 2D and 3D computations of rock scour related to hydraulic structures, using a wide variety of computational methods, all within an encrypted and secured user-specific environment. Data results are automatically stored on user-specific space allocated on the cloud, or on the user's local IT infrastructure by downloading csv result files.
- **X_ch@nge** is a free-access digital database platform for storage of any publicly available case study of rock scour related to hydraulic structures. The platform promotes to enhance parametric pertinence and experience. It aims widespread distribution and exchange of relevant data, so that shared benefits may be expected for our engineering community.

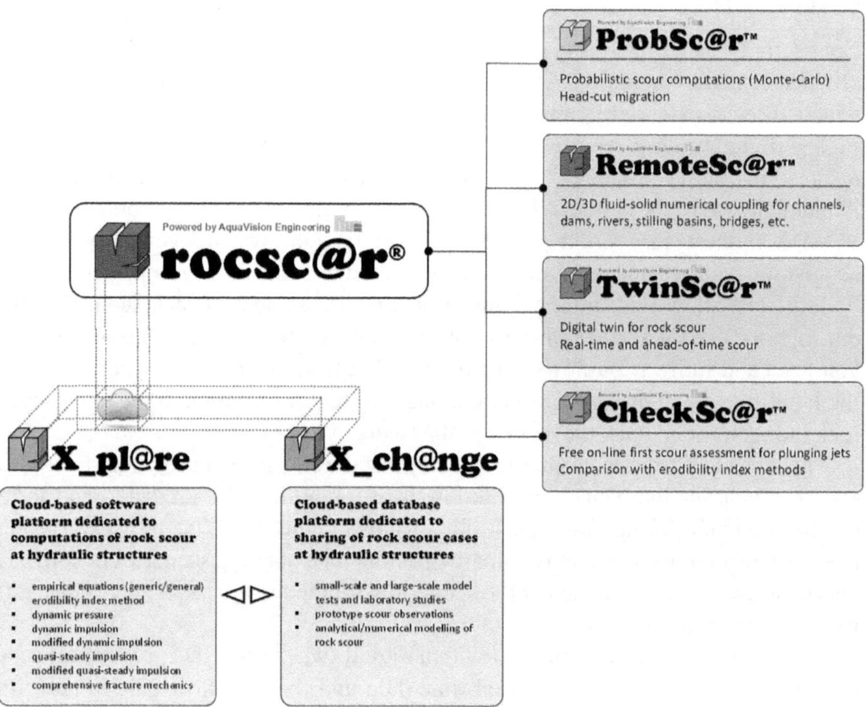

FIGURE 4.3 Sketch of the rocsc@r digital environment and its cloud-based platforms (Bollaert 2021).

The X_pl@re Software Platform

The X_pl@re digital platform comprises the rocsc@r software, i.e. a suite of modules and computational methods for numerical prediction of rock scour downstream of dam spillways and in unlined channels, rivers and stilling basins.

The range of methods covers impinging turbulent jets, as well as any other turbulent flow at the water-rock interface, such as free surface channel flows and hydraulic jumps. Users may apply and compare proven as well as new methods within one single numerical environment, based on user-defined parametric settings of the dam, the turbulent flow and the rock mass.

Topology

From the topological point of view, the software is subdivided into three main parts: the geometry, the flow and the rock mass.

First, the user must implement the geometry of the dam spillway and the downstream plunge pool, channel or stilling basin. This is done by using a set of 2D vertical profiles in the *X–Z* plane, as shown in Figure 4.4 (upper left graph), which shows 2D vertical profiles that have been determined in a 3D geometrical situation of an arch dam and its downstream plunge pool. The profiles can be oriented freely, for example, following the river axis (profile 1), or following the dam and spillway concentric curvature and geometry (profile 2). The dam and the spillway are introduced in a simplified manner, i.e. by determining the coordinates of the invert at jet issuance. The downstream rock bottom is introduced by user-defined *X–Z* coordinates. The scour computations are performed along each of these profiles separately. By assembling these 2D results, a quasi-3D result may be obtained.

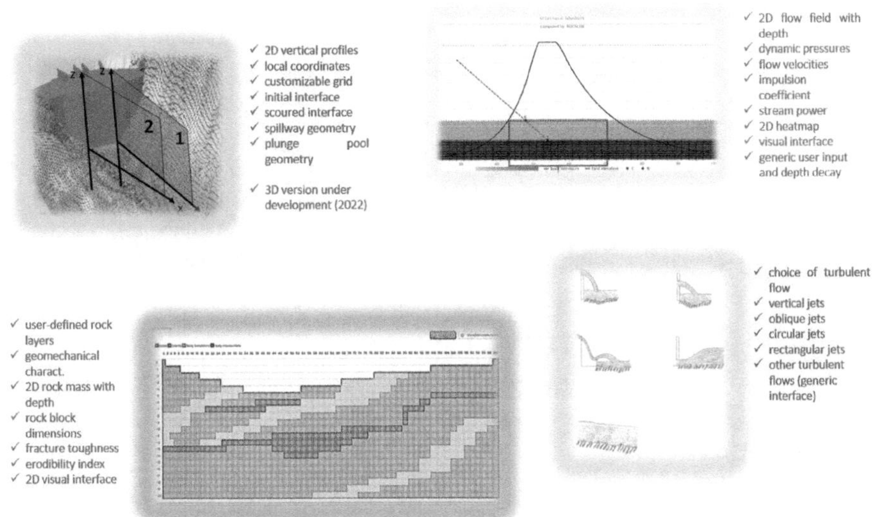

FIGURE 4.4 View of 2D vertical profiles and visual interfaces implemented in the rocsc@r digital environment.

Next, the software allows to implement the 2D turbulent flow in two ways. First of all, by making use of 2D flow matrices that use analytical expressions to determine the main hydraulic parameters with depth of scour for turbulent jets (i.e. mean and fluctuating dynamic pressures, impulsion coefficients, average flow velocities along the interface, stream power). Second, a generic interface is available, allowing the user to determine the hydrodynamic parameters at the water-rock interface by means of a csv-formatted exchange file, based for example on a HEC-RAS model, or by defining a customizable vertical decay of the hydrodynamic parameters with increasing scour depth.

Furthermore, this generic interface can be directly coupled to the FLOW-3D® CFD software by means of the RemoteSc@r API, in which the former continuously updates the hydraulic parameters based on scour progression computed by rocsc@r.

As shown in Figure 4.4, the visual interface of rocsc@r allows to present and check the hydraulic parameters by use of 2D heatmaps and curves. Another topologic choice is the type of turbulent flow, from vertical and oblique jets to differently shaped jets and finally any type of turbulent flow, by using the generic interface.

Furthermore, a 2D rock mass user interface has been set up allowing the user to determine the different rock mass layers and lithologies, and to set up a rock mass model with depth in each 2D vertical profile (Figure 4.4). For each lithology, the relevant geomechanical characteristics can be defined, such as the UCS strength, the density, the initial degree of fracturing, the typical block shape, the erodibility index and so on.

Physics Implemented

Aeration issues are accounted for by using analytical equations for the hydrodynamic parameters that take into account flow aeration of jets and plunge pools. Second, by using the FLOW-3D® coupling, automatic numerical coupling between air and water is obtained during the modelling for all types of turbulent flows (Figure 4.5).

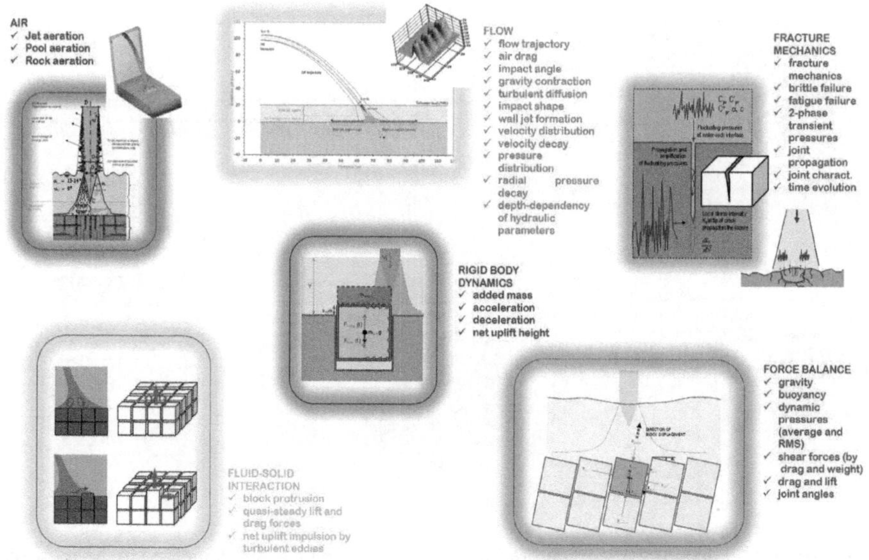

FIGURE 4.5 Main physics implemented in the rocsc@r digital environment.

The implemented flow physics account for basic issues like jet trajectory and air drag, impact angle of jets, gravitational contraction and turbulent diffusion, formation of wall jets along the rock interface, with velocity and pressure decay functions, both radially outwards and with increasing scour depth.

In terms of rock break-up mechanisms, fracture mechanics is accounted for, by both brittle and fatigue failure of rock joints (Figure 4.5).

Second, a force balance is performed, accounting for the main forces of influence, such as gravity, buoyancy, dynamic water pressures, shear forces in joints and drag and lift forces for inclined joints. This is directly coupled with rigid body dynamics, accounting for added mass, and block acceleration and deceleration phases during movement, allowing to define the net uplift height of the block while moving.

Finally, fluid-solid interaction is implemented by accounting for block protrusion using quasi-steady and fluctuating drag and lift forces on protruding blocks, depending on their local geometrical situation.

Digitalized Computational Methods

The currently available computational methods are illustrated in Figure 4.6 and briefly described hereafter. These methods have been entirely digitalized, offering full user customization of their relevant parameters. A more detailed description of the basic equations and digital capabilities of each of these methods can be found in Chapter 6.

Empirical Equations

Empirical equations (EMP) are mainly based on the state-of-the-art work performed by Castillo and Carrillo (2017), and contains general and simplified scour formulae, as well as a fully customizable generic formula.

Erodibility Index Method (EIM)

The Erodibility Index Method (EIM) has been integrated based on the work by Annandale (2006), but by making the scour threshold equation fully customizable, so that users are free to test and develop their own scour threshold equation.

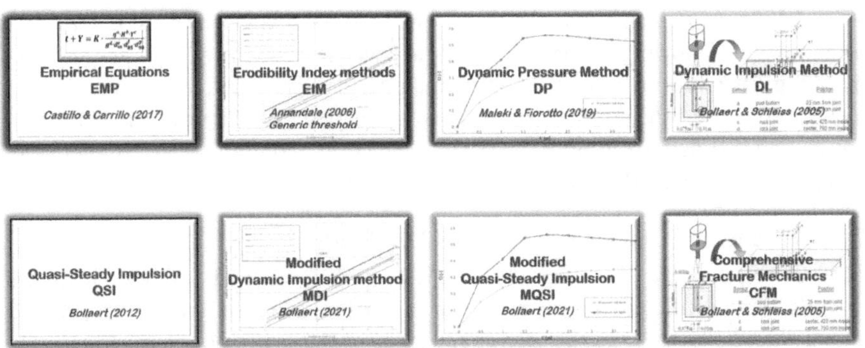

FIGURE 4.6 Summary of computational methods digitalized and implemented in the rocsc@r digital environment.

Dynamic Pressure (DP) Method

The Dynamic Pressure (DP) method is based on the work by Maleki and Fiorotto (2019), computing the uplift forces on rock blocks based on small-scale laboratory measurements of dynamic pressures generated by a jet impinging onto a pool bottom. The software proposes a 2D implementation of the block uplift model developed by the authors.

Dynamic Impulsion (DI) Method

The Dynamic impulsion (DI) method developed by Bollaert and Schleiss (2005) considers the maximum net uplift impulsion on a rock block, based on high-velocity aerated circular jets vertically impinging onto an artificial 2D rock joint. The net impulsion is obtained by time integration of net forces on the block and expressed by means of an impulsion coefficient. The critical uplift displacement of the block is determined as a function of the total block height.

Modified Dynamic Impulsion (MDI) Method

The Modified Dynamic Impulsion (MDI) method developed by Bollaert (2021) considers the maximum uplift impulsion on a rock block, based on high-velocity jets impinging onto an artificial 3D rock block (Federspiel 2011). The net impulsion is expressed by means of a multiplication coefficient times the RMS-value of pressures at the block surface, similar to the approach by Liu et al. (1998). The uplift of the block is subdivided into an acceleration phase and a deceleration phase, and accounts for the added mass during block acceleration (Federspiel, 2011). Critical uplift is determined as a function of the total block height.

Quasi-Steady Impulsion (QSI) Method

The Quasi-Steady Impulsion (QSI) method developed by Bollaert (2012) considers scour along the water-rock interface outside of the turbulent shear layer of the impacting jet. Following 2D jet diffusion through the pool, the water-rock interface deflects the jet towards up- and downstream. Quasi-steady wall jet flow is deviated by protruding rock blocks, generating quasi-steady net lift forces.

Modified Quasi-Steady Impulsion (MQSI) Method

The MQSI model is an enhanced version of the QSI developed by Bollaert (2021), by adding the following elements: additional shear forces in lateral joints, additional pulsating lift forces, rectangular and circular-shaped jets, obliquely impacting jets. Furthermore, the wall jets are expressed as a function of the jet thickness at impact (Beltaos 1976). Fluctuating uplift pressures are determined similar to the MDI method, i.e. by considering the RMS values of the pressure fluctuations at the block upper face.

Comprehensive Fracture Mechanics (CFM) Method

The Comprehensive Fracture Mechanics (CFM) method allows computing time-dependent fracture propagation based on linear elastic fracture mechanics, assuming a perfectly linear elastic, homogeneous and isotropic material, and relating the stress intensity generated by pressure pulsations in rock joints to the fracture resistance of the rock mass. The cyclic character of the pressures generated by the impact of a high-velocity jet makes it possible to describe joint propagation by fatigue stresses occurring at the tip of the joint. Bollaert and Schleiss (2005) developed a practical approach of the underlying theory, called the Comprehensive Fracture Mechanics (CFM) method and applicable to partially jointed rock.

The X_ch@nge Database Platform

The X_ch@nge digital platform offers a free-access database platform for any type of publicly available rock scour cases at hydraulic structures. For each case, specific geometric, flow and rock mass parameters are provided, in parallel to the observed scour for the related flow event. As such, this platform aims at offering a practical and continuously updated database with the following objectives:

- Feedback on the most relevant results, parameters and calibrations that have been used for each of the available scour computational methods
- Development of new parametric settings (refinement) and objective standardization of the most relevant parameters in rock scour
- Facilitate data-driven intelligence and machine learning algorithms in rock scour

The platform distinguishes between the different types of turbulent flow that are covered by the software: free overfall jets, pressurized jets, ski-jump jets, stilling basin flows and finally open channel flows.

Furthermore, X_ch@nge contains a database with basically the same topological structure as the X_pl@re platform, and allows the implementation of all input data in a way that rock scour computations can be easily set up. Inversely, when performing rock scour computations on the software platform, the case in question can be easily added to the database if wanted (Figure 4.7).

The RemoteSc@r API connection

RemoteSc@r is a dedicated API (Application Programming Interface) digital connection that offers a computational interface between the FLOW-3D® CFD software and the rocsc@r scour software. RemoteSc@r allows to perform sequentially coupled fluid-solid scour computations during which the flow parameters are sequentially updated as a function of generated scour and, vice versa, the scour formation accounts for sequentially updated flow parameters (Figure 4.8).

This is performed using an API connection between FLOW-3D (running on a local PC) and rocsc@r (running on the cloud). A user-friendly visual interface allows fast, secure and personalized access, as well as determination of all relevant parameters, such as the computational method to use, the time step or layer height during the iterations, the number of iterations to perform or also the time duration of FLOW-3D computations (Figure 4.9).

This sequential fluid-solid coupling is performed along 2D vertical profiles that are defined in rocsc@r. By covering the relevant bathymetric domain in the *Y*-direction using a series of 2D vertical profiles, a full 3D rock mass is modelled.

X_ch@nge

Database types of turbulent flows

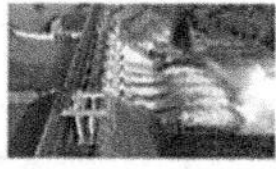

Free overfall jets

Free surface overfall jets, typically generated by dam crest spillways. This type of turbulent flow is characterized by jets that are impinging rather close to the dam toe.

Pressurized jets

Pressurized jet flows typically generated by intermediate or low-level (bottom) gated outlets through the dam body.

Ski-jump jets

Jets from spillways that flow over the downstream face of the dam and end up being ejected towards downstream by means of a flip bucket structure.

Stilling basin flows

Jets that start as spillway generated chute flow or gated pressurized flow and that are dissipated downstream by means of a stilling basin structure.

Channel flows

Turbulent flows that are generated by spillways and that flow in lined or unlined channels conveying the water towards downstream.

KEY FEATURES

- ✓ Easy cloud-access
- ✓ Continuously updated
- ✓ Same topology than X_pl@re – Projects are interchangeable between both platforms
- ✓ Data-driven intelligence - Parametric exchange and **optimisztion**

FIGURE 4.7 Types of turbulent flows and key features of the X_ch@nge database platform.

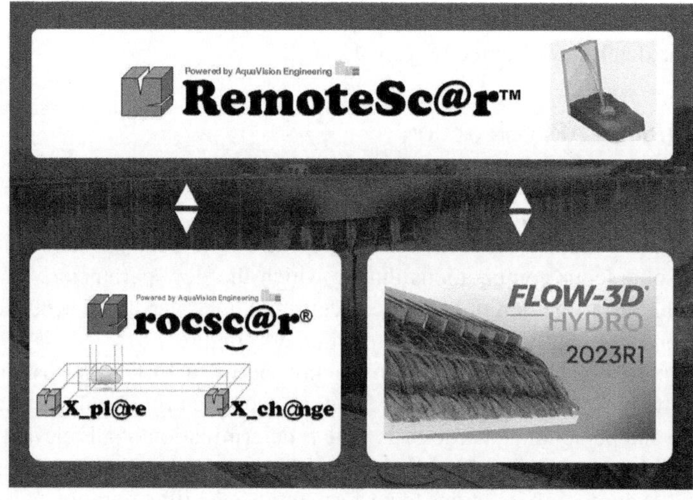

FIGURE 4.8 The RemoteSc@r API connection between FLOW-3D and rocsc@r digital platform.

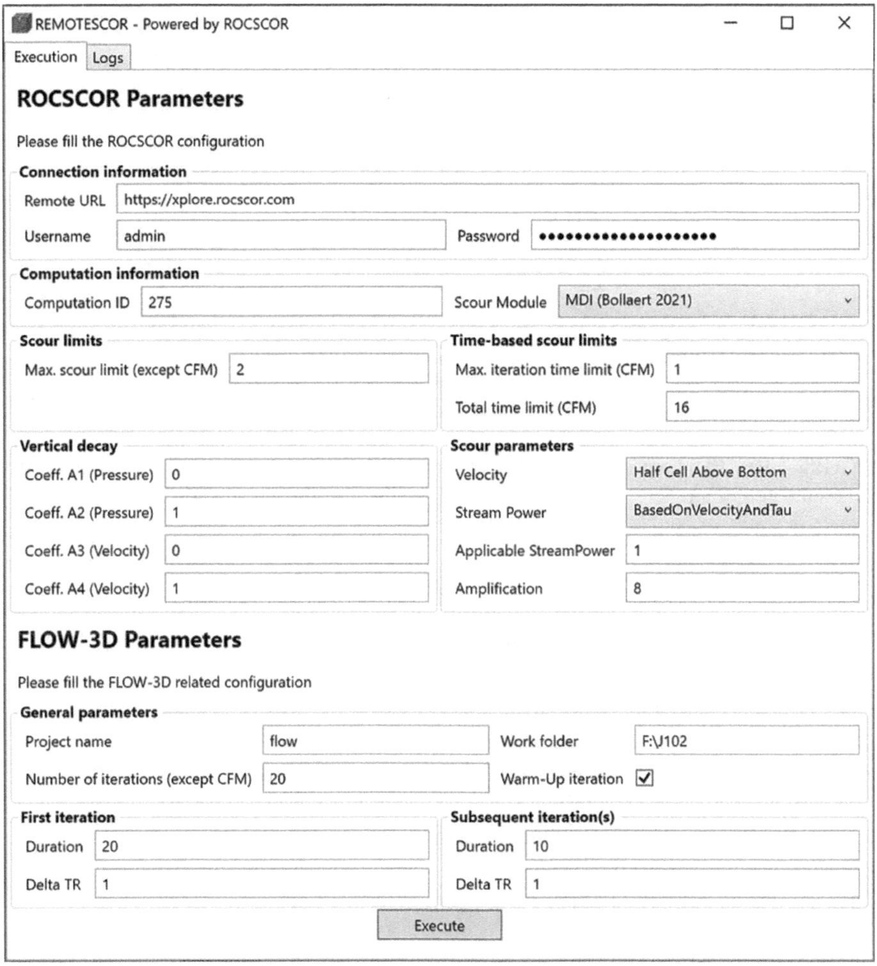

FIGURE 4.9 The RemoteSc@r visual interface allowing to determine the main parameters of the fluid-solid coupling.

This 3D bathymetry is transmitted to FLOW-3D as an STL file at each iteration of the coupling. FLOW-3D then adjusts the 3D hydrodynamic solution and transmits the relevant hydrodynamic parameters, such as flow velocities at the water-rock interface, average dynamic pressures, RMS values of dynamic pressure fluctuations, flow depths, shear stress and so on, to rocsc@r to prepare the next scour iteration.

Fluid-solid coupling comes to an end when subsequent iterations do not alter anymore the scoured water-rock interface (i.e. for rock block dynamic uplift, for rock block peeling off or for erodibility index methods), or when the total time duration of scour formation has been reached (i.e. for the time-dependent fracturing process) (Figure 4.10).

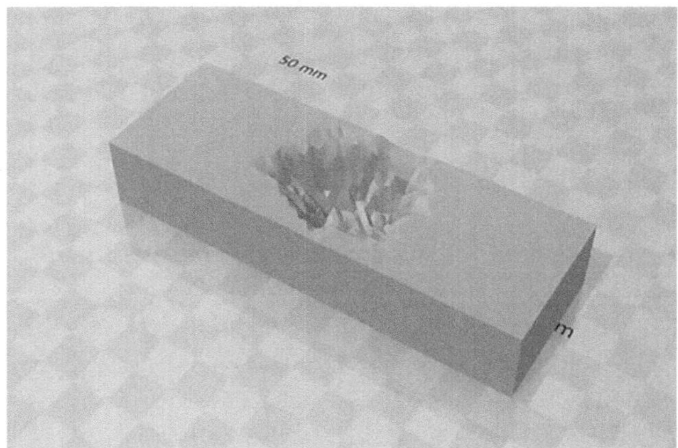

FIGURE 4.10 Example of an STL file of a scoured plunge pool as used by the RemoteSc@r fluid-solid coupling.

The key features of this digital connection can be summarized as follows:

- Choice of four different computational methods (erodibility index methods and physics-based methods)
- User-customizable parametric settings and scour thresholds
- Choice between 2D or 3D scour evolution with time
- Covers any type of turbulent flows that can be simulated in FLOW-3D
- Implementation of detailed 2D or 3D rock mass model with depth, including all relevant geomechanical characteristics

Case studies of fluid-solid coupling are provided in Chapter 7.

Key Features of the Rocsc@r Environment

It is interesting to confront the features of this novel digital environment with the key features and types of digital platforms as established under section "Types of Digital Platforms". Figure 4.11 summarizes the key features of both the X_pl@re and the X_ch@nge digital platforms.

The figure distinguishes between rocsc@r digital scour modelling with and without fluid-solid coupling. Some essential differences become apparent and are worthwhile mentioning.

First, without coupling, only plunging turbulent jets are dealt with in rocsc@r, and this only in 2D. A quasi-3D estimate may be obtained by multiplying 2D profiles, but the hydraulics are only available in 2D. Other turbulent flows must be computed in another software and implemented through use of a generic exchange (CSV) file.

Second, coupling allows to numerically simulate air entrainment through the FLOW-3D air-water flow mixture model, while rocsc@r uses constitutive equations

Key feature	X_pl@re rocsc@r software platform	X_pl@re rocsc@r - FLOW-3D coupling	X_ch@nge database platform
Modelling process			
precise and pertinent	O	O	X
2D / 3D geometries	2D vertical profiles only	O	X
time evolution	fracturing phase	fracturing phase	X
rock mass model	conceptual	conceptual	X
main physics implemented	O	O	X
Phase coupling process			
air-water	constitut.	air entrainment model	X
water-water	constitutive eqs.	constitutive eqs.	X
water-rock	conceptual	conceptual	X
Computational methods			
digitalization	O	O	X
reliable and approved	O	O	X
diversity of flow situations	turbulent jet flows only	all flows (coupling)	X
Adaptability / updatability			
easy, quick and universal access	O	O	O
uniqueness results	O	O	O
portability with other systems	not developed yet	not developed yet	X
data storage	on cloud / csv-files	on cloud and on PC	on cloud / csv-files
Security			
secure access	needs log-in and subscr.	needs log-in and subscr.	needs log-in
data encryption	O	O	X
backup redundancy	O	auto on cloud / ? on PC	O
intellectual property	O	O	O
data ownership	O	O	O
Connectivity			
API compatible	O	O	X
digital twin-ready	O	O	X
real-time and ahead-of-time runs	O	OK 2D / grid-depend. 3D	X
risk-informed decision-making	O	OK 2D / grid-depend. 3D	X
Data-driven intelligence			
parameter definition	by private users	by private users	O
parameter refinement	by private users	by private users	O
artifical intelligence-ready	not developed yet	not developed yet	not developed yet
Added value leveraging			
free access to all users	X	X	O
updatable by all users	X	X	O
feedback from all users	X	X	O
user participation	X	X	O

LEGEND	
O	feature available without constraints
constraint	feature available but with constraints
X	feature not available

FIGURE 4.11 Summary of key features of the X_pl@re and the X_ch@nge digital platforms.

for air entrainment and jet break-up. At first sight, this seems to be an advantage for coupled computations. However, it should be noted that, in contrast to air entrainment, full break-up of plunging jets is still difficult to model in CFD software. This aspect will be dealt with in Chapter 7.

Furthermore, a security issue might arise using coupling, in case the FLOW-3D result files, which are stored locally on the PC or hard disk, are not backed up automatically by the user. The resultant files on the cloud benefit from an automatic and redundant back-up on multiple data centres.

Also, for real-time and ahead-of-time scour computations, i.e. in case of risk-informed decision-making or digital twin connectivity (see Chapter 5), applying rocsc@r software without coupling will provide the fastest (quasi-instantaneous) responses. Fluid-solid coupling remains feasible in quasi-real time provided that the coupling is 2D, or in case the full 3D coupling is performed by a coarse numerical grid for the CFD (i.e. 1 m or more).

It is believed that the rocsc@r digital environment is the first attempt in the field of rock scour engineering to offer digitalized computational methods, automated 3D sequential fluid-solid coupling and a freely accessible database of published case studies of rock scour, all on the same cloud.

Added value leveraging becomes feasible by using the X_ch@nge free-access database platform, while private users on the X_pl@re software platform benefit from dedicated privacy, security and data encryption. Nevertheless, the topological structure of both digital platforms is similar, such that private users may choose to make public part or all of their computations on the database platform. This may happen typically at the end of a project, when private users decide that scour results may become publishable and the relevant data and feedback may become available to the engineering community.

Figure 4.12 determines the type of platforms for both X_pl@re and X_ch@nge. The software platform is service-oriented, while the database platform is both

Type of platform	X_pl@re		X_ch@nge
	rocsc@r software platform	rocsc@r - FLOW-3D coupling	database platform
Business model			
Social media			
Knowledge			•
Content and media sharing			
Service-oriented	•	•	•
Ownership			
Proprietary	•	•	• (free)
Open-source			
IT-framework			
Locally hosted server	• (twin)	• (CFD, twin)	
Cloud-based server	•	•	•
User intervention			
active service platform	•	•	
passive service platform			•
Data connectivity			
low connectivity	•	•	•
high connectivity	• (twin)	• (twin)	
live connectivity			

FIGURE 4.12 Summary of platform types for the X_pl@re and the X_ch@nge digital platforms.

knowledge-based and service-oriented, because the relevant parametric settings will aid users in developing their scour projects and computations. The platforms are proprietary-based, with free access to the database platform and subscription access to the software platform. The platforms are cloud-based, but may connect and exchange with locally hosted elements, such as CFD software for the fluid-solid coupling, or also a SCADA or local server for digital twin applications.

User intervention is actively implemented on the software platform, while a priori no user intervention is required on the database platform. Finally, data connectivity with third parties and services is a priori low, but can become high in case of automatic implementation of the software platform into a digital twin of a hydraulic structure. The latter aspect is dealt with in more detail in the next chapter.

REFERENCES

Annandale, G.W., *Scour technology: mechanics and engineering practice,* McGraw-Hill Professional, New York, 1 ed., 2006.

Asadullah, A., Faik, I. and Kankanhalli, A., "Digital Platforms: A Review and Future Directions", in *22nd Pacific Asia Conference on Information Systems*, Japan, 2018.

Beltaos, S., "Oblique Impingement of Circular Turbulent Jets", *Journal of Hydraulic Research, IAHR*, 14, 17–36, 1976.

Bollaert, E.F.R., "Wall Jet Rock Scour in Plunge Pools: A Quasi-3D Prediction Model", *International Journal on Hydropower & Dams*, 20, 1–9, 2012.

Bollaert, E.F.R., "The rocsc@r Cloud: An Innovative Digital Platform to Compute and Record Rock Scour", *International Journal on Hydropower & Dams*, 20, 5, 60–70, 2021.

Bollaert, E.F.R. and Schleiss, A.J., "Physically Based Model for Evaluation of Rock Scour Due to High-Velocity Jet Impact", *Journal of Hydraulic Engineering*, 131, 3, 153–165, 2005. https://doi.org/10.1061/(ASCE)0733-9429(2005)131:3(153)

Castillo, L.G. and Carrillo, J.M., "Comparison of Methods to Estimate the Scour Downstream of a Ski Jump", *International Journal of Multiphase Flow*, 92, 171–180, 2017.

Federspiel, M.P.E.A., "Response of an embedded block impacted by high-velocity jets", *PhD Thesis*, LCH-EPFL, 2011.

Koh, T.K. and Fichman, M., "Multi-Homing Users' Preferences for Two-Sided Exchange Networks," *MIS Quarterly*, 38, 4, 977–996, 2014.

Liu, P.Q., Dong, J.R. and Yu, C., "Experimental Investigation of Fluctuating Uplift on Rock Blocks at the Bottom of the Scour Pool Downstream of Three-Gorges Spillway", *Journal of Hydraulic Research, IAHR*, 36, 1, 55–68, 1998.

Maleki, S. and Fiorotto, V., "Blocks Stability in Plunge Pools Under Turbulent Rectangular Jets", *Journal of Hydraulic Engineering*, 145, 4: 04019007, 2019. https://doi.org/10.1061/(ASCE)HY.1943-7900.0001573

Spagnoletti, P., Resca, A. and Lee, G., "A Design Theory for Digital Platforms Supporting Online Communities: A Multiple Case Study", *Journal of Information Technology*, 145, 1–17, 2015.

5 Digital Twins in Rock Scour

DEFINITION OF DIGITAL TWINS

As stated in Chapter 1, a key element in reaching the goal of complete digital transformation is undoubtedly the development of *Digital Twins*. A digital twin (DT) is thereby defined as

> a digital replication of a physical entity or process, accounting for all the essential elements (geometry, material, physics, etc.) and allowing to accurately and realistically simulate all the relevant aspects throughout the entity/process lifecycle, whereby the physical-to-digital connection should be continuous and automatised.

The physical-to-digital connectivity is synchronized following a plausible frequency and fidelity and uses data sampled by sensor networks and generally transmitted through the Internet of Things (IoT).

Following Bado et al. (2022), the origin of digital twins is not fully clear but seems to go back to a presentation held at the University of Michigan by Dr Michael Grieves in 2002, and by a strategic technology roadmap published by NASA in 2010 (Shafto et al. 2012). In fact, the first ever digital twin was the simulator developed by NASA during the Apollo 13 mission, allowing to simulate the conditions inside the spacecraft following an explosion in the oxygen tanks that critically damaged the main engine. Instead of today's telemetry, NASA made use of the then available telecommunication technology to continuously update the simulator (digital twin) to understand the problem in space (physical twin) and help the astronauts in solving it.

In general, a digital twin groups the following key aspects:

- Digital twin systems transform engineering and management by accelerating holistic understanding, optimal decision-making and effective action.
- Digital twins use real-time and historical data to represent the past and present and simulate predicted futures.
- Digital twins are motivated by outcomes, tailored to use cases, built on data, guided by domain knowledge and implemented in IT systems.

Furthermore, once a digital twin is created, artificial intelligence (AI) and machine learning (ML) algorithms may continuously drive the digital twin to simulate future possible states and scenarios of the real system, allowing predictive or prescriptive maintenance, or to become more pertinent and accurate in digital replication and engineering decisions, using real-world data sets, all within the relative safety of a virtual world. Data can be imagined as sensor-provided engineering data, but also operational data, inspection reports, maintenance reports, and so on.

DOI: 10.1201/9781003319610-5

As outlined in Chapter 1, DTs and AI are already implemented today in dam engineering. Examples are smart operation and maintenance systems of power houses, allowing to perform automated remote inspection by robots and advanced transmission technologies. This allows condition-based maintenance, with suggestions for start-up and shut-down sequences of machines, and lifetime predictions based on neural network modelling.

Another example is energy companies that use digital twins to forecast and optimize the performance and health of their hydraulic or wind turbines (Tao and Qi 2019). A digital twin enables more effective asset design, project execution, and asset operations by integrating data and information throughout the asset lifecycle. As a data resource, it can improve the design of a new asset, understand an existing asset, run simulations and scenarios, or provide a digital snapshot for future works. Within this context, it is substantially different from BIM for example, which merely corresponds to a picture of an asset at a specific moment.

Following the digital twin consortium (https://www.digitaltwinconsortium.org), digital twins are recognized as an essential enabler of digital transformation and are the cornerstones of the metaverse, a quickly emerging virtual shared space in the future digital economy.

TYPES OF DIGITAL TWINS

The digital twin concept is composed of three main parts: a real or physical object, a virtual or digital object, and data that provide the connection between the physical and the digital objects. The physical object collects and stores real-time data that is sent to the digital object for processing. Vice versa, the digital object applies embedded engineering models and AI subjecting the received data to transformations and processing information. The types of digital twins can be based on different degrees of data exchange, or based on areas of application.

Twins Based on Data Exchange

Depending on the different degrees of data exchange and integration, a classification of the digital twin into three subcategories was proposed by Kritzinger et al. (2018). In reality, however, the terminology is fuzzy and all subcategories are most often referred to as Digital Twins in literature.

Digital Model

A Digital Model has the lowest level of data integration. The term indicates a digital representation of a physical object characterized by the absence of automated data flow between the physical and digital object. This suggests that the data flow from a physical object to a digital object and vice versa is provided manually. Consequently, any change that occurred in the physical element does not automatically impact the digital element and, similarly, any modification of the digital element does not automatically affect the physical element (Kritzinger et al. 2018).

Applied to rock scour engineering, these digital models typically represent numerical models of a real physical hydraulic structure and/or rock mass, which do not use any form of automatic data integration. Related to digital platforms described in Chapter 4, they would correspond to low-connectivity digital platforms.

Digital Shadow

A Digital Shadow has an intermediate level of data exchange and integration by an automated one-way data flow between the state of an existing physical object and a digital object. A change in state of the physical object leads to a change of state in the digital object, but not vice versa (Kritzinger et al. 2018).

Applied to dam engineering, a digital shadow typically represents a digitalized replicate of a real physical hydraulic structure, i.e. a dam for example, that has been developed and may be automatically updated by UAV or satellite investigations, and that is generally used for condition monitoring and predictive maintenance of the concrete of the structure. Frequency of data flow and updates would be typically on the order of max. several times per year.

To the author's knowledge, no digital shadow currently exists in the field of rock scour engineering. Related to digital scour platforms as described in Chapter 4, digital shadows may correspond to lowly or highly connected digital platforms.

Digital Twin

The highest level of data exchange and integration is reserved to a digital twin (Kritzinger et al. 2018). The data flow is automatic in both directions between the physical and digital object. In this context, a modification in the state of a physical object determines a modification in the state of a digital object and vice versa.

Unlike the digital shadow, the digital twin allows to verify physical processes and activities prior to their execution, in order to optimize performance and minimize failures. The digital twin is able to point out in real time any differences between the physical and simulated performances to optimize and predict the behaviour of the physical object. As such, ahead-of-time simulations and optimizations become possible and this information may be sent back to the physical object to adapt its functioning.

Furthermore, since data within the digital twin is derived not only from the physical environment but also from virtual models with data elaborated through processes such as statistics and regression, the digital twin is more abundant in data than the digital shadow.

To the author's knowledge, no digital twin currently exists in the field of rock scour engineering. Related to digital scour platforms as described in Chapter 4, digital twins correspond to highly connected or live connected digital platforms.

Twins Based on Application Areas

Digital twins may also be subdivided based on their application areas as outlined hereafter (source: https://vidyatec.com/blog/the-3-levels-of-the-digital-twin-technology-2).

Component/Part Twins

This type represents the model of an individual part of a system. Although it is more common to use the technology on a larger scale, it is possible to apply the tool only to the most relevant elements of a system. In rock scour engineering, this would be the case for a twin that only deals with bedrock bathymetric evolution, or that only deals with the outflows generated by the spillway.

Asset Twins

By combining the application of the technology in different components that are connected, the asset twin is formed. In this case it is possible to understand how different parts of a system interact and generate more concrete actions from the processing of the data provided by the analysed elements. In rock scour engineering, this would be the case for a twin that deals with both bedrock bathymetric evolution and outflows generated by the spillway, and that makes the link between these essential components.

System/Unit Twins

The next level of application is represented by the union of different assets that enable the analysis of a complete process within the operation. The insights generated at this level are much more complete, since it is possible to understand the consequences of a situation on a larger scale across the industry. In rock scour engineering, this would mean for example a link between a twin for rock scour and a twin for gate operations and power house functioning. The occurrence of rock scour might have consequences on gate operations or turbine functioning. As such, an asset twin would cover more than just the field of rock scour.

Process Twins

Finally, we have the most complete digital twin model, which covers all the systems of an industry to represent its operation from start to finish. The Process Twin allows you to integrate the data generated all the time in the industry so that you can make decisions more assertively and make processes more efficient. In rock scour engineering, this would imply that the twin is part of much larger-scale ensemble of twins that optimizes the entire functioning of the dam and its power house throughout their lifecycle.

LEVEL OF TECHNOLOGY OF DIGITAL TWINS

Different levels of technology can be defined for digital twins, based on the degree of connectivity and the importance the twin takes in decision-making processes (source: https://afry.com/en/newsroom/news/how-digital-twins-improve-efficiency-industry).

Level 0: Unconnected Model

This level corresponds to the development and unconnected use of a digital or virtual model. Depending on the project state, this model can be a numerical replicate of a physical reality (i.e. in case of an existing dam), or can also be a virtual prototype of a not-yet-existing reality (i.e. in case of a preliminary project before dam construction). In any case, no connection exists between the digital and the physical elements.

In rock scour engineering, this level would correspond to the set-up and initial calibration of a numerical scour model, based on historic flood events and related in-situ scour measurements. Also, it may be used to simulate how the real object would respond to different scenarios.

Level 1: Physical-to-Digital Connection

This level represents the (quasi) real-time connection of the previously developed digital model to the physical counterpart, such that the digital model can be soundly updated at each relevant change on the physical model. Depending on the system, this update can be manual or automated. The former corresponds to a digital model, while the latter corresponds to a digital shadow.

In rock scour engineering, this level would correspond to sequential real-time updates of the digital scour model during flood events, such that the current state of the turbulent flows and related scour can be visualized to the user.

Level 2: Diagnostics and Problem Identification

This level represents the diagnostics and problem identification that can be made following real-time experiences made with the physical-to-digital connection.

In rock scour engineering, this level would correspond to sequential recalibration of the digital scour model after each flood event, together with sound real-time rock scour analysis during flood events by the dam owner or operator.

Level 3: Prognostics of Different Future States

This level represents the prognostics that can be made and updated of different future states of the system that is being twinned. It uses the technology of level 2 to make ahead-of-time predictions for different potential scenarios.

In rock scour engineering, this level would correspond to recomputing of design flood events after each recalibration of the digital scour model following flood events. Also, it would correspond to ahead-of-time rock scour prognosis for different near-future spillway outflow hydrographs and/or gate opening scenarios. Comparison of several outflow scenarios and/or spillway triggering allows human analysis and optimization of actions to be taken in view of asset operation and management.

This level is currently proposed by the TwinSc@r API application for rock scour, as outlined in more detail further on in this chapter.

Level 4: AI and Active Decision-Making

This level represents artificial intelligence (AI) and active risk-informed decision-making processes performed by the digital twin, based on a real-time highly connected coupling with the physical twin, allowing the digital twin to perform real-time and ahead-of-time analyses and optimization of actions to be taken in view of asset operation and management.

In rock scour engineering, this level would correspond to an automatization of the level 3 process of optimization of actions to be taken in view of asset operation and management. The digital twin makes the computations, performs the analyses and finally proposes an optimal outflow scenario for each of the spillway gates. The twin proposes the actions but the dam operators may still actively judge and choose to apply the proposed scenario or not.

Level 5: Fully Operational Digital Twin – No Human Intervention

This level represents the highest level of artificial intelligence and active risk-informed decision-making processes performed by the digital twin. The digital twin takes its own decisions and handles the object itself. At this level, it replaces operators and employees.

In rock scour engineering, this level would correspond to a full automatization of the level 3 process of optimization of actions to be taken in view of asset operation and management, i.e. dam operators do not intervene anymore in the process.

EXAMPLES OF DIGITAL TWIN DEVELOPMENTS IN DAM ENGINEERING

As mentioned in Chapter 1, digital twins start to be developed in dam engineering. The twins under development generally focus on asset health management, i.e. checking the state and related structural integrity of the dam and its appurtenant structures, and rarely aim at a continuous real-time connection with the physical asset. As such, following the terminology outlined before, they are more like digital models or shadows than digital twins. For simplicity, the term digital twin is used throughout the remaining of this chapter.

A first example is the digital twin of Diablo Dam in Seattle, USA, generated by HDR Inc. for Seattle City Light (SCL) Power Company (Camus 2022) and making use of Bentley System's iTwin digital platform (https://www.bentley.com/software/itwin-platform). The twin is visualization-based and has been created by making use of drone-based remote sensing, allowing to set up a realistic virtual copy of the dam and surrounding rock mass by taking more than 10,000 pictures and using photogrammetry.

An ensemble of 82 million data points, with a topographical precision of less than 2 cm, allows to automatically detect the state of the surface concrete of the dam. Features such as cracks, weaknesses and so on can be detected on the digital twin, without any real-life inspection. Artificial intelligence and machine learning together with local visual verifications on the physical twin then allow to continuously enhance detection of spots on the surface concrete that need repair or maintenance (Camus, 2022).

The twin only concerns the dam surface, and even if the surrounding rock mass may be part of the detailed topography being remotely sensed, no rock mass model is present or automatically updated in this twin. Figure 5.1 illustrates a general view of Diablo Dam (courtesy of Seattle City Light).

FIGURE 5.1 General view of Diablo Dam. Courtesy of M. A. Firoozfar, Seattle City Light Power Company.

Seattle City Light Power Company used the digital twin in their most recent Federal Energy Regulatory Commission Part 12 inspection, as a complement to their traditional rope-access inspection. Specifically, they used it for their potential failure modes analysis. Experts in different locations reviewed the digital twin to figure out possible ways the dam could fail. The digital twin was very useful as a discussion aid, and enabled people to review the dam together, without requiring to travel for a site visit (M. Firoozfar, SCL, personal communication 2022). .

A second example is presented by Park and You (2023) who made a digital twin of Sumjin dam and its entire river basin for comprehensive digitalized smart water resource management and data-driven decision-making. Their platform includes a GIS-based high-precision twin with topography and facility information of both dams and rivers. The platform synchronizes real-time data such as rainfall, dam and river water levels, flow rate and closed-circuit television (CCTV), and incorporates hydraulic and hydrological simulation models. AI technology is used to suggest optimal dam discharge scenarios. It is not clear, however, if this is performed automatically or manually.

Additionally, the platform includes a geotechnical safety evaluation module for levees, advanced drone monitoring, and an AI CCTV video surveillance function. The digital twin-based platform supports efficient decision-making for smart flood responses (Figure 5.2).

In relation with digital twin developments, Bentley Systems Inc. developed health assessment of a wide range of infrastructure assets from telecommunication towers and bridges to rail and dams. They offer a wide portfolio of technologies, from photogrammetry, machine learning and 3D reality meshes to software for Internet of Things (IoT) applications that allow digital twins to incorporate real-time sensor data.

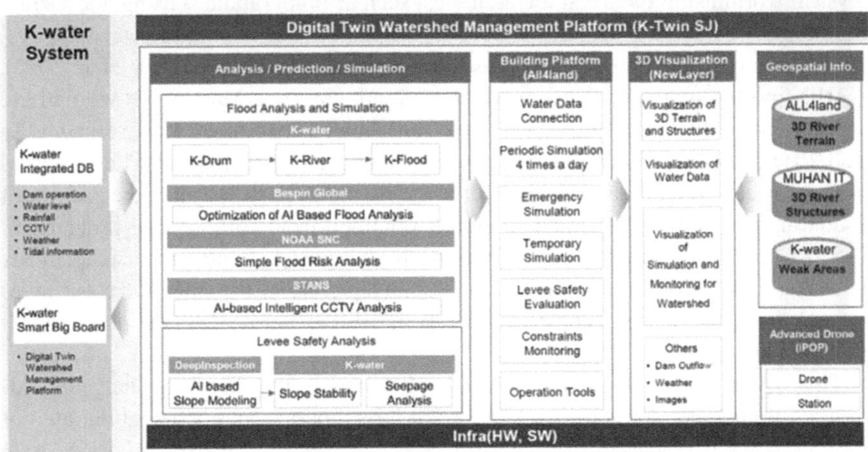

FIGURE 5.2 Digital twin of Sumjin Dam: UP: digital twin watershed platform, DOWN: architecture of digital twin platform for dam and river basin. Courtesy of Park and You.

These tools are integrated into Bentley's iTwin Open-API platform consisting of iTwin IoT – used to monitor in real time a series of physical parameters, such as displacements, vibrations, settlements, etc., and of iTwin Capture for creating high-resolution 3D models of assets. The corresponding products for real-time health monitoring are AssetWise Bridge Monitoring and AssetWise Dam Monitoring (https://www.bentley.com/software/itwin-platform/).

These solutions are not necessarily designed to eliminate in-situ physical infrastructure inspections, but help consultants keep a close eye on the asset and develop a more informed inspection plan from the office.

Furthermore, Bentley Systems Inc. collaborates with Niricson Software Inc., who developed an AI-based predictive damage assessing cloud platform, AUTOSPEX™, designed to verify the structural integrity of concrete structures. Niricson's technology uses optical, thermal and acoustic sensors on UAVs to check deep into the concrete where the rebar is, then applies artificial intelligence (AI) and machine learning (ML) to figure out exactly where the damage is happening (https://niricson.com).

Combining Niricson's software with Bentley's asset health monitoring product allows to set up a digital twin using IoT sensor data to see cracking analysis superimposed with temperature, displacement, vibrations and other metrics.

DIGITAL TWIN DEVELOPMENTS IN ROCK SCOUR ENGINEERING

INTRODUCTION

To the author's knowledge, no digital twin exists in the field of rock scour engineering at the time of printing of the present work. Given the ongoing developments in dam engineering, they will probably become available soon.

In rock engineering, Discrete Fracture Network (DFN) models may be considered as a sort of virtual rock mass that is generated based on digital data acquired from the physical world using the latest technologies, such as holographic sensing and virtual or mixed reality (VR/MR). The connection or updating between the digital and physical twins is not yet automatized, but is foreseen in the near future (Yang et al. 2022).

DFN models represent the foundation of Synthetic Rock Masses (SRM), and are stochastic by nature, thus requiring a probabilistic approach to rock engineering. As such, they do not provide an authentic replicate of the real rock mass. The data used by these recent models are not part of standard guidelines for characterization of discontinuities (ISRM 1981) and cannot be obtained from core logging procedures. The SRM approach fully accounts for anisotropy, scaling effects and displacements between blocks. Also, it provides a quite detailed picture of the state of fracturing of the rock mass and does not consider the rock mass as an equivalent continuum model with uniquely defined strength.

This is particularly of relevance to rock scour modelling, where knowledge regarding the initial state of fracturing (number of joints, persistency, orientation, etc.) of the rock mass is essential. Elmo et al. (2021) developed in this regard a new rock mass quality indicator, the Network Connectivity Index or NCI, which is a first step towards a more quantitative rock mass characterization. NCI is thereby determined based on DFN models and supposed to better reflect the degree of natural connectivity of the rock mass's fracture network. A sensitivity analysis for the NCI using DFN modelling can be found in Fogel (2022).

Furthermore, building a realistic digital replica of the real rock mass is not sufficient for a rock scour digital twin to come alive. The main challenges to develop rock scour digital twins have been discussed in Chapter 2. Among others, a continuous real-time digital update of the physical scour formation in the rock mass during flood events would be needed.

In the following, the relevant physical elements and digital building blocks needed to develop such twins are explained more in detail, together with the workflows connecting these blocks under the form of flowcharts. Finally, a first-hand version of a rock scour digital twin is developed and presented through the TwinSc@r application that connects the rocsc@r digital environment to real-life data measured on a hydraulic asset.

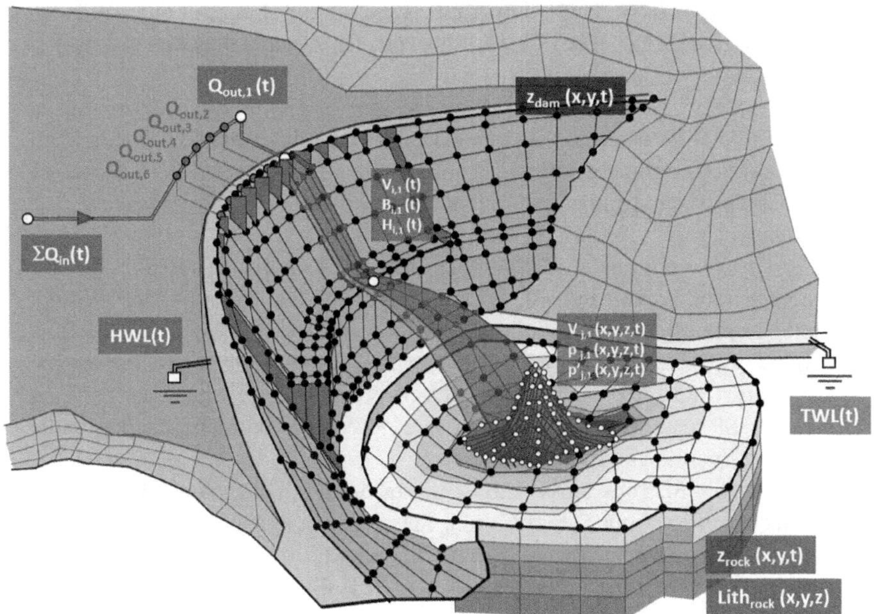

FIGURE 5.3 The complete digital twin in rock scour at an arch dam.

Physical Elements

A complete digital twin for rock scour would need continuous real-time implementation of topographic, bathymetric, hydrodynamic and rock mechanics parameters in a complete, realistic and accurate way. This would ask for automated digital assessment of the rock mass with depth, as well as real-time automatic updates of the essential bathymetric, hydrodynamic and geomechanic parameters, based on continuous monitoring of the physical twin. AI and ML techniques might then allow automated model calibration and validation, as well as future scour predictions as a function of incoming flow events.

The main physical elements of a rock scour digital twin as applied to a ski-jump jet issuing from an arch dam are illustrated in Figure 5.3 and may be summarized as follows:

- *Topography*: 3D detailed real-time topography of all elements involved, i.e. the dam and its appurtenant structures (spillway, chute, flip bucket, etc.) (z_{dam} (x,y,t)), the surrounding valley and downstream riverbed (z_{valley} (x,y,t));
- *Bathymetry*: 3D detailed real-time bathymetry of downstream plunge pool and rock mass, riverbed, etc. (z_{rock} (x,y,t));
- *Hydrology*: detailed functioning of catchment area (rainfall, snowfall, storage, evapotranspiration, etc.) and resulting reservoir inflow hydrographs ($\Sigma Q_{in}(t)$);
- *Reservoir routing*: detailed reservoir level changes as a function of in- and outflows with time, HWL-volume relationship, HWL(t);

- *Spillway (per bay)*: detailed hydraulic functioning of bays of dam spillway, outflow discharge per bay $Q_{out}(t)$, flow velocity $V_i(t)$ and flow dimensions per bay (shape, jet flow depth $h_i(t)$, jet thickness $B_i(t)$) with time;
- *Jet trajectory (per bay)*: point of jet issuance, angle of issuance, turbulence intensity at issuance, aeration, air drag, shape deformation, gravitational contraction, turbulent diffusion, point/shape of impact in TWL, degree of break-up of jet at impact, jet velocity $V_j(t)$, flow depth $h_j(t)$ and width $B_j(t)$ at impact;
- *Jet diffusion/Plunge pool flow*: type of jet (rectangular or circular shaped, compact or broken up), 3D turbulent diffusion of air-water jet through plunge pool depth, 3D spatial distribution of dynamic pressures and flow velocities $V_j(x,y,z,t)$ at water-rock interface, time-averaged and fluctuating pressure values $p_j(x,y,z,t)$ and $p'_j(x,y,z,t)$, shear stress $\tau_j(x,y,z,t)$ and stream power $SP_j(x,y,z,t)$, flow depth $h_j(x,y,z,t)$, TWL(t); automatic adaptation of these values during scour formation;
- *Rock mass*: 3D rock mass with depth, including 3D lithology, 3D detailed discontinuities (joints, faults, etc.), 3D rock block shapes and protrusion, roughness, degree of weathering, UCS strengths, degree of jointing, in-situ stresses, anisotropy, heterogeneity, etc.; ideally determined by Discrete Fracture Network modelling;
- *Rock scour*: 3D rock scour with depth, including all possible types of break-up (i.e. abrasion, weathering, quarrying and plucking, dissolution, peeling off, brittle and fatigue fracturing, cavitation) and accounting for all potential degrees of movement (translation, rotation, sliding, etc.), including detailed shape of all distinct scoured blocks and remaining water-rock interface, downstream mounding, automatic 3D and 3-phase interaction between scoured interface and 3D jet hydrodynamics.

Building Blocks

Digital twins are composed of building blocks that are essential to their correct functioning. Each building block thereby represents a distinct physical process or action. The main types of building blocks are:

- *Metadata*: Diverse types of information needed to describe the twinned object, including its physical components, how they're assembled, the object's behaviour and its main characteristics and specifications.
- *Generated data*: IoT or other sensor-based time-series data, external data and whatever other data that is used by the analytical models of the twin. Data can be readily available as real-time input data to the twin or can be numerically reproduced in real time by the twin.
- *Analytical models*: Software algorithms and procedures that ingest generated data and produce output results.
- *Software components*: Functional core elements of the twin containing one or more analytical models and interconnected such that, by using input data, output results are generated, such as digital platforms, user interfaces, automates, etc.

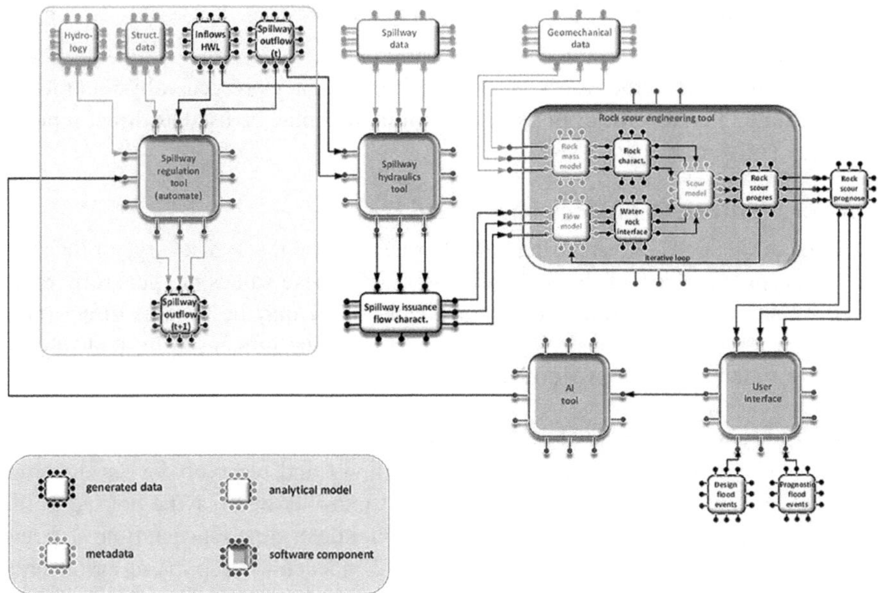

FIGURE 5.4 Flowchart of digital twin for rock scour at hydraulic structures.

The main building blocks needed to twin rock scour occurring at hydraulic structures are described hereafter. Some of these blocks may already exist as part of the automated regulation of hydraulic structures, while other blocks must be created ad hoc. The connectivity and main workflow needed between the blocks is outlined in Figure 5.4 and discussed in detail in the next subchapter.

Hydrology Block (Metadata)

The physically most upstream situated block manages the hydrology of the catchment area that feeds the hydraulic structure in question. Typical elements composing this block may be catchment areas, rainfall-runoff relationships, surface coverages, inflows, precipitation intensity and spatial distribution, pluviometry, infiltration and evaporation, evapotranspiration, and so on.

Typical output generated by such blocks are time-varying flow discharge values into a reservoir or in the river upstream. In case of reservoirs or any other type of water storage facility, these serve as direct input to make computations to determine the reservoir level evolution and reservoir outflows based on pre-defined regulations.

This building block is generally only available as a digital engineering model, without direct connection with the physical reality. In case a real-time connectivity with the hydraulic structure digital regulation exists, it may be used for real-time inflow and outflow prognosis.

Structural Elements/Spillway Block (Metadata)

These building blocks define the structural elements of the hydraulic structure, i.e. dams, spillways, concrete elements, walls, slabs, abutments, piers, baffle blocks, flip buckets, aprons, end walls, and so on, or also the specific design and dimensions of spillway bays.

In-Situ Measured Data Block (Generated Data)

This building block defines the in-situ measured data, such as for example discharge inflows of tributaries, the reservoir level and so on. They are generally recorded at distinct time intervals, for example, every 5 min, and may be used as direct input to model or software blocks.

Spillway Outflow Data Block (Generated Data)

This building block defines the current (t) or near-future ($t+1$) discharge of the river or of the spillway bays of the hydraulic structure. These values are generally regulated by an automate and are readily available. They may be used as direct input to models or software blocks. Their regulation may be influenced by real-time or ahead-of-time rock scour prognoses.

Spillway Regulation Block (Software Component)

In case of hydraulic reservoirs, discharge outflows and reservoir levels are often regulated by an automatic (PID or equivalent) controller that may be housed in the SCADA of the hydraulic structure. This controller determines at each time step and for each separate spillway bay the outflow that is needed to keep the reservoir level within tolerable or optimal limits. Rules generally account for the availability of spillway bays and a pre-defined set of logic rules distributing the total outflow over the different available bays.

The controller makes use of real-time measurements of the reservoir level and of upstream inflows into the reservoir to determine the total outflows during the next time step. Dispatching of these outflows over the available spillway bays then allows determining the gate opening ratio that is needed for each bay. This building block is generally implemented in an existing digitalized part of the hydraulic regulation.

Spillway Outflow Hydraulics (Software Component)

This building block is the physical suite of the spillway regulation block and computes the hydraulic characteristics at issuance for each spillway bay, i.e. the average flow velocities, the flow depths, the flow widths, the flow initial turbulence intensity and finally the orientation and angle of issuance. Flow hydraulics may be, for example, free surface or pressurized plunging jets or free surface or gated chute flows.

These parameters can be determined based on constitutive equations such as stage-discharge relationships for free surface flows or for different gate opening ratios, and can be stored as pre-defined 1D or 2D matrices. Another option is to directly derive them from detailed CFD modelling, in which the spillway structure itself is directly modelled.

This building block is generally not part of existing spillway regulators on hydraulic structures and thus must be developed to create a digital twin for rock scour.

Spillway Flow Issuance Characteristics Data Block (Generated Data)

This building block defines the detailed flow characteristics of the spillway bays of the hydraulic structure. These values are generated by the spillway outflow hydraulics tool and may be used as direct input to models or software components.

Rock Scour Engineering Tool (Software Component)

This tool represents the main engine for rock scour computations. It makes use of both hydraulic and rock mass data, and of numerical models to compute rock scour by fluid-solid coupling. The tool may implement different computational models and data that are interconnected. The main output is the rock scour prognosis for the given time step of operation of the system.

Geomechanics Data Block (Metadata)

This building block groups the geomechanical input data needed to set up a rock mass model with depth, such as the rock bed level and shape, lithologies, layers, joints, faults, bathymetry, topography, UCS strengths and so on.

Rock Mass Model (Analytical Model)

This block represents the model of the rock mass with depth, by using the information from the geomechanics data block. These models may be conceptual, i.e. based on a numerical grid in which each grid point incarnates rock mass characteristics, or also physics-based, i.e. 2D or 3D more or less accurate representations of the real mass with its joints, faults and blocks of different shapes.

Flow Model (Analytical Model)

This block represents the model of the flow issuing from the hydraulic structure and down into the rock mass. It distinguishes between free surface turbulent flows above the water-rock interface and pressurized flows inside the rock mass joint network. Corresponding models may be 1D, 2D or 3D, analytical or numerical (CFD), and time-dependent or not. Flows acting at the water-rock interface serve as boundary conditions for flows inside the rock mass discontinuities.

Rock Mass Data Block (Generated Data)

This building block defines the geomechanical characteristics of the rock mass with depth, as defined by the rock mass model. This data manages and transforms the input data provided by the geomechanics data block and is used as direct input by the scour model.

Rock Interface Data Block (Generated Data)

This building block defines the hydrodynamic characteristics of the flow at the interface between the water and the rock (i.e. flow velocity, stream power, RMS pressure fluctuations, etc.). This data is used as direct input by the scour model and generally changes during progressive scour formation.

Scour Model (Analytical Model)

This block represents the model used for rock scour computations and may implement multiple digitalized scour computational methods. The generated output is the scour formation and progression within the computed time step of operation of the hydraulic structure. Within this time step computation, whenever the flow characteristics at the water-rock interface significantly change during scour formation, an iterative loop is needed to compute this scour progression correctly, by means of sequential fluid-solid coupling. This also holds for example for cases of scour regression, where the point of issuance of the flow changes because of head-cut migration.

Scour Prognosis Data Block (Generated Data)

This building block defines the scour progression generated within the computed time step of operation of the hydraulic structure. It may be consulted by the user in a visual interface and/or be used as direct input into the spillway regulation block to influence near-future discharge outflows.

User Interface Block (Software Component)

This building block is the user interface through which the user may visualize scour computations in real time and ahead-of-time, consult ancient results, compare results, define prognostic flood events, launch computations and so on. It functions as a digital control room, allowing to implement rock scour potential into asset management and risk-informed decision-making processes during flood events at a hydraulic structure.

Flood Events Block (Metadata/Generated Data)

This building block bundles three types of flood events occurring at hydraulic structures, depending on the timeline of computation compared to reality (UTC time):

- *Design flood events*: composed of synthetic flood hydrographs based on return periods or a PMF event; such scenarios may be pre-defined, computed and stored as reference results in the twin environment.
- *Real-time flood events*: composed in real time based on the sequentially transmitted values of spillway outflows – note: rock mass scour is only determined in real time in the digital twin, not in the physical twin.
- *Prognostic flood events*: composed ahead-of-time based on prognosis of the near-future spillway outflows. Such prognosis may be provided by the hydrology building block, or also made manually.

Artificial Intelligence (AI) Block (Software Component)

This building block bundles different spillway scenarios that may be imagined at a hydraulic structure. These scenarios depend on the number and type of spillway bays and their availability, and may be determined and triggered in real time during incoming flood events. Main parameters of influence are generally the gate availability and the gate opening ratios.

Flowchart of Digital Twin for Rock Scour at Hydraulic Structures

The connectivity between the building blocks is illustrated by the flowchart in Figure 5.4 and determines the workflow process of a digital twin for rock scour at hydraulic structures.

Generally, part of these building blocks exists already on a hydraulic structure, i.e. the closed-loop outflow regulation or controller, which is typically a PID (Proportional-Integral-Derivative) controller or equivalent system using the outflowing discharge as set point to regulate. This controller makes use of real-time measured data as well as of system data (set points, rules, models) to make a spillway outflow proposal for the next time step. It allows maintaining the reservoir level within acceptable or optimal limits. The corresponding building blocks are isolated by the upper left rectangular area in Figure 5.4.

The digital twin for rock scour adds an additional loop to this existing flow regulating system, by modelling of spillway outflows, of turbulent flows at the rock interface and of the rock mass with depth. An appropriate scour model allows connecting these different models in an iterative manner such that a real-time prognosis can be made regarding the rock scour potential of the current time step of the flow regulation.

The scour model estimates the current situation of the water-rock interface based on an initial situation that may be derived for example from bathymetric measurements at a prescribed date. The outflows that occurred since that date are then used by the model to reconstitute the current water-rock interface. After each bathymetric measurement, a reset is made of the water-rock interface, allowing eventually to recalibrate the scour model.

Furthermore, depending on the feedback of the digital twin prognosis on rock scour potential, the spillway outflows as planned by the controller for the next time step may be corrected manually or automatically.

This scour prognosis may be visualized in a customizable user interface and compared with available design flood and prognostic flood events. While the real-time flood event only computes scour potential for the historic and instantly chosen spillway outflow scenario, prognostic and design flood events may be numerically reproduced for a range of different virtual scenarios.

This allows the dam owner or operator to compare scenarios between them and choose the optimal one in view of scour potential and risk, allowing real-time risk-informed decision-making. In such a case, the dam owner or operator manually adjusts the closed-loop outflow controller, such that rock scour potential is accounted for.

By use of artificial intelligence (AI), however, the scour prognosis may be directly used as input to the spillway regulation building block, to correct the controller following a set of rules. Without any human intervention, clearly defined criteria should be made for scenario optimization, such as for example max. scour depth, scour proximity with dam, etc.

Development of a Digital Twin for Rock Scour at Hydraulic Structures

The aforementioned digital technology and workflows are under development within the framework of the rocsc@r digital environment. Part of these developments is discussed in the next subchapter.

THE TWINSC@R™ APPLICATION

Introduction

TwinSc@r is a new feature that allows connecting the rocsc@r software and platform to third-party digital systems, such as for example the SCADA, or any locally hosted or cloud-based server related to a hydraulic structure. In this way, a digital twin for rock scour may be generated, allowing real-time and even ahead-of-time rock scour assessment on the twinned infrastructure.

This dedicated API (Application Programming Interface) application has a specific user interface that allows easy, fast and universal access to past, current or future scour states of the hydraulic structure being twinned. These different scour states are stored on a database from which the user can retrieve historic information upon demand.

Hydraulic structures often dispose of a spillway outflow regulating system or controller, which makes use of a set of in-situ measured variables, such as the water surface at a distinct location or the inflow discharge, following a pre-defined regulating time step. The in-situ measured variables are generally available in (quasi-) real time through a dedicated user interface in the control room of the hydraulic structure (from the SCADA for example).

Eventually, a digital twin may even exist of the hydraulic structure. In such a case, the twin groups multiple types of data and/or engineering services that allow the dam owner or operator to dispose of a virtual (numerical) copy of the physical reality, with automated monitoring and transmission of all relevant data in the field, using IoT or any other type of sensors.

As discussed in Chapter 2, digital twins of hydraulic structures are currently strongly developing, aiming primarily other issues than rock scour, such as for example predictive or prescriptive maintenance. Nevertheless, most of these twins are based on hydraulic, geometric or even geomechanical data that could also serve digitalized rock scour predictions.

Once TwinSc@r is connected to a server or SCADA, or embedded into an existing global digital twin, the rocsc@r computational environment becomes (and remains) continuously connected with the physical reality of the infrastructure. As such, the virtual replicate of the rock mass may be continuously kept updated during the lifetime of the infrastructure, based on a real-time numerical implementation of flood events.

As the rock mass cannot be monitored in real time during flood events, only the hydraulics of flood events may be twinned in real time. The scour formation of the rock mass is only measurable in between consecutive flood events and, as such, the digital system is forced to autonomously twin this scour formation without real-time feedback during events.

However, sound calibration following accurate rock mass data, as well as systematic feedback and recalibration following past flood events, progressively increases the replicability and performance of the rock scour digital twin.

As flood events generally do not occur continuously on hydraulic structures, TwinSc@r exploits the physical-to-virtual connection continuously and in real time, but relevant results and feedback allowing recalibration only become available after each distinct flood event.

A number of relevant hydraulic parameters, continuously or sequentially monitored on the physical structure, are thereby systematically transmitted to the scour platform, which remains in stand-by for real-time updates. Whenever a new set of hydraulic data is transferred, for example every 10 min, the virtual twin of the rock mass is being updated for that time step in question.

Generated Output

The output that is generated by TwinSc@r corresponds to time-dependent scour states of the rock mass that is impacted by flows from the hydraulic structure that is being twinned.

When making use of the rocsc@r 2D vertical profiles and the in-house hydraulic model for plunging turbulent jets, a 2D scour assessment is obtained, i.e. each TwinSc@r model is related to a specific spillway bay or overflow section.

Ongoing developments will allow to run TwinSc@r by coupling rocsc@r with the FLOW-3D® CFD software. As such, a full 3D scour assessment may theoretically be obtained, per spillway bay or for an ensemble of bays. However, depending on the complexity and size of the structure, the coupling option may need considerable computational time and may not offer a real-time or ahead-of-time prognosis. Therefore, for practice, 2D CFD coupling seems a priori more appropriate.

The different scour states computed by rocsc@r may be visualized by TwinSc@r in different manners, depending on the time-span of the computations compared to the real time (UTC-based):

- *Design flood events*: Scour results generated by synthetic flood hydrographs, i.e. prescribed return periods, PMF event and so on. Such scenarios are independent from the real-time and physical situation, and may be pre-defined and stored as reference results in the twin environment.
- *Real-time flood events*: Scour tracking in real time (UTC-based) based on sequentially transmitted values of monitored variables or spillway outflows – note: real-time rock scour is only tracked digitally, not physically.
- *Prognostic flood events (Forecast)*: Scour forecasting based on prognosis of the near-future spillway outflows, using the real-time scour as initial situation. Such prognosis may be provided automatically by the hydrology building block, or may be made manually through the user interface specifically developed for TwinSc@r.

User Interface

The current user interface of TwinSc@r is embedded into the secured user-specific digital environment of rocsc@r and makes use of widgets. A more enhanced user interface is currently under development. This interface will also be available on-line, as a stand-alone application.

The interface has been developed to track real-time scour and to assess and visualize scour forecasts, allowing to make risk-informed decisions related to scour safety of the twinned asset. These tasks are performed by the user without having to worry about the underlying scour computations and twinning issues, which continuously run in the background on the rocsc@r digital environment.

The user interface is illustrated in Figure 5.5 and makes use of user-definable and customizable widgets. For each 2D profile of a hydraulic structure, a widget may be created by the user. This widget is identified by a Key that is unique and points towards a 2D profile in the rocsc@r environment. By defining this 2D profile as a TwinSc@r profile upon its creation, a Key ID (GUID) is automatically generated for this profile. As such, the profile is in stand-by to be linked to a physical profile of a hydraulic structure, and cannot be used anymore for other scour computations in parallel.

Once TwinSc@r triggers use of this Key ID, the profile becomes intrinsically and definitively twinned with a real physical profile. In other terms, the profile becomes the

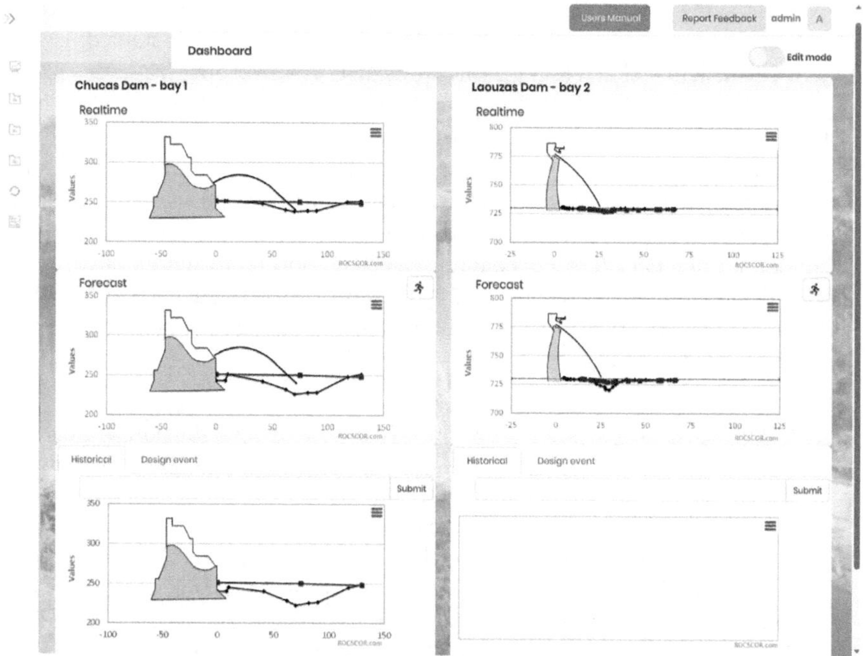

FIGURE 5.5 User interface of the TwinSc@r digital twin for rock scour, showing a widget for historic, real-time or ahead-of-time scour estimates, and a virtual SCADA for data transmission.

virtual counterpart of a physical profile and its continuous evolution, and the related computation is flagged as a TwinSc@r computation that is unique to the profile.

Figure 5.5 contains two widgets. These represent different hydraulic structures of the user's portfolio, but may also be different spillway sections of the same hydraulic structure. Each widget contains three graphs. The first graph illustrates the real-time (current) situation of jet trajectory and rock scour along the 2D profile of the hydraulic structure. The graph makes use of UTC (Universal Time Coordinated) time convention and is generally updated every 5–10 min, depending on the frequency of data transmission performed by the SCADA.

The second graph allows the user to make ahead-of-time projections of scour formation (i.e. Scour Forecasting) by manually inserting estimates of upcoming hydraulic values (i.e. spillway functioning) as a function of time. As shown in Figure 5.6, this corresponds to hydrographs defined as a series of time durations of constant discharge, determined by the flow velocity, flow depth, flow thickness and tailwater level.

The third graph allows the user to visualize both historical scour data and design flood events. The former are obtained by choosing from a scroll-down list the UTC date and time at which the scour state of the rock is requested. The latter allows to visualize the scour potential for a range of pre-computed and stored design flood events. These are typically representing scour potential for given flood return periods, or for the PMF event, and are independent from the real-time and forecast events. They point to standard scour computations that may

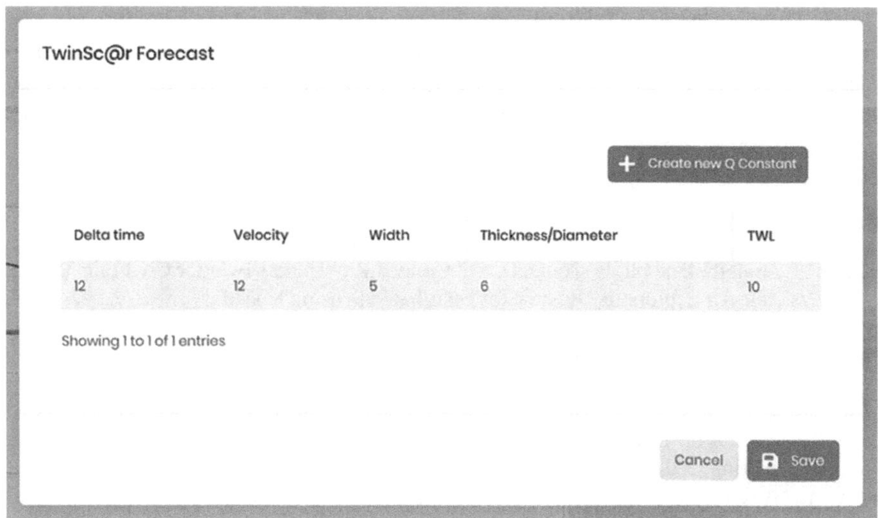

FIGURE 5.6 User interface and main parameters of TwinSc@r for Scour Forecasting.

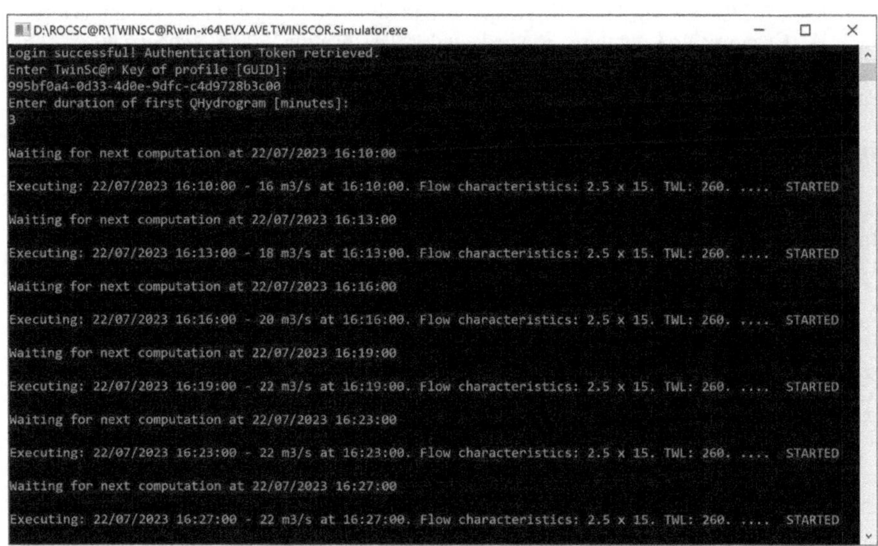

FIGURE 5.7 Spillway hydraulics software component for hydraulic data transmission from SCADA to rocsc@r software.

be adapted whenever plausible, for example, due to a recalibration of the rock mass scour resistance following a flood event.

Figure 5.7 presents an example of a spillway outflow hydraulics software component (see Figure 5.4). This component receives the monitored hydraulic variables from the SCADA (spillway regulator tool in Figure 5.4) and, if needed, computes the input hydraulic parameters needed for the scour modelling, such as flow velocity, flow depth, flow width, angle of issuance, flow turbulence and tailwater level.

Each time a new set of hydraulic input parameters has been received, the software component computes the hydraulic data needed for rock scour and transmits it to rocsc@r. By doing this, it automatically triggers an additional computational time step in the rocsc@r software environment at each time a new set of parameters has been received.

REFERENCES

Bado M.F., Tonelli, D., Poli, F., Zonta, D. and Casas, J.R., "Digital Twin for Civil Engineering Systems: An Exploratory Review for Distributed Sensing Updating", *Sensors*, 22, 3168, 2022. https://doi.org/10.3390/s22093168

Camus, A., "Digital Twin Ensures Safe Operations at Diablo Dam", International Journal on Hydropower and Dams, 29, 4, 36–37, 2022.

Elmo, D., Stead, D., Yang, B., Marcato, G., and Borgatti, L., "A New Approach to Characterize the Impact of Rock Bridges in Stability Analysis", Rock Mechanics and Rock Engineering, 2021. https://doi.org/10.1007/s00603-021-02488-x

Fogel, Y., "A sensitivity analysis for the network connectivity index (NCI) using discrete fracture networks (DFN)", M.Sc., The University of British Columbia, Vancouver, 2022.

ISRM, "Suggested Methods for Geophysical Logging of Boreholes", *International Journal of Rock Mechanics and Mining Sciences & Geomechanics*, 18, 1, 67–84, 1981.

Kritzinger, W., Karner, M., Traar, G., Henjis, J., and Sihn, W., "Digital Twin in Manufacturing: A Categorical Literature Review and Classification", *IFAC-PapersOnLine*, 51, 1016–1022, 2018.

Park, D. and You, H., "A Digital Twin Dam and Watershed Management Platform", Water, 15, 2106, 2023. https://doi.org/10.3390/w15112106

Shafto, M., Conroy, M., Doyle, R., Glaessgen, E., Kemp, C., LeMoigne, J., and Wang, L., "Modeling, Simulation, Information Technology & Processing Roadmap", *National Aeronautics and Space Administration*, 32, 1–38, 2012.

Tao, F. and Qi, Q., "Make More Digital Twins", *Nature*, 573, 490–491, 2019.

Yang, B., Mitelman, A., Elmo, D., and Stead, D., "Why the Future of Rock Mass Classification Systems Requires Revisiting Their Empirical Past", Quarterly Journal of Engineering Geology and Hydrogeology, 55, 1, 2022. https://doi.org/10.1144/qjegh2021-039

6 Digital Models for Rock Scour

INTRODUCTION

This chapter presents models that allow a digitalization of rock scour engineering, and ultimately allow the creation of rock scour digital twins. The models are discussed from both the physical and the digital point of view. The basic physical processes or characteristics are discussed, followed by the implementation of these processes and characteristics into a numerical grid or a numerical model, such that the corresponding modelling workflow may become digitalized.

Some of the presented models have been specifically developed within the framework of the present work and related digital environment, others are existing models that are freely or commercially available and that may be easily exploited for digital rock scour and digital twins.

To create a complete digital engine for rock scour, the following building blocks are essential:

- a digitalized flow model,
- a digitalized rock mass model,
- a digitalized model for computation of rock scour by coupling the flow and the rock mass models in space and time, based on a relevant rock break-up mechanism.

The flow model describes the main parameters of the flow action at the water-rock interface, accounting for flow turbulence and flow aeration. The rock mass model implements the characteristics of the intact rock mass, the rock joint network and the characteristics of the joint surfaces. The model for rock scour computations makes use of physics and related constitutive equations to express different rock break-up mechanisms by connecting the flow and rock mass models in space and time.

Hence, lots of different combinations of these building blocks are possible. For instance, a digital rock scour model may focus on the constitutive equations for scour formation, but may simplify the way the flow and/or the rock is represented. Another digital rock scour model may be very refined in flow and rock mass modelling, but implements the scour computations in a basic manner.

The following subchapters present some of the currently available or recently developed flow and rock mass models, as well as a series of rock scour computational models. The strengths and weaknesses of these models are pointed out.

DOI: 10.1201/9781003319610-6

DIGITALIZED FLOW MODELS

This subchapter presents digitalized flow models that are suitable to be implemented in digital rock scour analysis and on digital platforms. Some models may simulate free surface flows, such as encountered in plunge pools and in rivers or spillway channels, while other models simulate transient pressurized flows, such as occurring inside rock joints.

THE 2D PLUNGING JET MODEL

The 2D plunging jet model represents a digitalization of the theoretical equations governing the trajectory and the diffusion of aerated turbulent jets in a vertical (*X–Z*) plane, from their point of issuance (i.e. spillway) to the water-rock interface.

The 2D Vertical Profiles

The digital scour computations are performed within 2D geometrical profiles that have local coordinates in the *X–Z* plane, i.e. vertically oriented profiles that contain the dam and spillway geometry as well as the downstream rock mass, under the form of a longitudinal section. These profiles must be generated before start of the computations, by determination of minimum and maximum *X*-coordinates and *Z*-coordinates, as well as by determination of the initial (and eventually scoured or observed) water-rock interface and the related numerical grid with depth in the rock mass.

Figure 6.1 illustrates examples of 2D vertical profiles that have been determined in a 3D geometrical situation of an arch dam and its downstream plunge pool. The profiles can be oriented freely, for example, following the river axis (profile 1), or following the dam and spillway concentric curvature and geometry (profile 2).

Furthermore, a coordinate system must be defined. For the *Z*-axis, the most convenient way is to adopt the *Z*-altitudes following the national coordinate system. For the *X*-axis, local coordinates should be used. The origin of the *X*-axis can be freely chosen, for example, at the crest or the toe of the dam spillway. Moreover, the following elements must be defined:

- Computational grid size in *X* and *Z* directions
- Point of issuance of the turbulent flow
- Initial water-rock interface at each 2D vertical profile, based on the chosen local coordinates
- Scoured water-rock interface at each 2D vertical profile (optional)
- Non-erodible ranges of points in *X*-, *Y*- and *Z*-coordinates respectively
- Rock mass lithology in each of the grid points

The 2D Flow Matrix Model

The analytical equations for plunging turbulent jets impacting a flat bottom through a water body at rest allow assessment and storage of the main hydraulic parameters in 2D matrices. These matrices respect the network of computational nodes representing the rock mass in the 2D vertical profiles, along which the scour computations are

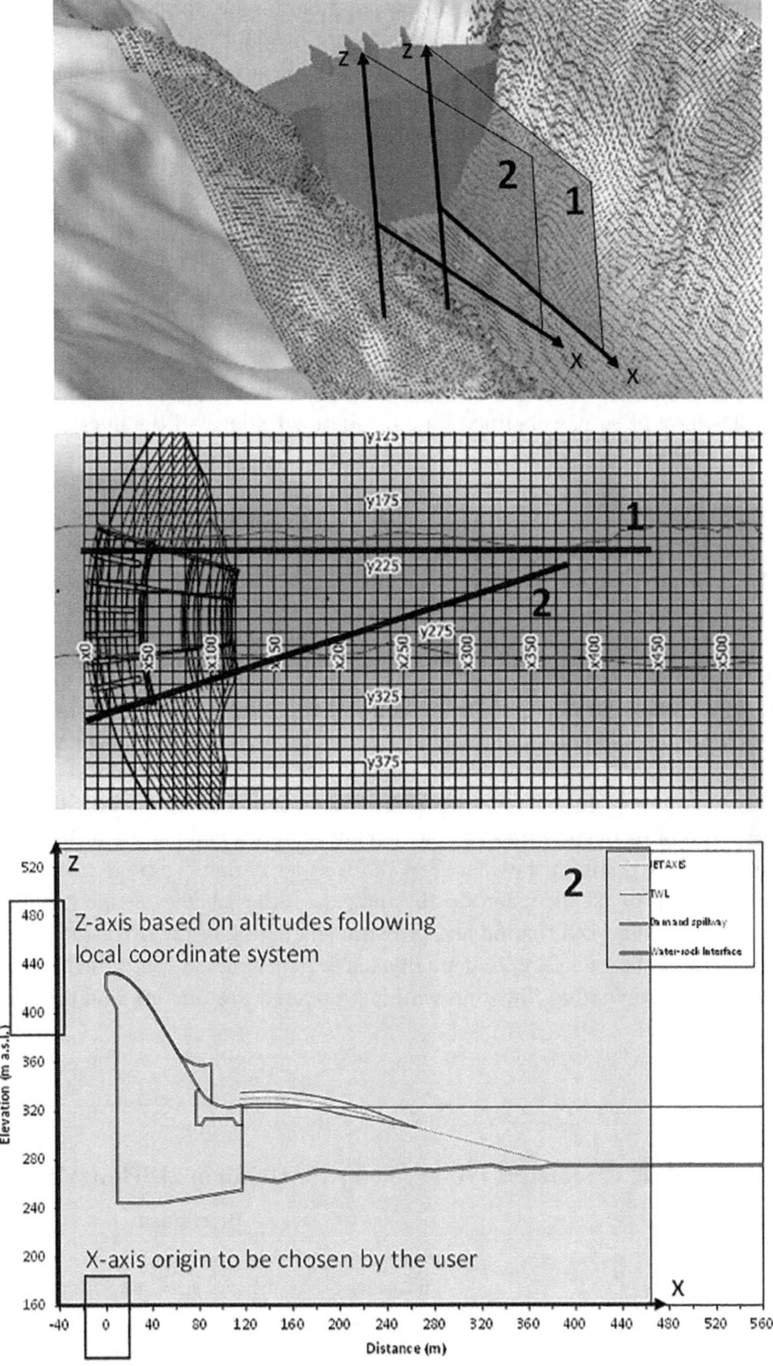

FIGURE 6.1 The 2D plunging jet model: 2D vertical profiles to be defined: perspective view, plan view and longitudinal profile view.

being performed. As such, in every computational node, the hydraulic parameters that are needed for the scour computations are available at start. The values automatically account for the progressively increasing flow depth during scour formation and do not change during the computations.

Table 6.1 presents the hydraulic parameters that are pre-defined along a computational grid and stored in 2D matrices, based on analytical solutions of the flow environment.

Figure 6.2 illustrates the mean dynamic pressure values generated by an impacting turbulent jet. These values are computed following a 2D network of nodes, and the corresponding values are stored in a 2D matrix. As shown, the graphical interface allows to visualize the values in two different ways:

- 2D heatmap of values illustrating the 2D matrix using a colour code
- 2D curve of values obtained for a pre-defined Z-level of the interface

These values are based on parametric settings to be made for the different analytical equations behind the computations. These settings and their theoretical background are discussed hereafter.

The flow model computes the 2D matrices and visualizes them for verification through the interface.

Plunging Turbulent Flows

Plunging turbulent flows issuing from dam spillways impact the water-rock interface by developing a ballistic flow trajectory under an angle of ~20°–90°. Whenever a downstream water cushion is present, these flows generate a diffusive shear layer by progressive mixing of the impacting high-velocity flow with the fluid at (quasi) rest.

This type of turbulent flows has been extensively studied and described in literature and mainly distinguishes between the free jet region (i.e. region in pool with no perturbation of jet diffusion by the interface), the jet impingement region (i.e. region in pool with perturbation and pressure stagnation of the diffusive shear layer by presence of the interface) and finally the wall jet region (i.e. transformation of impinging and stagnating flow into wall jets oriented towards up- and downstream) (see Figure 6.3).

TABLE 6.1
List of Hydraulic Parameters Pre-defined with Depth in 2D Flow Matrices

Parameter	Description
C_p	Mean dynamic pressure coefficient
C'_p	Fluctuating (RMS) dynamic pressure coefficient
C_{max}	Maximum dynamic pressure coefficient
C_I	Dynamic impulsion coefficient
V	Average flow velocity quasi-parallel to interface (for MQSI)
SP	Average stream power at interface

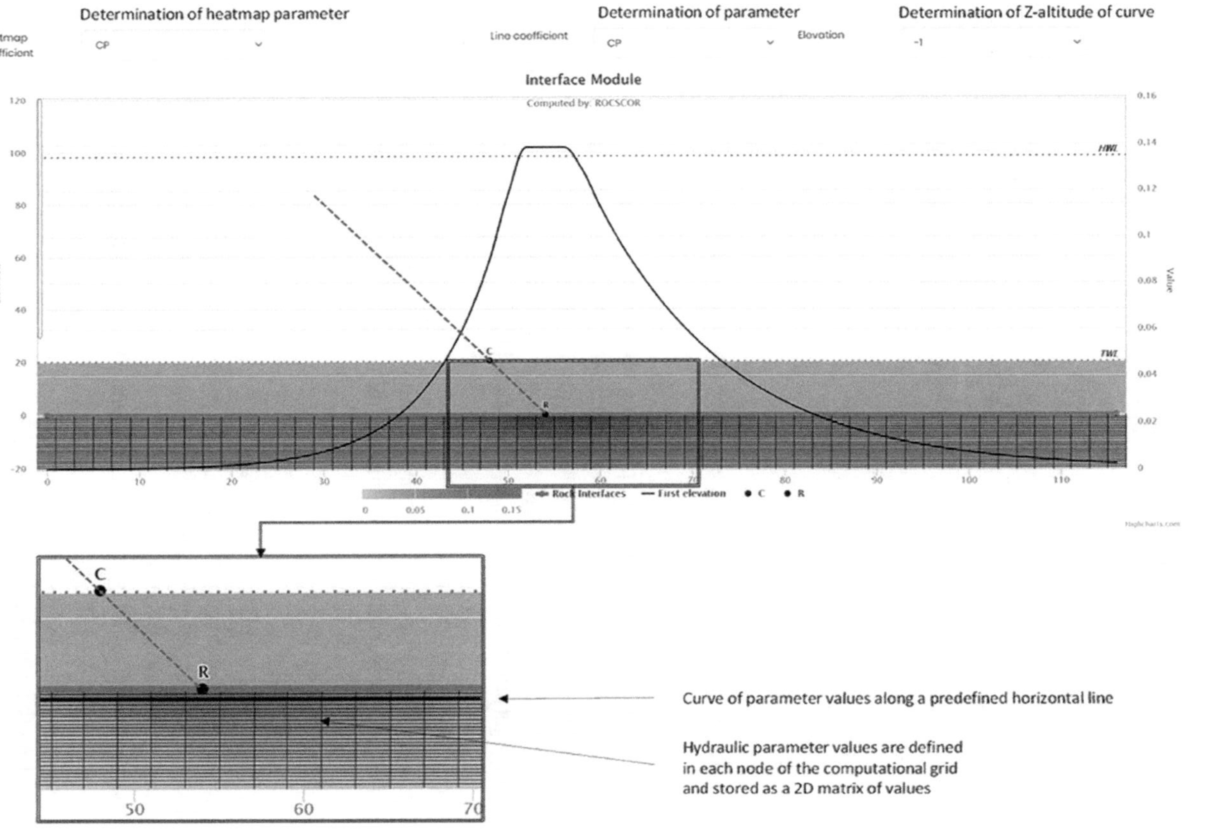

FIGURE 6.2 Interface of 2D flow model showing the 2D matrix of average dynamic pressures generated by a jet plunging into a pool at rest.

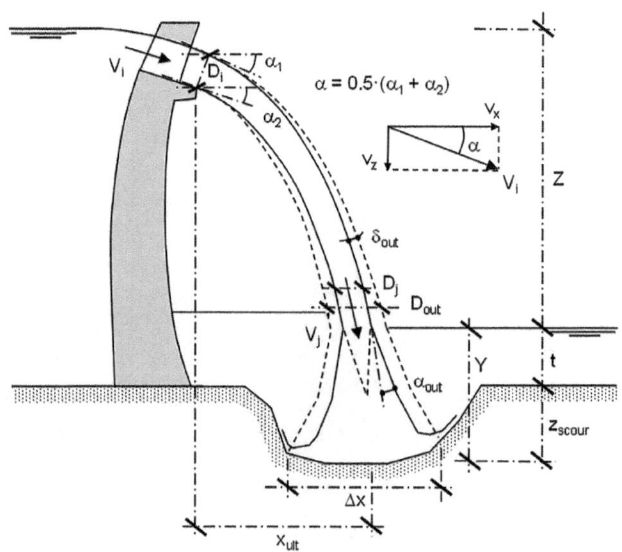

FIGURE 6.3 Main parameters of plunging jet flow (Bollaert 2002).

In the remainder of this document, plunging turbulent flows are termed as "jets". The flow model distinguishes between circular-shaped jets and rectangular-shaped jets, because their diffusion and deformation through the air and through the plunge pool water depth are significantly different. Nappe flows are considered as rectangular jets with a jet width that is very large compared to the jet thickness.

Jet Issuance from Dam Spillways

The issuance of jets from dam spillways depends on the type of spillway structure (free overfall jet, pressurized jet, ski-jump jet) and considers the following parameters:

- average jet velocity
- average jet angle
- jet shape (rectangular, circular, nappe)
- average jet diameter (circular) or jet thickness (rectangular, nappe)
- average jet width (for 3D modelling)
- initial turbulence intensity
- coordinates of spillway crest (lower boundary of jet circumference)

These values are thus not computed by the flow model but have to be determined before starting the computations.

Jet Trajectory through the Air

The trajectory of a free jet through the air is mainly influenced by the following aspects:

- jet issuance (velocity, diameter or thickness, angle, turbulence intensity, shape)
- air drag
- jet deformation (gravitational contraction, turbulent spread, change in shape)

The jet trajectory is typically described by ballistic equations, together with the incorporation of an air drag coefficient K that allows to shorten the trajectory due to air resistance:

$$z = x \cdot \tan(\alpha_i) - \frac{x^2}{4 \cdot K \cdot H \cdot \cos^2(\alpha_i)} \tag{6.1}$$

in which H stands for the head at issuance and α_i for the jet angle at issuance. This equation solely considers the velocity head at the brink of the spillway crest to compute the trajectory, as suggested by Wahl (2008). Typical values for K are between 0.8 and 1.0.

Nevertheless, a slightly different trajectory equation is sometimes used by professionals to describe the jet trajectories at a free overfall, based on both the pressure head h (flow depth) and the velocity head H at the brink:

$$z = x \cdot \tan(\alpha_i) - \frac{x^2}{4 \cdot K \cdot (h + H) \cdot \cos^2(\alpha_i)} \tag{6.2}$$

Based on Wahl (2008), this equation should in principle not be used because generating flat jet trajectories and introducing an error for flows where the water depth at the brink exhibits a static pressure head profile, which should normally be the case for free overfalls. For a static pressure head profile, the pressure head does not contribute to the jet trajectory. The error becomes particularly important at low Froude numbers, i.e. between 1 and 4. Under certain circumstances, however, the pressure head may have an influence as pointed out in Chapter 7. Hence, this factor is available in the model and can be freely defined.

During its fall, the jet is submitted to gravitational acceleration and contracts (Figure 6.3). Jet contraction reduces the initial diameter D_i for circular jets or thickness B_i for rectangular jets as follows:

$$\frac{D_j}{D_i} = \sqrt{\frac{V_i}{V_j}} \tag{6.3}$$

$$\frac{B_j}{B_i} = \frac{V_i}{V_j} \tag{6.4}$$

in which D_j stands for the contracted jet diameter (for circular jet) at impact in the pool, B_j for the contracted jet thickness (for rectangular jets and nappe flows) at impact in the pool and V_j the average jet velocity at impact in the pool.

Second, the jet spreads and gets aerated during its fall. Based on Ervine et al. (1997), the jet spread can be computed based on the initial jet velocity, turbulence intensity, Froude number and jet trajectory length. A similar equation for rectangular jets has been developed by Castillo et al. (2015). Outer jet spread δ_{out} is typically of 3%–4% and is circumferential for circular jets and purely lateral for rectangular jets. Inner jet spread is only about 15%–20% of the outer jet spread (0.5%–1.0%) and may be neglected for engineering purposes. This means that, to determine the diameter or thickness of the core of the jet at impact in the pool, only gravitational acceleration is being considered.

The total (outer) jet diameter D_{out} or jet thickness B_{out} at impact in the pool is obtained by superposing the contracted jet diameter or thickness to the jet spread at impact (Figure 6.4). The area situated in between the outer jet diameter/thickness and the core of the jet is generally highly aerated, while the core of the jet consists of water.

During the modelling, only the contracted jet diameter or thickness is considered relevant to the diffusion of the jet through the plunge pool water depth.

Jets may also deform their shape during their fall. Depending on issuance conditions and initial shape and velocity, compact jets (i.e. shape that is best represented by a circular or square shape) may exhibit a lateral diffusion that is much higher than the spread generated by jet turbulence. This diffusion generates a change in shape towards a much less compact one, i.e. a rectangular flat or curved shape.

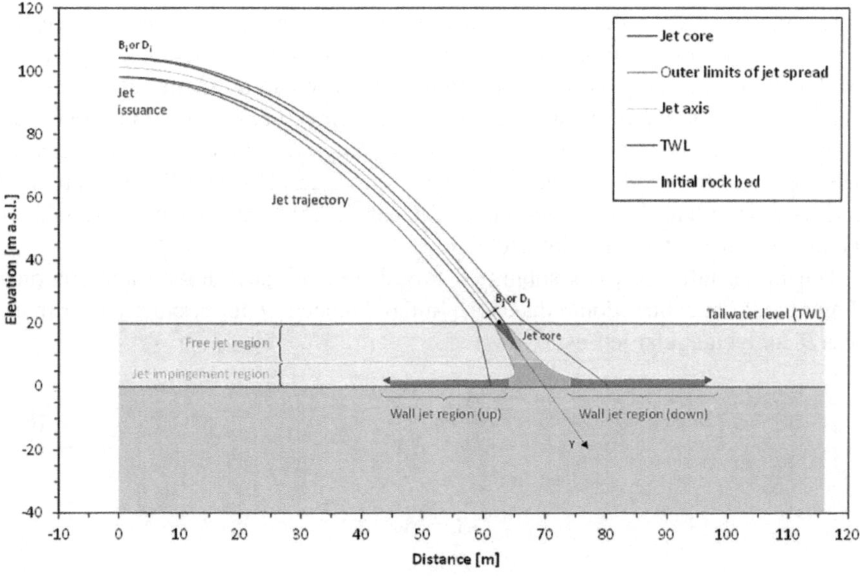

FIGURE 6.4 Sketch of plunging jet diffusion through pool at rest.

The jet diameter at issuance is then transformed during fall into a jet thickness at impact valid for rectangular jets, and the jet contraction should be computed following the equation for rectangular jets and not circular jets.

Finally, the jet core may disappear in the air, before impacting the tailwater. In such a case, the jet is said to be completely disintegrated or broken up into distinct droplets and corresponds to an ensemble of water and air pockets that impact the plunge pool. The trajectory length at which the jet disintegrates is called the jet break-up length L_b. This jet break-up length has been extensively studied and described in literature. The equations used here for circular and rectangular jets are based on the work by Ervine et al. (1997) and Castillo (1989) and express the jet break-up length as a function of the initial turbulence intensity Tu and the Froude number of the jet at issuance Fr:

$$L_b = 1.05 \cdot \frac{D_i \cdot Fr^2}{1.11 \cdot Tu^{0.82} \cdot Fr^{1.64}} \text{[m] for circular jets} \tag{6.5}$$

$$L_b = 0.85 \cdot \frac{B_i \cdot Fr^2}{1.057 \cdot Tu^{0.82} \cdot Fr^{1.64}} \text{[m] for rectangular jets}$$

$$Fr = \frac{V_i}{\sqrt{g \cdot D_i}} [-] \tag{6.6}$$

The degree of break-up of a jet at impact is expressed by the ratio of the jet trajectory length L_j to the jet break-up length L_b for circular jets, or by the ratio of the jet fall height H to the jet break-up length L_b for rectangular jets. For ratios ≥ 1, the jet is fully broken up.

For rectangular jets, Castillo et al. (2015) and Castillo and Carrillo (2017) have determined the dynamic pressures at the water-rock interface as a function of the degree of break-up of the jet. Both compact and broken-up jets are described by the equations. For circular jets, however, only compact jets are covered by the developments made by Ervine et al. (1997) and by Bollaert (2002). As such, in case of broken-up circular jets, the equations developed for broken-up rectangular jets are used during the modelling.

Jet Diffusion through Plunge Pools

Upon impact in the downstream plunge pool tailwater level, jets diffuse through the water cushion of the pool. This diffusion corresponds to a progressively increasing mixing by turbulent shear generated between the high-velocity air-water mixture of the jet and the water at rest in the pool. An aerated turbulent shear layer develops around the core of the jet, following an inner jet diffusion angle of about 8° and an outer jet diffusion angle of 13°–14°.

This progressively reduces the size of the core of the jet. If part of the initial jet core remains, the jet is considered as a developing jet. Once the core completely vanishes, the jet is fully developed in the pool. The longitudinal distance through the pool Y [m] at which the jet becomes fully developed in the pool has been extensively studied and described in literature, and corresponds generally to 4–6 times the initial diameter D_j or thickness B_j at impact (Figure 6.4).

This Y/D_j ratio determines the degree of development of the jet through the plunge pool water depth. The hydrodynamic action of the jet at the water-rock interface is highly dependent on this ratio. This action may be expressed by the time-averaged and fluctuating dynamic water pressures and by the time-averaged velocity or stream power of the jet.

In the following, these hydraulic parameters are assessed separately for rectangular and circular jets, and for compact and broken-up jets. The parameters are determined both along the jet axis and radially outwards from the jet axis.

Jet Parameters at Water-Rock Interface

Mean Dynamic Pressure Along Jet Axis The mean dynamic pressure coefficient C_{pa} [–] is the mean dynamic pressure value P_{mean} [m] along the jet axis (i.e. total pressure head H_{mean} minus the static pressure head Y) divided by the incoming kinetic energy head of the jet $V_j^2/2g$ [m] (simplified by taking $\phi = 1$).

$$C_{pa} = \frac{H_{mean} - Y}{\frac{V_j^2}{2g}} \; [-] \tag{6.7}$$

For compact circular jets impacting vertically, Ervine et al. (1997) defined C_{pa} along the jet axis as a function of Y/D_j and of the air concentration at impact in the pool α_j:

$$\begin{aligned} C_{pa} &= 38.4 \cdot \left(1 - \alpha_j\right)^{1.345} \cdot \left(\frac{D_j}{Y}\right)^2 && \text{for } Y/D_j > 5 \\ C_{pa} &= 0.85 && \text{for } Y/D_j < 5 \end{aligned} \tag{6.8}$$

The air concentration at impact α_j is defined as a function of the volumetric air-to-water ratio β:

$$\alpha_j = \frac{\beta}{1 + \beta} \tag{6.9}$$

Prototype values for α_j are typically of 40%–60% for aerated jets. A physical upper limit for β values is estimated around 2–3. Following equation (6.8), the mean dynamic pressure decreases with increasing air content in the plunge pool. These expressions, however, do not account for jet break-up, which generates unsteady dynamic pressures and additional turbulence at very low frequencies. This may have a significant impact on the mean and fluctuating pressure values.

For broken-up circular jets, no theoretical developments exist. Therefore, the expressions proposed for broken-up rectangular jets, as discussed hereafter, are recommended for such cases. For rectangular jets and nappe flows impacting quasi-vertically, expressions have been developed by Castillo and Carrillo (2017) that cover both compact and broken-up jets, for effective (i.e. $Y/B_j \geq 5$) and non-effective (i.e. $Y/B_j < 5$) water cushions:

$$C_{pa} = a \cdot e^{-b \cdot \frac{Y}{B_j}} \quad \text{for } Y/B_j \geq 5 \; [-] \text{(effective water cushion)}$$

$$C_{pa} = 1 - 0.0037 \cdot e^{5.2484 \cdot \left(H/L_b\right)} \quad \text{for } Y/B_j < 5 \text{ and } H/L_b < 0.85 \; [-] \text{(compact jets)}$$

$$C_{pa} = 0.456 \cdot \left(\frac{H}{L_b}\right)^{-2.393} \quad \text{for } Y/B_j < 5 \text{ and } H/L_b \geq 0.85 [-] \text{ (broken-up jets)} \tag{6.10}$$

in which "a" and "b" are parameters partially defined by Castillo and Carrillo (2017) and defined in Table 6.2 as a function of the degree of break-up of the jet H/L_b.

The expressions developed by Ervine et al. (1997) and Castillo et al. (2015) and Castillo and Carrillo (2017) are based on laboratory experiments for vertically or quasi-vertically impacting jets only. For both circular and rectangular jets under oblique impact, the development length of the jet under water Y should be computed following the angle at impact in the pool. Also, for rectangular jets under oblique impact, it is suggested to replace the fall height H in equation (6.10) by the jet trajectory length L_j, similar to the equation developed for circular jets.

Mean Dynamic Pressure Radially Outwards from the Jet Axis The mean dynamic pressures at points of the water-rock interface that are located radially outwards from

TABLE 6.2
Parameters of Mean Dynamic Pressure for Rectangular Jets

H/L_b	a	≻b
≤0.75	1.74	0.155
0.75–0.85	1.85	0.180
0.85–0.95	1.55	0.175
0.95–1.05	1.40	0.170
1.05–1.15	1.20	0.160
1.15–1.25	1.20	0.175
1.25–1.30	1.13	0.185
1.30–1.35	0.76	0.185
1.35–1.45	0.39	0.190
1.45–1.55	0.29	0.170
1.55–1.60	0.20	0.145
≥1.60	0.16	0.135

Source: Partially based on Castillo and Carrillo (2017).

the jet axis can be described by an exponential decay function of the mean pressure value valid along the jet axis, at the same altitude (i.e. and thus for the same Y/D_j ratio). Two methods are proposed to determine this decay function:

1. *METHOD 1*: decay function based on the oblique distance between the node along the interface and the point of jet impact in the pool (c_j). This function has been developed and calibrated based on case studies of rock scour.
2. *METHOD 2*: decay function based on a range of small-scale laboratory experiments and pressure recordings involving obliquely impinging jets in laboratory plunge pool bottoms.

The proposed methods are valid for both circular and rectangular jets, and for both compact and broken-up jets at impact.

METHOD 1: Radial Pressure Decay Based on Case Studies

For a given altitude Z of the computational node under consideration, corresponding to a fixed ratio Y/D_j along the jet axis, the radial decay from the jet axis is expressed as the horizontal distance between the node and the jet axis divided by a reference distance r_{ref}. This reference distance r_{ref} is determined as follows:

$$r_{ref} = 0.5 \cdot D_j + \tan(\gamma) \cdot Y_{local} [\text{m}] \tag{6.11}$$

in which D_j is the jet diameter (or jet thickness for rectangular jets) at impact in the pool, γ is a calibration parameter and Y_{local} is the oblique distance between the node under consideration and the jet axis at the point of impact of the jet in the pool. Typical values for ($\tan \gamma$) are between 0.5 and 0.8, corresponding to γ values between 27° and 40°.

METHOD 2: Radial Pressure Decay Based on Laboratory Jets

The second method is based on pressure recordings performed during laboratory experiments with obliquely impinging jets in realistically shaped small-scale plunge pools. For a given altitude of the node under consideration, corresponding to a fixed ratio Y/D_j along the jet axis, the radial decay from the jet axis is expressed as a function of the horizontal distance r between the node and the jet axis divided by a reference distance r_{ref} determined as the jet diameter (or jet thickness for rectangular jets) at impact in the pool D_j.

The radial decay is exponential and makes use of four parameters that depend on the angle of impact of the jet in the pool: n_1 and n_2 for radial decay towards upstream and n_3 and n_4 for radial decay towards downstream.

$$\frac{C_{pr}}{C_{pa}} = e^{-n_1 \cdot \left(\frac{r}{D_j} - n_2\right)} [-] \text{upstream of jet impact} \tag{6.12}$$

$$\frac{C_{pr}}{C_{pa}} = e^{-n_3 \cdot \left(\frac{r}{D_j} - n_4\right)} [-] \text{downstream of jet impact} \tag{6.13}$$

Table 6.3 summarizes the range of values to be used for these parameters, for different jet angles. The higher the n_1 and n_3 values, the higher the radial decay of pressure.

Figure 6.5 illustrates an example of radial decay of the mean dynamic pressure, for a jet impinging the plunge pool under an angle of 60°. For method 1, the parameter $\tan(\gamma)=0.50$. For method 2, the parameter values are $n_1=0.30$, $n_2=1$, $n_3=0.10$, $n_4=1$. Due to the oblique impact, the radial decay is asymmetric and biased towards downstream.

Fluctuating (RMS) Dynamic Pressure Along the Jet Axis The fluctuating dynamic pressure coefficient C'_{pa} [−] is the RMS (root-mean-square) value of the pressure deviations from the mean dynamic pressure ($P–P_{mean}$) [m] along the jet axis divided by the incoming kinetic energy head of the jet [m].

$$C'_{pa} = \frac{\text{RMS}(P - P_{mean})}{\frac{V_j^2}{2g}} [-] \quad (6.14)$$

For compact circular jets, Bollaert (2002) defined C'_{pa} along the jet axis as a function of Y/D_j, equation only valid for $Y/D_j < 14$:

$$C'_{pa} = 0.000215 \cdot \left(\frac{Y}{D_j}\right)^3 - 0.0079 \cdot \left(\frac{Y}{D_j}\right)^2 + 0.0716 \cdot \left(\frac{Y}{D_j}\right) + offset [-] \quad (6.15)$$

in which *offset* is a parameter to be chosen by the user between 0.00 and 0.20, based on the initial turbulence intensity of the jet at issuance (Bollaert 2002). Values can be found in Table 6.4. For highly turbulent prototype conditions, an offset of minimum 0.10 is recommended.

An extension of this expression that is valid for $Y/D_j \geq 14$ is proposed here as follows (Bollaert 2021):

$$C'_{pa} = a \cdot e^{-b \cdot \left(Y/D_j\right)} [-] \quad (6.16)$$

TABLE 6.3
Parameter Values to Be Used for Radial Decay Function of Mean Dynamic Pressures

	C_p Radial Decay Parameters							
	n_1		n_2		n_3		n_4	
α [°]	min	max	min	max	min	max	min	max
30–45	0.2	0.4	1	3	0.05	0.10	1	3
45–60	0.2	0.4	1	3	0.10	0.20	1	3
60–75	0.2	0.4	1	3	0.15	0.30	1	3
75–90	0.2	0.4	1	3	0.20	0.40	1	3

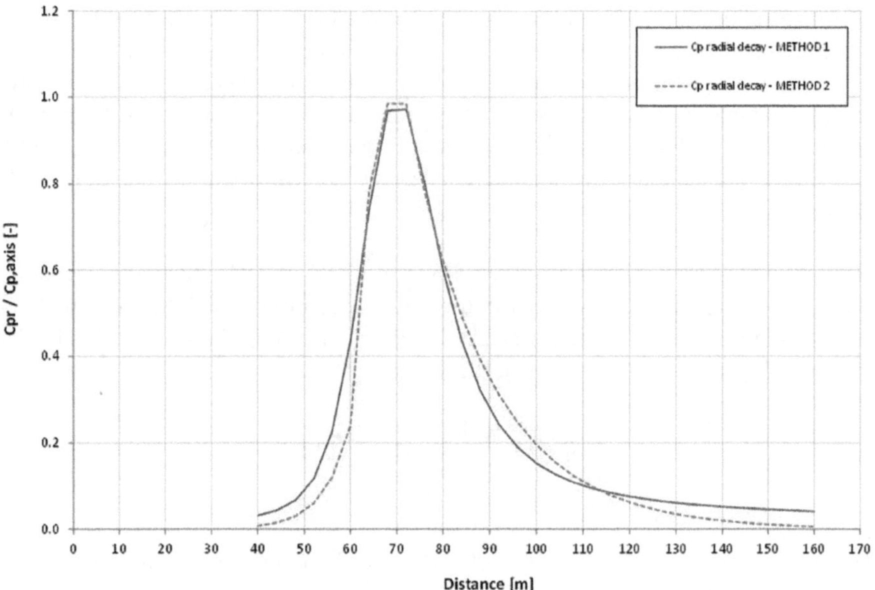

FIGURE 6.5 Example of radial decay for mean dynamic pressure, based on the two methods described, for a jet impinging the plunge pool under an angle of 60°.

TABLE 6.4
Parameter Values to Be Used for Fluctuating Dynamic Pressures of Compact Circular Jets (Bollaert 2021)

Offset in Equation (6.15)	a	b	Tu
[–]	[–]	[–]	[%]
0.00	0.80	0.20	<1
0.05	1.55	0.20	1–3
0.10	1.15	0.15	3–5
0.15	0.78	0.10	5–8
0.20	0.96	0.10	>8

in which the parameters a and b are determined as a function of the chosen offset in equation (6.15), as defined at Table 6.4.

For broken-up circular jets, no theoretical developments currently exist. Therefore, the expressions proposed for broken-up rectangular jets are recommended for such cases.

For rectangular jets, both compact and broken-up, similar expressions have been developed by Castillo and Carrillo (2017) that cover both compact and broken-up jets, for effective and non-effective water cushions. The equation becomes:

$$C'_{pa} = a \cdot \left(\frac{Y}{D_j}\right)^3 + b \cdot \left(\frac{Y}{D_j}\right)^2 + c \cdot \left(\frac{Y}{D_j}\right) + d\,[-] \text{ for } Y/B_j \leq 14 \quad (6.17)$$

$$C'_{pa} = a \cdot e^{-b \cdot \left(Y/D_j\right)}\,[-] \text{ for } Y/B_j > 14 \quad (6.18)$$

in which the parameters a to d, respectively a to b, are determined as a function of H/L_b and Y/B_j as illustrated in Figure 6.6 and given in Table 6.5 (partially based on Castillo and Carrillo (2017) and completed here).

Fluctuating Dynamic Pressure Radially Outwards from the Jet Axis The fluctuating dynamic pressures along water-rock interface nodes that are located radially outwards from the jet axis can be described by an exponential decay function of the fluctuating pressure value valid along the jet axis, at the same altitude and thus for the same Y/D_j ratio. The same two methods than for the mean pressures are proposed to determine this decay function:

1. *METHOD 1*: decay function based on the oblique distance between the node along the interface and the point of jet impact in the pool. This function has been developed and calibrated based on case studies involving prototype jets.
2. *METHOD 2*: decay function based on a range of small-scale laboratory experiments and pressure recordings involving obliquely impinging jets in laboratory plunge pool bottoms.

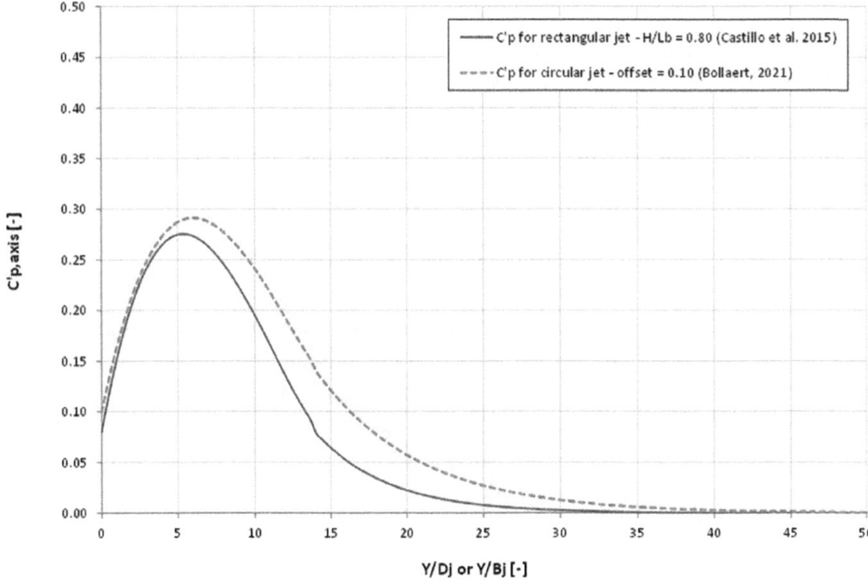

FIGURE 6.6 Example of fluctuating dynamic pressures along jet axis for circular or rectangular jets.

TABLE 6.5
Parameter Values to Be Used for Fluctuating Dynamic Pressures of Rectangular Jets

	$0 \le Y/B_j \le 14$				$Y/B_j > 14$	
H/L_b	a	b	c	d	a	b
≤0.75	0.000300	−0.0086	0.0590	0.1000	0.300	0.120
0.75–0.85	0.000340	−0.0090	0.0500	0.2000	0.400	0.130
0.85–1.05	0.000345	−0.0100	0.0625	0.2300	0.600	0.140
0.95–1.05	0.000258	−0.0077	0.0471	0.2319	0.600	0.140
1.05–1.15	0.000185	−0.0057	0.0334	0.2326	0.700	0.150
1.15–1.25	0.000089	−0.0033	0.0183	0.2290	0.600	0.140
1.25–1.30	0.000060	−0.0025	0.0124	0.2198	0.600	0.150
1.30–1.35	0.000060	−0.0025	0.0124	0.2198	0.600	0.150
1.35–1.45	0.000030	−0.0017	0.0070	0.2100	0.500	0.150
1.45–1.55	0.000030	−0.0017	0.0070	0.2000	0.420	0.150
1.55–1.60	0.000100	−0.0033	0.0165	0.1750	0.340	0.150
≥1.60	0.000200	−0.0054	0.0290	0.1400	1.050	0.240

Source: Partially based on Castillo and Carrillo (2017).

The proposed methods are valid for both circular and rectangular jets, and for both compact and broken-up jets at impact.

METHOD 1: Radial Decay Function Based on Case Studies

For a given altitude Z of the computational node under consideration, corresponding to a fixed ratio Y/D_j along the jet axis, the radial decay from the jet axis is expressed as the horizontal distance between the node and the jet axis divided by a reference distance r_{max}. This reference distance r_{ref} is determined as follows:

$$r_{ref} = 0.5 \cdot D_j + \tan(\gamma) \cdot Y_{local} [\mathrm{m}] \tag{6.19}$$

in which D_j is the jet diameter (or jet thickness for rectangular jets) at impact in the pool, γ is a calibration parameter and Y_{local} is the oblique distance between the node under consideration and the jet axis at the point of impact of the jet in the pool. Typical values for tan γ are between 0.5 and 0.8, corresponding to γ values between 27° and 40°.

METHOD 2: Radial Decay Function Based on Laboratory Jets

The second method is based on pressure recordings performed during laboratory experiments with obliquely impinging jets in realistically shaped small-scale plunge pools. For a given altitude Z of the computational node under consideration, corresponding to a fixed ratio Y/D_j along the jet axis, the radial decay from the jet axis is expressed as a function of the horizontal distance between the node and the jet axis divided by a reference distance r_{ref} determined as the jet diameter (or jet thickness for rectangular jets) at impact in the pool D_j.

The radial decay is exponential and makes use of four parameters that depend on the angle of impact of the jet in the pool: n_1 and n_2 for radial decay towards upstream and n_3 and n_4 for radial decay towards downstream.

$$\frac{C'_{pr}}{C'_{pa}} = e^{-n_1 \cdot \left(\frac{r}{D_j} - n_2\right)} [-] \text{ upstream of jet impact} \tag{6.20}$$

$$\frac{C'_{pr}}{C'_{pa}} = e^{-n_3 \cdot \left(\frac{r}{D_j} - n_4\right)} [-] \text{ downstream of jet impact} \tag{6.21}$$

Table 6.6 hereafter summarizes the range of values to be used for these parameters, for different jet angles. The higher the n_1 and n_3 values, the higher the radial decay of pressure.

Figure 6.7 illustrates an example of radial decay of the fluctuating dynamic pressure, for a jet impinging the plunge pool under an angle of 60°. For method 1, the parameter $\tan(\gamma)=0.50$. For method 2, the parameter values are $n_1=0.30$, $n_2=1$, $n_3=0.10$, $n_4=1$. Due to the oblique impact, the radial decay function is asymmetric and biased towards downstream.

Maximum Dynamic Pressure Along Jet Axis and Radially Outwards The maximum dynamic pressure coefficient $C_{max,a}$[-] is the sum of the time-averaged dynamic pressure, and a Γ^+ amplification of the RMS (root-mean-square) values of the pressure deviations from the mean dynamic pressure ($P-P_{mean}$) along the jet axis, divided by the incoming kinetic energy head of the jet $V_j^2/2g$.

$$C_{max,a} = C_{pa} + C'_{pa} = \frac{(H_{mean} - Y)}{\frac{V_j^2}{2g}} + \Gamma^+ \cdot \frac{RMS(P - P_{mean})}{\frac{V_j^2}{2g}} [-] \tag{6.22}$$

TABLE 6.6
Parameter Values to Be Used for Radial Decay Function of Fluctuating Dynamic Pressures

	C'_{pr} Radial Decay Parameters							
	n_1		n_2		n_3		n_4	
α [°]	min	max	min	max	min	max	min	max
30–45	0.2	0.4	3	3	0.05	0.10	3	3
45–60	0.2	0.4	3	3	0.10	0.20	3	3
60–75	0.2	0.4	3	3	0.15	0.30	3	3
75–90	0.2	0.4	3	3	0.20	0.40	3	3

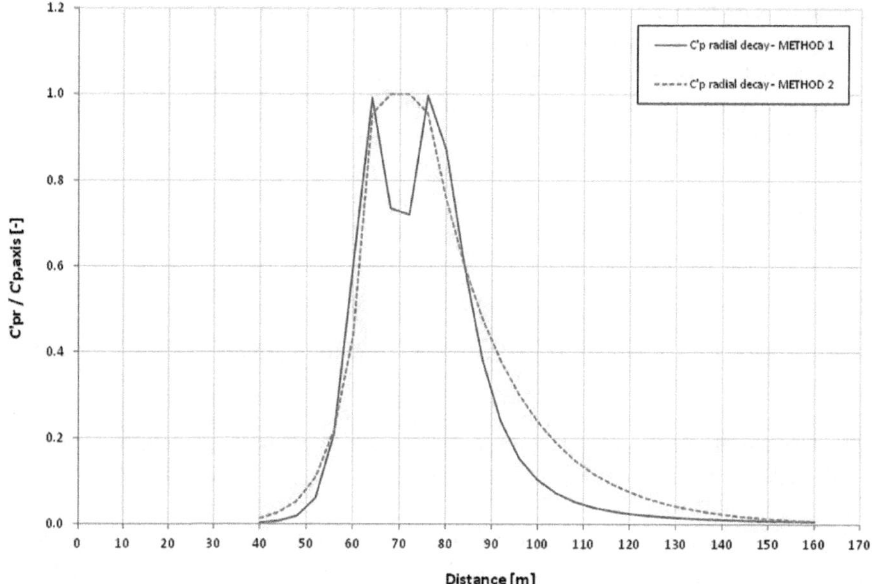

FIGURE 6.7 Example of radial decay for fluctuating dynamic pressure, based on the two methods described, for a jet impinging the plunge pool under an angle of 60°.

This maximum pressure value is used as input to the Comprehensive Fracture Mechanics method. Γ^{+} is computed based on the *gamma* parameter that can be freely chosen and is mathematically determined later on in this chapter, as a function of the Y/D_j ratio of the impacting jet.

Dynamic Impulsion Coefficient Along the Jet Axis and Radially Outwards The non-dimensional dynamic uplift impulsion coefficient C_I is based on large-scale laboratory recordings of pressures over and under a simulated 2D rock block. It allows to determine the net uplift impulsion I_{up} on a single rock block.

A first-hand mathematical analysis as defined by Bollaert (2002) for maximum recorded net uplift pressures and related time durations resulted in the following expression for C_I, valid along the jet centreline and for Y/D_j ratios < 14:

$$C_I = 0.0035 \cdot \left(\frac{Y}{D_j}\right)^2 - 0.119 \cdot \left(\frac{Y}{D_j}\right) + 1.22 \text{ for } Y/D_j < 14 \tag{6.23}$$

in which Y represents the plunge pool water depth [m] and D_j the jet diameter at impact [m].

According to a recent refinement of this mathematical analysis (Bollaert 2021), this equation has been modified and extended in order to incorporate C_I values for $Y/D_j > 14$ in the following manner:

$$C_I = 0.0023 \cdot \left(\frac{Y}{D_j}\right)^2 - 0.105 \cdot \left(\frac{Y}{D_j}\right) + 1.22 \text{ for } Y/D_j \leq 22 \tag{6.24}$$

$$C_I = 0.02 \text{ for } 22 < Y/D_j \leq 25$$

$$C_I = 0.01 \text{ for } 25 < Y/D_j$$

These equations are only valid along the jet's centreline of impact in the plunge pool. For corresponding C_I values at computational nodes located radially outwards from the jet centreline, the C_I values at the jet centreline for the same Y/D_j ratio (same node depth) are being multiplied by the radial reduction factor for RMS pressure coefficients C'_{pr} / C'_{pa} at the radially distant node.

Average Flow Velocity with Depth Through Pool (Along Jet Axis, Vertical and Oblique Impinging Jets) Because of diffusion of the jet through the plunge pool, the average velocity of the jet along its centreline progressively decays. Most studies relate the axial jet velocity decay to the local jet thickness B_j or jet diameter D_j, and to the inverse of the water depth Y. Following Hartung and Hausler (1973), the relationship for V_Z as a function of the jet velocity at impact in the pool V_j is written as:

$$V_z = \left(\frac{Z_{\text{core}} \cdot D_j}{Y}\right) \cdot V_j \text{ [m/s] for circular-shaped jets} \tag{6.25}$$

$$V_z = \sqrt{\frac{Z_{\text{core}} \cdot B_j}{Y}} \cdot V_j \text{ [m/s] for rectangular-shaped jets} \tag{6.26}$$

in which Z_{core} stands for the distance necessary for the jet to diffuse its core through the pool depth. This distance generally corresponds to 4–6 times the jet diameter at impact D_j. An example is provided in Figure 6.8 for a jet with an impact velocity V_j = 40 m/s and a jet diameter or thickness at impact D_j (or B_j) = 1.0 m. The plunge pool water depth is set at an altitude of 210 m a.s.l. The axial velocity decay is much more pronounced for a circular-shaped jet than for a rectangular-shaped jet, because of the axial-symmetric diffusion in the first case.

Based on Bohrer et al. (1998) and experimental and practical evidence, for broken-up rectangular jets, the velocity decay according to circular jets should be used.

Average Flow Velocity at Water-Rock Interface (Vertical and Oblique Impinging Jets) Once the jet deflected, the wall jets generated parallel to the interface may be characterized by their initial flow velocity $V_{Z\,\text{bottom}}$ and their initial thickness $h_{\text{up/down}}$ at the point of deflection. Initiating from this singular location, wall jets develop radially outwards following self-preserving velocity profiles. The resulting local average velocity V_{local}, that is valid at the computational nodes that are situated outwards from the jet axis, is described mathematically as a function of the shape and the angle of impact of the jet.

First, the following equation developed by Beltaos and Rajaratnam (1973), as used by the QSI method, is valid for 2D wall jets that are generated by vertically impinging rectangular-shaped jets:

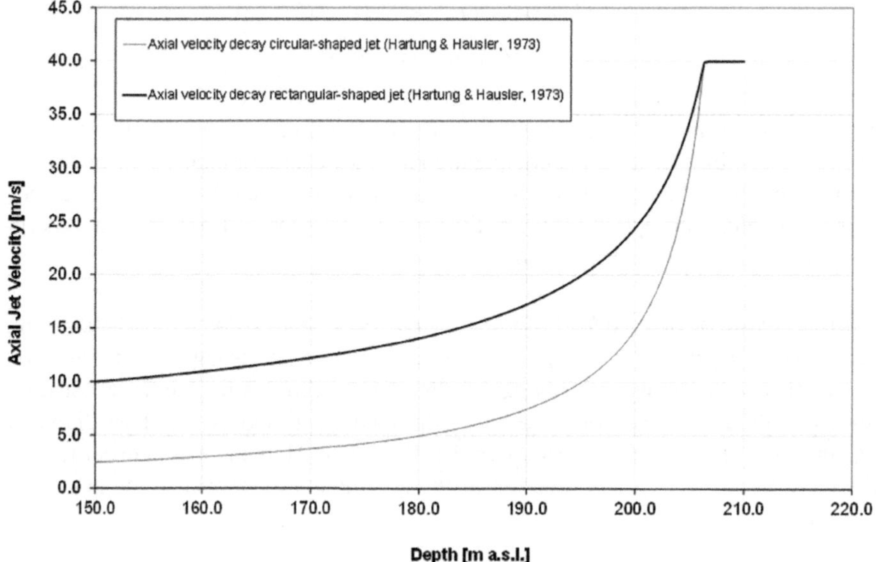

FIGURE 6.8 Axial velocity decay for a circular/rectangular-shaped jet at impact in a plunge pool.

$$\frac{V_{\text{local}}}{V_z} = \min\left(V_z, \frac{3.5}{\sqrt{\dfrac{X}{h_{\text{down}}}}}\right)[-] \tag{6.27}$$

V_z stands for the jet velocity along the jet axis and depends on the development length through the water depth, based on equations (6.25) and (6.26). V_z thus continuously changes (reduces) during scour formation. V_{local} expresses the decay of the cross-sectional jet velocity with the relative distance from the start of the wall jet. For small lateral distances from the impingement point, the wall jet velocity equals the maximum possible value of V_z. Also, with increasing distance from the jet deflection point, the jet velocity profile progressively flattens.

Second, a more generally applicable equation, used by the MQSI method, has been developed by Beltaos (1976a) for both vertical and oblique impinging rectangular jets as follows:

$$\frac{V_{\text{local}}}{V_z} = \frac{\sqrt{5.5\cdot\left(1+\cos(\alpha')\right)}}{\sqrt{ABS\left(\dfrac{r-s}{D_j}\right)}} \tag{6.28}$$

in which α' equals the angle of jet impact in the plunge pool α for downwards-oriented wall jets, and equals $(90° + (90° - \alpha))$ for upwards-oriented wall jets. Furthermore,

ABS $(r-s)$ represents the absolute value of the eccentricity towards upstream that may occur when determining the pressure stagnation point of the jet at impingement onto the water-rock interface. Such an eccentricity typically occurs for oblique impinging jets and is expressed as follows:

$$s = -2.15 \cdot \left(1 - \frac{\alpha}{90}\right) \cdot D_j \tag{6.29}$$

Figure 6.9 illustrates the radial velocity decay with lateral distance from the point of jet impact at the water-rock interface, for vertical (90°) and oblique (60°) rectangular impinging jets.

A similar relationship has been developed by Beltaos (1976b) for circular oblique impinging jets as follows:

$$\frac{V_{\text{local}}}{V_z} = \frac{h_{1,2}}{ABS\left(\dfrac{r-s}{D_j}\right)} \tag{6.30}$$

in which the eccentricity "s" is defined following equation (6.29) and h_1 and h_2 are written as:

$$h_1 = 1.10 \cdot \frac{1+\cos\alpha}{\sqrt{\sin\alpha}} \text{ for downwards oriented wall jets} \tag{6.31}$$

$$h_2 = 1.10 \cdot \frac{1-\cos\alpha}{\sqrt{\sin\alpha}} \text{ for upwards oriented wall jets} \tag{6.32}$$

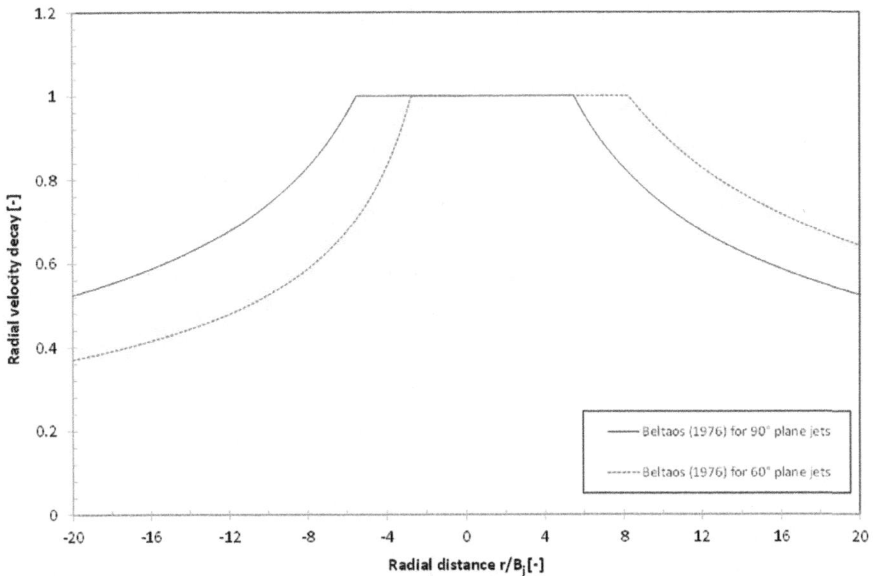

FIGURE 6.9 Radial velocity decay for rectangular-shaped jets: comparison of vertical jet impingement with oblique 60° jet impingement.

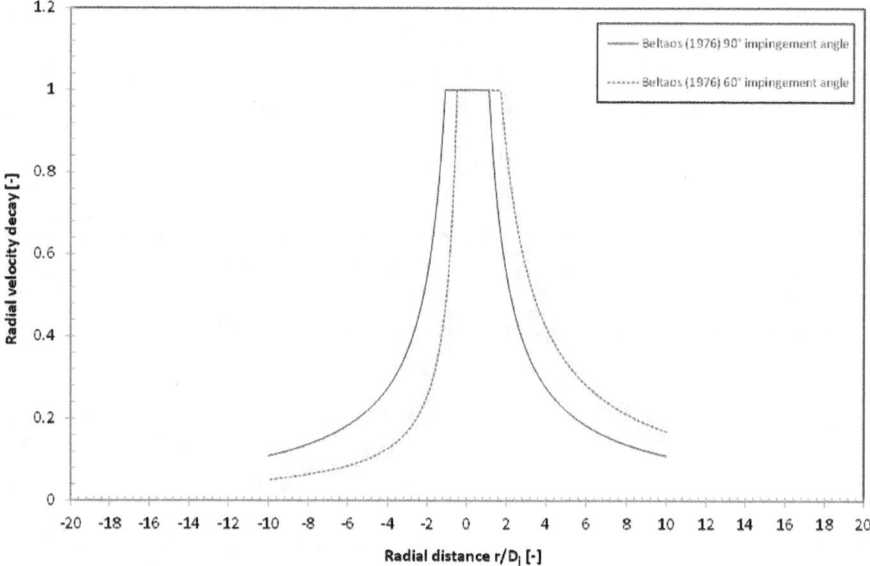

FIGURE 6.10 Radial velocity decay for circular-shaped jets: comparison of vertical jet impingement with oblique 60° jet impingement.

Figure 6.10 illustrates the radial velocity decay with lateral distance from the point of jet impact at the water-rock interface, for vertical (90°) and oblique (60°) circular impinging jets.

It can be observed that, similar to the velocity decay along the jet axis, the radial decay is much more pronounced for circular jets than for rectangular jets.

Applicable Stream Power Along the Water-Rock Interface The last hydraulic parameter to be stored in 2D flow matrices is the applicable stream power generated by the impinging jet at the water-rock interface. Based on the work by Bollaert (2002) and Bollaert and Schleiss (2005), Annandale (2006) related the rate of energy dissipation of the flow to the mean and fluctuating dynamic pressures and average flow velocities acting locally at the water-rock interface. By consideration of the applicable stream power (ASP), i.e. the part of the available stream power that is acting near the water-rock interface due to turbulence production, the following equation can be derived for impinging jets:

$$\text{ASP} = C_{\text{sp}} \cdot \frac{\rho_w \cdot g}{1{,}000} \cdot V_{\text{local}} \cdot \left(C_p + C_p'\right) \cdot \frac{V_j^2}{2g} \ [\text{kW/m}^2] \tag{6.33}$$

in which C_{sp} is a stream power coefficient, ρ_w the density of the water, g the gravitational acceleration, C_p and C_p' the mean and fluctuating dynamic pressure coefficients, V_j the jet velocity at impact in the plunge pool, and V_{local} the local flow velocity acting along the water-rock interface following lateral deviation of the jet. An example is illustrated in Figure 6.11.

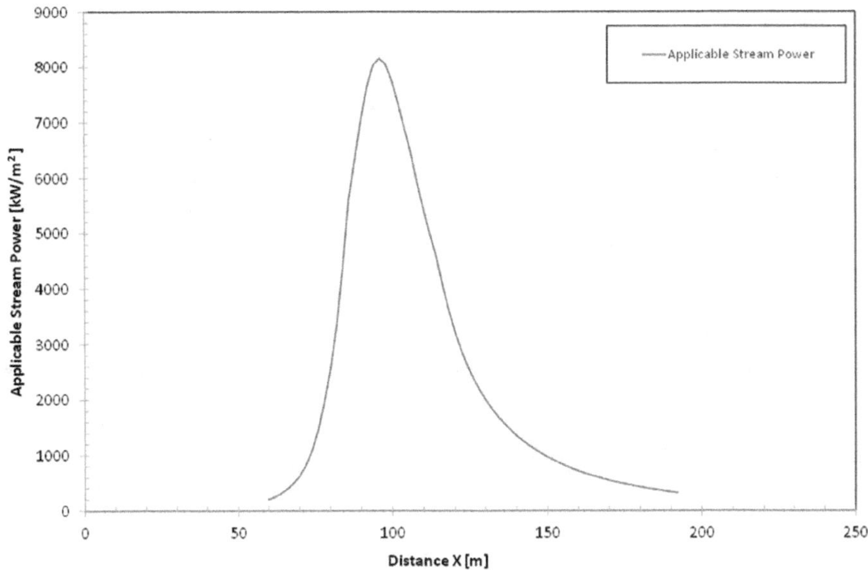

FIGURE 6.11 Applicable stream power at water-rock interface for a jet with a diameter of ~3.5 m and an impact velocity of ~44 m/s into a plunge pool with 20 m of water depth.

The value of the stream power coefficient depends on the ratio of the flow depth to the absolute roughness k_s [mm] of the interface. For rough turbulent flow conditions with a flow depth that is min. 50–100 times the absolute roughness, a typical value for C_{sp} is ~0.40 (Annandale 2006). The ASP decay with increasing scour depth is governed by the corresponding decays with depth of the local flow velocity and the dynamic pressure coefficients.

The Generic Interface Flow Model

The generic interface flow model allows determining the flow parameters at the interface between the water and the rock based on hydraulic computations made by a third-party software, such as for example HEC-RAS, FLOW-3D or any other numerical model for hydraulics.

Initial flow parameters are determined at the water-rock interface by use of a generic exchange file in csv format. Moreover, the flow model proposes customizable mathematical decay functions to automatically account for a change with depth of the flow parameters during the scour process.

The Flow Types

As the flow modelling is performed by third-party software, in principle any type of turbulent flow situation can be handled by the generic flow model, as long as the syntax of the generic exchange file is respected. For practice, the following flow situations are often encountered when dealing with scour at hydraulic structures:

- Turbulent flows under the form of plunging jets impacting the water-rock interface, with or without a plunge pool cushion.
- Turbulent flows over lined or unlined stilling basins, downstream of horizontal/sloped chutes or pressurized gates. This mainly involves submerged or free hydraulic jump flows, as well as wall jet flows.
- Turbulent flows in unlined channels, chutes or rivers, with low to high slopes.

The hydraulic parameters that are available to feed the generic exchange file directly depend on the third-party software used to compute the flow.

The Generic Interface Concept

The generic interface allows to define in a csv formatted file the main hydraulic parameters that represent the turbulent flow at the initial water-rock boundary and that are used by the different computational scour methods, i.e. the mean and fluctuating dynamic pressures, the flow depths, the average flow velocities and the stream power (Figure 6.12).

Furthermore, the generic interface offers the possibility to determine the vertical decay with depth of these hydraulic parameters, based on user-customizable exponential decay functions. As such, the generic interface flow model allows

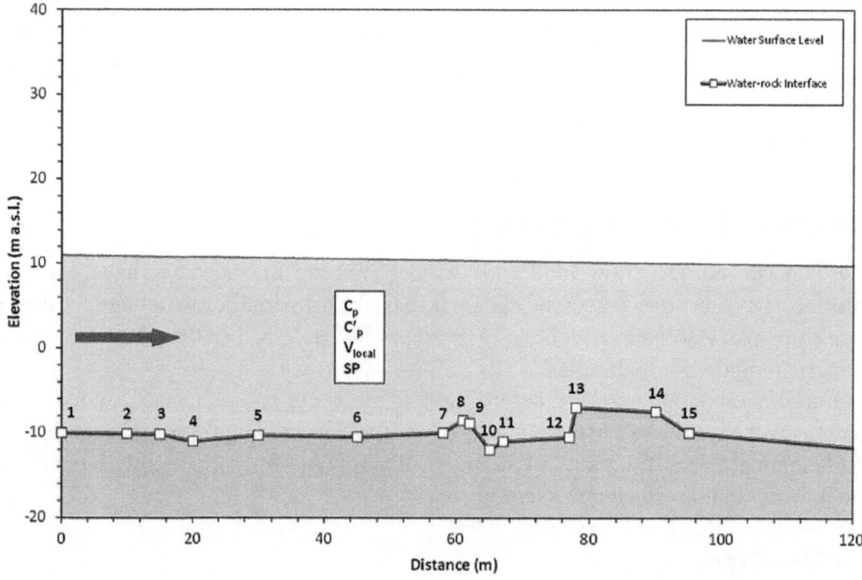

FIGURE 6.12 Generic interface flow model: example showing user-defined interface nodes and hydraulic parameters to be provided at each node.

users to perform scour computations with depth applying one of the following procedures:

1. *STAND-ALONE (NO COUPLING)*: single iteration for constant discharge, based on a csv file containing the hydraulic parameters at the water-rock interface valid at the start of scour formation. Automatic determination of hydraulic parameter evolution with depth based on user-customizable decay functions.
2. *MANUAL COUPLING*: multiple iterations at constant discharge, with manual preparation of csv files containing hydraulic parameters at the water-rock interface for each iteration separately. Limitation of scour formation per iteration. No parametric evolution with depth during scour computations. Stepwise scour computations by repetitive manual feedback and insertion of the scoured rock bottom into a third-party model and rerun of flow hydraulics.
3. *CFD COUPLING*: multiple iterations at constant discharge, with automatic determination of csv files containing hydraulic parameters at the water-rock interface based on third-party CFD software. Limitation of scour formation per iteration. Automatic stepwise scour computations by direct coupling of scour model and CFD model. Manual coupling in case of multiple blocks of constant discharge.

The Generic Interface Application

In the following, an example of application of the generic interface flow model within the rocsc@r digital scour model is presented.

Upon launch of a scour computation with user's choice of an external csv file to define the hydraulic parameters, the computation is held in stand-by mode and is waiting to receive a first iteration to compute. Start of the computations can then be performed in two ways:

1. *External mode based on user-defined csv file (stand-alone or manual coupling)*: The user defines the input file with the hydraulic parameters that are needed for the computations. When clicking the *Run next iteration* button, a window pops up asking the user to load the input (csv) file.
2. *External mode based on CFD defined csv file (FLOW-3D CFD coupling)*: The user leaves rocsc@r in stand-by mode and uses the RemoteSc@r application to pilot FLOW-3D and rocsc@r.

When only performing one iteration, the computation is performed in *stand-alone mode*, i.e. the initial hydraulic values in the csv file are applied with depth using vertical decay functions to be defined by the user. When performing multiple iterations, however, the computation is performed in *manual coupling mode*. The initial hydraulic values at each iteration are in principle not affected by a vertical decay, and the user applies small scour changes per iteration for this hypothesis to remain valid.

For both options based on a user-defined csv file, the following parameters have to be defined to perform the scour computations (Figure 6.13):

1. *Vertical decay coefficient* a_1: Coefficient used in the exponential equation that expresses the vertical decay with depth of average dynamic pressures and RMS pressure fluctuations at the water-rock interface.
2. *Vertical decay coefficient* a_2: Coefficient used in the exponential equation that expresses the vertical decay with depth of average dynamic pressures and RMS pressure fluctuations at the water-rock interface.

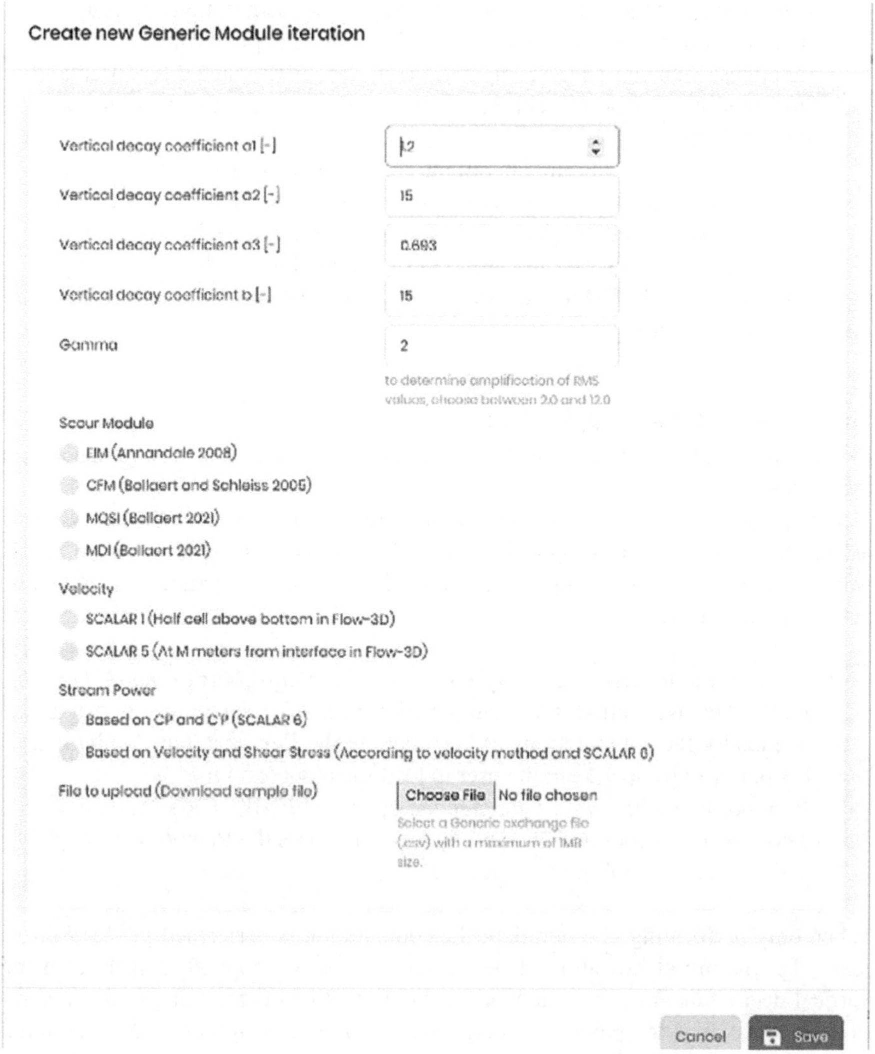

FIGURE 6.13 User interface in rocsc@r for creation of generic interface computation.

3. *Vertical decay coefficient* a_3: Coefficient used in the exponential equation that expresses the vertical decay with depth of average flow velocities at the water-rock interface.
4. *Vertical decay coefficient* a_4: Coefficient used in the exponential equation that expresses the vertical decay with depth of average flow velocities at the water-rock interface.
5. *Gamma*: Coefficient used to amplify the RMS pressure fluctuations inside rock joints, typical values are between 2 and 12.
6. *Scour Module*: Choice of computational module to be used during the scour computations (EIM, CFM, MQSI, MDI).
7. *Velocity*: Choice of average velocity defined as a function of the distance from the water-rock interface. NOTE: this option is for CFD couplings, for other couplings the user should choose the column in the generic exchange file where the average flow velocities have been stored.
8. *Stream Power*: Choice of computational procedure for the Stream Power, defined as a function of C_p and C_p' pressure coefficients or based on product of velocity and shear stress. NOTE: this option is for CFD couplings, for other couplings the user should choose the column in the generic exchange file where the average flow velocities have been stored.

When choosing the computational method, the following parameters must be defined by the user:

1. For fracture mechanics (CFM) method: *maximum time limit for iteration* (hours) = Time duration to be considered for each iteration separately.
2. For the other methods: *maximum scour limit* (m): Value that limits the maximum scour during the computation.

Vertical decay of the hydraulic parameters with increasing scour depth has been determined based on available research and based on real-life calibrated case studies. A summary of values to use for decay parameters is presented in Table 6.7.

For *time-averaged and fluctuating dynamic pressure vertical decay*, from C_{p0} to C_{pz} at a scour depth of z m, a typical value for a_1 is 1 (low decay), with a maximum value of about 10 (for very high decay). The value of a_2 is defined at 15.

$$\frac{C_{pz}}{C_{p0}} = e^{-a_1 \cdot \left(\frac{z}{a_2}\right)^2} \text{ [–] for both time – averaged and RMS values} \tag{6.34}$$

For *flow velocity vertical decay*, from $V_{\text{local},0}$ at start to $V_{\text{local},z}$ at a scour depth of z m, the decay parameters depend on the longitudinal distance from the point of origin upstream of the scour hole. The currently implemented decay model is user-customizable, but does not allow accounting for a point of origin upstream and a longitudinal distance. Based on physical model experiences, for long unlined channels, a typical value for a_3 is of 0.30–0.40, with a minimum value

TABLE 6.7
Parameter Values for Vertical Decay of Dynamic Pressures and Velocities

	Vertical Decay Parameters							
	a_1		a_2		a_3		b	
Case	min	max	min	max	min	max	min	max
No decay	0	0	<>0	<>0	0	0	<>0	<>0
Decay	1	10	15	15	0.2	0.8	5	5

Notes:
1. The decay function assumes that scour forms a local hole compared to the downstream water-rock interface, which locally increases the water depth and generates progressive flow submergence from downstream. For scour cases where generalized deepening of the water-rock interface occurs, such that the hydraulic parameters are not significantly affected and no submergence from downstream occurs, the decay functions are not applicable.
2. The above values have only been validated on a limited number of cases and should be used with caution.

of 0.20 (for very low decay) and a maximum value of 0.80 (for high decay), for an a_4 value of 5.

$$\frac{V_{\text{local},z}}{V_{\text{local},0}} = e^{-a_3 \cdot \left(\frac{z}{b}\right)^2} \; [-] \qquad (6.35)$$

In case of no vertical decay, for example, when performing multiple subsequent iterations with depth during the computations, or in case there is no progressive flow submergence from downstream, the a_1 and a_3 values equal 0, for any a_2 and b values different from 0.

Format of Generic Exchange File

Figure 6.14 shows the format to provide the hydraulic parameters in geometrical points. The model will automatically interpolate at grid points that are not covered by the interface nodes.

The format of the generic exchange file to be provided by the user is a csv file using semicolons, and is explained hereafter.

The first line of the file contains the headings of the different values. From the second to the last line, the following values are provided in columns:

- *Cycle*: Computational cycle used during coupling with FLOW-3D, of no relevance to generic module with external input file generated by the user.
- *Computational time [sec]*: Time at which the values are determined from a FLOW-3D computation, of no relevance to generic module with external input file generated by the user.
- *Longitudinal coordinate X [m]*: Longitudinal coordinate along the *X*-axis, following the water-rock interface, generally taken from the dam toe towards

X distance towards downstream [m]

Comput. cycle [-]	Comput. time [sec]	Longitud. coord. [m]	Transverse coord. [m]	Altitude [m a.s.l.]	Flow Velocity_1 [m/s]	Average pressure [N/m2]	RMS pressure [N/m2]	Flow depth [m]	Flow Velocity_M [m/s]	Shear stress [N/m2]	Depth_Avg Flow Vel [m/s]
cycle	time	x	y	z	sclr_1	sclr_2	sclr_3	sclr_4	sclr_5	sclr_6	sclr_7
1801	4.00E+01	5.00E-01	0.00E+00	5.00E-01	1.82E+01	7.33E+04	2.53E+05	2.44E+00	1.82E+01	7.00E+03	1.32E+01
1801	4.00E+01	1.50E+00	0.00E+00	5.00E-01	1.78E+01	7.31E+04	2.45E+05	2.48E+00	1.78E+01	7.00E+03	1.28E+01
1801	4.00E+01	2.50E+00	0.00E+00	5.00E-01	1.73E+01	7.33E+04	2.37E+05	2.52E+00	1.73E+01	7.00E+03	1.23E+01
1801	4.00E+01	3.50E+00	0.00E+00	5.00E-01	1.69E+01	7.36E+04	2.32E+05	2.53E+00	1.69E+01	7.00E+03	1.19E+01
1801	4.00E+01	4.50E+00	0.00E+00	5.00E-01	1.64E+01	7.41E+04	2.27E+05	2.55E+00	1.64E+01	7.00E+03	1.14E+01
1801	4.00E+01	5.50E+00	0.00E+00	5.00E-01	1.60E+01	7.46E+04	2.23E+05	2.65E+00	1.60E+01	7.00E+03	1.10E+01
1801	4.00E+01	6.50E+00	0.00E+00	5.00E-01	1.56E+01	7.51E+04	2.19E+05	2.78E+00	1.56E+01	7.00E+03	1.06E+01
1801	4.00E+01	7.50E+00	0.00E+00	5.00E-01	1.52E+01	7.56E+04	2.15E+05	2.83E+00	1.52E+01	7.00E+03	1.02E+01
1801	4.00E+01	8.50E+00	0.00E+00	5.00E-01	1.48E+01	7.61E+04	2.11E+05	3.81E+00	1.48E+01	7.00E+03	9.80E+00
1801	4.00E+01	9.50E+00	0.00E+00	5.00E-01	1.44E+01	7.65E+04	2.08E+05	3.85E+00	1.44E+01	7.00E+03	9.41E+00
1801	4.00E+01	1.05E+01	0.00E+00	5.00E-01	1.40E+01	7.70E+04	2.04E+05	3.88E+00	1.40E+01	7.00E+03	9.04E+00
1801	4.00E+01	1.15E+01	0.00E+00	5.00E-01	1.37E+01	7.74E+04	2.01E+05	3.90E+00	1.37E+01	7.00E+03	8.67E+00
1801	4.00E+01	1.25E+01	0.00E+00	5.00E-01	1.33E+01	7.78E+04	1.98E+05	3.92E+00	1.33E+01	7.00E+03	8.32E+00
1801	4.00E+01	1.35E+01	0.00E+00	5.00E-01	1.30E+01	7.83E+04	1.96E+05	3.94E+00	1.30E+01	7.00E+03	7.97E+00
1801	4.00E+01	1.45E+01	0.00E+00	5.00E-01	1.26E+01	7.87E+04	1.93E+05	4.95E+00	1.26E+01	7.00E+03	7.64E+00
1801	4.00E+01	1.55E+01	0.00E+00	5.00E-01	1.23E+01	7.91E+04	1.92E+05	4.95E+00	1.23E+01	7.00E+03	7.30E+00
1801	4.00E+01	1.65E+01	0.00E+00	5.00E-01	1.20E+01	7.95E+04	1.90E+05	4.96E+00	1.20E+01	7.00E+03	6.98E+00
1801	4.00E+01	1.75E+01	0.00E+00	5.00E-01	1.17E+01	8.00E+04	1.89E+05	4.96E+00	1.17E+01	7.00E+03	6.67E+00
1801	4.00E+01	1.85E+01	0.00E+00	5.00E-01	1.14E+01	8.04E+04	1.88E+05	4.97E+00	1.14E+01	7.00E+03	6.36E+00
1801	4.00E+01	1.95E+01	0.00E+00	5.00E-01	1.11E+01	8.09E+04	1.88E+05	4.97E+00	1.11E+01	7.00E+03	6.05E+00
1801	4.00E+01	2.05E+01	0.00E+00	5.00E-01	1.08E+01	8.13E+04	1.87E+05	4.98E+00	1.08E+01	7.00E+03	5.76E+00
1801	4.00E+01	2.15E+01	0.00E+00	5.00E-01	1.05E+01	8.18E+04	1.87E+05	4.98E+00	1.05E+01	7.00E+03	5.47E+00
1801	4.00E+01	2.25E+01	0.00E+00	5.00E-01	1.02E+01	8.22E+04	1.88E+05	4.99E+00	1.02E+01	7.00E+03	5.18E+00
1801	4.00E+01	2.35E+01	0.00E+00	5.00E-01	9.91E+00	8.27E+04	1.88E+05	5.99E+00	9.91E+00	7.00E+03	4.91E+00
1801	4.00E+01	2.45E+01	0.00E+00	5.00E-01	9.64E+00	8.32E+04	1.89E+05	5.99E+00	9.64E+00	7.00E+03	4.64E+00
1801	4.00E+01	2.55E+01	0.00E+00	5.00E-01	9.37E+00	8.36E+04	1.90E+05	5.99E+00	9.37E+00	7.00E+03	4.37E+00
1801	4.00E+01	2.65E+01	0.00E+00	5.00E-01	9.12E+00	8.41E+04	1.91E+05	6.99E+00	9.12E+00	7.00E+03	4.12E+00
1801	4.00E+01	2.75E+01	0.00E+00	5.00E-01	8.87E+00	8.45E+04	1.92E+05	6.99E+00	8.87E+00	7.00E+03	3.87E+00
1801	4.00E+01	2.85E+01	0.00E+00	5.00E-01	8.62E+00	8.50E+04	1.93E+05	6.99E+00	8.62E+00	7.00E+03	3.62E+00
1801	4.00E+01	2.95E+01	0.00E+00	5.00E-01	8.38E+00	8.54E+04	1.94E+05	6.99E+00	8.38E+00	7.00E+03	3.38E+00
1801	4.00E+01	3.05E+01	0.00E+00	5.00E-01	8.15E+00	8.59E+04	1.95E+05	6.99E+00	8.15E+00	7.00E+03	3.15E+00
1801	4.00E+01	3.15E+01	0.00E+00	5.00E-01	7.91E+00	8.63E+04	1.96E+05	7.00E+00	7.91E+00	7.00E+03	2.91E+00
1801	4.00E+01	3.25E+01	0.00E+00	5.00E-01	7.69E+00	8.68E+04	1.97E+05	7.00E+00	7.69E+00	7.00E+03	2.69E+00
1801	4.00E+01	3.35E+01	0.00E+00	5.00E-01	7.46E+00	8.72E+04	1.98E+05	6.00E+00	7.46E+00	7.00E+03	2.46E+00
1801	4.00E+01	3.45E+01	0.00E+00	5.00E-01	7.24E+00	8.77E+04	1.99E+05	6.00E+00	7.24E+00	7.00E+03	2.24E+00
1801	4.00E+01	3.55E+01	0.00E+00	5.00E-01	7.01E+00	8.81E+04	2.00E+05	6.00E+00	7.01E+00	7.00E+03	2.01E+00

FIGURE 6.14 Format of generic exchange file.

upstream. The density of the grid points does not necessarily have to be the same than the grid size of the interface.

- *Transversal coordinate Y [m]*: Transversal coordinate along the *Y*-axis, following the water-rock interface (NOTE: not relevant for 2D profile computations). The density of the grid points does not necessarily have to be the same than the grid size of the interface.
- *Altitude Z [m a.s.l.]*: Altitude of the water-rock interface points generally expressed in m a.s.l.
- *Flow velocity [m/s]*: Time-averaged flow velocity defined at the water-rock interface. Velocity taken positive from upstream to downstream.
- *Average dynamic pressure [N/m²]*: Average dynamic pressure of the turbulent flow at the water-rock interface.
- *RMS dynamic pressure [N/m²]*: RMS values of the dynamic pressure fluctuations at the water-rock interface.
- *Flow depth [m]*: Water depth at the water-rock interface grid points.
- *Flow velocity 2 [m/s]*: Time-averaged flow velocity defined at a distance *M* from the water-rock interface. Velocity taken positive from upstream to downstream.
- *Shear Stress [N/m²]*: Shear stress along the water-rock interface.
- *Depth_Avg Flow velocity [m/s]*: Depth-averaged flow velocity above the interface.

The file can be prepared in an Excel spreadsheet and should then be saved as a csv file using semicolons.

The 1D Transient Flow Model

Introduction

The 1D transient flow model describes the propagation of air-water pressure waves through rock joints. Bollaert (2002) mathematically described flows in joints based on the transient flow equations for homogeneous two-component flow mixtures with varying air content. This flow model uses the instantaneous pressure values at the entrance of joints at the water-rock interface as a weak boundary condition. The model provides the instantaneous pressure field around blocks or in a closed-end rock joint. Accounting for the pressure field over the water-rock interface, this allows to define net uplift pressures and impulsions on the blocks. Together with other relevant forces, such as the submerged weight, the friction along the joints and the added mass during block acceleration, block displacements are computed.

Figure 6.15 illustrates the transient pressure approach compared to a time-averaged or instantaneous approach. Time-averaged pressures do not account for time-dependent pressure fluctuations, while instantaneous pressures do. The latter assume a (quasi-) instantaneous propagation of pressures through rock joints, without deformation of

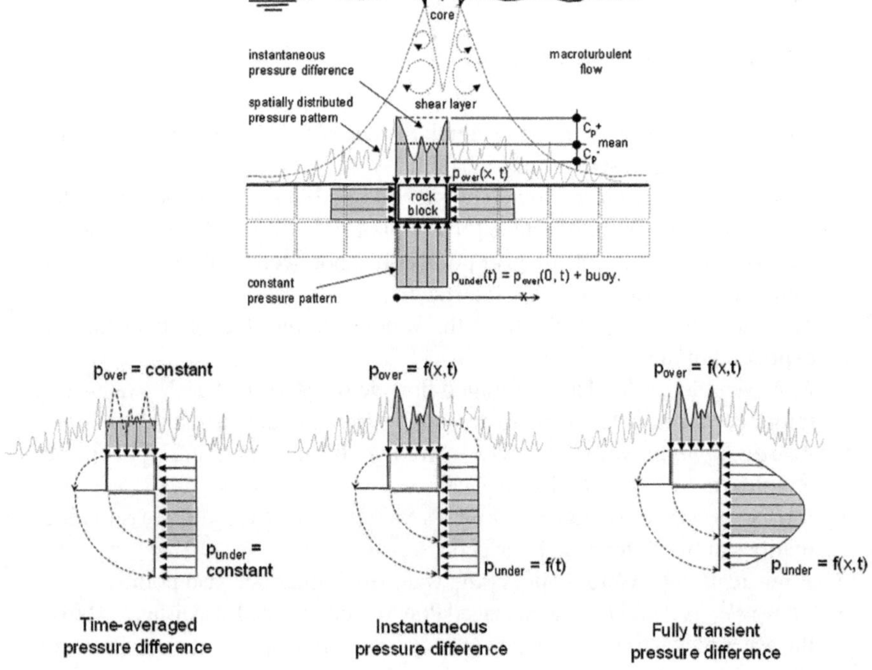

FIGURE 6.15 Dynamic pressure field around a rock block: comparison of time-averaged, instantaneous and transient pressure concepts.

the values. The transient pressure approach uses varying instantaneous pressures at joint entrances to compute pressure wave propagation and superposition. This may become relevant in case of air presence, for which pressure waves travel at very low celerities through the joints (<100 m/s).

The 1D transient flow model is applicable to closed-end or open-end rock joints around single rock blocks, or to 1D joint networks, as illustrated in Figure 6.16 (Bollaert 2002).

The model allows to compute transient pressurized flow phenomena such as wave superposition and amplification, and resonance conditions. Turbulent flow impact on a joint has all elements of a resonator system: the flow provides a periodic excitation and the joint acts as a resonance chamber. The boundaries are formed by the rock. The periodic nature of the flow excitation is defined by its spectral content.

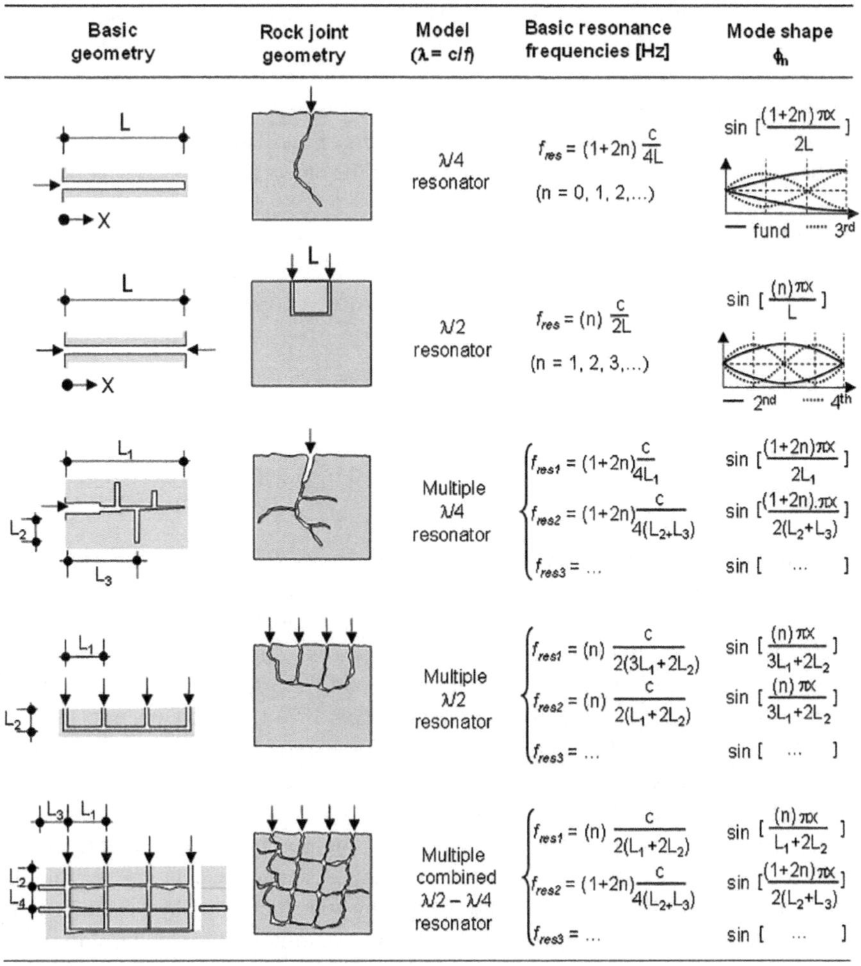

Basic geometry	Rock joint geometry	Model ($\lambda = c/f$)	Basic resonance frequencies [Hz]	Mode shape ϕ_n
L, x		λ/4 resonator	$f_{res} = (1+2n)\frac{c}{4L}$ (n = 0, 1, 2,...)	$\sin[\frac{(1+2n)\pi x}{2L}]$ — fund 3rd
L, x	L	λ/2 resonator	$f_{res} = (n)\frac{c}{2L}$ (n = 1, 2, 3,...)	$\sin[\frac{(n)\pi x}{L}]$ — 2nd 4th
L_1, L_2, L_3		Multiple λ/4 resonator	$f_{res1} = (1+2n)\frac{c}{4L_1}$ $f_{res2} = (1+2n)\frac{c}{4(L_2+L_3)}$ $f_{res3} = \ldots$	$\sin[\frac{(1+2n)\pi x}{2L_1}]$ $\sin[\frac{(1+2n).\pi x}{2(L_2+L_3)}]$ $\sin[\ \ldots\ ]$
L_1, L_2		Multiple λ/2 resonator	$f_{res1} = (n)\frac{c}{2(3L_1+2L_2)}$ $f_{res2} = (n)\frac{c}{2(L_1+2L_2)}$ $f_{res3} = \ldots$	$\sin[\frac{(n)\pi x}{3L_1+2L_2}]$ $\sin[\frac{(n)\pi x}{3L_1+2L_2}]$ $\sin[\ \ldots\ ]$
L_3, L_1, L_2, L_4		Multiple combined λ/2 – λ/4 resonator	$f_{res1} = (n)\frac{c}{2(L_1+2L_2)}$ $f_{res2} = (1+2n)\frac{c}{4(L_2+L_3)}$ $f_{res3} = \ldots$	$\sin[\frac{(n)\pi x}{L_1+2L_2}]$ $\sin[\frac{(1+2n)\pi x}{2(L_2+L_3)}]$ $\sin[\ \ldots\ ]$

FIGURE 6.16 Closed-end and open-end rock joint configurations, resonance frequencies and mode shapes (Bollaert 2002).

This may cause resonance effects whenever part of the spectral content (turbulent eddies) of the flow is situated near the natural frequencies of the joint. At each cycle, additional energy may so be injected into the system. When this periodical energy injection is higher than the periodical energy dissipation, resonance conditions might occur.

Such conditions typically happen at or near the theoretical natural frequencies or eigen-frequencies of the system in question. For joints in rock, two main boundary systems can be distinguished in Figure 6.16:

- the *open-closed boundary system*, relevant to joints that are not fully fractured, $\lambda/4$ – resonator, with resonance frequencies at $f_{res} = (1+2n).(c/4L)$, $n = 0, 1, 2, \ldots$
- the *open-open boundary system*, for joints forming distinct rock blocks or concrete slabs, $\lambda/2$ – resonator, with resonance frequencies at $f_{res} = (n).(c/4L)$, $n = 1, 2, 3\ldots$

in which c is the wave celerity and λ is the wavelength ($=c/f$). Due to the compressibility of the flow mixture, an infinite number of modes of oscillation exist. The first mode of vibration is the fundamental or first harmonic; the others are higher harmonics.

Transient Flow Equations

The one-dimensional transient flow equations are presented in their derivative form for a homogeneous two-component air-water mixture (Bollaert et al. 2001).

$$\frac{\partial p}{\partial t} + \frac{c^2}{g} \cdot \frac{\partial V}{\partial x} = 0 \text{ mass conservation} \tag{6.36}$$

$$\frac{\partial(\eta V)}{\partial t} + \frac{\partial}{\partial x}\left(\beta V^2\right) + g \cdot \frac{\partial p}{\partial x} + \frac{1}{2} \cdot \frac{\lambda}{D} \cdot V \cdot |V|^e = 0 \text{ momentum conservation} \tag{6.37}$$

in which p is the pressure head (m), V the mean velocity (m/s), c the pressure wave celerity (m/s) and D the hydraulic diameter of the rock joint. For a small joint, the hydraulic diameter D may be taken equal to twice the joint thickness. The terms λ, η and β account for steady, unsteady and non-uniform velocity distribution friction losses. They are three parameters to be optimized. The steady friction factor λ is calculated based on the Colebrook-White formula. The unsteady friction factor η depends on the cyclic behaviour of the flow inside the joints. For a uniform velocity distribution, β equals 1 in the convective term of equation. This coefficient has been neglected in the remainder of the analysis.

It is assumed that these friction terms also incorporate other possible energy losses, such as friction due to heat or momentum exchange between the air and the water phase. As such, they cannot be compared with the Darcy-Weisbach friction term that is usually applied for one-phase steady-state flow. Their values are often quite different, due to the particular damping effect generated by the two-phase transient character of the flow (Ewing 1980; Martin and Padmanabhan 1979). For turbulent flow conditions inside the joint, the exponent e has to be taken equal to 1.

However, as a result of the narrow geometry, the Reynolds numbers can be very low (~O (10^2)) and laminar flow might be more plausible under certain circumstances. The corresponding exponent e has then to be taken equal to 0.

The two-component air-water mixture inside the joint is simulated as a pseudo-fluid with average properties and, thus, only one set of conservation equations. The density is hardly modified by the gas and, at relatively small gas contents, may be approximated by the density of the liquid. This means that any possible mass or momentum transfer between the two components is excluded.

Furthermore, no slip velocity or heat transfer between the two phases is considered, so the energy equation is omitted. According to Wylie and Streeter (1978), this simplified approach is valid for air contents of at least 2%. Martin and Padmanabhan (1979) numerically verified the homogeneous flow assumption for air contents of up to 30%, and found correct wave celerities. Therefore, in the approach presented here, the homogeneous flow model has been applied. No further assumption is made regarding the distribution of air throughout the joint, so the wave celerity c is dependent on both time and space.

Air Content Inside Rock Joints

During the Bollaert (2002) large-scale laboratory experiments, determination of the instantaneous local wave celerity inside the modelled joints showed that the instantaneous local volumetric air concentration inside the joints depends on the pressure. The deduced celerity-pressure relationships did not only point out changing air bubble volumes as a function of pressure, as dictated by the ideal gas law, but were also found to be of changing mass of free air as a function of pressure, following Henry's law.

Hence, a constitutive relationship between the instantaneous local celerity $c(x,t)$ and pressure $p(x,t)$ has been defined and optimized by numerical modelling. This relationship replaces any kind of transfer (heat, mass or momentum) that could occur between the air and the water and has the advantage of simplicity. It is dependent on both space and time. A quadratic form matches quite well with the measured data points and is written as follows:

$$c(x,t) = k_1 + k_2 \cdot p(x,t) + k_3 \cdot p^2(x,t) \qquad (6.38)$$

in which k_1, k_2 and k_3 are three numerical parameters. In some rare cases, a double quadratic form revealed to be more appropriate.

Numerical Scheme and Discretization

The numerical scheme that is used to solve a weak formulation of this set of three equations is a second-order finite-volume scheme. As the experimental pressure measurements revealed the appearance of violent transient and highly non-linear wave phenomena, it is obvious that a shock-capturing scheme, introducing a fit amount of numerical dissipation without excessive smearing of the peak pressures, is preferable (Bollaert et al. 2001).

The numerical code defines an unsteady pressure signal as weak upstream condition and imposes a zero flow velocity as weak downstream condition (at the end of the joint). The upstream pressure signal has been taken from the experimental measurements made at the entrance of the rock joint. As the transfer of this pressure signal from the pool bottom to the joint cannot be fully assessed and probably needs some length to be introduced into the joint, a weak formulation of the upstream boundary condition has been chosen. This implies that the upstream boundary condition $p_{up}(t)$ is only applied as outer condition on the upstream finite volume, and not as condition over the whole volume directly. It is believed that this smoothing of the boundary condition is most plausible for the physical situation. The boundary conditions have been presented in Figure 6.17 together with the numerical grid.

Furthermore, an adaptive time stepping has been applied. The criterion that has been used to determine the critical time step is a classical Courant condition, in which the Courant number $C_t = (V + c) \cdot (\Delta t / \Delta x)$ is taken equal to 0.5. This condition is checked at every node of the system, and the most restrictive one is retained for the next time step. The calculations revealed numerical time steps on the order of $1 \cdot 10^{-5}$ to $5 \cdot 10^{-5}$ seconds, i.e. one to two orders of magnitude smaller than the time step of the experimental tests (= 0.001 s at 1,000 Hz acquisition rate).

Celerity-Pressure Relationships for Closed-End Rock Joints

The adjustment of the friction losses parameters λ, η, and of the three k-parameters, is based on the following criteria: mean pressure value, root-mean-square value, maximum and minimum pressure values and finally the histogram and the power spectral density of the computed pressure values. The computed values are systematically compared with the corresponding measured values for different test runs. For the histogram and the power spectral density, a least-square criterion has been applied.

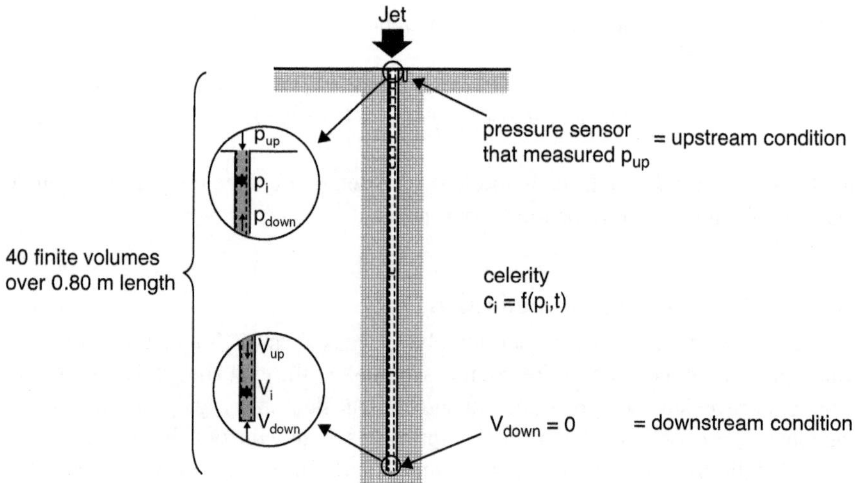

FIGURE 6.17 Definition of the numerical grid and of the upstream and downstream boundary conditions (Bollaert 2002).

For closed-end rock joints, the optimization process has been performed for laboratory pool depths ranging from 0.20 m (core jet impact) to 0.67 m (developed jet impact), and for laboratory jet velocities V_j between 10 and 30 m/s. Test run periods were of 10 s per optimization. This time period was found to procure an appropriate balance between correct numerical analysis and an acceptable computation time.

The optimization is characterized by two stages. In a first stage, a trial-and-error process has been applied. Based on the measured data, an appropriate range of values could be found for the three k-parameters. Within this range, an optimum was then searched for by performing several consecutive numerical runs and by comparing the mean, root-mean-square, maximum and minimum pressure values, as well as the obtained histogram. Although this approach is of subjective character, it allows defining the major tendencies of the celerity-pressure relationships. The second stage of the optimization process involves an automated process based on genetic algorithms. All the parameters could so be optimized.

The best-fit relationships for closed-end rock joints and different jet velocities and plunge pool depths are illustrated and summarized in Table 6.8 (Bollaert 2002).

For prototype closed-end rock joints, the k-parameters equal 20, 100 and 3 respectively. The corresponding relationship is presented in Figure 6.18. The η value is generally close to 1 and may be neglected. The λ value increases with increasing jet velocity and fluctuates between 0.20 and 1.0.

Figure 6.19 presents computed and measured pressures at the entrance and the closed-end of the rock joint, together with the corresponding histograms and power spectral densities of the values. The numerical power spectral content overestimates the higher frequencies.

This is probably due to the absence of thermal dissipation by high-frequency compression and expansion of air bubbles. Such effects have been investigated in the field of wave impact pressures on rock cliffs (Muller et al. 2003). As a result of these particular damping effects, the comparison of the spectral contents of the measured pressures and the numerically calculated pressures has only been performed for Fourier coefficients of up to 200 Hz maximum. The presented numerical solution is in good agreement with the measured power spectral content.

TABLE 6.8
Numerically Derived Celerity-Pressure Relationships (Bollaert 2002)

V_j	Y	k_1	k_2	k_3	λ	η	criterion
m/s	m	–	–	–	–	–	–
14.7	0.20	65	10.0	0.000	0.20	1.00	Histo
14.7	0.67	125	9.0	0.071	0.35	1.00	Histo
14.7	0.67	77.1	12.0	0.021	0.47	1.00	Spec
19.7	0.67	22	10.2	0.047	0.53	1.14	Histo
19.7	0.67	70	5.0	0.143	0.58	1.00	Spec
24.6	0.20	−10.7	0.5	0.173	1.00	0.80	Spec
24.6	0.67	55.7	11.4	0.000	0.71	1.00	Histo
24.6	0.67	10.7	6.4	0.147	0.78	1.00	Spec

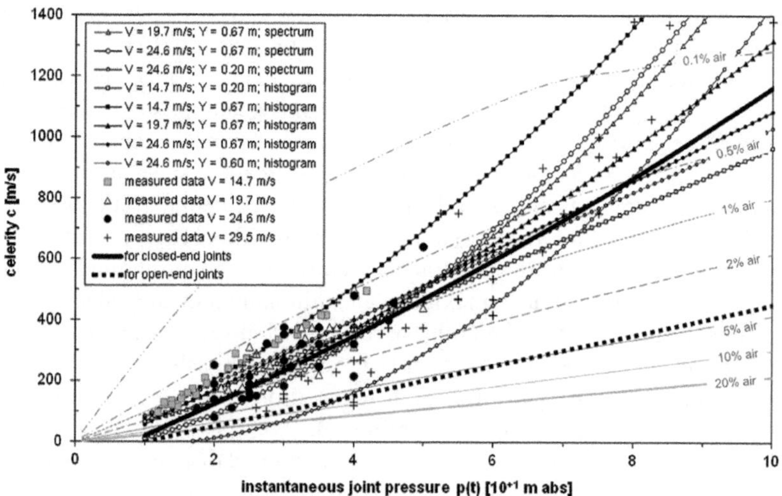

FIGURE 6.18 Numerically derived celerity-pressure relationships and comparison with measured data points for different jet velocities and plunge pool depths (Bollaert 2002).

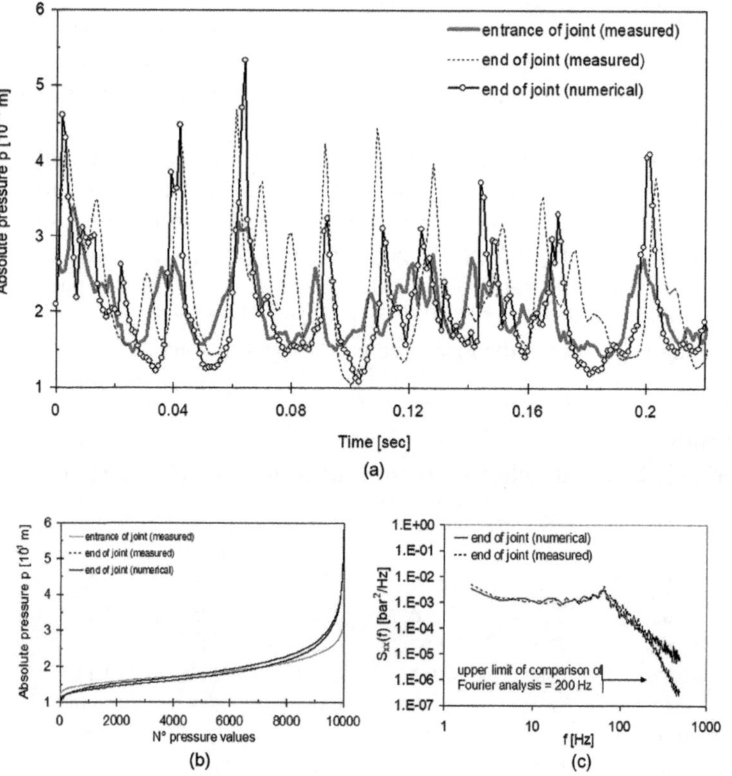

FIGURE 6.19 Comparison of experimental and numerical derived pressures at closed-end joint, for a jet velocity of 19.7 m/s and a plunge pool depth of 0.67 m: (a) pressure signals in the time domain; (b) corresponding histograms; (c) corresponding power spectral densities (Bollaert 2002).

Celerity-Pressure Relationships for Open-End Rock Joints

Similar relationships have been measured and numerically reproduced for open-end rock joints, i.e. joints consisting of two entrances at the water-rock interface and composing a distinct rock block (Bollaert 2003). During these measurements, the pressure underneath the simulated block has been recorded simultaneously with the surface pressures, allowing a direct comparison between measured and computed values.

The measured *c*–*p* relation for this case is given in equation (6.39), involving a lot of free air in the joint (10%–20%). The jet impact velocity is 25 m/s and the friction factor $\lambda = 0.25$. Significant air damping is occurring, reducing the peak pressures that may be observed in closed-end joints. Figure 6.20 compares computed and laboratory-measured pressure fluctuations in a 1-mm-thick open-end joint around an artificial rock block. For prototype cases, the following relationship holds (Bollaert 2003):

$$c(x,t) = 0 + 50 \cdot p(x,t) + 0 \cdot p^2(x,t) \tag{6.39}$$

Figure 6.20 shows that the computed pressures are in quite good agreement with the measured data. The general shape and values of the peak pressures are correctly generated. However, not all pressure peaks are apparent or at the right moment, because the measured pressure signals at the joint entrances are not made exactly at the entrances but somewhat aside (for technical reasons). This automatically induces errors and time lags between measurements and computations.

Turbulent Flow – Pressurized Flow Interface Coupling

The 1D transient flow model accounts for time-varying boundary conditions at the rock joint entrances. These are generally determined based on a free surface flow

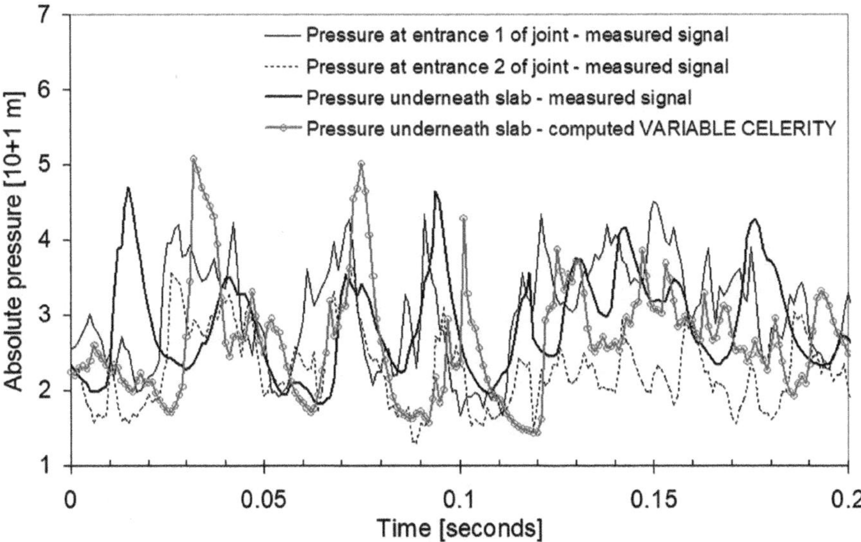

FIGURE 6.20 Measured and computed transient pressures at entrances and underneath distinct rock block (Bollaert 2003).

model of the plunge pool or the river. These flow models allow determining the mean and fluctuating (RMS) values of dynamic pressures at the water-rock interface. Based on these characteristics and together with the spectral content obtained through numerical or physical modelling, representative time-domain pressure signals can be generated as boundary conditions at the rock joint entrances.

Figure 6.21 presents an example of such a coupling for a rock foundation of a bridge pier on a river (Bollaert 2010). It shows the 1D transient pressure signals computed over and under a 0.4-m-high and 1.2-m-long rock block. The block has a protrusion of 0.1 m and is impacted by a turbulent flow in a 1% sloped river with a unitary discharge of $10\,m^2/s$ and an approach flow velocity of 5.1 m/s.

By combining both quasi-steady pressures and turbulent pressure fluctuations, the total dynamic pressure signal at the entrances of the rock joints can be defined. For simplicity, a sinusoidal pressure shape has been used, defined as follows (Figure 6.22):

$$p(t) = \frac{1}{2} \cdot B \cdot \sin(\omega \cdot t) + C \tag{6.40}$$

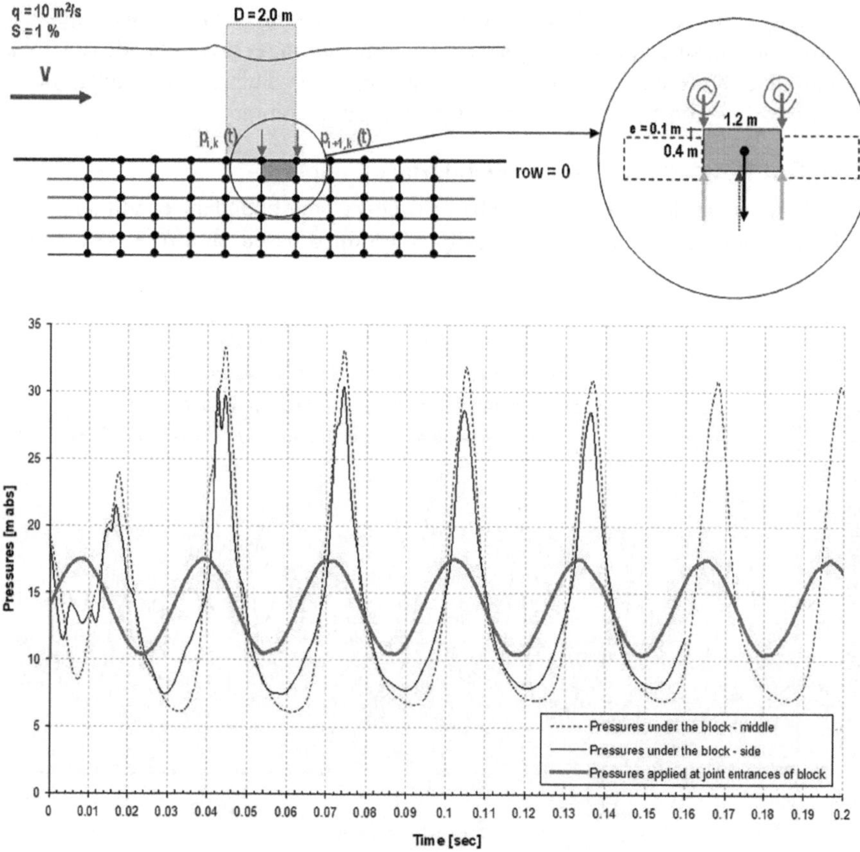

FIGURE 6.21 Sinusoidal (relative) pressure signal applied at joint entrances of rock block.

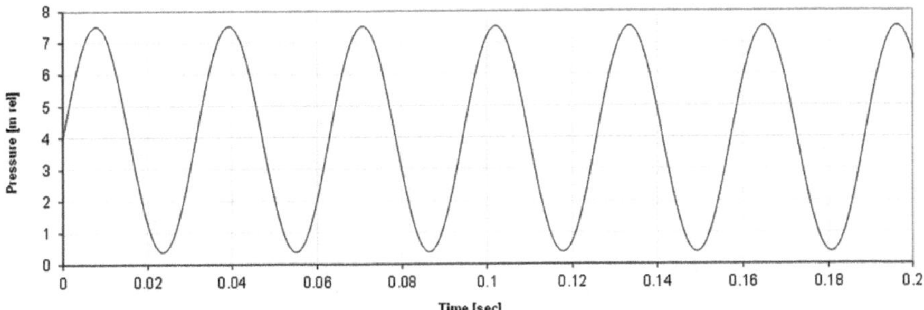

FIGURE 6.22 Example of 1D transient pressures computed around rock block at bridge pier (Bollaert 2003).

in which:

p = relative pressure signal varying with time

t = time duration

$B = p^+$ = maximum positive deviation from quasi-steady pressure value

$C = C_p\, V^2/2g + 0.5 \cdot p^+$

$C_p = 0.29$

$\omega = 2\pi f$, with $f = 32\,\text{Hz}$

The sinusoidal pressure signal has been applied to both joint entrances separating the rock block from the adjacent blocks, without any time lag between both pulses (simultaneous action). Finally, the surface pressure field acting at the surface of the block (in between both joints) has been neglected.

As such, the modelled 1D transient pressure situation may be considered as the most critical one that might be encountered in practice.

2D and 3D Networks of 1D Joints

The 1D transient flow model may be applied on 2D and 3D networks of 1D rock joints, such as illustrated in Figure 6.23.

A 2D network of 1D joints allows describing transient flows in between 2D rock blocks within a vertical plane. Such a model accounts for multiple pressure waves, depending on the characteristics of the blocks. Figure 6.23 illustrates such a network for prismatic blocks (on the left-hand side).

An example is provided in Chapter 3 (Figure 3.28) for a 2D block network where the joints are determined based on DFN modelling (Li and Liu 2010). Applying a Monte-Carlo stochastic method, a DFN was developed, together with 1D transient flow modelling to generate pressure wave propagation within the joints. The authors stochastically generated a fluctuating pressure signal by use of 40 sin and cosine functions with different frequencies and amplitudes. This signal was then locally applied at the entrance of one of the joints along the rock surface. Li and Liu (2010) found significant amplification of pressure fluctuations inside joints, especially at closed-end joints.

A 3D network of 1D joints describes transient flows and wave propagation in between blocks in a 3D space. This is illustrated in Figure 6.23 (on the right-hand

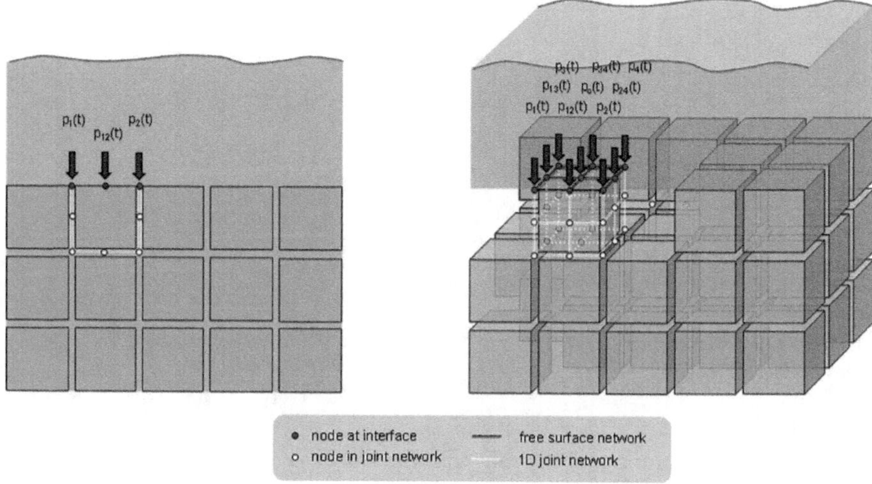

FIGURE 6.23 1D joint networks: LEFT: 2D network of 1D joints (for 2D rock mass); RIGHT: 3D network of 1D joints (for 3D rock mass).

side). Depending on the density of the joint network compared to the block dimensions, such a modelling may become representative for a joint network of 2D joints in the 3D space of the rock mass.

The Computational Fluid Dynamics (CFD) Model

This chapter provides a brief overview of CFD modelling aspects that are directly relevant to coupled rock scour computations, and is based on the work performed by Chesterton et al. (2019) and Vonkeman (2019). The basic equations and assumptions used by the large number of freely or commercially available CFD models are out of the scope of the present work, they can be found for example in Versteeg and Malalesekera (2007).

Introduction

Computational fluid dynamics is a branch of fluid mechanics that uses numerical analysis and algorithms to obtain approximate solutions for problems related to motion of fluids. CFD modelling techniques involve dividing a domain into a grid or mesh and aim at finding a numerical solution throughout this domain, by solving the Navier–Stokes equations with various simplifications. For time-dependent problems the simulation is progressed by solving the governing equations at incremental time steps.

CFD approaches that use a numerical grid or mesh are referred to as mesh-based approaches. On the other hand, meshless CFD methods include particle simulations, such as the Smoothed Particle Hydrodynamics (SPH) method which considers the interactions of large numbers of particles and does not require a computational mesh.

CFD models are frequently used in hydraulic engineering to solve 3D fluid flow problems of various complexity, and provide the ability to simulate full-scale prototype flow conditions which cannot be achieved by physical scale models. Typical hydraulic engineering problems are flows around hydraulic structures, such as dams, spillways, weirs, bridge piers, abutments, etc. As such, CFD models are particularly relevant when it comes to predicting local scour and erosion around such structures.

The codes that are most frequently used today for flow and scour problems at hydraulic structures are predominantly mesh-based and have the capability to reliably predict the location of a free water surface. Codes most commonly in use include Flow Science FLOW-3D, ANSYS Fluent and CFX, Siemens StarCCM++, and OpenFOAM, among others.

Turbulence Models

Turbulence defines the unsteady, aperiodic motions in which all three velocity components fluctuate at high Reynolds numbers. Turbulence models aim at closing the Navier–Stokes equations and solve the turbulent structures by making different assumptions. No single turbulence model is suited for solving all problems, but each has its advantages and associated limitations over the others. The most frequently used turbulence models in hydraulic engineering are briefly discussed hereafter (Versteeg and Malalasekera 2007).

Reynolds Averaged Navier–Stokes (RANS)

RANS equations approximate time-averaged solutions to describe the mean flow field. Commonly used examples are:

- The k–ε model is the most widely used for its simplicity, stability and reasonable accuracy. However, it generally predicts circulating or separated flows with less accuracy. The 2-equation model allows for the turbulent kinetic energy (k) and the turbulent dissipation (ε) to be determined independently. Three variants exist which differ in the way that the turbulent viscosity, Prandtl number and the generation and dissipation terms of ε are calculated.
  - The standard k–ε is robust and economical, giving reasonable results for a wide range of situations, but mediocre results for complex flows with severe pressure gradients and strong streamline curvature. The k–ε model fails to differentiate between, for example, the spreading rate of a circular and rectangular jet (Morgans et al. 1999). Furthermore, it is only applicable to fully turbulent flows.
  - The Renormalization Group (RNG) k–ε gives improved results for streamline curvature and transitional flows and has wider applicability. The RNG model is known to describe low-intensity turbulence flows and flows having strong shear regions more accurately, and converges twice as fast than the standard k–ε model (Karim and Ali 2000).
  - The realizable k–ε further improves complex secondary and separated flows with strong recirculation. It performs better than the other k–ε models for complex flow separation simulations (Vonkeman 2019).

- The k–ω model is another 2-equation model based on the k–ε model whereby the specific dissipation (ω) is determined directly. The model suffers from the same drawbacks with the exception that adverse pressure gradients are better modelled. The model is considered superior to k–ε or RNG models near wall boundaries or flows with streamwise pressure gradients, like plunging jets for example.
- The Shear Stress Transport (SST) k–ω model performs well for adverse pressure gradients and separated flow. The model applies the k–ω approach in boundary layers near walls and a k–ε approach in free streams, making it more robust. The SST k–ω turbulence model was developed by Menter (1994) and simulates flow separation and free shear flows more accurately compared to the other turbulence models. The k–ω model outperforms the k–ε model for boundary layer flow, but is overly sensitive to the free stream value, whereas the k–ε is not (Bosman 2021).
- The Reynolds Stress Model (RSM) requires seven differential equations to be solved, making it more efficient for flows with streamline curvature, circulation, separation or rapid changes in high strain rates. It is described as the most physically complete and superior model for complex flows but requires 2–3 times more CPU effort and time (Vonkeman 2019).

Large Eddy Simulation (LES)

- The Large Eddy Simulation (LES) model is based on space-filtered, time-dependent equations. Large eddies are explicitly solved while smaller eddies are accounted for by subgrids. While the model is highly accurate, it is often described as uneconomic. Furthermore, fluctuations must be initialized and/or input at the inflow boundaries. Despite larger CPU efforts and smaller mesh sized than RANS models, LES models generally provide more information.
- The Detached Eddy Simulation (DES) is a hybrid combination of the RANS and LES turbulence models that attempts to alleviate the uneconomic near-wall meshing requirements imposed by LES by switching to RANS where the turbulent length scale is less than the prescribed maximum grid size.

Direct Numerical Simulation (DNS)

- The Direct Numerical Simulation (DNS) model solves the full unsteady Navier–Stokes equations and can therefore model even the smallest meaningful eddy. The DNS model is the most accurate turbulence model. The DNS model is therefore generally used for small Reynolds number flow simulations, but is also described as uneconomical.

Applicability to Spillway Hydraulics

Many numerical studies have evaluated these different turbulence models on their applicability to different types of structures and flow situations. Researchers that used FLOW-3D to simulate spillways hydraulics are Castillo and Carrillo (2014; 2016; 2017), Savage and Johnson (2001), Avila and Pitt (2008), Chanel and Doering

(2007), Kamanbedast and Aghamajidi (2013) and Epely-Chauvin et al. (2014). The studies that used ANSYS FLUENT to simulate the flow at dam spillways are Karim and Ali (2000), Boroomand et al. (2007), Avila and Pitt (2008), Dey and Eldho (2009), Dasgupta et al. (2011) and Bosman (2021).

In general, the simplest and most widely used turbulence model to simulate flow and scour at spillway hydraulic structures is the k–ε model. The main reasons are because it's fast and robust. Nevertheless, drawbacks are its weakness in modelling strong pressure gradients and streamline curvature, such as typically encountered in aerated jets plunging into a pool and being deviated by the water-rock interface. Castillo and Carrillo (2017) successfully used the RNG k–ε model for the 59-m-high concrete dam of Toachi in Ecuador. Furthermore, Badas et al. (2020) recently used the k–ε model to simulate ski-jump jets and plunge pool flow conditions for Sa Stria Dam in Italy. Following the authors, given the remarkable good agreement between present numerical results and measurements from the physical model, RANS modelling provides a reasonably accurate and robust modelling framework from the engineering perspective.

As will be shown in Chapter 7, k–ε turbulence models can be used for plunging jets, but show some limitations in simulating highly aerated jets and in particular fully broken-up jets.

According to Bosman (2021), from the many numerical studies that have evaluated the applicability of the different turbulence models to simulate plunging jets, the SST k–ω model appears to be the best turbulence model to simulate rock scour due to a plunging jet. Castillo and Carrillo (2017) also made use of the SST k–ω model for the 135-m-high double arch dam of Paute-Cardenillo in Ecuador.

LES models are much less used in spillway hydraulics because of their computational cost and their relative inability to correctly reproduce air entrainment for high-velocity plunging jets. Nevertheless, Chapter 7 will present a successful application of LES to 2D fluid-solid coupled digital scour modelling of plunging jets.

Applicability to Bridge Pier Hydraulics (Vonkeman 2019)

In the field of bridge pier modelling and scour formation, Richardson and Panchang (1998) and Ali et al. (1997) recommended the RNG k–ε model above the standard k–ε model because it requires less reliance on the empirical constants and produces improved results in high shear stress problems. On the other hand, Ali and Karim (2002) state that both provide similar results for the velocity profile. Salaheldin et al. (2004) agreed, despite the commonly perceived weakness of the k–ε model, that it performs satisfactorily in reproducing the velocity profile near the bed, albeit slightly underestimated. However, the k–ε models show some discrepancy with the measured bed shear stress and generally overestimate the area of scour initiation.

Furthermore, even though the realizable k–ε model is considered superior among the k–ε models, it performs the most poorly and substantially overestimates the flow velocity. Salaheldin et al. (2004) recommended that the realizable k–ε model should not be used in bridge pier scour modelling. Salaheldin et al. (2004) concluded that the RSM is the most accurate model in simulating the velocity distribution and shear stress on flat beds and in scour holes.

The selection of an optimum turbulence model requires some degree of compromise between accuracy and economy. Thus, the two equation models are favoured above the RSM model. Menter (1993) evaluated the standard k–ε model, k–ω model and the SST model and established that the k–ε model did not yield as accurate results as the others. Furthermore, he recommended the SST model for its ability to handle adverse pressure gradient flows. Mendoza and Cabrales (1993), Khosronejad et al. (2012) and Xiong et al. (2014) could not resolve the horseshoe vortex which they attributed to the inadequacy of the k–ε model.

Additionally, Constantinescu et al. (2004) believe that RANS models are not suitable for the prediction of junction flows if a detailed study of the coherent structures and frequency spectra inside the horseshoe vortex system is of interest. Khosronejad et al. (2012) acknowledged that the 2-equation turbulence models tend to underpredict the intensity of the horseshoe vortex. To overcome the limitations associated with URANS, they propose the LES model which has been shown to accurately reproduce vortices at bridge piers. Furthermore, LES does not employ wall functions or adjustable constants, and is capable of resolving very fine mesh sizes.

LES models have led to an improved understanding of the flow field around bridge piers. However, their application to coupled hydrodynamic and sediment transport models is challenging due to the excessive computational resources required. Owing to the discrepancy in temporal scales, hydro-morphodynamic simulations using LES or DES are impractical for engineering applications (Khosronejad et al. 2012).

Only RANS models have thus far been used to simulate bridge pier scour because they are generally capable of resolving the primary horseshoe vortex structure. The influence of the secondary horseshoe vortices (captured by LES) on the maximum scour depth is still largely undetermined. Consequently, the less CPU intensive RANS models are considered sufficient.

DIGITALIZED ROCK MASS MODELS

This subchapter presents rock mass models that are particularly suitable for digital rock scour analysis. Some of these models exist since quite some time, and others have been specifically developed within the framework of the rocsc@r digital environment (Bollaert 2021).

Types of Rock Mass Models

According to Lemos (2013), two fundamental options exist for the representation of a jointed or fractured rock mass. The essential difference lies in the manner of representing the discontinuities (faults, joints, fractures, interfaces) and their effects on rock mass behaviour.

On the one hand, we have the equivalent continuum approach, in which a continuum constitutive model is employed to represent in an average manner the effects of the discontinuities. The alternative is the discontinuum approach, in which the discontinuity surfaces are explicitly represented individually. The rock mass is then viewed as a jointed medium, or as an assembly of interacting blocks (or particles).

One of the criteria in deciding whether to use an equivalent continuum or blocky model is based on the scale of the jointing with respect to the size of the excavation or foundation under study (Wittke 1990). As such, closely spaced joint sets may be easily included in a continuum rock matrix behaviour, while major, singular features and faults may have to be considered individually.

In practice, hybrid models are possible combining continuum and discontinuum in the way that best fits the needs of the problem in question. For example, in dam or tunnel failure analysis, it may be sufficient to include a few joint planes of each set to represent the major potential mechanisms. However, the deformability of the block material, if important, should include the contribution of the joints not explicitly modelled.

The Continuum Models

These models represent the rock mass and its discontinuities as an equivalent continuum responding to constitutive equations to determine its behaviour. The most common way to solve this problem is to scale the intact rock properties down to the rock mass properties by using empirically defined relationships.

For example, in tunnel engineering, use is made of the Hoek-Brown criterion for describing the rock mass behaviour. The starting point of the scaling process is the definition of intact rock material parameters, such as the UCS strength (uniaxial compressive strength) which can be obtained by laboratory testing. Then, by using correlations with rock mass indices (i.e. Q, RMR, GSI), the rock mass parameters such as m_b and s_b (rock mass constants according to the Hoek-Brown criterion) or c and ϕ (rock mass cohesion and friction angle respectively) can be estimated. The next step is the adoption of an appropriate constitutive relation for the rock mass. A yield function is often chosen that coincides with the Hoek-Brown failure criterion: (1) the rock mass response is elastic, if the state of stress is within the bounds defined by the yield function; (2) the rock mass response is plastic, once the state of stress is such as to reach the yield function.

As summarized by Startfield and Cundal (1988), numerical methods have been developed over the past few decades for obtaining appropriate solutions to tunnel engineering problems in the framework of the equivalent continuum approach. These numerical methods can be divided into two classes: boundary and domain methods.

The boundary methods comprise several types of boundary element methods (BEM) and imply the subdivision of the boundary of the excavation into elements, as the interior of the rock mass is represented mathematically as an infinite continuum. The domain methods, which include the finite element (FEM) and finite difference methods (FDM) imply that a physical problem is modelled numerically by discretizing (i.e. dividing into zones or elements) the problem region, i.e. the rock mass in which the excavation is to be created.

Continuum models are of interest for problems where the deformability of the rock mass is the main issue, or when the complexity of joint patterns may be difficult to simulate by a block pattern. An application at the engineering scale is the modelling of dam-rock interactions at dam foundations. For normal operating conditions, the aim is that the behaviour of structure and foundation remain essentially in the elastic, reversible domain. Therefore, equivalent continuum models are usually sufficient to represent the rock mass. The main requirement is to represent correctly the deformability of the various rock mass zones, and possibly major singular features. These models are typically sufficient to predict structural displacements during normal operation, and may be validated and calibrated with monitoring data (Lemos 2013).

The Discontinuum Models

Discontinuum models represent the rock mass as an assembly of distinct blocks that interact with each other. These models have primarily been developed to compute failure mechanisms of a rock mass.

The first solution to this problem was the joint element proposed by Goodman et al. (1968), which allowed the explicit representation of individual discontinuities in a FEM of a rock mass. This is a special type of finite element, which is assumed to have zero thickness and employs a joint constitutive model relating the joint normal and shear stresses with the differences in nodal displacements across the elements. FEMs with joint elements can represent a discontinuous medium by use of the conceptual model of a continuum crossed by a few discontinuities where failure may take place. Nevertheless, failure mechanisms involving block rotational modes, with large movements and/or total block separation, are not accounted for.

In a next step, the Distinct Element Method (DEM) has been developed (Cundall 1971). This model views the rock mass as an assembly of distinct blocks or particles in mechanical interaction. These blocks may behave as rigid bodies, a simplification that is not possible with continuum-based FEMs.

The Bonded Particle Model (BPM) is a simplified version of the DEM model and has been developed to study rock fracture propagation paths (Lemos 2013). The basic idea is to represent the rock material as an assembly of disks (in 2D) or spheres (in 3D), to simplify the contact detection between the elements and make faster computations.

Next, Discrete Fracture Network Models (DFN) simulate the joint network of a real rock mass based on a wide range of in-situ monitored data that is analysed and transformed into a 2D or 3D joint network that is generated based on stochastic procedures.

Finally, Synthetic Rock Mass (SRM) models extend the range of particle models to engineering scale, by considering the presence of the discontinuities (Mas Ivars et al. 2011). A discrete fracture network (DFN) is overlaid on a particle assembly, thus partitioning it into a system of grains or blocks formed by bonded circular particles.

Models Including Flow Through Joints

In the field of hydraulic structures, it is essential to simulate the flows and water pressures inside rock joints. The analysis of flow and pressure distributions under dams was a challenge for a long time, and has mainly been implemented based on Darcy's law (seepage flows) applied to continuum rock mass models. Finite difference methods were initially used (e.g., Serafim 1968), followed by finite element models, first assuming uncoupled flow, and later coupled flow-stress analysis. Cases of gravity and arch dams are presented, for instance, in Wittke (1990). Multilaminate models were also used to account for the joint structure (Lamas and Sousa 1993).

The experimental study of flow in fractures by various researchers provided the background for discontinuum numerical representations, either using joint finite elements or within a DEM framework (Damjanac and Fairhurst 2000). Several authors have applied 2D fracture flow models to the study of the water flow under gravity (e.g., Lemos 1999; Gimenes and Fernández 2006), using idealized joint patterns or random joint generation (Barla et al. 2004).

Fracture flow models require much more information than a continuum idealization. For the latter, the average permeability of the various foundation zones is sufficient, while a discontinuous representation of flow implies the specification of joint apertures, as well as realistic values of the joint normal stiffness. The representation of the joint network also requires a good knowledge about the rock mass structure. For dams, this necessary data is often not available.

Application of discontinuum models to dam failure analysis asks for water pressures to be specified in all discontinuities, so that the possibility of sliding is properly checked on the basis of effective stresses. However, these water pressures most often are not the result of fracture flow analysis. Either standard design hypotheses are assumed, or a less demanding continuum flow calculation may be performed, from which the water pressures can be transferred to the mechanical model for safety assessment. It should also be stressed that a fracture flow model requires a much finer representation of the jointing, in particular its connectivity, while for a stability analysis only the main discontinuities that may contribute to define a failure mechanism are necessary.

Application of discontinuum rock mass models to rock scour computations at hydraulic structures has been developed by Bollaert (2002) for single rock blocks with open-end or closed-end joints. The difference with the previously discussed models is that the flow and water pressures inside the rock joints are considered highly fluctuating and transient, instead of constant and/or based on Darcy's law. Furthermore, air entrainment has to be accounted for because of the pressurized flow conditions. This is logic because the main break-up mechanisms of rock layers situated at or near the surface of the rock mass, and impacted by turbulent flows issuing from spillways, are governed by dynamic pressures propagating through the rock joints and fracturing the rock into distinct blocks.

The Comprehensive Scour Model (CSM) (Bollaert 2002; Bollaert and Schleiss 2005) is applicable to plunging turbulent jets and computes 2D scour of fractured rock as a function of the intensity and duration of flows issuing from dam spillways. As will be detailed hereafter, this model is conceptual and based on a series of constitutive equations, but nevertheless considers the representative shape and dimensions of distinct rock blocks during the computations. Furthermore, a modified version, applicable to plucking of regularly shaped rock blocks in channel flows and flows around bridge piers, has been developed in 2010 (Bollaert 2010).

This has triggered the development of other scour models using differently shaped rock blocks over the last decades, such as the BTRE (George 2015) model, the BS3D model (Asadollahi and Thonon 2010), the BSS model (Kieffer and Goodman 2012), the cumulative plucking model (Pan et al. 2014) and so on (see Chapter 3).

Over the last decade, DEM and DFN rock mass models start to be coupled with more or less advanced hydraulic models, such as LBM-DEM coupling (Gardner and Sitar

2019; Gardner 2023), CFD-DEM coupling (PFC^{3D}, Pan et al. 2014), FLUENT-DEM (UDEC, Dasgupta et al. 2012) or DFN-transient flow coupling (Li and Liu 2010).

Nevertheless, these attempts are mostly research-based and limited by computer power and time constraints. No rock scour engineering models currently exist allowing professionals to numerically simulate rock break-up by automated coupling between a flow model and a rock mass model, such that a quasi-instantaneous 2D or 3D result may be obtained for practical engineering purposes.

The Conceptual Model (CM)

The conceptual rock mass model (CM) simulates a real rock mass in a conceptual manner, i.e. by simplifying the physical reality of rock mass and joint networks to a network of nodes following a numerically defined grid in the *XYZ* space. Each node thereby represents the necessary information on the intact rock mass, the rock joint network and finally the rock joint main characteristics. Such a conceptual model is particularly suited for digitalization and allows fast numerical scour computations.

Description

The framework of the conceptual model is illustrated in Figure 6.24. The user first determines the grid size and the grid coordinates in the *X*, *Y* and *Z* dimensions.

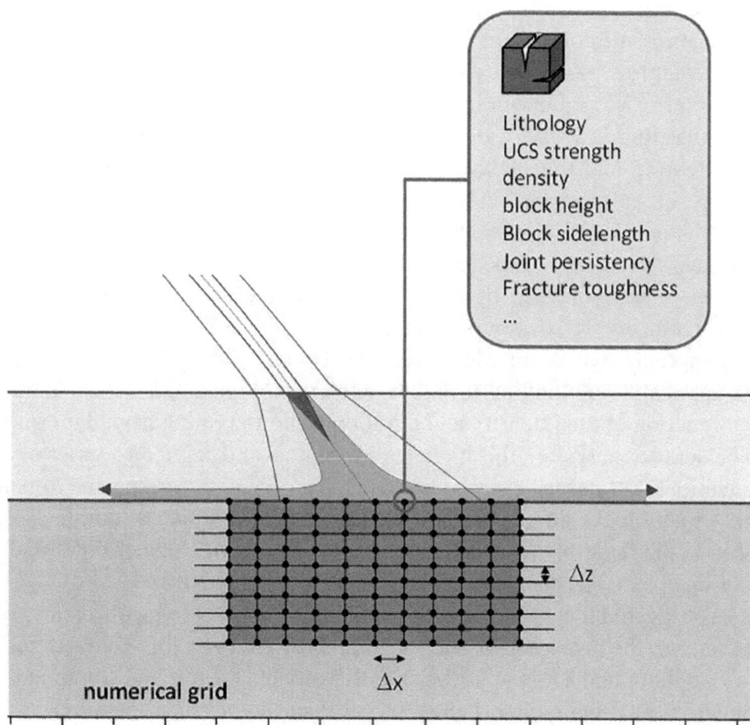

FIGURE 6.24 Framework of the conceptual rock mass model (CM).

This orthogonal Cartesian system fixes a numerical grid of nodes or computational points that will be used during the rock scour modelling process.

Each node thereby represents the local rock mass in that area, by implementation of the necessary information on the intact rock mass, on the rock joint network, and finally on the rock joint main characteristics.

There is a priori no direct relationship between the numerical grid size and shape and the joint network (or rock block dimensions). Nevertheless, knowing that scour modelling obeys the numerical grid and progresses from one grid point to another in a vertically downward direction, it is advisable to choose the vertical grid size in adequacy with the expected rock block vertical dimension (i.e. height), i.e. the vertical grid size should not be larger than the block height, the latter ideally being equal to or a multiple of the vertical grid size.

In case of multiple rock layers and block heights, the smallest block height should be considered to determine the vertical grid size. Also, in case of non-orthogonal and/or inclined joint orientations and/or rock lithologies, a sound estimate may be obtained by considering the horizontal and vertical distances between the centres of mass of the different blocks.

Water-Rock Interface

The water-rock interface is determined based on a 3D topography or bathymetry of the channel or plunge pool bottom. Based on the deployed measurement technique and the chosen sizes of the numerical grid, this may allow the implementation of a certain degree of detail regarding the initial water-rock interface of the modelled rock mass.

Nevertheless, the conceptual model does not integrate these geometrical details. Whenever scour formation progresses with depth through such a model, eroding from one grid point to another, any detailed information on the blocky shape of the interface is lost.

This means that the geometrical interface elements that are relevant to scour formation, such as surface roughness of the rock mass and local protrusion of rock blocks compared to their surroundings, must be implemented as grid point parameters in order to be accounted for during the scour modelling. Furthermore, the relevance of these interface elements on rock scour should be incorporated in the applied rock break-up methods.

Hydrodynamic Action

The hydrodynamic action of the turbulent flow is defined by the hydraulic parameter values valid at the node, and by the global shape of the water-rock interface that connects the water-rock interface (surface) nodes by straight lines (Figure 6.25).

Hence, in case of scour of neighbouring nodes (or blocks), the resulting turbulent flow will be adapted following these straight lines and will not follow the real physical shape of the rock blocks. Detailed information regarding the interface thus gets lost, such as angularity and flow deviating effects (protuberances, protrusions).

Rock Mass Implementation

The geomechanical parameters that are implemented in the nodes are directly related to one of the rock break-up mechanisms as described by the rock scour computational

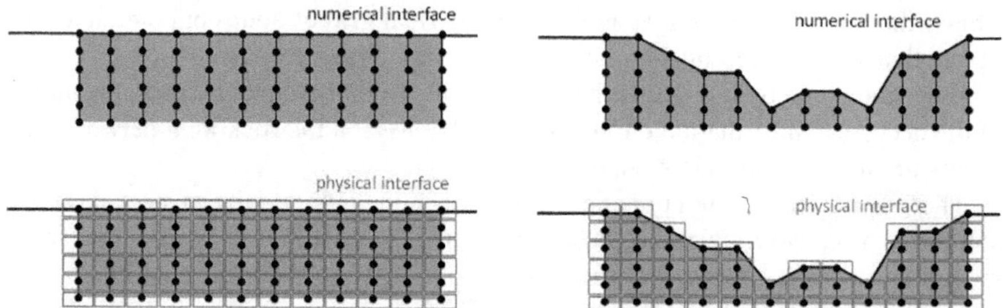

FIGURE 6.25 Hydrodynamic action on nodes of the conceptual rock mass model (CM).

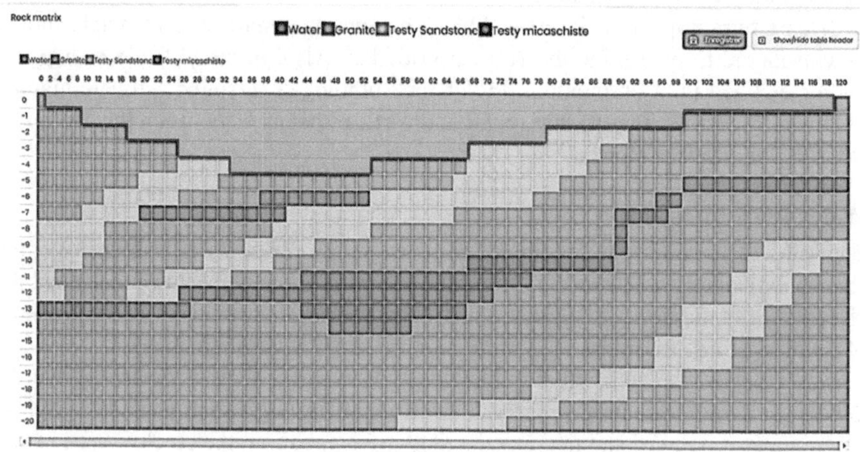

FIGURE 6.26 Grid points of the conceptual rock mass model (CM).

models discussed later on in this chapter. As stated before, the rock mass model is conceptual in the sense that no physical rock blocks are being directly modelled, only their main dimensions and their relevant geomechanical characteristics are being accounted for in the computational nodes. Each of the nodes corresponds to a certain pre-defined lithology. An unlimited number of rock lithologies can be created and their geomechanical parameters pre-defined.

The list of geomechanical parameters that have to be defined for each rock mass layer separately is discussed hereafter. The implementation of the so defined rock lithologies is then performed by attribution of a rock lithology to each grid point of the computational network.

Figure 6.26 illustrates the rock mass lithologies chosen for each of the grid points of the computational network in the *X–Z* plane of a typical computational 2D profile, consisting of 120 columns and 21 rows. Each grid point is represented here by a prismatic block of a certain colour, similar to the physical interface blocks in Figure 6.26.

The different rock mass parameters that must be defined for each of the rock lithologies, as well as the different computational methods that make use of these parameters, are listed in Tables 6.9 and 6.10 and are further described hereafter.

TABLE 6.9
Main Geomechanical Parameters and Relevant Computational Methods (Part 1)

Parameter	Unit	Description	DI/MDI	QSI/MQSI	DP	EIM	CFM
Name	–	Name of layer					
Description	–	Short description of layer (lithology, etc.)					
ρ_r	kg/m^3	Density of rock mass	√	√	√	√	√
UCS	MPa	UCS strength				√	√
M_s	MPa	Mass strength				√	
RQD	%	Rock quality designate				√	
J_n	–	Joint set number				√	
K_b	–	Block size number = RQD/J_n				√	
K_d	–	Discontinuity bond shear strength number = Jr/J_a				√	
J_s	–	Relative ground structure number				√	
z_b	m	Height of block	√	√	√	√	√
x_b	m	*X*-length of block	√	√	√	√	
y_b	m	*Y*-length of block					
L_f	m	Total length of joint around block	√	√			√
EL/SE	–	2D shape of joint					√

TABLE 6.10
Main Geomechanical Parameters and Relevant Computational Methods (Part 2)

Parameter	Unit	Description	DI/MDI	QSI/MQSI	DP	EIM	CFM
Pe	%	Initial degree of fracturing of joint					√
Γ	–	Pressure amplification inside joint					√
m_r	–	Joint fatigue sensitivity					√
C_r	–	Joint fatigue coefficient					√
α_b	°	Joint/block angle with horizontal	√	√		√	
c	m/s	Joint wave celerity	√	√			√
f	°	Joint friction angle	√	√	√		

Name

The name given to the rock lithology in question. It informs about the type of rock and its physical characteristics. Physical characteristics include colour, texture, grain size, and composition. Lithology may refer to either a detailed description of these characteristics or a summary of the gross physical character of a rock. Examples of lithologies in the second sense include sandstone, slate, basalt, or limestone.

Description

A short description of further main particularities of the rock lithology in question, such as for example the main lithological background, mineralogy, schistosity, weathering, etc.

Specific Weight

The density of the rock mass ρ_r is expressed in kg/m³ and relates to the bulk volume, i.e. including the porosity. Table 6.11 illustrates the most often encountered rock densities for different rock lithologies and for mass concrete.

Uniaxial Compressive Strength

The Unconfined or Uniaxial Compressive Strength (UCS) of the rock mass in MPa, as typically obtained from laboratory testing on rock cores. This is one of the most important mechanical properties of the rock mass, i.e. the maximum axial compressive stress that a right cylindrical sample of rock material can withstand before failing. It is used by the Comprehensive Fracture Mechanics (CFM) method during scour modelling.

Typical UCS values range from 20 to 100 MPa for softer rock like shale, mica, sandstones and limestones, to 100–300 MPa for harder rock such as gneiss, granite or quartzite and basalt.

Mass Strength

The mass strength M_s of the rock mass, derived from the UCS strength as follows:

$$M_s = C_r \cdot 0.78 \cdot UCS^{1.05} \text{ for UCS} \leq 10 \text{ MPa} \quad (6.41)$$

$$M_s = C_r \cdot UCS \text{ for UCS} > 10 \text{ MPa} \quad (6.42)$$

TABLE 6.11
Determination of Rock Density as a Function of Lithology

Rock Lithology	Density ρ_r [kg/m³]
Hard solid rock, intact compact rock, compact and dense quartz rock and basalt, other extraordinary hard rocks	2,800–3,000
Very hard granite rock, quartz porphyry, vary hard shale, quartzite, very hard sandy rock, very hard calcite	2,600–2,700
Granite, hard sandstone and calcite, hard conglomerate, hard limestone, marble, dolomite, pyrite	2,500–2,600
Sandstone, medium sandy shale	2,400
Hard mudstone, softer sand rock and calcite, chalky clay	2,300–2,400
Shale, soft limestone, chalk, anthracite, marl, soft conglomerate	2,200–2,600
Compact clay, black coal	1,800–2,200
Concrete	2,400

in which $C_r = \rho_r \cdot g/27{,}000$, expressing the influence of a reduced specific density. This parameter is used to determine the erodibility index method (EIM) following Annandale (2006).

Rock Quality Designate

The Rock Quality Designation (RQD) index was developed by Deere et al. (1967) to provide a quantitative estimate of rock mass quality from drill core logs. RQD is defined as the percentage of intact core pieces longer than 100 mm (4 inches) in the total length of the core (usually 1.5 m). The core should be at least 54.7 mm or 2.15 inches in diameter and should be drilled with a double-tube core barrel. RQD values range between 5 and 100. An RQD of 5 represents very poor quality rock, and an RQD of 100 represents very good quality rock.

Palmström (1982) suggested that, when no core is available but discontinuity traces are visible in surface exposures, the RQD may be estimated from the number of discontinuities per unit volume. The suggested relationship for clay-free rock masses is $RQD = 115 - 3.3\ J_v$, where J_v is the sum of the number of joints per unit length for all joint (discontinuity) sets known as the volumetric joint count. RQD is a directionally dependent parameter and its value may change significantly, depending upon the borehole orientation. The volumetric joint count can be quite useful in reducing this directional dependence. RQD is intended to represent the rock mass quality in situ.

The RQD value is used to determine the block size number of the erodibility index method (EIM) following Annandale (2006).

Joint Set Number

The Joint set number J_n is a direct function of the number of joint sets encountered in the in-situ rock mass, ranging from rock with no or few joints (essentially intact rock) to rock formations consisting of multiple joint sets. The classification accounts for both random discontinuities, that do not form regular patterns, and regular joint sets (Table 6.12).

TABLE 6.12
Determination of Joint Set Number

Number of Joint Sets	J_n
Intact, no or few joints	1.00
1 joint set	1.22
1 joint set plus random	1.50
2 joint sets	1.83
2 joint sets plus random	2.24
3 joint sets	2.73
3 joint sets plus random	3.34
4 joint sets	4.09
Multiple joint sets	5.00

Source: Kirsten (1982).

Block Size Number

The block size number is defined as a function of the rock joint spacing and of the number of joint sets. Joint spacing is estimated from borehole data by means of the RQD value. The block size number K_b is then expressed as follows:

$$K_b = \frac{RQD}{J_n} \tag{6.43}$$

Discontinuity Bond Shear Strength Number

The discontinuity bond shear strength number is a parameter representing the relative resistance offered by discontinuities in rock, determined as the ratio between the joint wall roughness J_r and the joint wall alteration J_a:

$$K_b = \frac{J_r}{J_a} \tag{6.44}$$

in which J_r represents the degree of roughness of opposing faces of a rock discontinuity, see Table 6.13, and J_a represents the degree of alteration of the materials that form the faces of the discontinuity. Values for J_a are given in Table 6.14.

Relative Ground Structure Number

The relative ground structure number J_r represents the relative ability of earth material to resist erosion due to the structure of the ground. It is a function of the dip and the dip direction of the least favourable discontinuity (i.e. the most easily eroded one) in the rock with respect to the direction of flow, and the shape of the material units (Table 6.15). The orientation and the shape affect the ease by which the stream can penetrate the ground and dislodge individual material units. For intact rock, J_s equals 1.0. Other values can be found in Annandale (2006).

TABLE 6.13
Determination of Joint Roughness Number

Joint Separation	Condition of Joint	J_r
Joints tight or closing during excavation	Stepped joints	4.0
	Rough or irregular, undulating	3.0
	Smooth undulating	2.0
	Slickensided undulating	1.5
	Rough or irregular, planar	1.5
	Smooth planar	1.0
	Slickensided planar	0.5
Joints open and remain open during excavation	Joints either open or containing relatively soft gouge of sufficient thickness to prevent joint wall contact upon excavation	1.0
	Shattered or micro-shattered clays	1.0

Source: Kirsten (1982).

TABLE 6.14
Determination of Joint Alteration Number

Description of Gouge	Joint Alteration Number J_a for Joint Separation in mm		
	1.0	**1.0–5.0**	**5.0**
Tightly healed, hard, non-softening impermeable filling	0.75	–	–
Unaltered joint walls, surface staining only	1.0	–	–
Slightly altered, non-softening, non-cohesive rock mineral or crushed rock filling	2.0	2.0	4.0
Non-softening, slightly clayey non-cohesive filling	3.0	6.0	10.0
Non-softening, strongly over-consolidated clay mineral filling, with or without crushed rock	3.0	6.0	10.0
Softening or low-friction clay mineral coatings and small quantities of swelling clays	4.0	8.0	13.0
Softening moderately over-consolidated clay mineral filling, with or without crushed rock	4.0	8.0	13.0
Shattered or micro-shattered (swelling) clay gouge, with or without crushed rock	5.0	10.0	18.0

Source: Kirsten (1982).

TABLE 6.15
Determination of Relative Ground Structure Number

Dip Direction of Closer Spaced Joint Set (Degrees)	Dip Angle of Closer Spaced Joint Set (Degrees)	Ratio of Joint Spacing, *r*			
		1:1	**1:2**	**1:4**	**1:8**
180/0	Vertical 90	1.14	1.20	1.24	1.26
	89	0.78	0.71	0.65	0.61
	85	0.73	0.66	0.61	0.57
	80	0.67	0.60	0.55	0.52
	70	0.56	0.50	0.46	0.43
	60	0.50	0.46	0.42	0.40
	50	0.49	0.46	0.43	0.41
	40	0.53	0.49	0.46	0.45
	30	0.63	0.59	0.55	0.53
	20	0.84	0.77	0.71	0.67
	10	1.25	1.10	0.98	0.90
	5	1.39	1.23	1.09	1.01
	1	1.50	1.33	1.19	1.10
0/180	Horizontal 0	1.14	1.09	1.05	1.02

(Continued)

TABLE 6.15 (*Continued*)
Determination of Relative Ground Structure Number

Dip Direction of Closer Spaced Joint Set (Degrees)	Dip Angle of Closer Spaced Joint Set (Degrees)	Ratio of Joint Spacing, *r*			
		1:1	1:2	1:4	1:8
	−1	0.78	0.85	0.90	0.94
	−5	0.73	0.79	0.84	0.88
	−10	0.67	0.72	0.78	0.81
	−20	0.56	0.62	0.66	0.69
	−30	0.50	0.55	0.58	0.60
	−40	0.49	0.52	0.55	0.57
	−50	0.53	0.56	0.59	0.61
	−60	0.63	0.68	0.71	0.73
	−70	0.84	0.91	0.97	1.01
	−80	1.26	1.41	1.53	1.61
	−85	1.39	1.55	1.69	1.77
	−89	1.50	1.68	1.82	1.91
180/0	Vertical −90	1.14	1.20	1.24	1.26

Source: Kirsten (1982).

Rock Block Dimensions

As only quadrilateral-shaped rock blocks are considered, the main rock block dimensions are the block height z_b, and the block side lengths x_b and y_b. The side length x_b is measured from up- to downstream, and y_b stands for the transversal dimension.

The ratios of z_b/x_b and z_b/y_b define the shape of the block and are relevant to dynamic block uplift. The higher this ratio, the more difficult it becomes for a rock block to be uplifted. These ratios are automatically computed and accounted for by the software.

Total Rock Joint Length

The total length L_f of the rock joint all around the block is defined as the sum of x_b (or y_b) + $2 \cdot z_b$. This length is used to compute the fundamental resonance frequency of the rock joints and thus also the time duration of net uplift pulses on rock blocks.

2D Shape of Rock Joint

The shape of the rock joints in the 2D vertical *X–Z* plane is of importance to the ease of fracturing of the joint by dynamic water pressures. The joints can be elliptical-shaped (EL), single-edge (SE) shaped or centre-cracked (CC) shaped (Figure 6.27).

Initial Degree of Fracturing of Joints

The persistency Pe stands for the initial degree of fracturing of rock joints. It is defined as the ratio of the initially fractured length of the joint to the max. possible

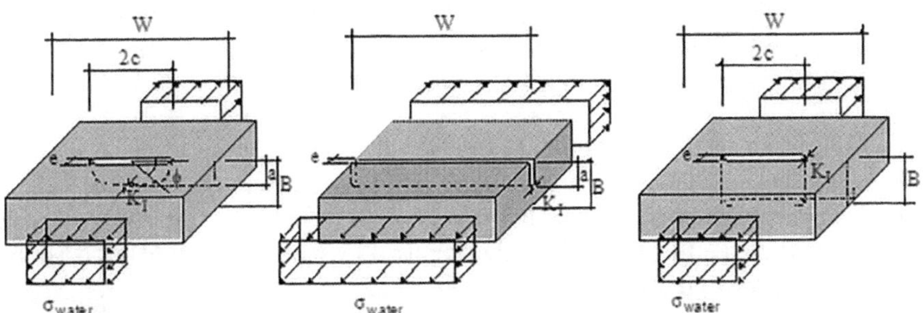

FIGURE 6.27 Principal shapes of rock joints (Bollaert 2002).

(totally fractured) joint length. It must be considered as a means to express the degree of development of the potential fracture network, given the existing state of the different joint sets.

Totally massive or intact rock corresponds to a persistency of 0.0, while totally broken-up rock (i.e. only distinct rock blocks without any massive bridges remaining in between the blocks) corresponds to a persistency of 1.0. Lowly broken-up rock typically has a persistency of 0.2–0.3, while highly fractured rock has a persistency of more than 0.5.

Pressure Amplification Inside Rock Joints

The maximum dynamic pressure $C^{max}{}_p$ is obtained through multiplication of the rms pressure C'_{pa} with an amplification factor Γ^+, and by superposition with the mean dynamic pressure C_{pa}. Γ^+ expresses the ratio of the peak value inside the rock joint to the RMS value of pressures at the pool bottom and has been determined experimentally (Bollaert 2002):

$$\Gamma^+ = \max\,(4, \mathit{gamma} - 8 + 2\,Y/D_j) \quad \text{for } Y/D_j < 8 \tag{6.45}$$

$$\Gamma^+ = \max\,(4, \mathit{gamma} + 8) \quad \text{for } 8 \le Y/D_j \le 10$$

$$\Gamma^+ = \max\,(4, \mathit{gamma} + 28 - 2\,Y/D_j) \quad \text{for } 10 < Y/D_j$$

in which *gamma* is close to 0 for joints with several side branches or joints that are not tightly healed, and up to maximum 12 for tightly healed joints. The former joints produce less significant pressure peaks, due to pressure diffusion and air dampening effects. Typical *gamma* values are of 2–6, with minimum Γ^+ values of 4 for both core and developed jets. The product of C'_{pa} times Γ^+ results in a maximum pressure, written as:

$$P_{max}\,[\mathrm{Pa}] = \gamma \cdot C_p^{max} \cdot \frac{V_j^2}{2g} = \gamma \cdot \left(C_{pa} + \Gamma^+ \cdot C'_{pa}\right) \cdot \frac{V_j^2}{2g} \tag{6.46}$$

It is interesting to notice that maximum pressures inside joints occur for Y/D_j ratios between 8 and 10. This means that the most critical flood situation may not be the PMF but rather the flow that results in a critical Y/D_j ratio.

Joint Fatigue Sensitivity and Fatigue Coefficient

The joint fatigue sensitivity m_r and the joint fatigue coefficient C_r define the fatigue law used by the Comprehensive Fracture Mechanics method. The fatigue law fits an equation of the type that was originally proposed by Paris et al. (1961) to describe fatigue crack growth in metals:

$$\frac{da}{dN} = C_r \cdot \left(\frac{\Delta K_I}{K_{Ic}} \right)^{m_r} \tag{6.47}$$

in which a is the joint length that fractures, and N the number of cycles per second of dynamic water pressure waves generated by the impinging jet at the tip of the joint. C_r and m_r are rock material parameters that can be determined by experiments. ΔK_I is the difference of maximum and minimum stress intensity factors at the joint tip, while K_{Ic} stands for the fracture toughness of the rock mass.

To implement time-dependent joint propagation into a comprehensive engineering model, m_r and C_r have to be known. They represent the vulnerability of rock to fatigue and have been derived from available laboratory data on the sensitivity of rock to quasi-steady break-up by water pressures in joints (Atkinson 1987).

The fatigue law coefficient C_r and exponent m_r may be defined as a function of the scour vulnerability of the rock, which is described in the fracture mechanics method later on in this chapter.

Rock Block Angle with the Horizontal

The rock block angle with the horizontal, or with the average slope of the water-rock interface, is illustrated in Figure 6.28. It is being used to define the potential direction of movement of the blocks by dynamic uplift, and to define the relevant forces on the blocks (Bollaert 2021).

Wave Celerity of Rock Joint

The wave celerity of the air-water mixture inside the rock joints highly depends on the % of air content out of the solution. Laboratory tests involving high-velocity aerated jets, define wave celerity c = 50–100 m/s for high air content, as typically encountered on turbulent plunging jets, and c = 200 m/s for low air content turbulent flows.

The wave celerity is used to define the number of pressure cycles per sec. travelling inside the rock joints, N, as follows (max. 100 Hz, consider no jet energy beyond) (Bollaert 2021):

$$N\ [-] = \frac{c}{4(a/B)L_f} \tag{6.48}$$

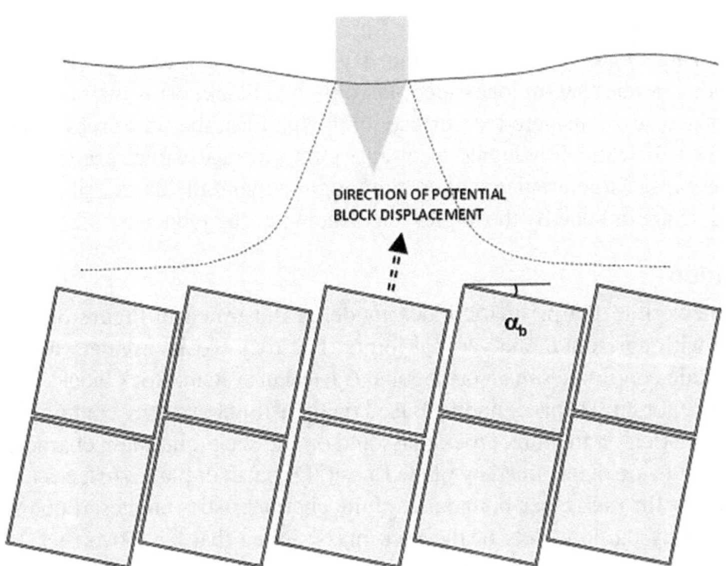

FIGURE 6.28 Rock block angle with the horizontal or with the average slope of the water-rock interface.

Joint Friction Angle

The basic friction angle of the rock joints depends on the roughness and asperities of the surfaces that form the joints. Typical values for most rock lithologies are between 25° and 35°.

The Prismatic Block Model (PBM)

The prismatic block model (PBM) simulates the real rock mass in a simplified manner by an ensemble of prismatic elements, for which the dimensions are determined by 2 (2D) or 3 (3D) series of joint sets. Each rock element or block is considered as an independent entity that may be scoured during the computations.

Introduction

Like for the modes in the conceptual model, each block represents the necessary information on the intact rock mass and the rock joint network. In contrast with the conceptual model, the PBM implements the geometry of each block by accounting for the block coordinates. Moreover, the water-rock interface is automatically defined based on the coordinates of the rock blocks at the surface of the rock mass.

The model implements the following main geometric elements:

1. *Nodes*: Nodes are determined by the coordinates of rock blocks. Nodes are not computational nodes but allow defining the rock joints around the blocks.
2. *Blocks*: Blocks are determined by 4 (2D) or 8 (3D) node coordinates that allow forming a closed prismatic shape. Blocks are the basic scouring elements when performing scour computations. They may be of different lithologies.

3. *Joints*: joints are defined by plane lines (2D) or surfaces (3D) that two neighbouring blocks have in common, based on the coordinates of the block nodes. Joints that are one-sided, i.e. only one block forms the plane line or surface, are considered as surface joints that form the water-rock interface. When different lithologies meet at a joint, average values are defined for the joint characteristics such as the roughness and the degree of alteration. Joints are defined by the angles and spacing of the joint sets.

Description

The framework of the prismatic block model is illustrated in Figure 6.29 for a 2D situation with horizontal and vertical joints. The rock blocks are determined based on their node coordinates in an orthogonal Cartesian system. Rock blocks have either 4 (2D space) or 8 (3D space) nodes. Based on their lithology, they contain the necessary information on the intact rock mass and on the rock joint main characteristics.

Rock joints are determined by plane lines (2D space) or plane surfaces (3D space) as defined by the rock block boundaries. Joint characteristics (angles, roughness, etc.) are defined by the joint sets of the rock mass. Joints that have two rock blocks in common are defined as internal joints, i.e. they are situated inside the rock mass

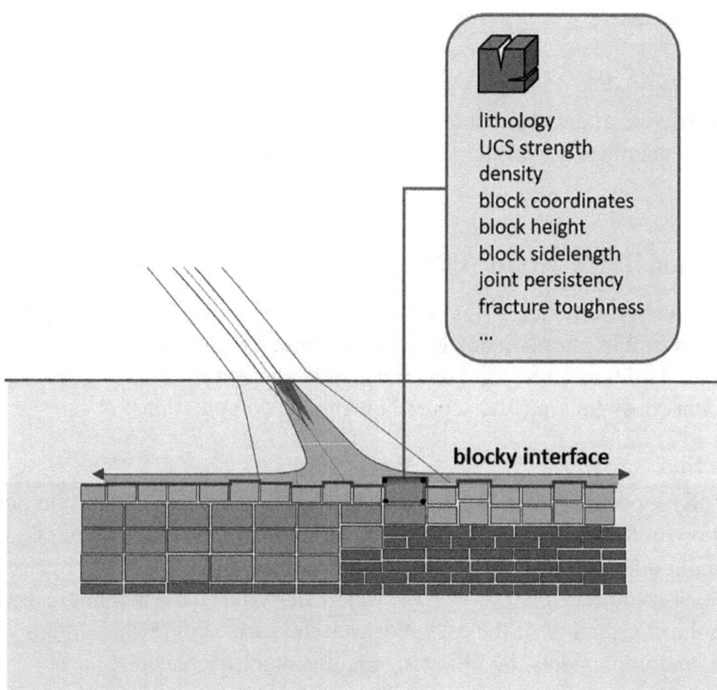

FIGURE 6.29 Framework of the prismatic block model of the rock mass model (PBM) containing different lithologies.

and transmit the flow by propagation of pressure waves. Joints that are formed by only one block are defined as surface joints, i.e. they are situated at the surface of the rock mass and form the water-rock interface. Depending on the dimensions and the position of the rock blocks, plane lines or surfaces as defined by blocks may be subdivided into different subjoints, allowing to generate both internal and surface joints if needed.

Rock blocks that have at least one surface joint are defined as surface blocks, while rock blocks without surface joints are defined as internal blocks.

The scour computations do not follow a pre-defined numerical grid like for the conceptual model. Scour is computed on a block-by-block basis. At each computational iteration, only the surface blocks are checked for scour. For a surface block to be scoured, two conditions must be satisfied:

- Complete fracturing of the joints surrounding the block
- Rock block displacement by uplift (plucking) or lateral displacement (sliding)

After each computational iteration, the water-rock interface and related surface blocks are redefined to prepare the next iteration.

Water-Rock Interface

Like for the conceptual model, the water-rock interface is determined based on a 3D topography or bathymetry of the river channel or plunge pool bottom. Based on the prismatic block approach, the interface may now be defined with a level of detail as provided by the single rock blocks at the surface.

Moreover, during scour progression with depth, the block shape, position and dimensions are continuously accounted for at the water-rock interface.

This means that the geometrical characteristics that are relevant to scour formation, such as surface roughness of the rock mass and local protrusion of rock blocks compared to their surroundings, are duly accounted for, provided that the relevance of these characteristics on rock scour is incorporated in the applied rock break-up methods (Figure 6.30).

Hydrodynamic Action

The hydrodynamic action of the turbulent flow is defined by the hydraulic parameters valid at the surface blocks. These parameters depend, among others, on the rock block shape, roughness and protrusion compared to neighbouring blocks. Especially the latter parameter may have a significant influence on the flow velocities and dynamic pressures acting at the water-rock interface. It may be accounted for computationally by using a series of pressure coefficients valid for different geometrical situations (protrusions), or also by a more detailed hydraulic modelling of the hydraulic parameters, accounting for the detailed shape and protrusion of the surface blocks.

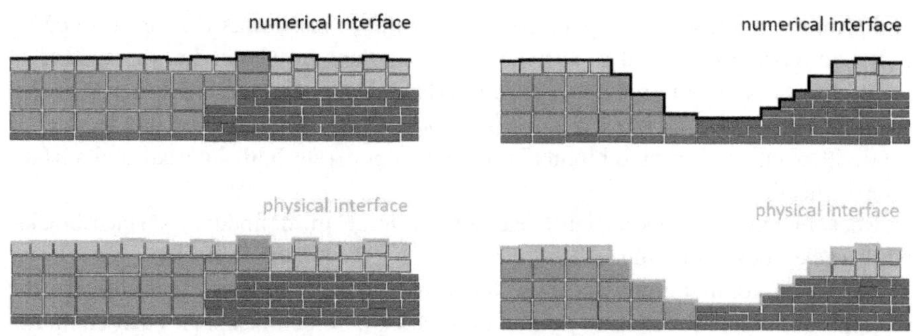

FIGURE 6.30 Hydrodynamic action on nodes of the conceptual rock mass model (CM).

In case of scour of a surface block, the resulting hydraulic parameters must be adapted following the new physical shape of the water-rock interface. Detailed information regarding the interface is thus accounted for during the modelling process.

Rock Mass Implementation

The different rock mass parameters that must be defined for each of the rock lithologies, as well as the different computational methods that make use of these parameters, are listed in Tables 6.9 and 6.10. Each rock block is attributed rock mass parameters by its lithology.

The Distinct Element Model (DEM) (Cundall 1971)

The Distinct Element Method (DEM) (Cundall 1971) considers the rock mass as an assembly of distinct blocks or particles in mechanical interaction. These blocks may behave as rigid bodies. The solution method relies on the integration of the equations of the motion using a time stepping explicit algorithm and is particularly suited for discrete systems undergoing large motions, for example, in the analysis of failure mechanisms, disaggregation processes or rock falls.

A distinctive feature of DEM is the use of contact points. Rock joints are treated as surfaces of interaction between neighbouring bodies, instead of being pre-defined as joint elements. This greatly enhances the flexibility of model generation and discretization, and simplifies the consideration of large movements (e.g., Lemos 2008). Assigning representative areas to these contacts allows standard rock joint constitutive models to be employed, such as the classical Mohr-Coulomb model or Barton's joint model (Barton et al. 1985).

Block deformability has been considered by dividing each block in an internal finite element mesh, which can be refined according to the stress analysis precision required. This has been implemented in the UDEC code (Cundall 1980) and its 3D counterpart 3DEC (Hart et al. 1988). Similar schemes, sometimes designated as discrete-finite elements, are now employed in various codes.

Because of the much simpler numerical procedures than for polygonal blocks, circular particle models have been developed. This implies that much larger systems can be simulated with reasonable run times, a major factor in the fast expansion of particle models that are very popular for rock fracture studies. Many DEM codes, based either on block or particles, are routinely employed in rock mechanics research and practice, as well as other numerical methods that share some of its essential concepts for the representation of a discontinuum, including Discontinuous Deformation Analysis (DDA) (Shi and Goodman 1988), Discrete-Finite Element Method (DFEM) (Munjiza 2004), Non-Smooth Contact Dynamics (NSCD) and others (e.g. Jing and Stephansson 2007; Zhao et al. 2012).

The Bonded Particle Model (BPM) (Potyondy and Cundall 2004)

The Bonded Particle Model (BPM) is a simplified version of the DEM model and has been developed to study rock fracture propagation paths (Lemos 2013). The basic idea is to represent the rock material as an assembly of disks (in 2D) or spheres (in 3D), randomly generated according to some size distribution curve. These rigid particles are linked by breakable bonds.

The BPM described by Potyondy and Cundall (2004) provides a formal procedure for representation of intact rock using Particle Flow Code (PFC, Itasca International Inc.). The model is conceptually simple, consisting of spheres bonded at their contacts, with the inputs restricted to stiffness and strength for the particles and bonds. Damage within PFC samples of intact rock is represented as broken bonds, which form and coalesce into macroscopic fractures when load is applied. The system deformation derives from the inter-particle movements. The contact normal and shear stiffness govern the deformability of the unbroken system. Bond strength is described by the tensile strength and cohesion terms. After bond breakage the interaction between the particles becomes purely frictional. The bond formulation employed, known as parallel bonds, offers the capability to transmit moments, in addition to forces, between the particles.

This simple structure allows very complex behaviours between the particles, such as brittle and ductile behaviour. Potyondy and Cundall (2004) demonstrate that the model reproduces many features of intact rock behaviour, including elasticity, fracturing, damage accumulation producing material anisotropy, dilation, strength increase with confinement and so on.

Use of macro-particles composed of groups of circular particles joined together was found to provide more interlocking in the assembly and enhanced frictional effects. The macro-particles may behave as rigid bodies ("clumps"), or the bonds between the component particles may be allowed to break, simulating intra-granular cracks. These models essentially provide a way to discretize the grain structure in more detail, accomplishing more complex grain shapes and behaviour, at the expense of increased computational efforts.

Polygonal block models resemble more closely the grain structure of many rocks. Very interesting results are obtained in fracture analysis. Kazerani and Zhao (2011) used both Voronoi and Delaunay block assemblies to study the fracture propagation in uniaxial and Brazilian test specimens. Fracture of blocks is possible in some codes (e.g., Eberhardt et al. 2004; Munjiza 2004), either through pre-defined potential crack paths or by means of some form of internal remeshing. The main disadvantage of polygonal models is the computational effort, as contact detection and particle interaction calculations are much slower than for circular particles. This fact limits the size of assemblies that can be handled with reasonable run times, particularly in 3D. The aim to create larger sized models has led to the development of simplified versions of these models, such as the flat-joint model (Potyondy 2012) or the lattice model (Cundall 2011).

The Discrete Fracture Network (DFN) model (Xu and Dowd 2010)

A Discrete Fracture Network (DFN) refers to a computational model that explicitly represents the geometrical properties of each individual fracture (e.g. orientation, size, position, shape and aperture), and the topological relationships between individual fractures and fracture sets. As such, DFN represents, in a deterministic or statistical sense, the discontinuities in the real rock mass.

According to Xu and Dowd (2010), the fundamental motivation of DFN is the recognition that at every scale, water transport in fractured rocks tends to be dominated by a limited number of discrete pathways formed by fractures and other discrete features.

The DFN model simulates the joint network of a real rock mass based on a wide range of in-situ monitored data that are analysed and transformed into a 2D or 3D joint network that is generated based on deterministic or stochastic procedures.

DFN focuses on identifying and geometrically describing the most relevant individual discrete features. The DFN approach provides the three-dimensional description of these features, which concentrate flow and transport.

It is not feasible to map accurately on an engineering scale the fractures and fracture networks in rock masses, because accurate field measurement of a single discontinuity is difficult and measurement of all discontinuities is impossible. Therefore, in practical applications one usually constructs a stochastic model informed by sparse data, which normally come from surveys of analogues, such as rock outcrops, or from direct or indirect observations of the rock mass such as drill cores, borehole imaging, geophysical surveys or seismic monitoring during fracture stimulation (Xu and Dowd 2010).

Stochastic modelling is a general approach in which the fracture characteristics, such as size and orientation, are treated as random variables with given probability distributions. In the simplest case, once the parameters of the distributions are inferred, the rock fracture model is constructed by a Monte Carlo simulation (Xu and Dowd 2010).

The Synthetic Rock Mass (SRM) model (Pierce et al. 2007)

SRM samples are three-dimensional and simulate rock as an assembly of bonded spheres (intact rock) with an embedded discrete network of disc-shaped joints. The technique brings together two well-established methods, Bonded Particle Modelling (BPM) with PFC^{3D} (Potyondy and Cundall 2004) and Discrete Fracture Network (DFN) simulation.

The methodology employs a sliding-joint model that allows for large rock volumes containing thousands of joints to be simulated in a rapid fashion. The inputs to the

SRM are from standard rock mass characterization methods while the outputs are in the form of rock mass properties that may be employed in empirical, numerical or analytical analyses.

Of particular interest is the ability to obtain predictions of rock mass brittleness. This is considered a significant step forward as there is no established method for quantifying this property.

Different properties are assigned to the bonds of the contacts between particles belonging to the same block, representing the intact rock material, and to the contacts between adjacent blocks, representing the joint behaviour. Joints may be persistent or terminate within blocks, thus allowing progressive crack development processes.

Creation of SRM Samples

SRM samples are three-dimensional and simulate rock assembly of bonded spheres (intact rock) with an embedded discrete network of disc-shaped flaws. The samples are constructed in PFC^{3D} and spherical in shape to permit application of strains in any direction and to rotate these during testing. The three main inputs required for construction of an SRM sample are the intact rock properties (UCS, modulus of elasticity and Poisson's ratio), a Discrete Fracture Network (DFN) and the joint properties (stiffness and shear strength).

Intact Rock

The Bonded Particle Model (BPM) described by Potyondy and Cundall (2004) provides a formal procedure for representation of intact rock using PFC. The intact rock properties can be obtained from standard uniaxial compression tests on core samples and then scaled to account for the average in-situ rock block size. According to the empirical relation developed by Hoek and Brown (1980), the laboratory-measured intact rock strengths are scaled to obtain a target rock block strength for the SRM samples. Using the procedures outlined by Potyondy and Cundall (2004), a series of uniaxial compression tests were conducted on cylinders of PFC^{3D} material to obtain a match to the target properties. The UCS, modulus of elasticity and Poisson's ratio values may be so obtained.

Discrete Fracture Network

Discrete Fracture Network (DFN) modelling techniques represent the most logical choice for representing joints within the SRM sample. They can be developed in external software from the measured fracture data in situ, for example, from borehole and scanline mapping, and then imported into the SRM samples to permit direct representation of the joint network.

Joint Characteristics

Joints are represented in PFC^{3D} by debonding contacts along a plane. This may result in unrealistic sliding on the plane because of the roughness or bumpiness induced by the particles. Small particles could be used to keep the thickness of the joint plane to realistic levels, but this is often not feasible because of computational constraints.

Hence, a sliding-joint model was developed. This model allows the different contacts to slide past one another without over-riding one another. The sliding-joint model overcomes some of the limitations of PFC in modelling jointed rock mass behaviour. More details can be found in Pierce et al. (2007).

DIGITALIZED SCOUR COMPUTATIONAL MODELS

This subchapter presents digital computational models to assess rock scour potential at hydraulic structures. Each of these computational models makes use of a flow sub-model and a rock mass sub-model, and mathematically relates both sub-models based on a well-defined scour criterion or rock break-up mechanism.

ERODIBILITY INDEX METHOD (EIM)

Background

The erodibility index method (EIM) is a semi-empirical computational method that is based on stream power, a geomechanical index and a threshold relationship to determine the scour potential in rock and earthen materials. The scour extent is determined by comparing the available stream power of the turbulent flow with the one that is required to scour the material.

Several scour threshold relationships using the erodibility index approach are available. Erodibility index methods are relatively easy to digitalize, because both the turbulent flows and the rock resistance to scour are represented using straightforward engineering parameters. Furthermore, scour thresholds are most of the time simple mathematical relationships easy to digitalize and customize.

The rock break-up mechanism implemented by these methods, however, is less straightforward. The continuum rock mass approach applied by these methods is a simplification that reduces the rock behaviour to a simple scour index, and no particular break-up mechanism is inherent. The EIM rather represents the most relevant break-up mechanisms, such as joint fracturing and block plucking, in a global manner.

Erodibility Index *K*

The erodibility index K is a geomechanical index that is used to quantify the relative ability of earth and engineered earth materials to resist erosive capacity of water. The erodibility index is defined as:

$$K = M_s \cdot K_b \cdot K_d \cdot J_s [-] \tag{6.49}$$

in which M_s stands for the mass strength number, K_b the block size number, K_d the discontinuity bond shear strength number and finally J_s the ground structure number. More details and corresponding values to be used for each number can be found in Tables 6.11–6.15.

Applied Stream Power (ASP)

The stream power (SP) that is available to a flow in a channel corresponds to the rate of energy expenditure and is defined in the following manner:

$$\text{SP} = \rho \cdot g \cdot q \cdot J_f \; [\text{kW/m}^2] \tag{6.50}$$

in which q [m²/s] stands for the unitary discharge of the flow and J_f [m/m] for the energy slope of the flow in the channel. It may also be expressed as the unitary stream power dissipation or expenditure rate, defined as the change in unitary stream power along the channel, when considering power as work or energy that is locally available along the channel per unit of time.

Based on the universal energy balance developed by Schlichting and Gersten (2000), the production of turbulence in the near-bed region is defined by Annandale (2006) as the applied stream power at the channel bed (ASP):

$$\text{ASP} = 7.853 \cdot \frac{\overline{\tau}_w^{3/2}}{\sqrt{\rho}} \left[\text{W/m}^2\right] \tag{6.51}$$

in which $\overline{\tau}_w = \rho \cdot g \cdot h \cdot J_f$ stands for the average wall shear stress of the flow [N/m²], with h [m] the flow depth. The applied stream power is generated by turbulence production in the near-bed region and corresponds to turbulent pressure fluctuations at the boundary, which is the principal factor leading to scour.

By expressing this applied stream power as a proportion of the total available stream power in the flow environment, it becomes possible to quantify it. This proportion depends on the boundary roughness and on the flow depth to absolute roughness ratio h/k_s, in which k_s is the equivalent sand roughness [m]. For rough turbulent flow, and h/k_s ratios > 50–100, this ratio equals more or less 0.40 (Annandale 2006).

Similarly, for channel flows, based on equation [51], the applied stream power (ASP) may be written as follows (George et al., 2022):

$$\text{ASP} = \frac{7.853}{1{,}000}\left(\frac{f}{8}\right)^{3/2} \cdot \rho \cdot V^3 \quad \left[\text{kW/m}^2\right] \tag{6.52}$$

in which the friction factor f [–] of the channel bottom is written as:

$$f = \left(\frac{1}{2 \cdot \log\left(12 \cdot \dfrac{R}{k}\right)}\right)^2 \tag{6.53}$$

where the hydraulic radius R [m] may be taken equal to the flow depth h [m] for large river channels compared to the flow depth, and k [m] is the equivalent roughness coefficient according to Darcy-Weisbach.

For plunging jets, the available stream power (SP) [W/m²] at impact in the pool is computed as follows:

$$\text{SP} = \rho \cdot g \cdot Q \cdot H \tag{6.54}$$

in which Q stands for the total jet discharge and H for the vertical plunge distance of the jet (headwater level – tailwater level). The available unitary stream power SP is computed by dividing the available stream power by the area of the footprint of the jet upon impact in the pool.

As for channel flows, the energy expenditure rate is of relevance. However, in case of plunge pools, this rate is difficult to quantify. When the jet traverses the plunge pool water depth, it progressively diffuses in three dimensions and modifies its available stream power at the water-rock interface.

The rate of energy expenditure at the water-rock interface is expressed here by making use of an applicable stream power coefficient C_{sp}, of coefficients that express the local dynamic pressures, and of the local velocities acting at the interface. By making a distinction between average and fluctuating dynamic pressures, the following equation is obtained for the applied unitary stream power ASP [kW/m²] of a jet impinging through a plunge pool onto a water-rock interface:

$$ASP = C_{sp} \cdot \frac{\rho_w \cdot g}{1{,}000} \cdot V_{local} \cdot \left(C_p + C_p'\right) \cdot \frac{V_j^2}{2g} \quad [kW/m^2] \tag{6.55}$$

in which C_{sp} [–] is a stream power coefficient, ρ_w the density of the water, g the gravitational acceleration, C_p [–] and C_p' [–] the mean and fluctuating dynamic pressure coefficients, V_j the jet velocity at impact in the plunge pool, and V_{local} the local flow velocity acting along the water-rock interface following lateral deviation of the jet.

As mentioned before, the value of the stream power coefficient C_{sp} depends on the ratio of the flow depth to the absolute roughness k_s of the interface (Annandale 2006). The ASP decay with increasing scour depth is governed by the corresponding decays with depth of the local flow velocities and the local dynamic pressures.

Scour Threshold

A general scour threshold relationship between the applied unitary stream power and the erodibility index can be mathematically defined as follows:

$$K = a \cdot ASP^b \text{ (for } K > 0.1\text{)} \tag{6.56}$$

Users may freely choose the a and b values. For example, by applying the Annandale (1995) threshold relationship, $a=1$ and $b=4/3$.

Based on multiple logistic regression analysis for spillway erosion, Wibowo et al. (2005) used the same dataset as Annandale (1995) and proposed an equation for the probability of occurrence of scour initiation. Figure 3.10 (Chapter 3) illustrates that the Annandale (1995) threshold corresponds more or less to a 20%–50% probability of erosion curve.

A 3D digitalization of the EIM method applies equation (6.55) all along the water-rock interface, based on the hydraulic parameter values C_p, C_p', V_j and V_{local} as defined in each of the computational nodes. The decreasing ASP with depth determines at which depth the scour formation ceases following the scour threshold relationship of equation (6.56).

Dynamic Pressure (DP) Method

Introduction

A computational method purely based on dynamic pressures around prismatic rock blocks has been digitalized, based on the research performed by Maleki and Fiorotto (2019).

Maleki and Fiorotto (2019) compute the uplift forces on prismatic rock blocks based on small-scale laboratory measurements of mean and fluctuating dynamic pressures generated by a jet impinging onto a pool bottom, for different Y/D_j ratios, and by analytical determination of the jet characteristics in the fully developed flow region. This is done by use of the length of the submerged jet in both the flow development and fully developed jet regions of the plunge pool, defined as ς, as well as use of the corresponding jet width $b(\varsigma)$ at those depths. Self-similar velocity profiles are assumed.

By accounting for appropriate pressure correlation functions in both the streamwise and transversal directions, they express the standard deviation of the net fluctuating uplift forces on a block as a function of the maximum standard deviation of the fluctuating pressure field on the block, by means of an Ω coefficient. Furthermore, they make use of the ratio "n" of the maximum net uplift pressure coefficient on a block ($C^+_p - C^-_p$) to the RMS coefficient of the fluctuating pressures C'_p on the block, defined as follows:

$$n = \frac{\left(C_p^+ - C_p^-\right)}{2 \cdot C_p'}[-] \tag{6.57}$$

Maleki and Fiorotto (2019) obtained mean values for n of 6.21, with a standard deviation equal to 0.57. Following the authors, this is in agreement with Castillo et al. (2014), who found values in the range of 5–7. Marson et al. (2007) found values of 6.0 for long-term small-scale laboratory experiments.

The authors finally multiply both Ω and n values with the RMS coefficient of pressure fluctuations at the block upper face to determine the max. net uplift pressure (force) on a block generated by dynamic pressures over and under the block.

Their approach is restricted to the determination of the net uplift force on a block and to the resulting block thickness needed to keep it stable, accounting for its submerged weight, in which stability is obtained when the submerged weight counteracts the net uplift force.

This approach has been digitalized by Bollaert (2021) in the rocsc@r software (see Chapter 7) by a 2D implementation of the related pressure coefficients at the water-rock interface. These coefficients automatically change with depth until user-defined blocks become stable.

Basic Input Parameters

The basic parameters that are used as input to the DP method are outlined in Table 6.16.

Forces on the Block

The net force $F_{net}(t)$ acting on a rock block at each time instant Δt is decomposed into fluctuating dynamic pressure forces acting under and over the block, buoyancy forces, gravity forces and finally shear forces along the vertical joints based on the weight and the buoyancy (for oblique joint sets) and based on differential pressures along the upstream and downstream vertical faces of the block.

$$F_{net}(t) = F_{under}(t) - F_{over}(t) + F_{buoy} - F_{grav} \tag{6.58}$$

TABLE 6.16
Basic Parameters Used as Input to the DP Computational Method

Parameter	Unit	Description
ς (= 0.65.Y)	m	Development length of jet through plunge pool
$b\varsigma$)	m	Jet width at development length ς
n	–	Ratio of maximum deviation from mean to RMS
	–	coefficient
L' (= x_b/b or y_b/b)	–	Non-dimensional block length
Ω	–	Reduction factor for net block uplift force
C'_p (Y/D_j)		RMS pressure coefficient at block surface, valid at jet centreline
C'_{pr} (Y/D_j)	–	$n \cdot C'_p$ (Y/D_j), RMS pressure coefficient at block surface, valid radially outwards from jet centreline
n'	–	$C'_{pr}/C'_{p,}$ RMS pressure reduction factor
V_j	m/s	Average jet velocity at impact in the pool
D_j or $\underline{B_j}$	M	Jet diameter or jet thickness at impact in the pool
H_j	M	$V_j^2/2g$, hydraulic head at impact of jet in the pool
x_b or $\underline{y_b}$	m	Rectangular block side lengths in the horizontal plane
z_b	m	Rectangular block vertical height
α_b	°	Angle of rock blocks/joints with the vertical
φ	°	Rock joint frictional angle
a_{am}	–	Coefficient of added mass
m_{am}	kg	Added mass of the block during its acceleration phase
MULT	–	Multiplication factor of the RMS pressure coefficient
Δt_{pulse}	sec	Time duration of net uplift pressure pulse
C_{UVF}	–	Average pressure coefficient along upper vertical face of block
C_{DVF}	–	Average pressure coefficient along downstream vertical face of block

$$F_{up,DP} = F_{under}(t) - F_{over}(t) = \Omega \cdot n \cdot C'_p \cdot \rho_w \cdot x_b \cdot y_b \cdot \frac{V_j^2}{2} \tag{6.59}$$

$$F_{buoy} = \rho_w \cdot g \cdot x_b \cdot y_b \cdot z_b \tag{6.60}$$

$$F_{grav} = \rho_r \cdot g \cdot x_b \cdot y_b \cdot z_b \tag{6.61}$$

where

F_{under} = dynamic pressure forces acting under the block [N]
F_{over} = dynamic pressure forces acting over the block [N]
$F_{up,DP}$ = net dynamic uplift pressure forces acting on the block [N]
F_{buoy} = buoyancy forces acting on the block [N]
F_{grav} = gravity forces acting on the block [N]
Ω = reduction factor of the RMS pressure coefficient at block upper face [–]
n = ratio of maximum deviation from mean coefficient to RMS coefficient [–]

Dynamic Pressures Over and Under the Block

For the present DP method, in accordance with available research on rock block uplift, the average value of the dynamic pressure field may be neglected in the analysis. Hence, the pressure field acting along the upper face of the block can be described by the RMS pressure fluctuation coefficient $C'_{p,\mathrm{UF}}$ of the pressure over the upper face (UF) of the block, and by assuming a Gaussian pressure probability distribution:

$$C'_{p,\mathrm{UF}} = \frac{\mathrm{RMS}\ (\text{spatial} - \text{average pressure over UF})(\text{in [m]})}{\dfrac{V_j^2}{2g}}[-] \quad (6.62)$$

The Maleki and Fiorotto (2019) experiments have allowed to express the net uplift force on the block as a function of this $C'_{p,\mathrm{UF}}$ coefficient, by multiplying this coefficient with an Ω factor and a n-factor. For a 3D quadratic surface prismatic block ($x_b = y_b$), and by neglecting the effect of the lateral (i.e. vertical) joints on the net uplift force, the Ω factor is reported as a function of the non-dimensional length L' of the block and depends on the covariance of the pressure field as a function of the integral scales of the pressure fluctuations in the longitudinal and transversal directions respectively. Close to the water-rock interface, their experiments showed practically the same trend for all different Ω curves:

$$\Omega = 0.35 \cdot \left(1 - \exp\left(\frac{-L'}{0.89}\right)\right)[-] \quad (6.63)$$

Figure 6.31 illustrates this Ω function for 3D prismatic blocks, as well as the 2D equivalent as computed by Maleki and Fiorotto (2019), which also provided values for triangular 2D block shapes. The n-factor between the maximum net uplift

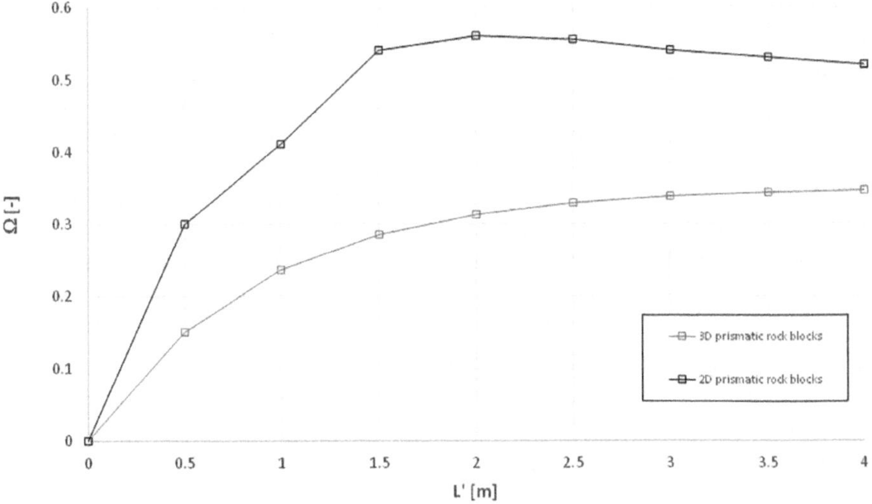

FIGURE 6.31 Ω values as a function of L', used as input to the DP computational method (adapted from: Maleki and Fiorotto 2019).

pressure deviation from the mean and the RMS coefficient of pressure fluctuations at the block upper face is defined following equation (6.57). This ratio is a value that has been calibrated in Chapter 7 based on different case studies. Following equation (6.59), once n and Ω are determined, the net uplift force $F_{up,DI}$ generated by dynamic pressures acting over and under the block is fully defined.

Block Stability Criterion

The stability of the rock block is expressed by the following equation:

$$F_{net} < 0 \tag{6.64}$$

This criterion does not incorporate a dynamic analysis of the movement of the block. As long as the net uplift force on the block remains positive, the block is considered to be uplifted.

Dynamic Impulsion (DI) Methods

The Dynamic Impulsion (DI) method has been initially developed by Dr E. Bollaert at the Swiss Federal Institute of Technology in Lausanne (Bollaert 2002; 2004; Bollaert and Schleiss 2005). The method determines the potential for vertical uplift of a rock block from its mass, due to a sudden pressure pulse. The method is valid in the turbulent diffusive shear layer of a plunging jet impacting a plunge pool, and is based on a series of near-prototype scaled laboratory experiments for a compact circular jet vertically impinging onto a 2D rock block, for which net uplift pressures and impulsions have been recorded. The net uplift impulsion is thereby defined based on an impulsion coefficient C_I, which depends on the pool-depth-to-jet-diameter ratio Y/D_j.

Based on more recent laboratory experiments and additional literature feedback, performed by multiple researchers such as Liu et al. (1998), Federspiel (2011), Bollaert et al. (2013) and Pells (2016), an enhanced version of the DI has been developed by Bollaert (2021) within the framework of the rocsc@r digital environment, called the Modified Dynamic Impulsion method or MDI.

This novel method is based on RMS pressure fluctuations at the rock block surface to define net uplift pressures and impulsions. In contrast with the initial DI method, it is considered applicable to any type of turbulent flows.

DI methods are a priori easy to digitalize. However, the success of these methods relies on sound knowledge of the RMS pressure fluctuations along the water-rock interface. This implies the need for an adequate modelling of turbulent flow.

Dynamic Impulsion Based on C_I values (DI, Bollaert and Schleiss 2005)

Forces and Impulsions on a Rock Block

The net dynamic impulsion on a 2D rock block is obtained by the integration of the different forces on the block during time periods Δt_{pulse} for which the net uplift force is positive. This generates an impulsion on the block that can be expressed as:

$$I_{net} = \int_0^{\Delta t_{pulse}} \left(F_u - F_o - G_b - F_{sh}\right) \cdot dt = m \cdot V_{\Delta t_{pulse}} \tag{6.65}$$

in which F_u and F_o are the forces under and over the block, G_b is the immerged weight of the block (i.e. accounting for buoyancy forces), F_{sh} represents the shear and interlocking forces, m the mass of the rock block (defined as the specific weight γ_r multiplied by the block volume) and V_{Dtpulse} the vertical displacement velocity that is given to the block following the impulsion. The different forces are sketched in Figure 6.32 for a prismatic rock block.

The immerged weight of the block G_b is determined based on the specific weight of the rock and on the block dimensions. As sketched in Figure 6.33, for rock blocks with a square horizontal section of side length x_b and for a block height z_b, the immerged weight of the block G_b is written as follows:

$$G_b = \forall \cdot (\gamma_r - \gamma_w) = x_b^2 \cdot z_b \cdot (\gamma_r - \gamma_w) \tag{6.66}$$

in which γ_r and γ_w represent the specific weight of the rock, respectively the water. For rock, typical values for γ_r vary between 22,000 and 28,000 N/m³.

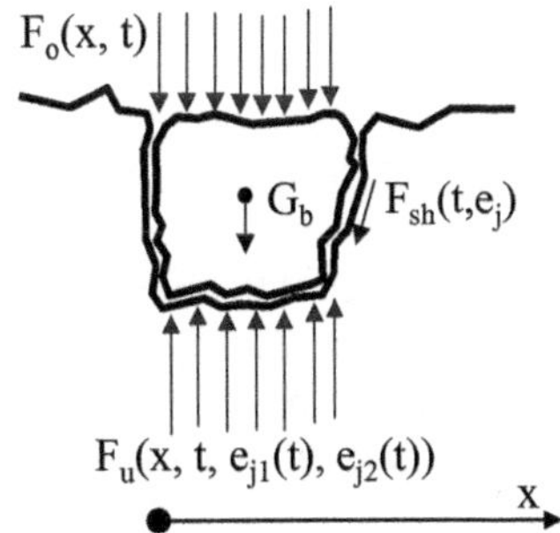

FIGURE 6.32 Force balance on a prismatic rock block (Bollaert 2002).

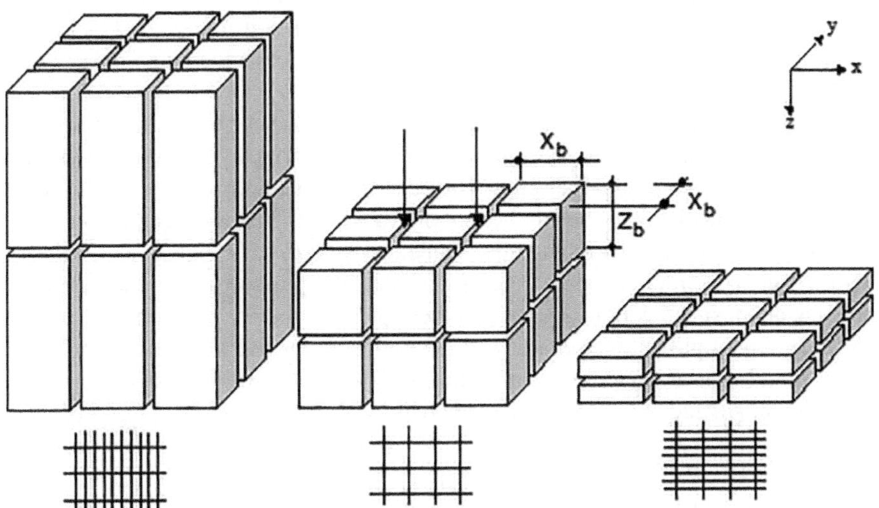

FIGURE 6.33 Three cases of form factors of a characteristic rock block used in the DI method. The influence of the y-direction is thereby neglected (Bollaert 2002).

Figure 6.33 shows that the unitary weight force (per m^2) depends on the form factor z_b/x_b of the rock block. Three cases of form factors can be considered: $z_b/x_b > 1$, $z_b/x_b = 1$ and $z_b/x_b < 1$. Block (a) will a priori be much more difficult to uplift than block (c). Moreover, the shear forces obviously will also be higher in case (a) than in case (c). As a result, knowledge about the joint sets is of importance to define the ultimate scour depth based on dynamic uplift of rock blocks.

The shear and interlocking forces F_{sh} of a rock block depend on the joint pattern and joint roughness, on contact points with neighbouring blocks and on the in-situ stress field of the rock mass. They are often neglected by assuming that progressive dislodgment and opening of the joints occurred during the break-up (preparatory) phase of the rock mass.

The immerged block weight remains constant during block movement and can be subtracted from the time integral calculated in equation (6.65) by multiplying this constant force by the time duration of the net positive pressure pulse, i.e. Δt_{pulse}.

Hence, by using equation (6.66) for the immerged weight of the block and by neglecting shear and interlocking forces, equation (6.65) becomes:

$$I_{\text{net}} = \int_0^{\Delta t_{\text{pulse}}} (F_u - F_o) \cdot dt - x_b^2 \cdot z_b \cdot (\gamma_r - \gamma_w) \cdot \Delta t_{\text{pulse}} = m \cdot V_{\Delta t_{\text{pulse}}} \ [\text{N} \cdot \text{s}] \quad (6.67)$$

The dynamic force exerted by the water at the block surface (F_o) is governed by the turbulent diffusive shear layer of the jet. The dynamic force of the water pressures underneath the block (F_u) is determined based on the propagation of transient pressure waves that enter the joint entrances at the block upper face. Pressure forces acting underneath the block will change once the block starts moving. For simplicity, these pressures underneath the block are assumed to be independent of its movement.

Equation (6.67) must be solved to express the net uplift impulsion I_{up} on a rock block generated by dynamic water pressures. This requires determination of the instantaneous spatially distributed pressure fields over and under the blocks as a function of time, during the duration Δt_{pulse}, in order to define the integral part of the equation.

It is obvious that determining both these forces and the corresponding time durations in detail is a complex task. Bollaert (2002) performed near-prototype scaled laboratory experiments with circular compact jets that impinge, in a small plunge pool, onto an artificially generated open-end rock joint. The test installation represented a 2D joint around a rectangular-shaped rock block. The experiments allowed high-frequency, simultaneous recordings of both the pressures at the entrances of the joint and inside the joint. Analysis of these pressure values during time durations of net positive uplift pressures allowed determining positive net uplift impulsions. These impulsions are mathematically expressed by a net uplift impulsion coefficient, for different jet velocities and plunge pool depths.

Computational Procedure

For a given 2D rock block, the specific weight γ_r and dimensional parameters x_b and z_b are known. The unknown in equation (6.67) is the net uplift impulsion I_{up} based on dynamic pressures over and under the block and defined by the time integral:

$$I_{up} = \int_0^{\Delta t \text{pulse}} (F_u - F_o) \cdot dt \quad (6.68)$$

NOTE: The original analysis of the Bollaert (2002) experiments has been performed per m², allowing to define the recorded net uplift pressures as the net forces $(F_u - F_o)$. Similarly, the recorded net uplift pressures were expressed in m of water column and not in N/m². The corresponding net uplift impulsion I_{up} thus had a dimension of [m·s]. In the following, for consistency and to avoid confusion, pressures will be expressed in [N/m²] and impulsions in [N·s], in accordance with international SI units.

The recorded net uplift pressures are averaged over the time durations of positive pulses and are then made non-dimensional by dividing them by the incoming kinetic energy $V_j^2/2g$ [m], in which V_j is the velocity of the jet at impact in the pool. For a squared block surface and accounting for the specific density of the water γ_w, this results in a net uplift pressure coefficient C_{up} [–]:

$$C_{\text{up}} = \frac{\dfrac{\overline{(F_u - F_o)}}{x_b^2 \cdot \gamma_w}}{\dfrac{V_j^2}{2g}} [-] \quad (6.69)$$

The recorded time durations of positive pulses Δt_{pulse} are non-dimensionalized by the travel period that is characteristic for pressure waves inside open-end rock joints, i.e. $T = 2 \cdot L_f/c$ [s], in which L_f is the total joint length and c the pressure wave celerity. This results in a time coefficient T_{up} [–]:

$$T_{\text{up}} = \frac{\Delta t_{\text{pulse}}}{\dfrac{2L_f}{c}} [-] \quad (6.70)$$

Next, a non-dimensional uplift impulsion coefficient C_I is defined as the product of C_{up} and T_{up} as follows:

$$C_I = C_{\text{up}} \cdot T_{\text{up}} [-] \quad (6.71)$$

The net uplift impulsion by water pressures over and under the block, I_{up} [*N·s*], defined as the time-averaged net force times the pulse duration, can then be written as:

$$I_{\text{up}} = \int_0^{\Delta t \text{pulse}} (F_u - F_o) \cdot dt = \overline{(F_u - F_o)} \cdot \Delta t_{\text{pulse}} = \gamma_w \cdot x_b^2 \cdot C_I \cdot \frac{V_j^2 \cdot L_f}{g \cdot c} [\text{N} \cdot \text{s}] \quad (6.72)$$

During the laboratory experiments, maximum measured C_{up} values varied between 0.30 and 0.40, while time duration coefficients T_{up} at near-prototype jet velocities varied between 0.5 and 2.0, with typical values ranging from 1.0 to 1.5, corresponding more or less to the characteristic resonance frequency (or travel time) of a pressure wave through the joint around the block, mathematically expressed as follows:

$$\Delta t_{\text{pulse}} \approx \frac{2L_f}{c} \quad (6.73)$$

A first-hand mathematical analysis performed by Bollaert (2002) for maximum recorded net uplift pressures and related time durations resulted in the following expression for C_I, valid along the jet centreline and for Y/D_j ratios < 14:

$$C_I = 0.0035 \cdot \left(\frac{Y}{D_j}\right)^2 - 0.119 \cdot \left(\frac{Y}{D_j}\right) + 1.22 \text{ for } Y/D_j < 14 \tag{6.74}$$

in which Y represents the jet travel distance through the water [m] and D_j the jet diameter at impact [m].

According to feedback from experience and a more recent mathematical analysis (Bollaert 2021), this equation has been modified and extended to incorporate C_I values for $Y/D_j > 14$ in the following manner:

$$C_I = 0.0023 \cdot \left(\frac{Y}{D_j}\right)^2 - 0.105 \cdot \left(\frac{Y}{D_j}\right) + 1.22 \text{ for } Y/D_j \leq 22 \tag{6.75}$$

$$C_I = 0.02 \text{ for } 22 < Y/D_j \leq 25$$

$$C_I = 0.01 \text{ for } 25 < Y/D_j$$

Equation (6.75) is only valid along the jet's centreline of impact in the plunge pool. For corresponding C_I values at computational nodes located radially outwards from the jet centreline, the C_I values at the jet centreline (i.e. for the same Y/D_j ratio (same node depth)) have to be multiplied by a radial reduction factor for RMS pressure coefficients C'_{pr}/C'_{pa} at the radially distant node.

Once the net uplift impulsion based on water pressures I_{up} is defined, equation (6.65) can be solved for the uplift velocity $V_{\Delta t\text{pulse}}$ that is given to the block. In a first step, the existence of a positive net uplift force is being checked for as follows:

$$\gamma_w \cdot x_b^2 \cdot C_I \cdot \frac{V_j^2 \cdot L_f}{g \cdot c} > x_b^2 \cdot z_b \cdot (\gamma_r - \gamma_w) \cdot \frac{2L_f}{c} \tag{6.76}$$

If the net uplift force on the block is negative, the block remains in place and no uplift height is computed. In case the net uplift force is positive, failure by ejection of a block is determined based on the vertical displacement height it receives due to the net impulsion I_{net}.

According to equation (6.67), the net impulsion I_{net} provides a displacement velocity V_{Dtpulse} [m/s] to a block of mass m. This momentum can be approximated by transformation of $V_{\Delta t\text{pulse}}$ into a net uplift displacement h_{up} [m] by conversion of kinetic into potential energy:

$$V_{\Delta t\text{pulse}} = \sqrt{2 \cdot g \cdot h_{up}} \tag{6.77}$$

In combination with equations (6.67) and (6.72), the net uplift velocity $V_{\Delta t\text{pulse}}$ and uplift displacement h_{up} can then be expressed as follows (Figure 6.34):

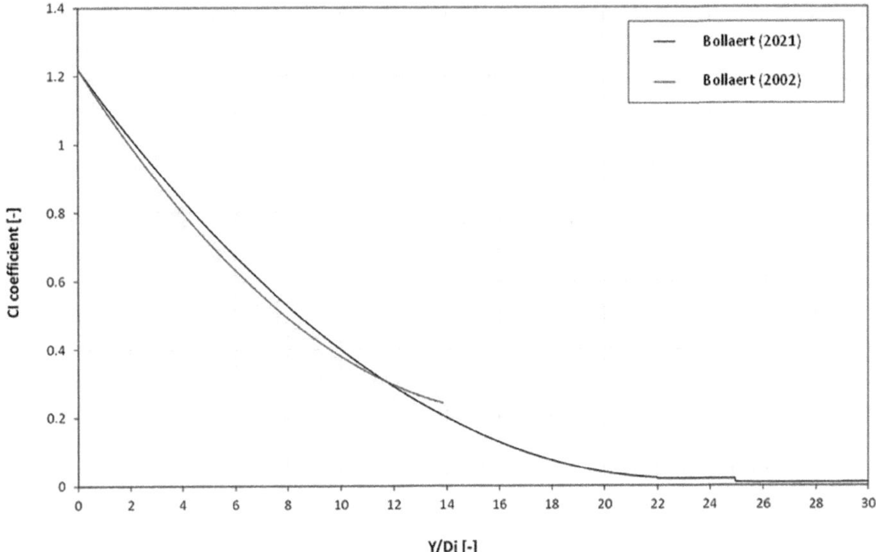

FIGURE 6.34 Net uplift impulsion coefficient C_I as a function of Y/D_j ratio for circular compact jets. Comparison between the initial Bollaert (2002) equation and the newly formulated extended equation (Bollaert 2021).

$$V_{\Delta t\text{pulse}} = \frac{I_{\text{net}}}{m} = \frac{1}{\rho_r \cdot x_b^2 \cdot z_b} \cdot \left[\gamma_w \cdot x_b^2 \cdot C_I \cdot \frac{V_j^2 \cdot L_f}{g \cdot c} - x_b^2 \cdot z_b \cdot (\gamma_r - \gamma_w) \cdot \frac{2 \cdot L_f}{c} \right] \tag{6.78}$$

$$h_{\text{up}} = \frac{1}{2g \cdot \rho_r^2 \cdot x_b^4 \cdot z_b^2} \cdot \left(\left[\gamma_w \cdot x_b^2 \cdot C_I \cdot \frac{V_j^2 \cdot L_f}{g \cdot c} - x_b^2 \cdot z_b \cdot (\gamma_r - \gamma_w) \cdot \frac{2 \cdot L_f}{c} \right] \right)^2 \tag{6.79}$$

The critical uplift displacement that is necessary to eject a rock block from its matrix is difficult to define. It depends on the initial protrusion and on the degree of interlocking of the blocks (shear and interlocking forces). These forces change during pressure pulses, depending on the pulse intensity and on the in-situ rock mass stress field. For practical situations, it is considered that blocks may get stuck in between neighbouring blocks following a first small displacement, increasing their protrusion and their potential for further uplift during subsequent pressure pulses.

The presented approach neglects both initial protrusion and shear and interlocking forces, and thus simplifies the problem to just one pressure pulse. Following this approach, a tightly jointed rock would theoretically need block displacements that are equal to or higher than the height of the block. Less tightly jointed rock, or protruding rock, would be uplifted more easily.

Several laboratory experiments available in literature point out that for protrusions that become significant compared to the total block height, i.e. around 50%, the displaced block will definitely be ejected. Block displacements of only half of the block height would thus be sufficient to eject the block from its matrix. To soundly account for the complex real block situations, the necessary or critical uplift displacement is a model parameter that needs to be calibrated according to site-specific parameters and is expressed in the following manner:

$$h_{\text{up,crit}} = n_b \cdot z_b \tag{6.80}$$

Based on practical feedback and calibrations performed with the DI model since its development, as discussed in Bollaert (2021), critical net uplift displacements of ~0.10–0.20 times the block height have been found.

Hence, for practice, a critical net uplift displacement factor of $n_b \sim 0.10$–0.20 should be used, together with wave celerity values of $c \sim 50$–$100\,\text{m/s}$ in case of highly aerated flow conditions, which is generally the case in prototype situations. In a global manner, plausible block behaviour as a function of n-value is presented in Table 6.17.

Modified Dynamic Impulsion Based on C'_p Values (MDI Method, Bollaert 2021)

Introduction

The DI method has shown its relevance for practice but nevertheless suffers from the following drawbacks that might restrict its applicability for real-life flow situations:

- The method is based on C_I values that have been recorded for circular compact jets that vertically impinge into a plunge pool.
- The method is based on C_I values determined at the point of jet impact only.
- The method deals with net uplift pressure pulses recorded on artificial 2D rock blocks, and not with the real pressure pulses acting at the water-rock interface.

The first drawback signifies that the DI method can strictly speaking not be used for other types of jets or other turbulent flow structures. The second drawback signifies that the DI method is not developed for rock blocks located radially outwards from the jet's centreline. Both drawbacks have been partially circumvented in the DI method by multiplying the C_I values at the jet centreline with radial reduction factors for the RMS pressure fluctuations at locations radially outwards. For turbulent flows other than plunging jets, however, no centreline exists and no C_I values are available.

TABLE 6.17
Block Behaviour as a Function of n_b-value

Ratio $n_b = h_{up}/z_b$	Rock block behaviour
$0.00 < n_b < 0.10$	Insignificant pulsations, no block ejection
$0.10 < n_b < 0.50$	Moderate pulsations, potential block ejection
$0.50 < n_b < 1.00$	Significant pulsations, block ejection very likely
$1.00 < n_b$	Major pulsations, definite block ejection

The third drawback excludes any influence of spatial pressure distribution or related block side length on the net uplift pulses on a real 3D rock block.

Hence, an enhanced version of the DI method has been developed within the framework of the rocsc@r digital environment, called the modified DI or MDI (Bollaert 2021) method. This method makes use of the dynamic pressure coefficient C'_p, expressing the RMS values of the pressure fluctuations at the joint entrances, instead of C_I values. Based on available literature data, among others the laboratory experiments on 3D blocks as described by Liu et al. (1998), Federspiel (2011), Bollaert et al. (2013) and Pells (2016), a direct relationship between these C'_p coefficients at the block surface and the related net uplift forces on a block may be established.

Furthermore, the method makes use of the time durations of the pressure pulses as recorded on these same 3D laboratory experiments to transform the net uplift forces into net uplift impulsions.

Finally, the method implements added mass forces exerted by water on a submerged rigid block during its upward acceleration, and accounts for the submergence of the block during accelerations and decelerations.

Basic Input Parameters

The basic parameters that are used as input to the MDI method are presented in Table 6.18.

TABLE 6.18
Basic Parameters Used as Input to the MDI Block Uplift Computational Method

Parameter	Unit	Description
$C'_p\ (Y/D_j)$	–	RMS pressure coefficient at block surface, valid at jet centreline
$C'_{pr}\ (Y/D_j)$	–	$n \cdot C'_p\ (Y/D_j)$, RMS pressure coefficient at block surface, valid radially outwards from jet centreline
N	–	$C'_{pr}/C'_{p,}$ RMS pressure reduction factor
V_j	m/s	Average jet velocity at impact in the pool
D_j or B_j	m	Jet diameter or jet thickness at impact in the pool
H_j	m	$V_j^2/2g$, hydraulic head at impact of jet in the pool
x_b or y_b	m	Rectangular block side lengths in the horizontal plane
z_b	m	Rectangular block vertical height
a_b	°	Angle of rock blocks/joints with the vertical
φ	°	Rock joint frictional angle
a_{am}	–	Coefficient of added mass
m_{am}	kg	Added mass of the block during its acceleration phase
MULT	–	Multiplication factor of the RMS pressure coefficient
Δt_{pulse}	sec	Time duration of net uplift pressure pulse
C_{UVF}	–	Average pressure coefficient along upper vertical face of block
C_{DVF}	–	Average pressure coefficient along downstream vertical face of block

One-Degree-of-Freedom Rigid Body Dynamics

The motion of a submerged rigid body subjected to turbulent flow may be described by a single-degree-of-freedom, flow-induced, damped vertical vibration of the body contained within the surrounding water mass. The corresponding mass-spring-dash-pot equation of motion equals (Bollaert 2013):

$$(m_b + m_{\text{add}}) \cdot \ddot{x} + (c + c_{\text{add}}) \cdot \dot{x} + k \cdot x = F_{\text{tot}}(t) \tag{6.81}$$

where

m_b = mass of the rock block [kg]
m_{add} = added mass of the fluid surrounding the block [kg]
c, c_{add} = viscous damping and added viscous damping around the block [kg/s]
k = stiffness (spring constant) of system during block displacement [N/m]
x = absolute vertical displacement (uplift height) of the block [m]
$F_{\text{tot}}(t)$ = net total external force applied to the block as a function of time [N]

The effective mass of the system is obtained by superposing the mass of the block m_b and the mass of the surrounding fluid that must be accelerated during block acceleration m_{add}. The added mass force is an important hydrodynamic quantity for rigid bodies undergoing an acceleration in a fluid, it opposes the motion of the block. The same holds for the added damping generated by the surrounding fluid.

Added mass forces generated during block acceleration, damping forces generated during block movement and spring forces depending on absolute block displacement have been estimated and described by Bollaert (2013) based on a numerical simulation of the Federspiel (2011) 3D block uplift experiments, by using equation (6.81) (see Figure 6.35).

For practical situations of rock blocks in plunge pools, the added mass may be expressed as follows, with typical α_{am} values situated between 2 and 3 (Federspiel 2011):

$$m_{\text{add}} = \alpha_{\text{am}} \cdot \rho_w \cdot \pi \cdot x_b \cdot y_b \cdot z_b \; [\text{kg}] \tag{6.82}$$

The viscous damping forces and the spring forces as determined by Bollaert (2013) are only of significance in the case of small vibrational block movements, which may occur during a quasi-instantaneous re-pressurization of the gap formed during sudden block uplift. These forces are not part of the digitalized engineering model, because they are considered specific to particular laboratory boundary conditions and not relevant for practical situations.

Forces and Impulsions on a Rock Block

The net force $F_{\text{net}}(t)$ acting on a prismatic block at each time instant Δt is decomposed into fluctuating dynamic pressure forces acting under and over the block, buoyancy forces, gravity forces, shear forces along the vertical joints based on the weight (for oblique joint sets) and finally shear forces based on differential pressures along the vertical faces of the block (Figure 6.36).

$$F_{\text{net}}(t) = F_{\text{under}}(t) - F_{\text{over}}(t) + F_{\text{buoy}} - F_{\text{grav}} - F_{\text{sh},G} - F_{\text{sh},P} \tag{6.83}$$

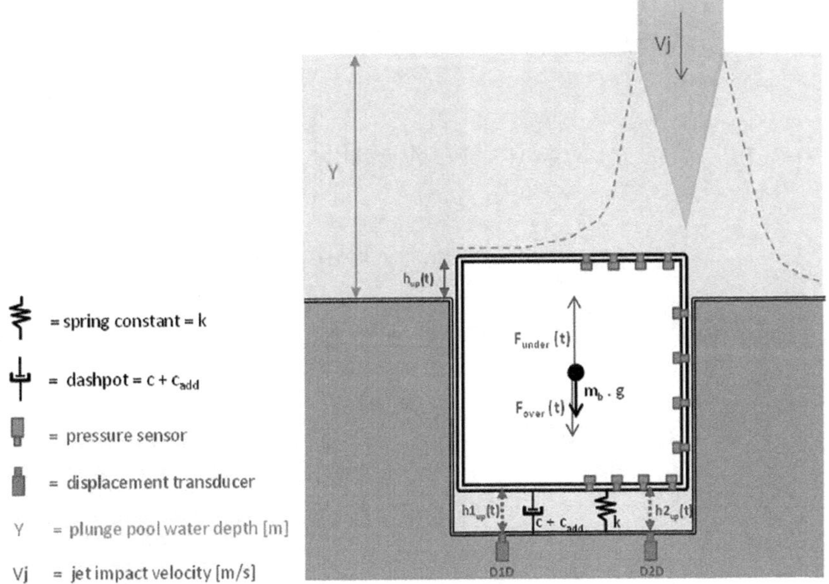

FIGURE 6.35 Main parameters of the single-degree-of-freedom mass-spring-dashpot system valid for a rigid submerged rock block (Bollaert et al. 2013).

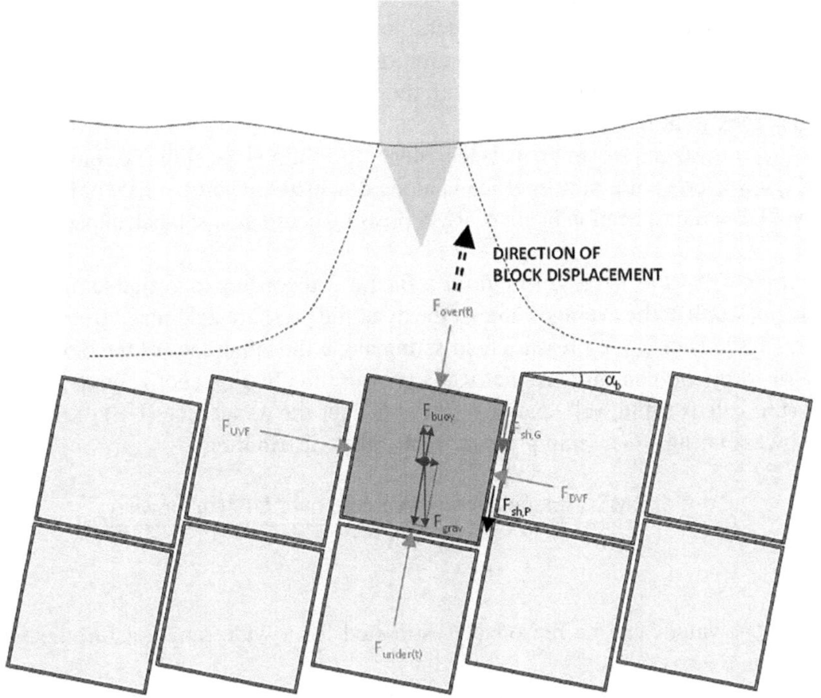

FIGURE 6.36 Main forces acting on a single rock block subjected to an impinging jet (Bollaert 2021).

$$F_{\text{up,DI}} = F_{\text{under}}(t) - F_{\text{over}}(t) = \text{MULT} \cdot C_p^{'} \cdot \rho_w \cdot x_b \cdot y_b \cdot \frac{V_j^2}{2} \tag{6.84}$$

$$F_{\text{buoy}} = \rho_w \cdot g \cdot x_b \cdot y_b \cdot z_b \cdot \cos(\alpha_b \cdot \pi / 180) - \text{ABS}\begin{pmatrix} \rho_w \cdot g \cdot x_b \cdot y_b \cdot z_b \cdot \\ \sin\left(\alpha_b \cdot \dfrac{\pi}{180}\right) \cdot \tan(\varphi) \end{pmatrix} \tag{6.85}$$

$$F_{\text{grav}} = \rho_r \cdot g \cdot x_b \cdot y_b \cdot z_b \cdot \cos(\alpha_b \cdot \pi / 180) \tag{6.86}$$

$$F_{\text{sh},G} = \text{ABS}\left((\rho_r - \rho_w) \cdot g \cdot x_b \cdot y_b \cdot z_b \cdot \sin\left(\alpha_b \cdot \frac{\pi}{180}\right) \cdot \tan(\varphi)\right) \tag{6.87}$$

$$F_{\text{sh},P} = \text{ABS}\left((C_{\text{UVF}} - C_{\text{DVF}}) \cdot \rho_w \cdot z_b \cdot y_b \cdot \frac{V_j^2}{2} \cdot \tan(\varphi)\right) \tag{6.88}$$

where

F_{under} = dynamic pressure forces acting under the block [N]
F_{over} = dynamic pressure forces acting over the block [N]
$F_{\text{up,DI}}$ = net dynamic uplift pressure forces acting on the block [N]
F_{buoy} = buoyancy forces acting on the block [N]
F_{grav} = gravity forces acting on the block [N]
$F_{\text{sh},G}$ = shear forces along block lateral joint based on weight and buoyancy [N]
$F_{\text{sh},P}$ = shear forces along block lateral joint based on differential pressures [N]
α_b = angle of rock blocks/joints with the vertical [°]
φ = rock joint frictional angle [°]
C_{UVF} = average pressure coefficient along upstream vertical face of block [–]
C_{DVF} = average pressure coefficient along downstream vertical face of block [–]
MULT = multiplication factor of RMS pressure coefficient at block upper face [–]

Dynamic Pressures Over and Under a Block According to available research on rock block uplift, the average value of the dynamic pressure field may be neglected in the analysis. Hence, the pressure field acting along the upper face of the block can be described by the non-dimensional RMS pressure fluctuation coefficient $C'_{p,\text{UF}}$ of the pressure values (in [mwc]) spatially averaged over the upper face (UF) of the block, and by assuming a Gaussian pressure probability distribution:

$$C_{p,\text{UF}}^{'} = \frac{\text{RMS (spatial-average pressure over UF)(in [mwc])}}{\dfrac{V_j^2}{2g}} [-] \tag{6.89}$$

Such $C'_{p,\text{UF}}$ values can be reasonably estimated for a wide range of turbulent flow situations.

The net dynamic uplift pressure force on a block is obtained by subtracting the down-force at the upper face of the block from the up-force at the lower face of the block. Figure 6.37 illustrates the net dynamic uplift pressure force $F_{up,DI}$ recorded on a 3D cubic block with a 0.20 m side length impacted by a vertically impinging jet at a velocity of 14.7 m/s.

The corresponding instantaneous net dynamic uplift pressure coefficient $C_{up,DI}$ can be written as $C_{p,LF} - C_{p,UF}$. The range of values recorded during the Federspiel (2011) tests for vertical impinging jets, as well as the values recorded during the Pells (2016) tests for parallel turbulent flow, are compared in Figure 6.38. For perfectly flat bottoms, turbulent jets generate net uplift values that are much higher than the ones for parallel turbulent flows.

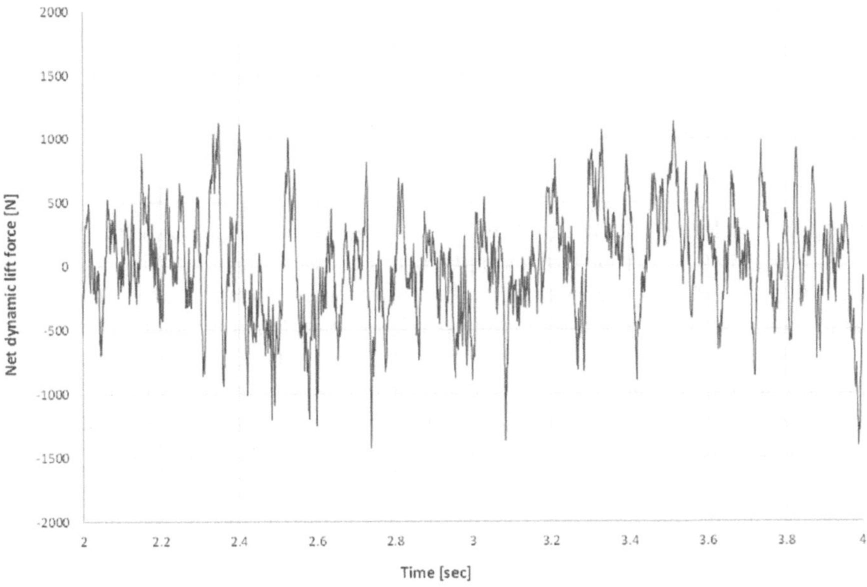

FIGURE 6.37 Net dynamic uplift pressure force $F_{up,DI}$ recorded on a 3D cubic block impacted by a vertical jet at a velocity of 14.7 m/s.

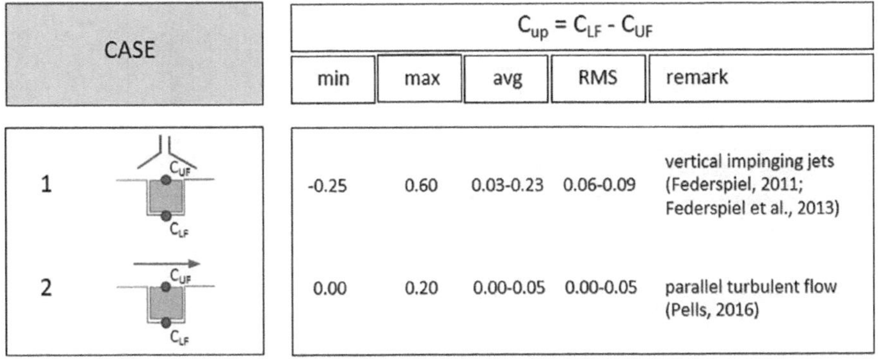

CASE	$C_{up} = C_{LF} - C_{UF}$				
	min	max	avg	RMS	remark
1 (C_{UF}, C_{LF})	-0.25	0.60	0.03-0.23	0.06-0.09	vertical impinging jets (Federspiel, 2011; Federspiel et al., 2013)
2 (C_{UF}, C_{LF})	0.00	0.20	0.00-0.05	0.00-0.05	parallel turbulent flow (Pells, 2016)

FIGURE 6.38 Instantaneous net dynamic uplift pressure coefficient recorded on 3D prismatic blocks impacted by vertical jets (Federspiel 2011) and parallel turbulent flows (Pells 2016).

Next, based on experiments described by Liu et al. (1998), Federspiel (2011) and Bollaert et al. (2013), the periods of positive net pressure forces have been studied more in detail. For a vertically impinging jet with a velocity of 14.7 m/s, Figures 6.39 and 6.40 present these periods or pulses as a function of their time duration (abscissa) and their average uplift pressure under the form of the uplift pressure coefficient $C_{\text{up,DI,avg}}$ (abscissa). At higher values, both parameters seem a priori uncorrelated.

The average value (per pulse) of the net dynamic uplift pressure coefficient $C_{\text{up,DI,avg}}$ may be determined by the RMS pressure fluctuations at the block surface $C'_{p,\text{UF}}$ multiplied by a factor MULT, as follows:

$$\text{MULT} = \frac{C_{\text{up},DI,\text{avg}}}{C'_{p,\text{UF}}} [-] \tag{6.90}$$

$$F_{\text{up},DI} = \rho_w \cdot x_b \cdot y_b \cdot \text{MULT} \cdot C'_{p,\text{UF}} \cdot \frac{V_j^2}{2} [\text{N}] \tag{6.91}$$

Next, by considering each net uplift pressure coefficient $C_{\text{up,DI,avg}}$ as a distinct value of a continuous random variable in a sample space of values, the relative likelihood that the value of the random variable would be within a prescribed range of values of the sample space may be expressed by a probability density function (PDF).

As the laboratory experiments are performed over short (1 min) time intervals, the extreme part of the sample space (extreme values) is most probably missing. Nevertheless, sound statistical analysis of the Federspiel (2011) experiments for

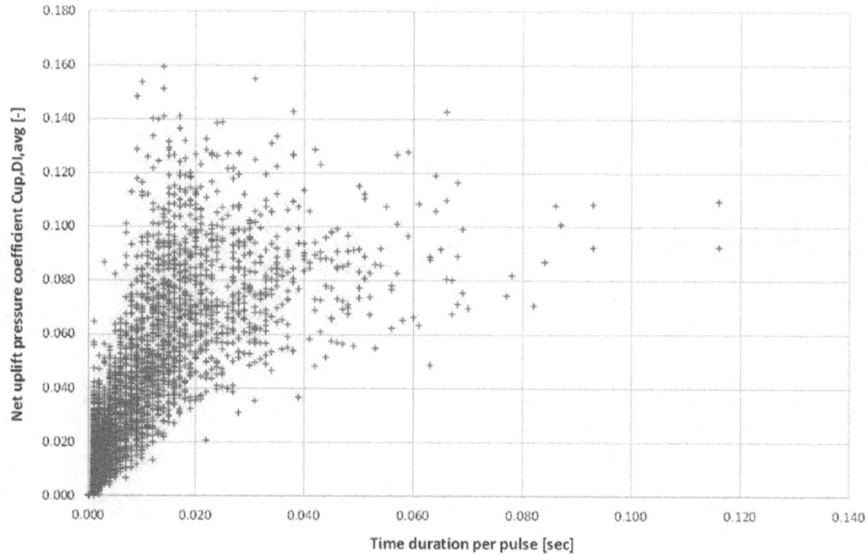

FIGURE 6.39 Ensemble of average pressure coefficient and time duration of net dynamic uplift pulses recorded on a 3D cubic block impacted by a vertical jet at a velocity of 14.7 m/s.

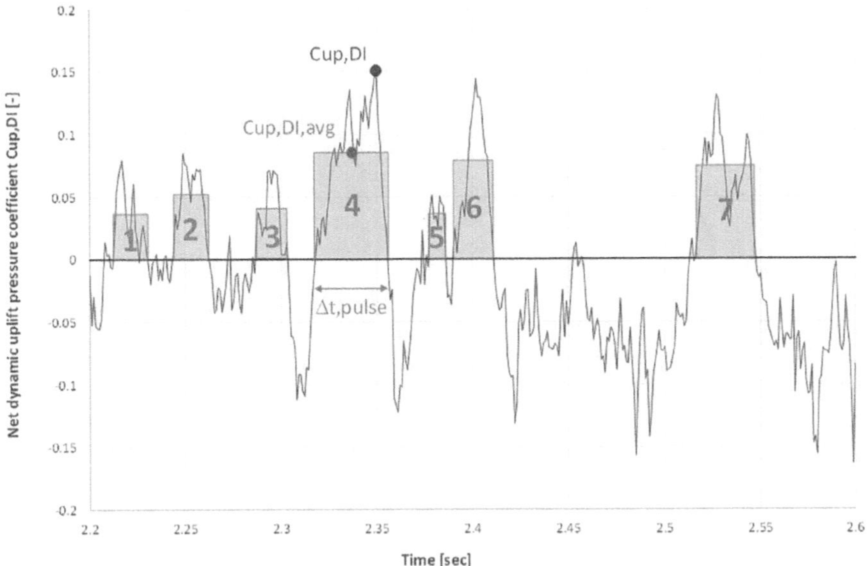

FIGURE 6.40 Determination of pressure coefficients of net dynamic uplift force pulses.

different near-prototype jet velocities and plunge pool depths has shown that the PDFs of the net uplift pressure pulses recorded over 1 min time intervals tend towards a Beta-distribution for low-probability (i.e. extreme) values. Such a Beta-distribution has been determined separately for the net dynamic uplift pressure forces, and for the time duration of the pressure pulses.

Figure 6.41 compares the cumulative sample distribution with a cumulative Beta-distribution for the net dynamic uplift pressure forces on a 3D cubic block impacted by a vertical jet at a velocity of 19.6 m/s. The Beta-distribution matches with the recorded probability of the sampled values, especially at extreme values with low probability. The Gaussian distribution is clearly not applicable.

The Beta-distribution is then used for determination of extreme values with very low probability of occurrence. This has been done for different jet impact velocities and a plunge pool depth of 0.60 m. By relating the maximum recorded values to the time duration of the tests, i.e. 1 min each, the Beta-distribution allows to estimate the extreme values for longer time durations. Corresponding MULT values are presented in Figure 6.42 as a function of the time duration of single flood events.

MULT values of 4.2 (i.e. the maximum recorded values by Liu et al. (1998)) correspond to uplift forces with a probability of occurrence of about once every 10–150 h. Similarly, MULT values of 4.5 correspond to pressure pulses with a probability of occurrence of about once every 50–800 h. For higher MULT values, the needed time duration of flood events rapidly becomes several thousands of hours. The MULT value is a parameter that must be defined as a function of the total duration of flood events for the problem in question.

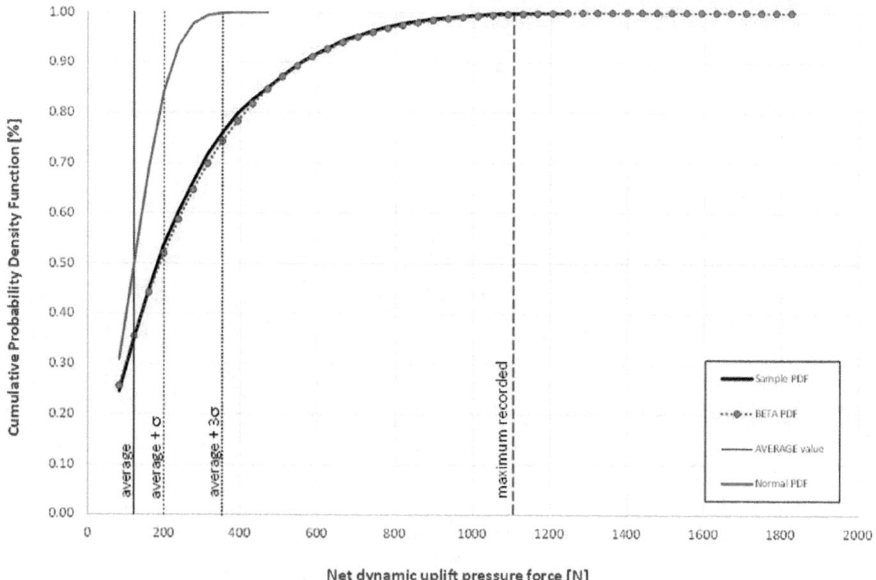

FIGURE 6.41 Cumulative sample distribution and Beta-distribution of net dynamic uplift forces recorded on a 3D cubic block impacted by a vertical jet at a velocity of 19.6 m/s.

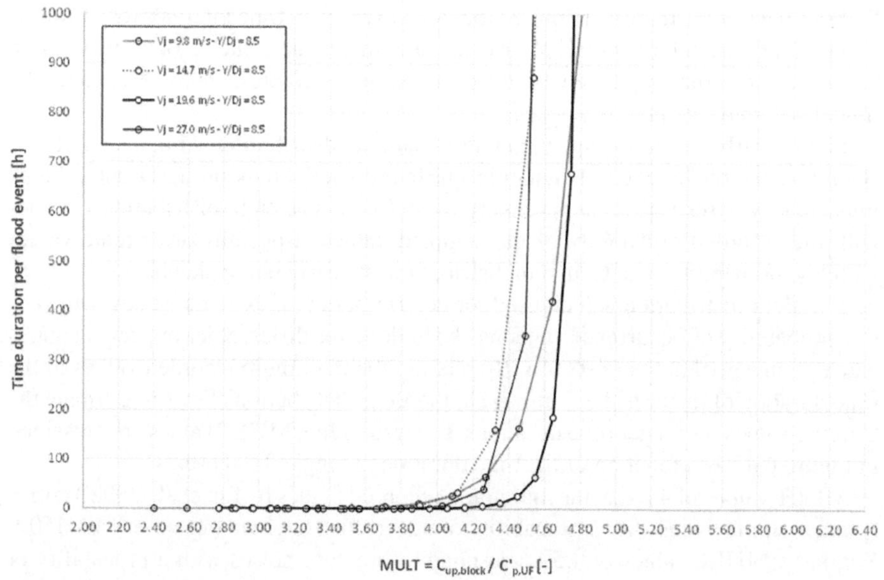

FIGURE 6.42 MULT values of net dynamic uplift forces as a function of the time duration of a flood event.

Time Duration of Pressure Pulsations The time duration to be considered for the net uplift pressure pulsations on rock blocks depends on two factors: the spectral content of the turbulent flow over the upper face of the block and the resonance frequency range of the joints around those blocks.

The spectral content of the turbulent flow depends on the flow velocities and the average sizes of the main turbulent eddies that govern that turbulence. On laboratory models, it is generally difficult to reproduce the entire prototype spectral content of the flow. Low frequencies are limited by the small dimensions of the model and high frequencies are limited by the low flow velocities of the model. Especially within the frequency range of macroturbulent flow, i.e. 0–20 Hz, part of the spectrum may be underestimated on models.

The Federspiel (2011) experiments that have been used to determine the present time durations of uplift pulsations have been performed by using near-prototype flow velocities but for a small-scale plunge pool basin. As such, the time durations of the pulses as measured may somewhat underestimate low-frequency pulsations, i.e. within the range of 0–5 Hz, for which a larger plunge pool would be needed. Typical time durations of pressure pulsations as recorded during the laboratory experiments are of 0.10 s.

Figure 6.43 illustrates the cumulative Beta-distribution for the time duration of the pressure pulses on a 3D cubic block impacted by a vertical jet at a velocity of 14.7 m/s. This Beta-distribution again matches with the recorded probability of the sampled values, especially at extreme values with low probability. The Gaussian distribution is clearly not applicable.

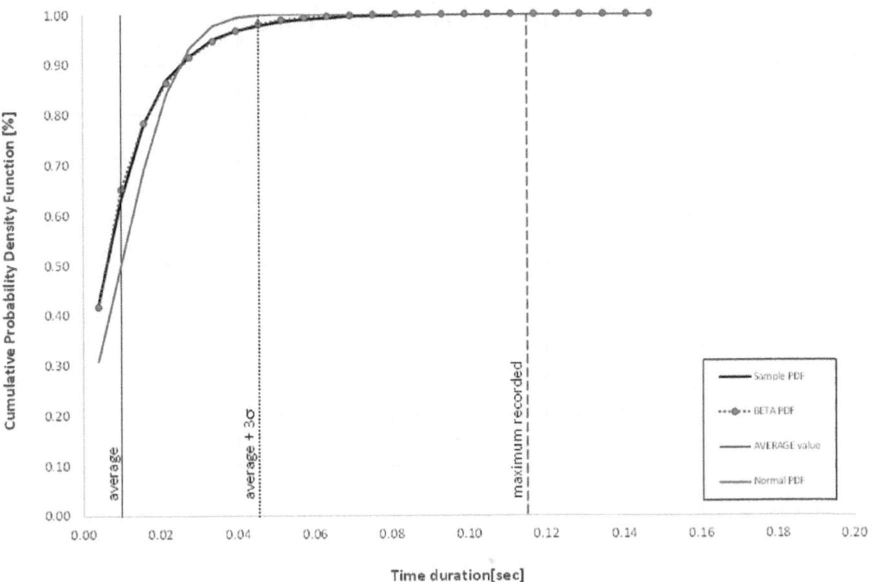

FIGURE 6.43 Cumulative sample distribution and Beta-distribution of time durations of pulses recorded on a 3D cubic block impacted by a vertical jet at a velocity of 14.7 m/s.

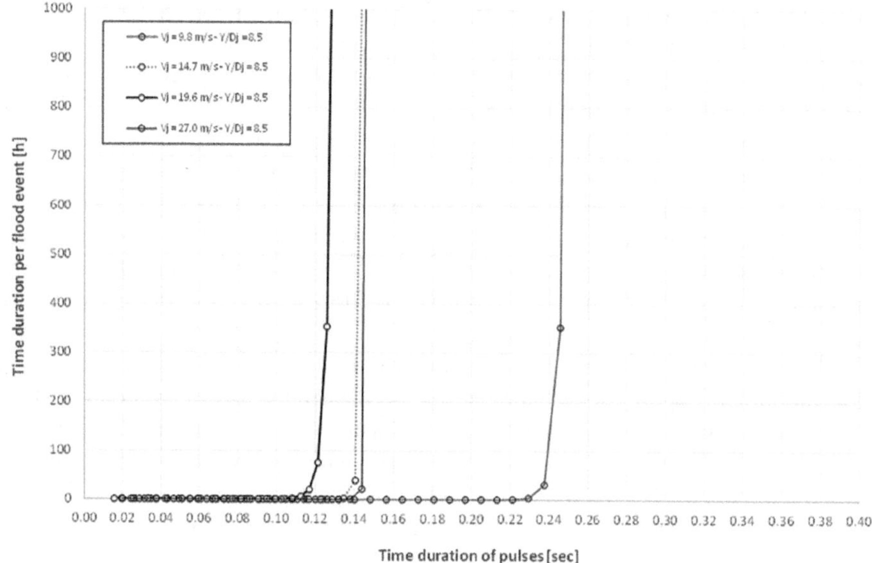

FIGURE 6.44 Time duration of distinct pressure pulsations on 3D prismatic blocks as a function of the time duration of a flood event.

The Beta-distribution is, again, used for determination of extreme values with very low probability of occurrence. Corresponding time periods of distinct pressure pulsations on 3D blocks are presented in Figure 6.44 as a function of the time duration of single flood events. Most pulse periods fall within the 0.10–0.15 s range of values, except for the recordings at a low jet velocity of 9.8 m/s, for which values of 0.20–0.25 are more appropriate.

For real-life rock joints at prototype scale, however, the time durations may become longer in case of longer joints, and may be determined by the resonance frequency of the joints around the blocks, based on the following equation valid for open-ended joints:

$$\frac{1}{f_{\text{res}}} = \Delta t = \frac{2L}{c} \tag{6.92}$$

For air-water wave celerities typically on the order of ~100 m/s, the time durations mainly depend on the total length of the joints. The larger the blocks, the larger the max. time durations of pulsations that may eject those blocks. For small blocks, however, this frequency content becomes too high and is not able to stimulate the joint to resonance conditions. Hence, for digital engineering applications, the following rules are applied:

- Determine the time duration following the resonance frequency of the joints around the blocks.
- If this duration becomes smaller than the values obtained on the Federspiel (2011) experiments, i.e. ~ 0.10 s, then use 0.10 s as a minimum value.

Dynamic Pressures along Upstream (UVF) and Downstream (DVF) Vertical Block Faces The dynamic water pressures also act in the joints that are located along the upstream and downstream vertical (or quasi-vertical) faces of the block (Figure 6.36), generating shear forces. Such shear forces are opposed to block movements, and occur when the pressure forces in the lateral joints are not equal, i.e. when a net lateral pressure difference is generated that laterally pushes the block towards its neighbouring block.

Accounting for a rock joint friction angle φ, this generates a shear force $F_{sh,P}$ in one of the lateral joints (equation 6.88), by expressing the lateral water pressures by means of average pressure coefficients C_{UVF} and C_{DVF}.

Based on laboratory experiments performed by Reinius (1986) and Pells (2016), eight different geometrical situations of a rock block as compared to the surrounding rock mass have been investigated and are shown in Figure 6.45.

For each case, the range of minimum, maximum and average differences between the lateral pressure coefficients has been defined. For some cases, like case number 4 and case number 7, a mathematical equation is available to express this net lateral pressure difference (Pells 2016).

It may be observed that, on average, these pressure differences are rather low to very low. Hence, during the computations, safe-side assumptions of low to quasi-zero values are recommended.

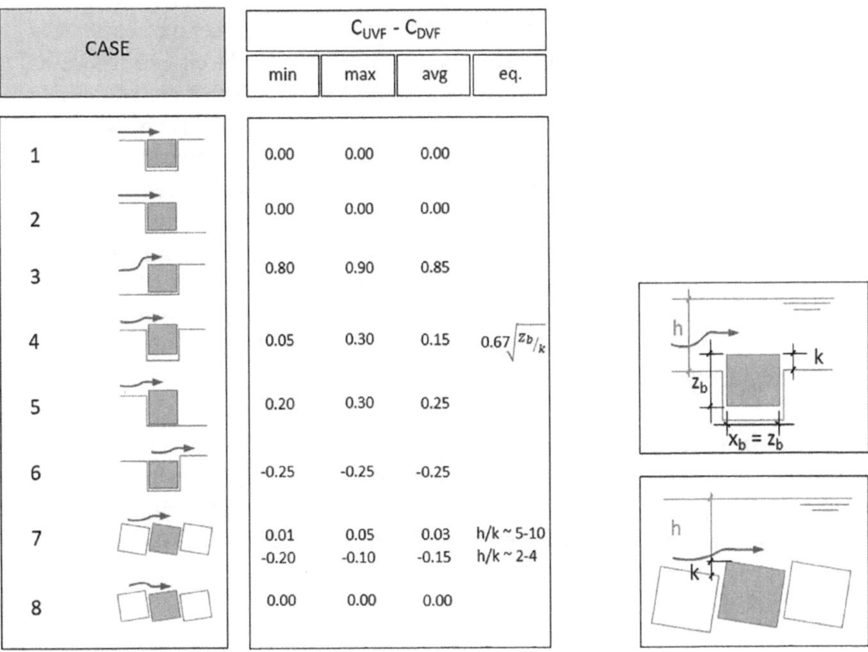

CASE	$C_{UVF} - C_{DVF}$			
	min	max	avg	eq.
1	0.00	0.00	0.00	
2	0.00	0.00	0.00	
3	0.80	0.90	0.85	
4	0.05	0.30	0.15	$0.67\sqrt{z_b/k}$
5	0.20	0.30	0.25	
6	-0.25	-0.25	-0.25	
7	0.01 -0.20	0.05 -0.10	0.03 -0.15	h/k ~ 5-10 h/k ~ 2-4
8	0.00	0.00	0.00	

(for cubical blocks and $0.01 < k/Z_b < 0.2$)
* based on quasi-steady lateral pressures/forces

FIGURE 6.45 Net lateral pressure differences along lateral (i.e. vertical face) joints of blocks in different geometrical situations (Bollaert 2021).

Shear Forces in Lateral Joints due to Gravity and Buoyancy Similar to the shear forces that may be generated by differential dynamic water pressures inside the lateral joints, an additional shear force may occur for blocks that have a non-zero angle of orientation with the horizontal. In such a case, the component of the gravity force that is oriented perpendicular to the lateral joint will add a shear force $F_{sh,G}$ in the joint by accounting for the rock joint friction angle φ (equation 6.87).

In the same way, the component of the buoyancy force that is oriented perpendicular to the lateral joint will reduce this added shear force because of counteracting gravity forces.

Movement of a Block

The movement of a submerged rigid block due to pressure pulses can be subdivided into different phases. The significance of each of the phases depends on the temporal structure of the pressure pulses. The following phases can be outlined and are illustrated in Figure 6.46 (Bollaert 2021):

- *Phase 1*: The block initially at rest receives a constant net uplift force during a time period Δt_{pulse}. This force provides momentum to the block, for a positive velocity and acceleration, and expressed by a final acceleration and a final uplift velocity valid at the end of the time duration of the net force acting on the block.
- *Phase 2*: The block in movement receives a subsequent constant net force, but downwards, during a time period $\Delta t_{downpulse}$. This force opposes the uplift movement of the block and will progressively decelerate the block and reduce its uplift velocity. For downward forces and/or time periods that are sufficiently important, the block will eventually reach zero values and start its inverse movement. In case the downward force is (quasi-) similar to the precedent upward force, then the time duration that is needed to inverse block movement will be somewhat smaller than the initial time duration of uplift, because the submerged weight now acts in the same direction than the downwards-oriented force.
- *Phase 3*: The block starts to move downwards, the net force is still negative and both the block velocity and the block acceleration are now negative. The block will eventually reach the bottom again and remain at rest, until a subsequent uplift pressure pulse is received.

As such, when the block movement inverses, the block reaches its maximum uplift displacement. This displacement consists of two parts: the displacement during phase 1 (acceleration), called h_1 in Figure 6.46, and the additional displacement because of remaining positive momentum during phase 2, called h_2 in Figure 6.46 (Bollaert 2021).

Acceleration Phase of a Block

The total net uplift force as defined by equation (6.83) is considered to act on the block for a time duration Δt_{pulse}. The net impulsion that is given to the block can then be written as:

$$I_{net,DI} = F_{net,DI} \cdot \Delta t_{pulse} \text{ [N} \cdot \text{s]} \tag{6.93}$$

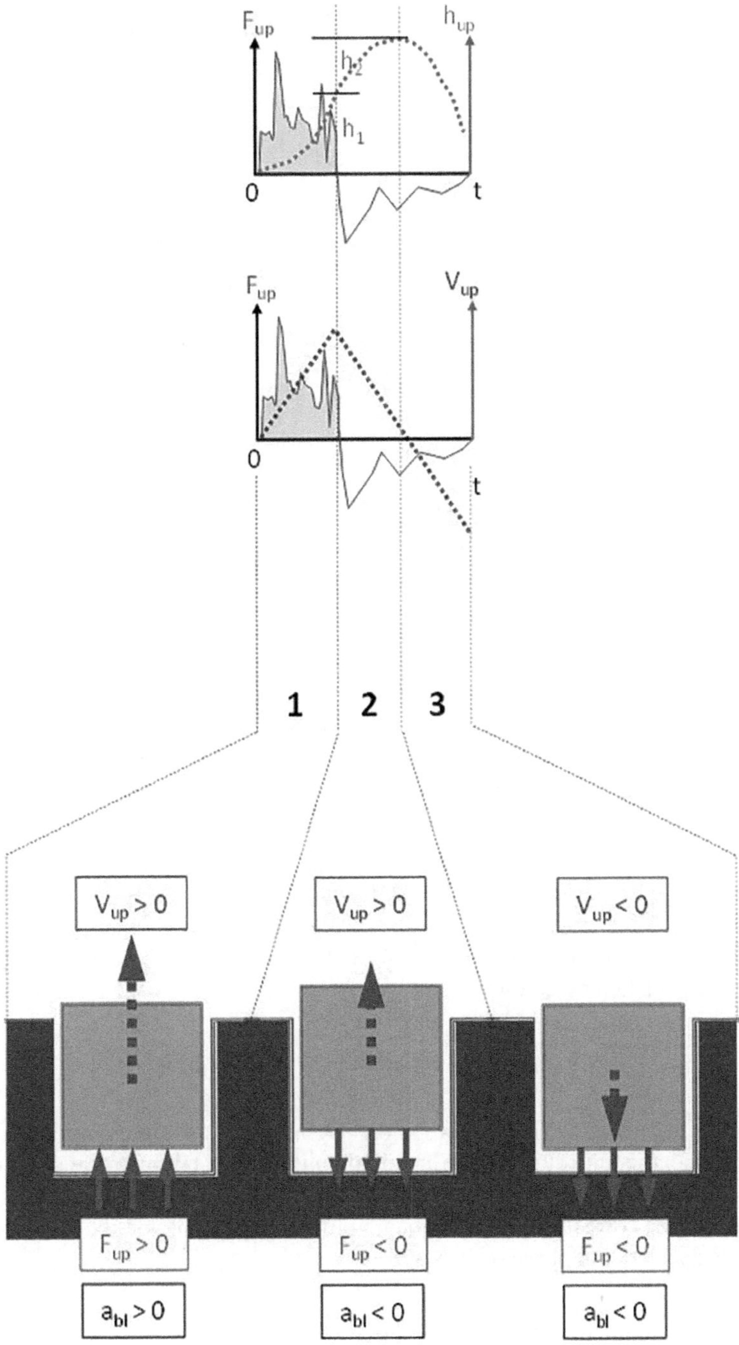

FIGURE 6.46 Different phases of dynamic rock block uplift (Bollaert 2021).

According to equation (6.81), which becomes simplified by neglecting the damping forces and the stiffness of the system, the net acceleration that is given to the block is defined as:

$$a_{\text{net,DI}} = \frac{F_{\text{net,DI}}}{\left(m_{b+m_{\text{add}}}\right)} \,[\text{m/s}^2] \tag{6.94}$$

A double temporal integration over Δt_{pulse} then allows computing the displacement $h_{\text{up,net,DI}}$ that the rock block exhibits during its phase of acceleration from 0 to $a_{\text{net,DI}}$:

$$h_{\text{up,net,DI}} = a_{\text{net,DI}} \cdot \frac{\left(\Delta t_{\text{pulse}}\right)^2}{2} \,[\text{m}] \tag{6.95}$$

During this same time interval, a final velocity $V_{\text{up,net,DI}}$ is given to the block after a time Δt_{pulse}:

$$V_{\text{up,net,DI}} = \frac{I_{\text{net,DI}}}{\left(m_b + m_{\text{add}}\right)} \,[\text{m/s}] \tag{6.96}$$

Deceleration Phase of a Block

The deceleration phase starts whenever the net total force on the block becomes negative, i.e. downwards oriented. This depends on the net dynamic water pressure force on the block, i.e. $F_{\text{under}}(t) - F_{\text{over}}(t)$. In the present MDI model, and based on the 3D near-prototype scaled experiments by Federspiel (2011), this force is considered equal to the precedent upwards-oriented dynamic water pressure force, but now becomes negative. Hence, the total net force on the block may still be written as in equation (6.81), but now becomes negative:

$$F_{\text{down,DI}}(t) = \left(F_{\text{under}}(t) - F_{\text{over}}(t)\right) + F_{\text{buoy}} - F_{\text{grav}} - F_{\text{sh},G} - F_{\text{sh},P} \;[\text{N}] \tag{6.97}$$

The time necessary for this downwards-oriented force to stop the momentum of the block is then written as follows:

$$\Delta t_{\text{downpulse}} = \frac{V_{\text{up,net,DI}} \cdot \left(m_b + m_{\text{add}}\right)}{F_{\text{down,DI}}} \,[\text{s}] \tag{6.98}$$

When considering equal shear forces in the lateral joints, this time duration is somewhat less than the time duration of the initial uplift period, because of the submerged weight that now acts in the same direction as the dynamic pressure force during the downpulse. The corresponding additional uplift displacement that the block exhibits during its deceleration phase is then written as:

$$h_{\text{decel,DI}} = \frac{V_{\text{up,net,DI}} \cdot \Delta t_{\text{downpulse}}}{2} \,[\text{m}] \tag{6.99}$$

Finally, for downpulse periods (deceleration phases) that are longer than $\Delta t_{\text{downpulse}}$, the block will inverse movement, and start moving and accelerating downwards, until it touches base again. This last phase, however, is not accounted for by the MDI model.

Total Vertical Displacement of a Block

The total (maximum) vertical displacement of the block during phase 1 and phase 2 corresponds to a superposition of the consecutive individual displacements:

$$h_{\text{up,tot,DI}} = h_{\text{up,net,DI}} + h_{\text{decel,DI}} \text{ [m]} \tag{6.100}$$

Similar to the DI method, to account for the more complex real block situations, the necessary or critical uplift displacement is a model parameter that needs to be calibrated according to site-specific parameters and is expressed following equation (6.80).

Based on real-life scour calibrations performed with the MDI method (see Chapter 7), critical net uplift displacements of ~0.10 times the block height were found adequate in conjunction with time durations of net uplift pulses of ~0.10 s and added mass coefficients of 2–3.

Quasi-Steady Impulsion (QSI) Methods

Introduction

The Quasi-Steady Impulsion (QSI) method has been developed by Bollaert (2012). The method determines potential detachment ("peeling off") of a rock block from its mass, due to a quasi-steady (time-averaged) net uplift force. This lift force is generated by a partial or total deviation of the flow velocity parallel to the boundary by a rock block that protrudes into the low compared to the surrounding blocks or rock mass interface.

In contrast with the DI method, the QSI method makes use of the local flow velocity V_{local} along the water-rock interface. This local flow velocity is based on the axial flow velocity decay of a plunging jet during diffusion through a water cushion, and based on the radial velocity decay of the deviated wall jet extending radially outwards from the point of impingement.

Also, the QSI method is not valid at the point of stagnation of the jet impacting the water-rock interface, but starts in the area of the interface where the impinging jet has been soundly transformed into a wall jet, i.e. oriented quasi-parallel to the interface. This area theoretically starts at a distance equal to about one time the jet impact diameter in the plunge pool, both upstream and downstream from the stagnation point of jet impingement.

Especially the scour computed outside of the turbulent diffusive shear layer of the jet impacting the interface is of importance, because the DI method is somewhat less relevant in these areas, while the QSI method is able to predict the shape of the scour hole in these areas. For practice, however, both DI and QSI methods are computed all over the interface.

Based on laboratory experiments and literature feedback for different geometrical situations of rock blocks that are protruding, performed by different independent research investigations (Reinius 1986; Otto 1989; USBR 2007), the flow deviations generated by protruding blocks have been transformed into time-averaged net uplift pressure coefficients. These are at the base of the QSI method developed by Bollaert (2012). Based on additional research investigations and laboratory experiments performed by Pells (2016), and recently by Wahl and Heiner (2023), both quasi-steady and fluctuating net uplift pressure coefficients are available for many protruding geometrical block situations.

An enhanced version of the QSI has recently been developed by Bollaert (2021) within the framework of the rocsc@r digital environment. This MQSI method is based on both quasi-steady and fluctuating net uplift pressures acting on blocks. The quasi-steady values define a critical net uplift force, while the additional pulsating values define a net uplift impulsion. The method is applicable to any type of impinging jets, from circular to rectangular jets, and from compact to broken-up jets.

Quasi-Steady Impulsion Based on Bollaert (2012)

The QSI method integrates a sequence of physical phenomena that are illustrated in Figure 6.47 and that are described as sub-models hereafter.

Free Falling Jet Model

Modelling of the free jet trajectory based on ballistics and air drag: The model needs the jet velocity, jet angle and jet diameter and jet shape at issuance from the spillway to compute these characteristics at the point of impact in the plunge pool, accounting for potential deformation of the shape of the jet during its fall.

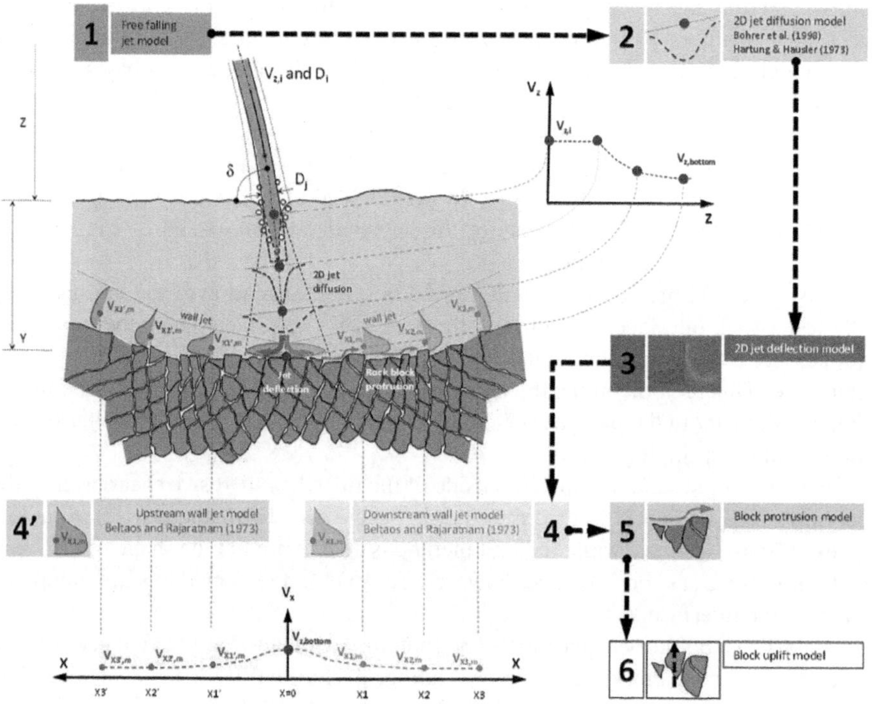

FIGURE 6.47 Step-by-step methodology of quasi-steady impulsion (QSI) computations, using a series of six sub-models (Bollaert 2012).

2D Jet Diffusion Model

During jet diffusion through the plunge pool water depth, the average velocity of the jet along its centreline progressively decays. This decay is illustrated in Figure 6.48 for a circular-shaped and a rectangular-shaped jet. Most studies relate the axial jet velocity decay to the local jet thickness B_j or jet diameter D_j, and to the inverse of the water depth *Y*. Following Hartung and Hausler (1973), the relationship for V_z as a function of the jet velocity at impact in the pool V_j is written as:

$$V_z = \left(\frac{Z_{core} \cdot D_j}{Y}\right) \cdot V_j \text{ [m/s] for circular-shaped jets} \tag{6.101}$$

$$V_z = \sqrt{\frac{Z_{core} \cdot B_j}{Y}} \cdot V_j \text{ [m/s] for rectangular-shaped jet} \tag{6.102}$$

in which Z_{core}. Dj stands for the distance necessary for the jet to diffuse its core through the pool depth. This distance is generally taken at 4–6 times the jet diameter at impact D_j.

An example is provided in Figure 6.48 for a jet with an impact velocity $V_j = 40$ m/s and a jet diameter or thickness at impact D_j (or B_j) = 1.0 m. The plunge pool water depth is set at an altitude of 210 m a.s.l. The axial velocity decay is much more pronounced for a circular-shaped jet than for a rectangular-shaped jet, because of the axial-symmetric diffusion in the first case.

2D Jet Deflection Model

Based upon the pool depth *Y* and the angle and location of impact of the jet onto the pool surface, the impact point of the jet near the bedrock may be defined. The angle

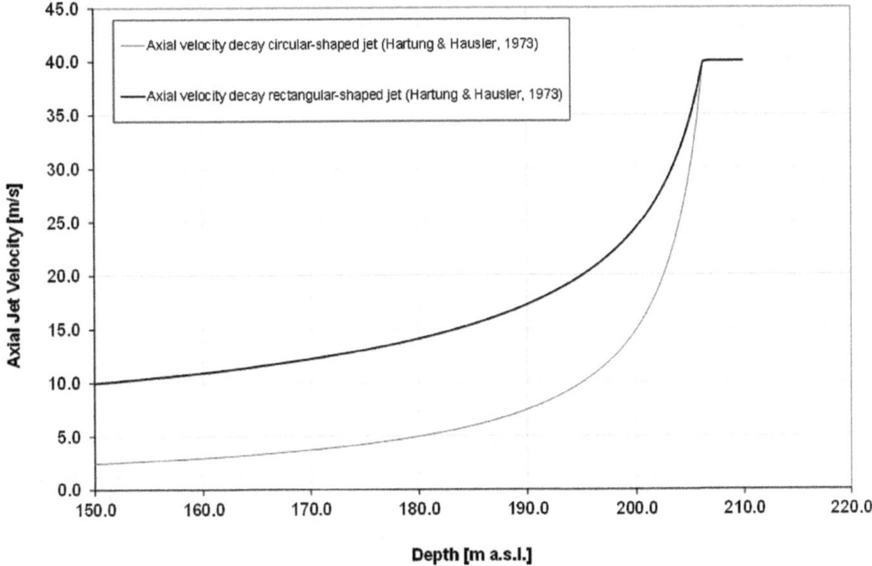

FIGURE 6.48 Axial velocity decay for a circular/rectangular-shaped jet at impact in a plunge pool.

of the jet through the pool depth is considered equal to the angle d of the jet at impact on the water-rock interface. This constant and linear trajectory is a simplification of reality, especially for confined 3D plunge pool volumes where significant returning flow currents exist that may alter the jet diffusion trajectory. This impact point is used as the starting point for the development of wall jets radially outwards, both towards up- and downstream.

Next, the parts of the total flow rate (q_{total}) deviated towards up- and downstream are defined as percentages of the total rate, based on equations (6.103) and (6.104). The deflection of the jet at the pool bottom occurs in both the up- and downstream directions. The importance of each of these deflections directly depends on the angle δ of the jet upon impact in the pool (Bollaert 2012).

As shown in Figure 6.49, based on Reich (1927), a theoretical approach for plane jets with initial discharge q_{total} and thickness D_j impinging on a flat plate relates the respective discharges q_{up} and q_{down} and thicknesses h_{up} and h_{down} by means of the cosinus of the jet angle δ with the horizontal.

$$\frac{q_{\text{up}}}{q_{\text{total}}} = \frac{h_{\text{up}}}{D_j} = \frac{1}{2} \cdot (1 - \cos\delta)[-] \tag{6.103}$$

$$\frac{q_{\text{down}}}{q_{\text{total}}} = \frac{h_{\text{down}}}{D_j} = \frac{1}{2} \cdot (1 + \cos\delta)[-] \tag{6.104}$$

Table 6.19 hereunder shows the up- and downstream deviated parts of the total flow for different jet angles δ.

2D Wall Jet Model

Once the jet deflected, the generated wall jets oriented parallel to the interface may be characterized by their initial flow velocity $V_{z,\text{bottom}}$ and their initial

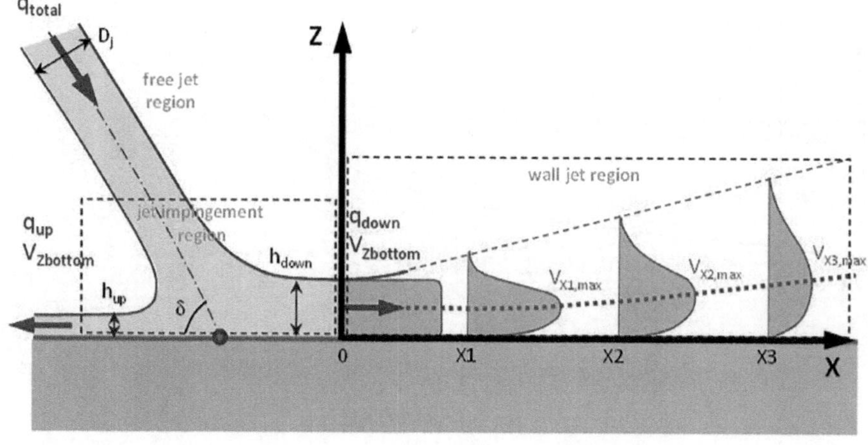

FIGURE 6.49 Plane jet deflection on a flat bottom according to Reich (1927) and wall jet velocity profiles following Beltaos and Rajaratnam (1973) (Bollaert 2012).

TABLE 6.19
Discharge Distribution Up- and Downstream of the Point of Jet Impingement

Jet Angle δ	10°	20°	30°	40°	90°
q_{up}	1.5%	6%	7%	12%	50%
q_{down}	98.5%	94%	93%	88%	50%

thickness $h_{up/down}$ at the point of deflection. Initiating from this singular location, the wall jets develop radially outwards following self-preserving velocity profiles (Beltaos and Rajaratnam 1973) as given by the following equation valid for 2D wall jets (i.e. for rectangular-shaped jets at impact in the pool):

$$\frac{V_{X,\max}}{V_{z,\text{bottom}}} = \min\left(V_{z,\text{bottom}}, \frac{3.5}{\sqrt{\dfrac{X}{h_{\text{down}}}}}\right) [-] \tag{6.105}$$

$V_{z,bottom}$ depends on the shape of the jet and on its development length through the water depth Z. $V_{z,bottom}$ thus continuously changes (reduces) during scour formation. $V_{X,max}$ expresses the decay of the maximum cross-sectional jet velocity with the relative distance from the start of the wall jet (i.e. lateral distance X divided by the initial thickness of the deflected jet $h_{up/down}$). At small lateral distances from the impingement point, the wall jet velocity equals the maximum possible value of $V_{z,bottom}$. Also, with increasing distance from the jet deflection point, the jet velocity profile progressively flattens.

The flow separation defines the initial heights of the wall jets towards up- and downstream. These are simply defined as the flow percentages applied to the jet diameter D_j or thickness B_j upon pool impact. Even if reality is more complex, this approach has the merit to respect the discharge distribution between both wall jets in a simple manner. These initial jet thicknesses represent the length scales $h_{up/down}$ in equation (6.105), and thus also the degree of decrease of the max. wall jet velocity with increasing distance X from the starting point.

It is assumed that this max. wall jet velocity applies at the bottom and is directly responsible for the generation of quasi-steady pressure gradients on protruding rock blocks.

Block Protrusion Model

Several researchers have defined the pressure gradient generated by sudden flow deviation at protruding blocks, by means of an uplift pressure coefficient C_{uplift} expressing the pressure as a percentage of the kinetic energy $V^2_{z,m}/2g$ of the quasi-parallel flow deviated by the block.

Reinius (1986) extensively studied potential net uplift stagnation pressures for different configurations of block protrusion and joint angles subjected to a high-velocity flow parallel to the pool bottom. The results are summarized in Figure 6.50 (Bollaert 2012).

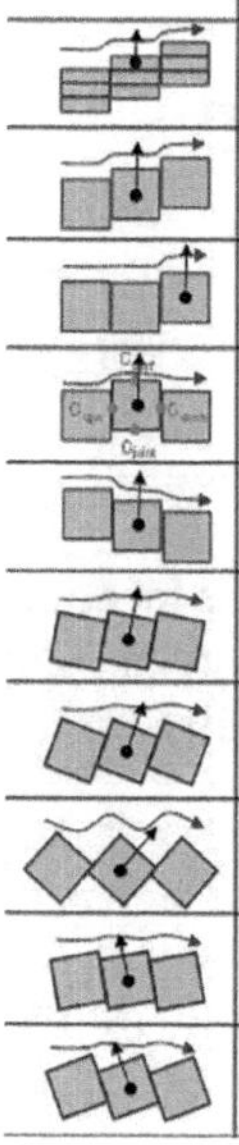

	h_{block}/e_{block}	β_{block}	C_{surf}	C_{joint}	C_{uplift}	Comment
	17-29	0°	0.030	0.250	0.220	
	17-29	0°	0.030	0.250	0.220	
	17-29	0°	0.020	0.105	0.085	
	17-34	0°	-0.010	0.145	0.155	
	4-9	0°	0.075	-0.110	-0.070	block stabilizing forces
	4-10	3°	0.030	0.350	0.310	
	2-4	9°	-0.10-0.17	0.36-0.55	0.37-0.47	
	1.0-2.5	18°	-0.15-0.00	0.23-0.40	0.25-0.45	
	4.2-8.7	-3°	0.02-0.13	-0.105	-0.070	block stabilizing forces
	1.0-2.3	-18°	-0.10-0.16	-0.075	-0.150	block stabilizing forces

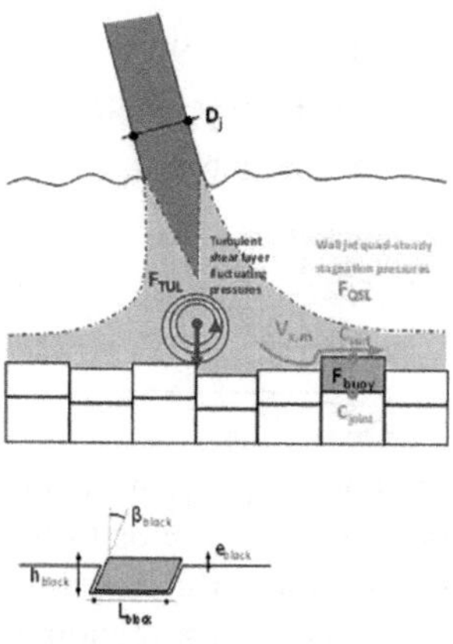

$$F_{QSL} = (C_{joint} - C_{surf}) \cdot L_{block} \cdot \frac{V_{x,m}^2}{2g}$$

$$V_{x,m} = f(X, h_{down}, V_{Zbottom})$$

$$h_{down} = D_j \cdot \frac{q_{down}}{q_{total}}$$

FIGURE 6.50 Summary of quasi-steady pressures over and under protruding rock blocks subjected to parallel high-velocity flow following Reinius (1986) (Bollaert, 2012).

By subtracting the surface pressure coefficient C_{surf} from the joint pressure coefficient C_{joint}, net uplift pressure coefficients C_{uplift} of up to 0.67 have been measured. For practice, sound values are situated between 0.10 and 0.50, depending on the importance of the protrusion of the block and on the joint angle. For joint angles that are oriented against the flow together with negative steps, negative or stabilizing coefficients are obtained.

For the computations, coefficients of 0.1–0.2 may be considered plausible for low to very low block protrusions, while coefficients of 0.3–0.5 correspond to moderate to significant block protrusions, i.e. for rough and irregular pool bottoms, typically encountered in fractured rock. The latter values are considered most plausible for a real water-rock interface. A series of calibrations for real-life scour formation at dams are provided in Chapter 7.

Block Uplift Model

Finally, based on site observations or available geomechanical characteristics of the rock mass, typical rock block shape (z_b/x_b) and dimensions are determined. The shape of the block is directly related to its ease of detachment. For example, a plate-like shaped flat block will be easily detached by the impacting flow, while a columnar prismoidal block is more profoundly embedded in the surrounding rock mass and will be detached with more difficulty, even in case of protrusion compared to the surrounding blocks.

Block stability under quasi-steady parallel flow is computed by determining the stabilizing and destabilizing forces on the block and by equating the force balance.

For rectangular-shaped blocks with horizontal and vertical joint orientations, the main destabilizing force is the time-averaged net uplift force that may act on a representative protruding rock block. This force can be defined as a function of a net uplift pressure coefficient C_{uplift}. Second, the buoyancy force must be accounted for as destabilizing force. The main stabilizing force that opposes these uplift forces is the weight of the block F_{grav}.

Furthermore, for rectangular-shaped blocks with joints that have an angle with the horizontal, the main gravity and buoyancy forces must be adapted, and additional shear forces are generated in one of the lateral joints due to the weight (stabilizing) and due to the buoyancy (destabilizing).

$$F_{net} = F_{up,AVG} + F_{buoy} - F_{grav} - F_{sh,G} \tag{6.106}$$

$$F_{up,AVG} = \rho_w \cdot g \cdot x_b \cdot y_b \cdot C_{uplift} \cdot \frac{V_{X,max}^2}{2g} \tag{6.107}$$

$$F_{buoy} = \rho_w \cdot g \cdot x_b \cdot y_b \cdot z_b \cdot \cos\left(\alpha_b \cdot \frac{\pi}{180}\right) - \text{ABS}\left(\begin{array}{c} \rho_w \cdot g \cdot x_b \cdot y_b \cdot z_b \\ \cdot \sin\left(\alpha_b \cdot \frac{\pi}{180}\right) \cdot \tan(\varphi) \end{array}\right) \tag{6.108}$$

$$F_{grav} = \rho_r \cdot g \cdot x_b \cdot y_b \cdot z_b \cdot \cos(\alpha_b \cdot \pi/180) \tag{6.109}$$

$$F_{sh,G} = \text{ABS}\left((\rho_r - \rho_w) \cdot g \cdot x_b \cdot y_b \cdot z_b \cdot \sin\left(\alpha_b \cdot \frac{\pi}{180}\right) \cdot \tan(\varphi) \right) \tag{6.110}$$

Block stability in the QSI method is obtained whenever the following relation of force balance holds:

$$F_{net} < 0 \tag{6.111}$$

Modified Quasi-Steady Impulsion (MQSI) Method Based on Bollaert (2021)

Introduction

The QSI model developed by Bollaert (2012) has been slightly enhanced within the framework of the rocsc@r digital environment by adding the following elements (Bollaert 2021):

- potential additional shear forces in the lateral joints generated by net pressure differences acting on the upper and lower vertical faces (i.e. lateral joints) of the block
- potential additional pulsating uplift forces generated by block protrusion, valid during a limited time duration and superposed to the time-averaged component of the uplift force
- wall jet velocity decay relationships valid for circular-shaped jets
- wall jet velocity decay relationships valid for jets impacting the plunge pool under an oblique angle, for both circular- and rectangular-shaped jets, and based on the jet diameter at impact in the plunge pool D_j

This enhanced version of the QSI method is called the modified quasi-steady impulsion or MQSI method. The MQSI method makes use of the average local wall jet velocity quasi-parallel to the water-rock interface. The main difference with the original QSI method lies in the additional consideration of shear forces in the lateral joints generated by differential water pressures acting in these joints, and by the additional consideration of net uplift pressure fluctuations generated by deviation of these velocities at protruding blocks, while the original model only considers quasi-steady (time-averaged) net uplift pressures.

These fluctuating uplift pressures are determined similar to the MDI method, i.e. by considering the dynamic pressure coefficient C'_p of the RMS values of the pressure fluctuations at the joint entrances. Furthermore, like the MDI method, the MQSI method makes use of the time durations of the pressure pulses as recorded on 3D laboratory experiments to transform the net uplift forces into net uplift impulsions.

Also, velocity decay relationships for circular-shaped wall jets have been added to the model, because the decay for such jets is significantly higher than for rectangular-shaped wall jets, and because such relationships are also representative for broken-up rectangular jets.

Finally, all velocity decay equations have been expressed as a function of the jet diameter at impact in the plunge pool D_j, and not as a function of the theoretical discharge distribution and related height of the wall jet after impingement, the jet diameter at impact being a parameter that is easy to estimate.

Verification of rock block stability is performed in two separate steps as outlined further on.

Basic Input Parameters

The basic parameters that are used as input to the MQSI (2021) method are outlined in Table 6.20.

Quasi-Steady Forces on a Block

The quasi-steady net force F_{net} ([N]) acting on a block is decomposed into net dynamic uplift pressure forces acting on the block, buoyancy forces, gravity forces and finally shear forces along the vertical joints based on weight and buoyancy (for oblique joint sets) and based on differential pressures along the upstream and downstream vertical faces of the block.

$$F_{net} = F_{under} - F_{over} + F_{buoy} - F_{grav} - F_{sh,G} - F_{sh,P} \tag{6.112}$$

$$F_{up,QSI,AVG} = F_{under} - F_{over} = C_{up,QSI} \cdot \rho_w \cdot x_b \cdot y_b \cdot \frac{V_{X,max}^2}{2} \tag{6.113}$$

$$F_{buoy} = \rho_w \cdot g \cdot x_b \cdot y_b \cdot z_b \cdot \cos(\alpha_b \cdot \pi / 180) - \text{ABS}\begin{pmatrix} \rho_w \cdot g \cdot x_b \cdot y_b \cdot z_b \\ \cdot \sin\left(\alpha_b \cdot \dfrac{\pi}{180}\right) \cdot \tan(\varphi) \end{pmatrix} \tag{6.114}$$

$$F_{grav} = \rho_r \cdot g \cdot x_b \cdot y_b \cdot z_b \cdot \cos(\alpha_b \cdot \pi / 180) \tag{6.115}$$

TABLE 6.20
Basic Parameters Used as Input to the MQSI (2021) Computational Method

Parameter	Unit	Description
$C'_p\ (Y/D_j)$	–	RMS pressure coefficient at block surface, valid at jet centreline
$C'_{pr}\ (Y/D_j)$	–	$n \cdot C'_p\ (Y/D_j)$, RMS pressure coefficient at block surface, valid radially outwards from jet centreline
n'	–	$C'_p/C'_p,$ RMS pressure reduction factor
V_j	m/s	Average jet velocity at impact in the pool
D_j or B_j	m	Jet diameter or jet thickness at impact in the pool
H_j	m	$V_j^2/2g$, hydraulic head at impact of jet in the pool
$V_{z,\text{bottom}}$	m/s	Axial average flow velocity along centreline of jet in pool
$V_{x,\max}$	m/s	Local average flow velocity at protruding rock block
x_b or y_b	m	Rectangular block side lengths in the horizontal plane
z_b	m	Rectangular block vertical height
α_b	°	Angle of rock blocks/joints with the vertical
φ	°	Rock joint frictional angle
α_{am}	–	Coefficient of added mass
m_{am}	kg	Added mass of the block during its acceleration phase
MULT	–	Multiplication factor of the RMS pressure coefficient
Δt_{pulse}	sec	Time duration of net uplift pressure pulse
C_{SURF}	–	Average pressure coefficient along surface of block
C_{BOTTOM}	–	Average pressure coefficient along bottom of block
C_{UVF}	–	Average pressure coefficient along upper vertical face of block
C_{DVF}	–	Average pressure coefficient along downstream vertical face of block

$$F_{\text{sh},G} = \text{ABS}\left(\left(\rho_r - \rho_w\right) \cdot g \cdot x_b \cdot y_b \cdot z_b \cdot \sin\left(\alpha_b \cdot \frac{\pi}{180} \right) \cdot \tan\left(\varphi\right) \right) \tag{6.116}$$

$$F_{\text{sh},P} = \text{ABS}\left(\left(C_{\text{UVF}} - C_{\text{DVF}}\right) \cdot \rho_w \cdot z_b \cdot y_b \cdot \frac{V_j^2}{2} \cdot \tan\left(\varphi\right) \right) \tag{6.117}$$

where

F_{under} = quasi-steady dynamic pressure force acting under the block [N]
F_{over} = quasi-steady dynamic pressure force acting over the block [N]
$F_{\text{up,QSI,AVG}}$ = quasi-steady net dynamic uplift force acting on the block [N]
F_{buoy} = buoyancy forces acting on the block [N]
F_{grav} = gravity forces acting on the block [N]
$F_{\text{sh,G}}$ = shear forces along block lateral joint based on weight and buoyancy [N]
$F_{\text{sh,P}}$ = shear forces along block lateral joint based on differential pressures [N]
α_b = angle of rock blocks/joints with the vertical [°]
φ = rock joint frictional angle [°]
$C_{\text{up,QSI}}$ = average net uplift pressure coefficient on block [–]

C_{UVF} = average pressure coefficient along upstream vertical face of block [–]
C_{DVF} = average pressure coefficient along downstream vertical face of block [–]

Quasi-Steady Pressures Over and Under the Block The quasi-steady pressure forces that act along the upper and lower faces of the rectangular block generate a lift force on the block. In contrast with the MDI method, the time-averaged value of this uplift pressure or force is essential to the analysis. It is determined by using a time-averaged net uplift pressure coefficient $C_{up,QSI}$ that is computed based on the time-averaged pressure coefficients on both the surface and the bottom of the block:

$$C_{up,QSI} = C_{BOTTOM} - C_{SURF} \text{ [–]} \quad (6.118)$$

Based on laboratory experiments performed by Reinius (1986) and Pells (2016), eight different geometrical situations of a rock block as compared to the surrounding rock mass are available. For each case, the range of minimum, maximum and average differences between the bottom and surface pressure coefficients have been defined in Figure 6.51 (Bollaert 2021). For the case of a perfectly flat bottom (case 1), theoretically no quasi-steady lift forces are generated by the wall jets.

Quasi-Steady Pressures along Upstream and Downstream Vertical Faces of Block The quasi-steady water pressures also act in the joints that are located along

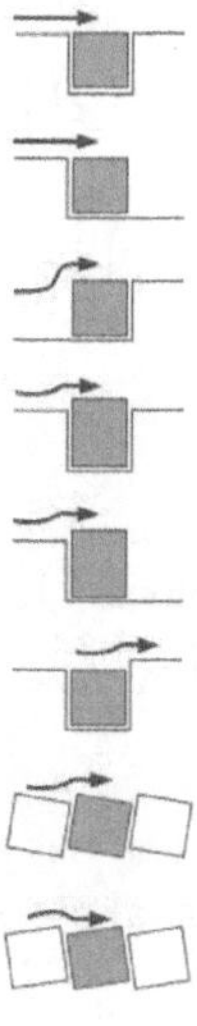

CASE	C_{up}, QSI				C'_{up}, QSI			
	min	max	avg	remark	min	max	avg	remark
1	0.00	0.10	0.05		0.00	0.00	0.00	
2	0.10	0.15	0.12		---	---	---	
3	0.70	0.90	0.80		---	---	---	
4	0.10	0.40	0.25	$0.9\sqrt{z_b/k}$	0.025	0.11	0.07	$0.24\sqrt{z_b/k}$
5	0.10	0.20	0.15	approx.	---	---	---	
6	0.08	0.09	0.08		---	---	---	
7	0.30 0.40	0.35 0.50	0.32 0.45	h/k ~4-9 h/k ~1-2	---	---	---	
8	-0.07 -0.12	-0.07 -0.20	-0.07 -0.16	h/k ~4-9 h/k ~1-2	---	---	---	

FIGURE 6.51 Quasi-steady net uplift pressures on blocks for different geometrical situations (Bollaert 2021).

the upstream and downstream vertical (or quasi-vertical) faces of the block. Such shear forces are opposed to block movements, and occur when the pressure forces in the lateral joints are not equal, i.e. when a net lateral pressure difference is generated that laterally pushes the block towards its neighbouring block. Accounting for a rock joint friction angle φ, this generates a shear force $F_{sh,P}$ in one of the lateral joints, by expressing the lateral water pressures by means of average pressure coefficients C_{UVF} and C_{DVF}.

Laboratory experiments performed by Reinius (1986) and Pells (2016) allow distinguishing eight different geometrical situations of a rock block as compared to the surrounding rock mass. For each case, the range of minimum, maximum and average differences between the lateral pressure coefficients is defined in Figure 6.45 (Bollaert 2021). For case number 4, a mathematical equation is available to express this net lateral pressure difference. It may be observed that, on the average, these pressure differences are rather low to very low. Hence, for practice, safe-side assumptions of low to quasi-zero values are recommended.

Shear Forces in Lateral Joints due to Gravity and Buoyancy Similar to the shear forces that may be generated by differential dynamic water pressures inside the lateral joints, an additional shear force may occur for blocks that have a non-zero angle of orientation with the horizontal.

In such a case, the component of the gravity force that is oriented perpendicular to the lateral joint will add a shear force $F_{sh,G}$ in the joint by accounting for the rock joint friction angle φ.

In the same way, the component of the buoyancy force that is oriented perpendicular to the lateral joint will reduce this added shear force because of counteracting gravity forces.

Block Stability Criterion

Like the original QSI model, block stability is obtained whenever the following relation of force balance holds:

$$F_{net} < 0 \tag{6.119}$$

Fluctuating Forces on a Block

The fluctuating part of the lift force on a protruding block can be described by the RMS pressure fluctuation coefficient $C'_{up,QSI}$ of the pressure acting at the point of block protrusion (i.e. over the upper face of the block):

$$C'_{up,QSI} = \frac{\text{RMS}\left(\text{local pressure at block protrusion}\right)}{\rho \cdot \frac{V_j^2}{2}} [-] \tag{6.120}$$

These $C'_{up,QSI}$ values have been recorded by Pells (2016) for rectangular-shaped blocks with or without protrusion and are presented in the last column of Figure 6.51.

Similar to the MDI method, and based on the Federspiel (2011) experiments, the fluctuating lift force has been related to the RMS pressure fluctuations at the block surface $C'_{up,QSI}$ by means of a multiplication factor MULT (Bollaert 2021):

$$\text{MULT} = \frac{C_{\text{up,QSI,RMS}}}{C'_{\text{up,QSI}}}[-] \tag{6.121}$$

The fluctuating lift force on the block $F_{\text{up,QSI,RMS}}$ can then be written as:

$$F_{\text{up,QSI,RMS}} = \text{MULT} \cdot C'_{\text{up,QSI}} \cdot \rho_w \cdot x_b \cdot y_b \cdot \frac{V_{X,\max}^2}{2} \text{ [N]} \tag{6.122}$$

Based on the statistical procedure outlined for the MDI method, it is observed that for MULT values of 4.2, the probability of occurrence of such a pulse is once every 10–150h. For higher MULT values, i.e. 4.5 and higher, the needed time duration of flood events rapidly becomes hundreds and even thousands of hours. The MULT value is a parameter that must be defined as a function of the total duration of flood events for the problem in question.

Total Lift Forces on a Block

By superposing the fluctuating lift force to the time-averaged lift force, the total lift force $F_{\text{net,QSI}}$ is then written as:

$$F_{\text{net,QSI}} = F_{\text{up,QSI,AVG}} + F_{\text{up,QSI,RMS}} = \left(C_{\text{up,QSI,AVG}} + \text{MULT} \cdot C'_{\text{up,QSI}}\right) \cdot \rho_w \cdot x_b \cdot y_b \cdot \frac{V_{X,\max}^2}{2} \tag{6.123}$$

Movement of a block

The block movements and the block stability can then be expressed in two different steps:

STEP 1: quasi-steady lift forces

Block stability criterion according to equation (6.119), i.e. a net force balance is checked for by only considering quasi-steady forces. In case the block is unstable, no second step is being checked for and the computation ends.

STEP 2: total lift forces

Block stability criterion according to the procedure of net uplift height is given to the block based on the net total lift force and its time duration. The total lift force is composed of a quasi-steady part (STEP 1) and of a single pressure pulse of duration Δt_{pulse} that is superposed to the quasi-steady lift force. In case the total net lift force is positive, there is application of a critical uplift height criterion to determine block detachment or stability, in different phases of movement following Figure 6.52 and as further detailed hereafter. In case the total net lift force is still negative, the computation ends and the block is stable.

Acceleration Phase of a Block

The total net uplift force as defined by equation (6.123) is considered to act on the block for a time duration Δt_{pulse}. The net impulsion that is given to the block can then be written as:

$$I_{\text{net,QSI}} = F_{\text{net,QSI}} \cdot \Delta t_{\text{pulse}} \text{ [N} \cdot \text{s]} \tag{6.124}$$

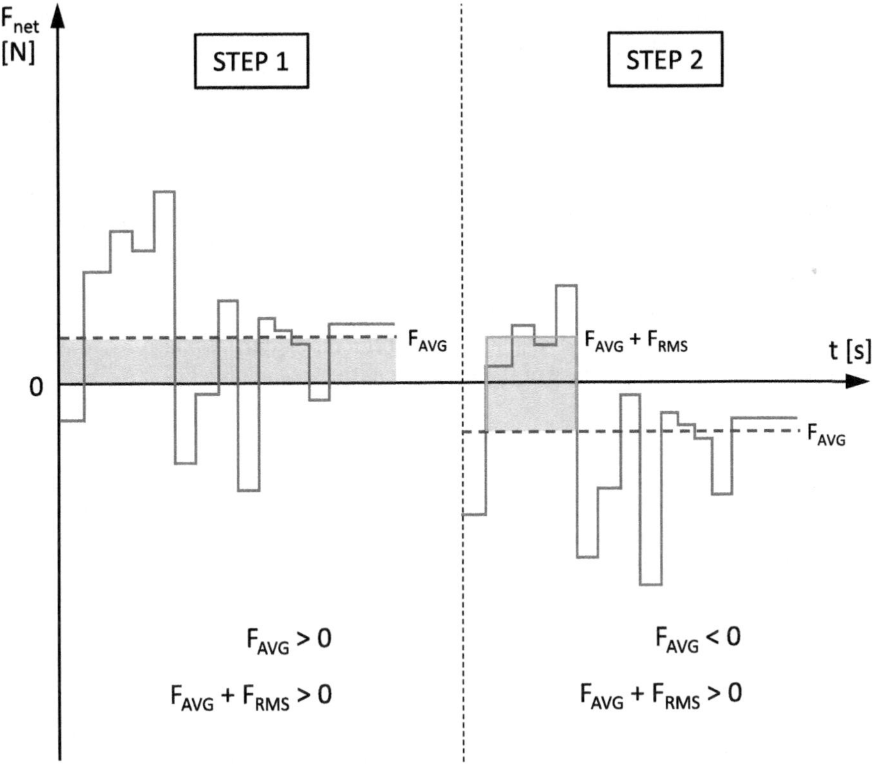

FIGURE 6.52 Block stability verification by MQSI method in two STEPS: STEP 1 = force balance based on average net lift force on block, STEP 2 = impulsion and block uplift height generated by single pulse model.

A double temporal integration over Δt_{pulse} then allows computing the displacement $h_{\text{up,net,QSI}}$ that the rock block exhibits during its phase of acceleration from 0 to $a_{\text{net,QSI}}$. During this same time interval, a final velocity $V_{\text{up,net,QSI}}$ is given to the block after a time Δt_{pulse}:

$$V_{\text{up,net,QSI}} = \frac{I_{\text{net,QSI}}}{(m_b + m_{\text{add}})} \text{ [m/s]} \tag{6.125}$$

Deceleration Phase of a Block

Like for the MDI method, the total net uplift force now becomes negative. The time necessary for this downwards-oriented force to stop the momentum of the block is written as follows:

$$\Delta t_{\text{downpulse}} = \frac{V_{\text{up,net,QSI}} \cdot (m_b + m_{\text{add}})}{-F_{\text{down,QSI}}} \text{ [s]} \tag{6.126}$$

The corresponding additional uplift displacement that the block exhibits during its deceleration phase is then expressed as:

$$h_{\text{decel,QSI}} = \frac{V_{\text{up,net,QSI}} \cdot \Delta t_{\text{downpulse}}}{2} \text{ [m]} \tag{6.127}$$

For downpulse periods (deceleration phases) that are longer than $\Delta t_{\text{downpulse}}$, the block will start moving and accelerating downwards, until it touches base again. This, however, is not accounted for by the model.

Total Vertical Displacement of a Block

The total (maximum) vertical displacement of the block during phase 1 and phase 2 corresponds to a superposition of the individual displacements:

$$h_{\text{up,tot,QSI}} = h_{\text{up,net,QSI}} + h_{\text{decel,QSI}} \text{ [m]} \tag{6.128}$$

Similar to the MDI method, to soundly account for the complex real block situations, the necessary or critical uplift displacement is a model parameter that needs to be calibrated according to site-specific parameters and is expressed similar to equation (6.80):

$$h_{\text{up,crit}} = n_b \cdot z_b \tag{6.129}$$

Based on real-life scour calibrations performed with the MDI method (see Chapter 7), critical net uplift displacements of ~0.10 times the block height were found adequate in conjunction with time durations of net uplift pulses of ~0.10 s and added mass coefficients of 2–3.

Comprehensive Fracture Mechanics (CFM) Model

Introduction

The Comprehensive Fracture Mechanics (CFM) method developed by Bollaert (2002) and Bollaert and Schleiss (2005) makes use of dynamic water pressures that may be generated inside rock joints. Based on principles directly derived from the field of Linear Elastic Fracture Mechanics (LEFM), the method transforms these water pressures into a stress field induced in the rock mass, at the tip of closed-end rock joints.

Depending on the resistance of the rock mass against fracture propagation, the method determines whether the closed-end rock joint remains stable or propagates along its tip. Propagation of rock joints can occur instantaneously or over a certain time duration. The former case is called instantaneous or brittle crack propagation. The latter is called time-dependent or fatigue crack propagation.

The dynamic water pressures have a cyclic character and travel as transient pressure waves inside rock joints. They can be described by an amplitude and a main frequency. LEFM models can handle such transient loadings by assuming a perfectly linear elastic, homogeneous and isotropic material. These models become quite complicated when accounting for all the relevant parameters. Therefore, a simplified application is used here. It is called the *Comprehensive Fracture Mechanics (CFM) method* and is applicable to partially jointed rock, i.e. to rock joints that are not fully formed yet and, as such, have a closed end.

The Stress Intensity

The fracture mechanics implementation of the hydrodynamic water loading in rock joints consists of a transformation of the water pressures into rock mass stresses. The intensity of these stresses is expressed by the stress intensity factor K_I. This factor depends on the water pressure distribution inside the joint, on the geometry of the joint, on the global stress field of the surrounding rock mass and on the loading rate of the water pressures.

The following simplifying assumptions are made: (1) the dynamic character of the loading has no influence; (2) the pressure distribution inside the joints is constant; (3) only simple geometrical configurations of rock joints are considered; (4) the joint surfaces are planar; and (5) the joints are without filling material.

Figure 6.27 (see rock mass models) presents three basic geometrical situations for partially jointed rock. The water pressure in the joints is applied from outside the rock elements. No geometries with multiple joints are considered. The choice of the most relevant geometry will depend on the type and the degree of jointing of the rock in question.

The first crack is semi-elliptical (EL) or semi-circular (CIRC) and, with regard to the laterally applied stress field, is considered partially sustained by the surrounding rock mass in two directions. As such, it is the geometry with the highest possible support of surrounding rock. Corresponding stress intensity factors should be used in case of low to moderately jointed rock (0 to 3 joint sets). The second crack is single-edge notched (SE) and of two-dimensional in nature. Support from the surrounding rock mass is only exerted perpendicular to the plane of the notch and, as a result, stress intensity factors will be substantially higher than for the first case. Thus, it is more appropriate for highly jointed and/or disintegrated rock (more than three joint sets, sheet joints). The third geometry is centre-cracked (CC) throughout the rock. Similar to the single-edge notch, only one-sided rock support can be accounted for. This support, however, should be slightly higher as for the single-edge notch. The second and third geometrical configurations correspond to a partial destruction of the first one and generate higher stresses at the tip of the joint.

Figure 6.53 illustrates the concept of partially jointed rock, by visualizing the different geometrical configurations. Only one layer of rock is presented in the z-direction.

The stress intensity at the tip of the joint is described by a stress intensity factor K_I. The stress intensity factor depends on the joint length a, on the boundary correction factor f and on the water pressure along the joint σ_{water}:

$$K_I = \sigma_{\text{water}} \cdot f \cdot \sqrt{\pi a} \tag{6.130}$$

K_I is expressed in MPa$\sqrt{\text{m}}$ and the water pressures are in MPa. It is interesting to notice that the stress intensity factor not only grows with the square root of the joint

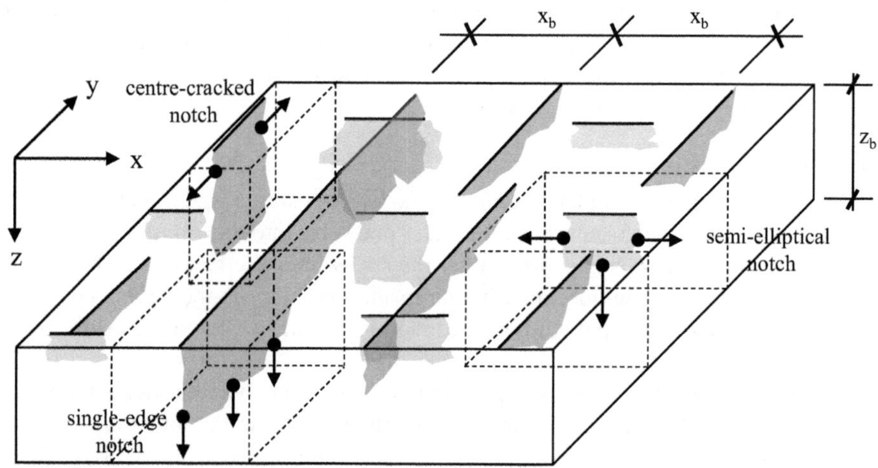

FIGURE 6.53 Examples of geometrical configurations in jointed rock mass. In the vertical (z) direction, only one layer is presented.

length a (or the half joint length c for a centre-crack) but also increases as a function of the geometrical ratio a/B (or c/W for a centre-crack). The f factor reflects the initial degree of break-up or persistency (a/B) of the rock joint. It reflects the influence of the surrounding rock support and is a function of the geometrical ratio a/B (for the elliptical and single-edge notches) or c/W (for the centre-cracked specimen).

Figure 6.54 presents a comparison of the f function for the three joint geometries. Following the choice of geometrical configuration and initial degree of break-up a/B, a tendency of fracture propagation enhancement can be distinguished. The more the rock is broken up, the higher the differences become between the configurations.

Following equation (6.130), the water pressure σ_{water} should not be taken exactly equal to the maximum possible dynamic water pressure $C^{max}{}_{p}$ at the tip of the joint, because the rock mass stresses at the tip of the joint are defined by the water pressure distribution throughout the whole joint length, and not only by the water pressure at the tip.

Hence, the pressure distribution over the joint length is needed to accurately assess the stress intensity at the tip. The pressures at the boundaries are defined by the pressures valid at the entry (= pressures at the water-rock interface) and the pressures valid at the end of the joint. In between those values, a sinusoidal mode shape is assumed, according to a $\lambda/4$ – resonator model. This simplification results in the following definition of the pressure distribution:

$$p(x) = p_0 + \left((p_{max} - p_0) \sin\left(\frac{\pi \cdot x}{2 \cdot L} \right) \right) \tag{6.131}$$

For $x = L$, this conducts to a maximum pressure at the tip of $(0.36) \cdot p_0 + (0.64) \cdot p_{max}$. Based on prototype scaled laboratory measurements, the maximum pressure at the

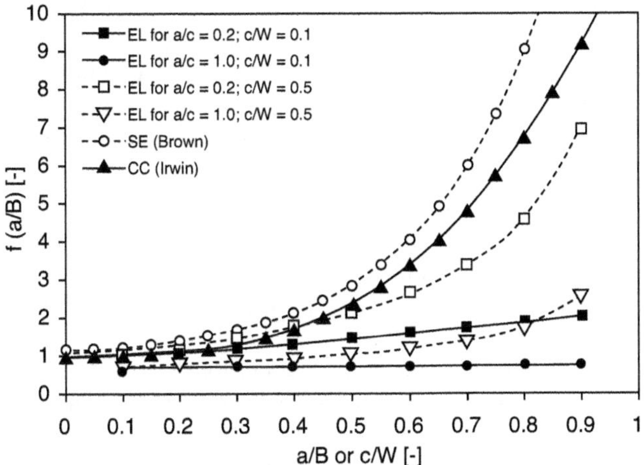

FIGURE 6.54 Influence of initial degree of break-up *a*/*B* of joint on the *f* factor.

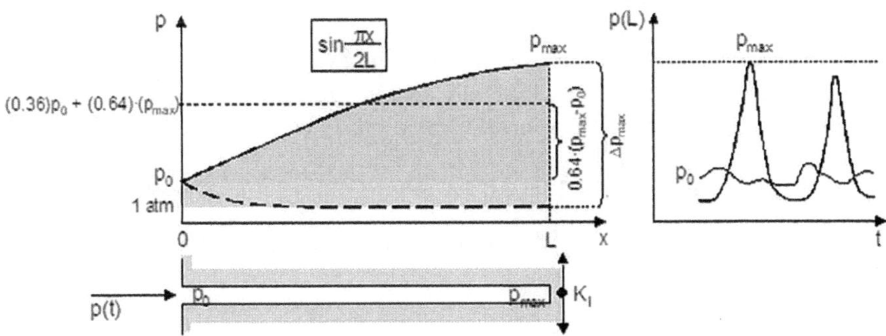

FIGURE 6.55 Sinusoidal pressure distribution over the whole joint length and definition of the average maximum pressure value to be used as σ_{water}.

tip is on the order of 2–5 times the pressure at the entrance. This results in an average pressure value over the whole joint length that is equal to 71%–82% of the maximum pressure p_{max}. Hence, a value of 80% of p_{max} ($= 0.5\rho V^2 C^{max}_p$) is used by the model for σ_{water} for practical applications. This allows rewriting equation (6.130) as follows:

$$K_I = 0.8 \cdot P_{max} \cdot f \cdot \sqrt{\pi a} \tag{6.132}$$

The Fracture Toughness

The resistance of the fractured rock against propagation of its joints is expressed by its fracture toughness K_{Ic}. The fracture toughness is directly related to the mineralogical type of rock and to the unconfined compressive strength UCS (or eventually the tensile strength T). Also, potential in-situ rock mass stresses can be of influence on its value. The fracture toughness is generally obtained through fracturing tests of rock core specimens in a laboratory or by performing in-situ hydraulic fracturing tests. For most types of rocks, a reasonable range of fracture toughness values may be estimated.

Similar to the stress intensity K_I, the implementation of the fracture toughness K_{Ic} of a partially jointed rock needs simplifying assumptions. The fracture toughness is depending on a vast range of parameters. In the following, the fracture toughness is directly related to the mineralogical type of rock and to the tensile strength T or the unconfined compressive strength UCS.

Furthermore, corrections are made to account for the effects of the loading rate and the in-situ stress field of the rock mass. The corrected fracture toughness is here called the in-situ fracture toughness $K_{Ic,\text{ins}}$. According to Bollaert (2002), the following general formulae are proposed for the in-situ fracture toughness value:

$$K_{Ic,\text{ins},T} = A(1.2 \text{ to } 1.5)T + (0.054\sigma_c) + B$$

$$K_{Ic,\text{ins},\text{UCS}} = C(1.2 \text{ to } 1.5)\text{UCS} + (0.054\sigma_c) + D$$

in which T, UCS and σ_c are expressed in MPa and $K_{Ic,\text{ins}}$ in MPa$\sqrt{\text{m}}$. The parameters A to D depend on the type of rock as indicated in Table 6.21.

For other types of rock, the general expressions to be used are:

$$K_{Ic,\text{ins},T} = (0.105 \text{ to } 0.132)T + (0.054\sigma_c) + 0.5276$$

$$K_{Ic,\text{ins},\text{UCS}} = (0.008 \text{ to } 0.010)\text{UCS} + (0.054\sigma_c) + 0.4200$$

Brittle Joint Propagation

Finally, instantaneous or brittle crack propagation will occur if the following expression is valid:

$$K_I \geq K_{I,\text{ins}} \tag{6.133}$$

If this is not the case, crack propagation can still occur within a certain time interval. This is outlined in the next section.

TABLE 6.21
Parameters for the In-Situ Fracture Toughness Value $K_{Ic,\text{ins}}$

Type of Rock	A	B	C	D
Silicate rocks	0.0648	0.8693	0.0023	1.3257
Carbonate rocks	0.3230	−0.0405	0.0145	−0.0190
Quartz rocks	0.1283	0.2747	0.0088	0.1429

Time-Dependent Joint Propagation by Fatigue

The comparison of the stress intensity with the fracture toughness defines whether brittle or instantaneous rock break-up will occur. When the stress intensity is less than the fracture toughness, a second type of rock break-up can happen. It is called time-dependent or subcritical crack propagation and distinguishes between two mechanisms: stress corrosion (or static fatigue) and cyclic (or dynamic) fatigue.

Stress corrosion or static fatigue happens when a steady stress is applied in a chemically active environment. Water pressures in rock joints represent such an environment. The steady stresses exerted by the water can progressively weaken the strained bonds of the rock mass by chemical reactions. A qualitative relationship between subcritical crack propagation velocities $v(K)$ and a normalized stress intensity factor ($= K_I/K_{Ic}$) can be found for a wide range of geologic materials in Atkinson (1987). The exact propagation velocities, however, are difficult to determine and only tendencies can be pointed out.

Failure by dynamic fatigue of partially jointed rock occurs when a considerable cyclic loading is applied inside the joints. Since the 1960s, a significant body of theoretical and experimental evidence has been generated. The most interesting studies focus on a combination of stress corrosion and cyclic fatigue. At small values of the amplitude of the stress cycle Δp_c, stress corrosion is the dominant mechanism, because of the very little cycling. At moderate and high values of Δp_c, the results can only be explained by cyclic fatigue. They are fit by an equation of the type that was originally proposed by Paris et al. (1961) to describe fatigue crack growth in metals:

$$\frac{dL_f}{dN} = C_r \cdot (\Delta K_I)^{m_r} \tag{6.134}$$

in which L_f is the joint length and N is the number of pressure cycles per sec, defined as follows (max. 50 Hz, consider no jet energy beyond):

$$N = \frac{c}{4\left(\frac{a}{B}\right)L_f} \tag{6.135}$$

Wave celerity c may be determined between 50–100 m/s (for high air content) and 200 m/s (for low air content). C_r and m_r are rock material parameters that can be determined by experiments and that express the vulnerability of the rock to fatigue. ΔK_I is the difference of maximum and minimum stress intensity factors.

To implement time-dependent joint propagation into a comprehensive engineering model, m_r and C_r have to be known. They represent the vulnerability of rock to fatigue and have been derived from available laboratory data on the sensitivity of rock to quasi-steady break-up by water pressures in joints, together with laboratory measurements of the sensitivity of a granite rock to fatigue break-up (Atkinson 1987).

Figure 6.56 illustrates the crack propagation rate in m/cycle for a Westerly granite as a function of ΔK. The fracture toughness is of 2.2 MPa$\sqrt{\text{m}}$, and the crack

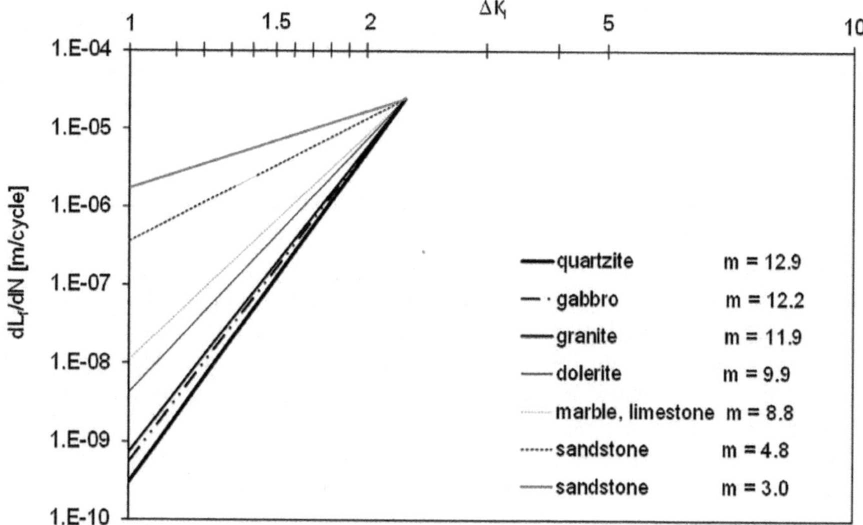

FIGURE 6.56 Crack propagation rate in *m*/cycle for a Westerly granite as a function of ΔK. Qualitative comparison with propagation rates for other types of rock.

propagation rate of 4.0E^{-5} m/cycle has been deduced from typically measured propagation rates of ~10^{-3}m/s based on Swanson (1984), and accounting for 50 pressure cycles per second. The m_r value represents the slope of the log-log relationship, i.e. 11.9 for granite, while the C_r value stands for the intersection of the relationship with the ordinate at $\Delta K = 1.0$, i.e. a value of 7.3E^{-10} m/cycle for granite. It can be clearly pointed out that, for decreasing stress intensities compared to the fracture toughness of the rock, the crack propagation rate strongly decreases.

For sake of comparison, Figure 6.56 also represents similar relationships for other types of rock, depending on their respective m_r value and starting these relationships from the same origin at a fracture toughness of 2.2 MPa$\sqrt{\text{m}}$ and a crack propagation rate of 4.0E^{-5} m/cycle. Most other types of rock seem to have higher crack propagation rates than granite.

As the fracture toughness K_{Ic} is highly dependent on the type of rock, however, the curves for other rocks should be adapted (shifted) following their respective fracture toughness values. A range of average fracture toughness values is provided in Figure 6.57 (Bollaert 2002).

Accounting for these average values, an m_r–C_r relationship for the different rock types discussed in Atkinson (1987) is presented in Figure 6.58 by a series of grey dots (log-lin relation). The relationship depends on the fracture toughness value of the rock and is obtained using equation (6.134), i.e. for absolute values of stress intensity amplitude ΔK.

However, for practice, a different version of the Paris' equation is privileged. This version considers m_r–C_r values as a function of relative values of stress intensity, i.e. $\Delta K/K_{Ic}$. This has the advantage not to depend on the fracture toughness values of the rock, and thus to offer a qualitatively equivalent comparison

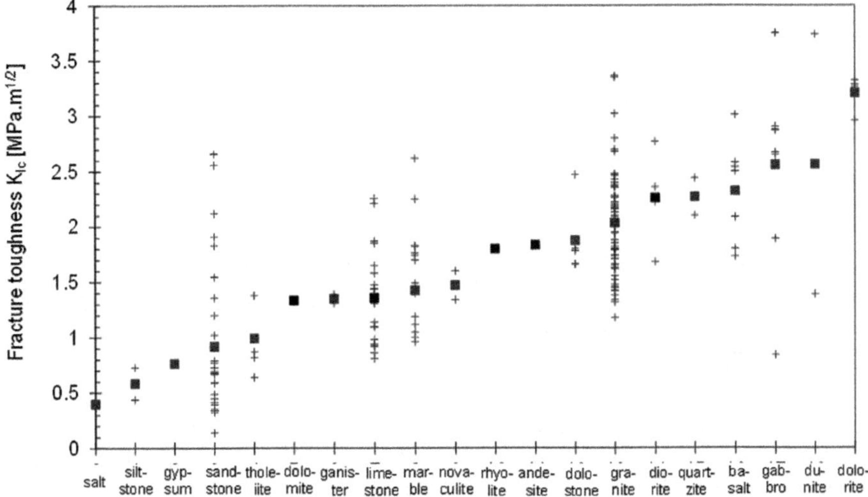

FIGURE 6.57 Average fracture toughness values for different types of rock (Bollaert 2002).

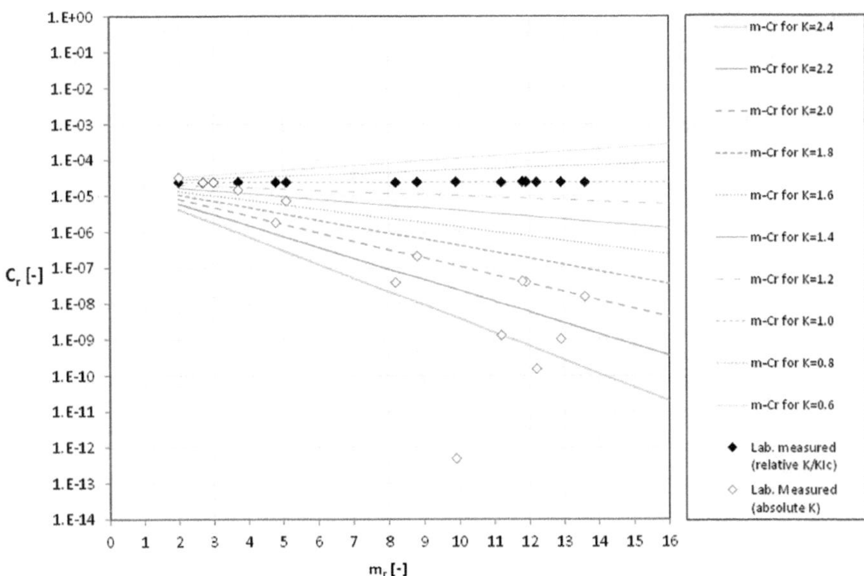

FIGURE 6.58 m_r–C_r relationships for the laboratory-investigated rock types based on Atkinson (1987) and valid for absolute values of stress intensity K.

between rocks. The slightly modified version of the Paris' equation is written as follows:

$$\frac{dL_f}{dN} = C_r \cdot \left(\frac{\Delta K_I}{K_{Ic}}\right)^{m_r} \tag{6.136}$$

The same different rock types are now presented by the black dots in Figure 6.58 and follow a constant line of $C_r = 4E^{-5}$, valid for $\Delta K/K_{Ic} = 1.0$ (i.e. max. possible relative value, start of brittle fracturing).

The fracturing speed is computed following equations (6.134) and (6.135) and by assuming $K_I = \Delta K_I$:

$$\frac{dL_f}{dt}\left[\frac{m}{h}\right] = 3600 \cdot N \cdot \frac{dL_f}{dN} \tag{6.137}$$

Equation (6.136) is the fatigue equation that is used in the CFM, and one might theoretically use the black dot (laboratory-measured) values in Figure 6.58 for practice. However, feedback from practice points out a slightly different behaviour for most rock types. This is explained hereafter in more detail.

For a given (i.e. measured) scour depth observed following a flood event of a certain time duration, in fact an infinite number of m_r–C_r couples satisfy equation (6.136). The resulting lines of equivalent m_r–C_r couples are illustrated in Figure 6.59 for both a granite, with a laboratory-measured m_r value of 11.9, and a sandstone with a laboratory-measured m_r value of 4.8. The corresponding lines depend on the $\Delta K/K_{Ic}$ ratio in equation (6.136), and results are presented for ratios of 0.40, 0.60 and 0.80.

The only difference between these m_r–C_r couples is the curvature and slope of the time evolution of scour formation, the end value of scour being exactly similar. As such, for high m_r values a strong curvature is observed, while for low m_r values the lines become quasi-linear, for a constant slope. This is illustrated

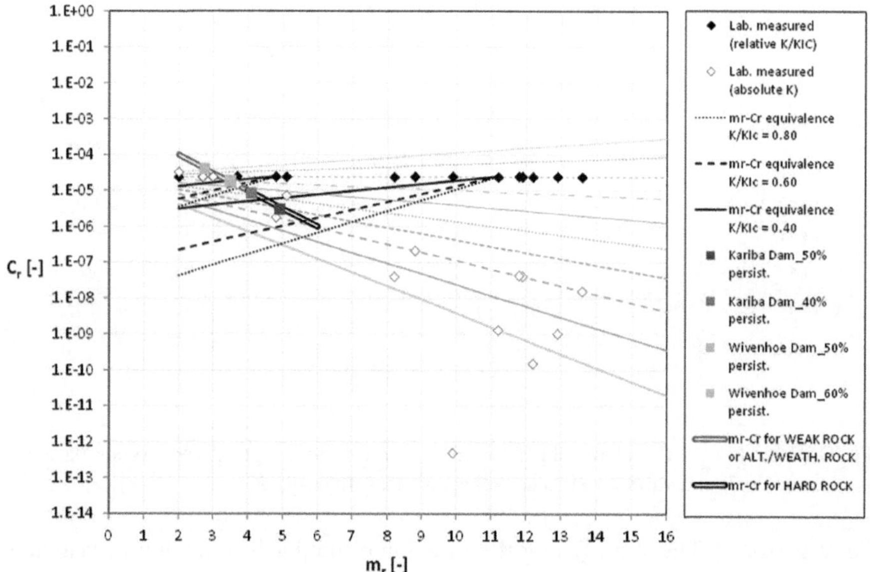

FIGURE 6.59 m_r–C_r relationships for the laboratory-investigated rock types based on Atkinson (1987) and valid for different $\Delta K/K_{Ic}$ ratios. Comparison with calibrations from practice.

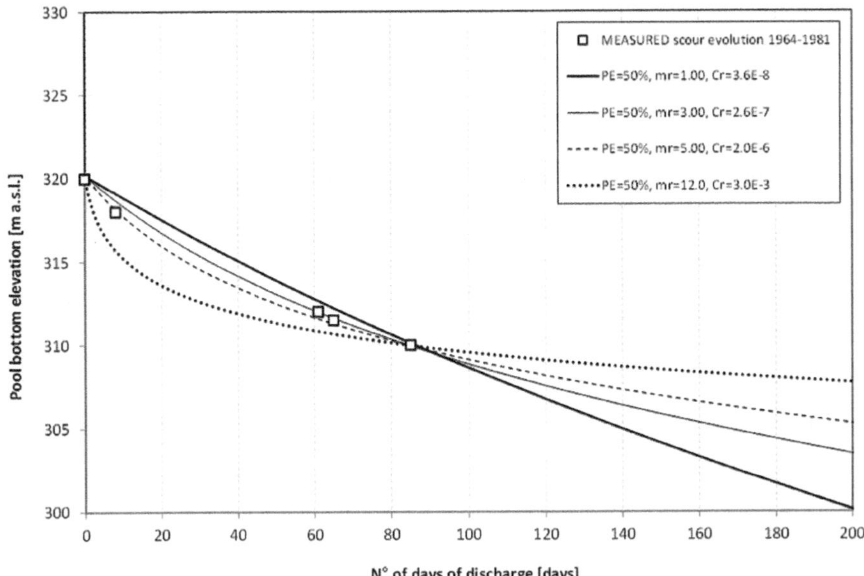

FIGURE 6.60 Scour evolution with time computed at Kariba Dam for different m_r–C_r couples and comparison with in-situ measured scour.

in Figure 6.60 for scour observed at Kariba Dam between 1970 and 1981, and comparing with computational results for different m_r–C_r couples and a persistency of 50%. The best fit of values is obtained for $m_r = 3$–5, while the laboratory values are estimated around $m_r = 10$–12. The results for these m_r values are added to Figure 6.59 and comply with the lines of equivalent m_r–C_r couples.

A second example is provided in Figure 6.59 for a sandstone rock downstream of Wivenhoe Dam in Australia, showing best-fit values very close to the laboratory-measured values.

The explanation for this deviating behaviour of rock in the field compared to laboratory rock samples is probably multiple and related to physical and mechanical weathering and alteration of rock in the field, together with a high degree of moisture, which strongly affects the fracture resistance of rock.

Figure 6.59 finally shows that the best-fit m_r–C_r couples obey to a line that is very similar to the couples obtained when expressing the laboratory-measured values using the original (or absolute ΔK) Paris' equation.

This log-lin relationship was tested on many other scour cases and found adequate for practice. As presented in more detail in Chapter 7, calibration of a significant number of scour cases for different rock types and flow conditions has allowed to define a line of best fit of m_r–C_r couples. This is presented in Figure 6.61. A first line of best fit corresponds to the grey line and covers m_r values from 2 to 4 and C_r values from 1.0E^{-4} to 1.0E^{-5}. It is valid for rock that may be considered as scour vulnerable, i.e.:

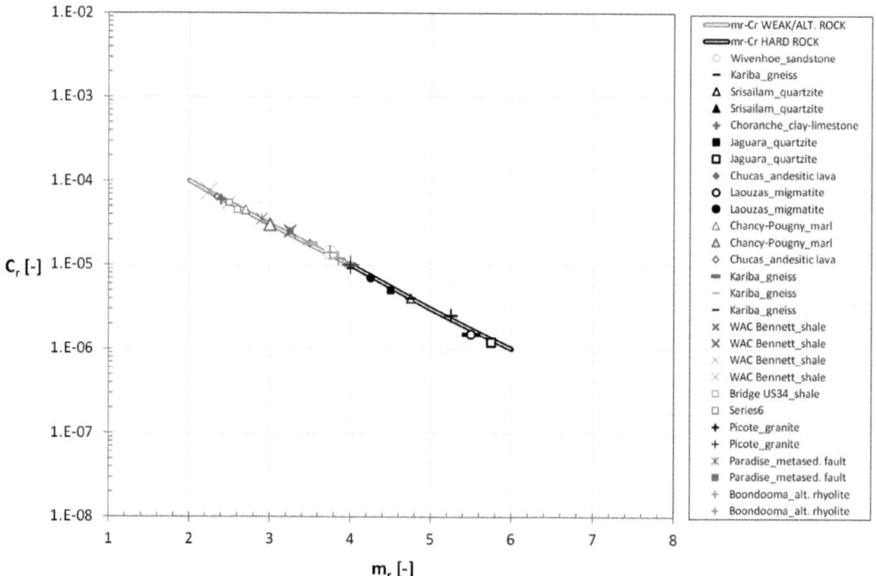

FIGURE 6.61 Summary of m_r–C_r best-fit relationships for weak or weathered/altered rock (in grey), and for strong competent rock (in black).

1. Weak rock, i.e. mainly sedimentary rock and/or rock with UCS strength <25 MPa (classes R0-R1-R2 according to ISRM (1978)).
2. Strong competent rock that is moderately to strongly weathered and/or altered, typically weathering grades W3 to W5 according to ISRM (1978).
3. Strong competent rock that is significantly tectonically disturbed, due to presences of folding, faulting and/or shearing planes, reducing the original structural integrity of the rock mass. Typically blocky/disturbed, disintegrated or sheared/laminated rock according to the GSI strength index criterion (Hoek 1994).

Typical examples of scour vulnerable rock are marls, mudstones, limestones, weathered sandstones, shales and so on.

A second line of best fit corresponds to the black line and covers m_r values from 4 to 6 and C_r values from 1.0E^{-5} to 1.0E^{-6}. It is considered valid for rock that may be considered as lowly to not scour vulnerable, i.e. strong competent rock, such as metamorphic and igneous rock types, with UCS strengths typically >25 MPa (classes R3 to R6 according to ISRM 1978), and with low to no weathering (W1 to max. W2 according to ISRM, 1978) or tectonic perturbations (i.e. no folding, no faulting, no bedding layers, no shear planes, etc.).

The rock may contain several joint sets and be partially fissured, typically intact/massive or blocky rock with very good to good surface conditions according to GSI strength index criterion (Hoek 2006). Typical examples of strong competent rock are quartzite, granite, dolerite, gneiss.

The line of best fit between m_r and C_r can be written as follows:

$$C_r = \frac{0.001}{(3.162278)^{m_r}} \tag{6.138}$$

The log-transformed version of this power function corresponds to a straight line in the log-lin graphic of the m_r–C_r relationship of Figure 6.61.

Furthermore, to implement time-dependent joint propagation into a comprehensive engineering model, the initial degree of fracturing of the rock, or persistency PE (in %), has to be estimated. This becomes obvious by combining equations (6.132) and (6.136) as follows:

$$\frac{d(PE \cdot L_{max})}{dN} = C_r \cdot \left(\frac{0.8 \cdot P_{max} \cdot f(PE) \cdot \sqrt{\pi \cdot PE \cdot L_{max}}}{K_{Ic}} \right)^{m_r} \tag{6.139}$$

As the persistency PE is a parameter of the stress intensity equation, joint propagation by fatigue is directly dependent on this parameter. In fact, the m_r–C_r couples described above are only valid for a given persistency value, and their values will slightly change with varying persistency. This is outlined in Figure 6.62, which directly relates the best-fit m_r values to the adopted persistency PE, by respecting equation (6.139), for the different scour cases under study.

Distinction is made between weak and strong competent rock, and lowly (PE < 40%), moderately (40% < PE < 60%) and highly (60% < PE < 80%) fractured

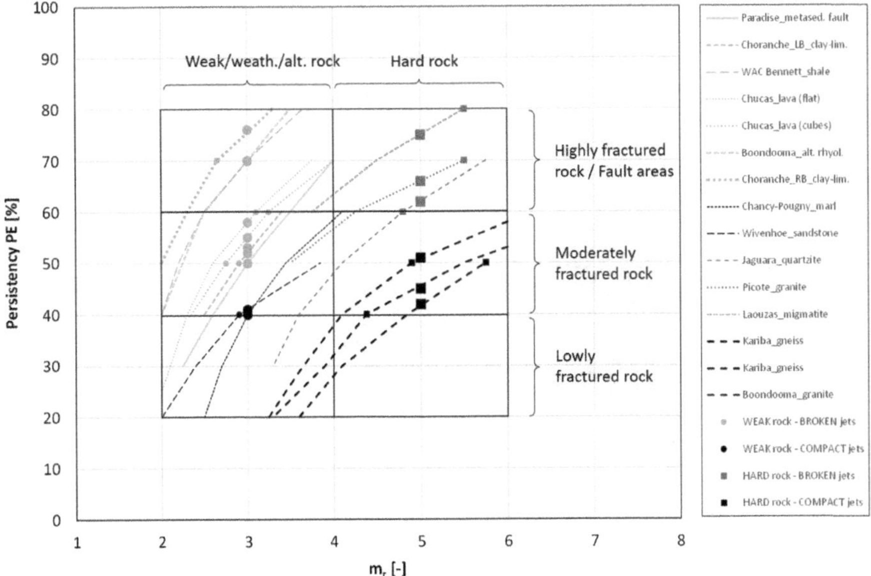

FIGURE 6.62 Summary of m_r–PE best-fit relationships for weak or weathered/altered rock (in left-hand side on figure), and for strong competent rock (in right-hand side on figure).

rock. Also, distinction is made between compact (dark lines) and broken-up (light grey lines) jet impact.

This shows m_r values that reduce with reducing persistency PE. In other words, for fissures that are more difficult and time-consuming to break open (i.e. low PE), a more steeply sloped fatigue law (i.e. low m_r) is needed to suit the scour evolution with time.

For practice, a simplification is made by determining a constant m_r value of 3 for weak rock and a constant m_r value of 5 for strong competent rock. When doing so, the parameters needed for joint propagation by fatigue are solely defined by the type of rock (weak or strong competent) and by the intensity of the initial degree of fracturing, i.e. lowly, moderately or highly fractured rock. A range of PE values is used for scour estimates.

It is worthwhile mentioning that most case studies that involved broken-up jets were characterized by highly fractured rock, or at least an initial persistency of 50%. Case studies involving compact jets generally had lower persistencies of 40–50%. Also, no studies were observed involving lowly fractured rock.

As a result, for practice, Figure 6.63 summarizes the best-fit values to use for m_r, C_r and PE. Distinction is thereby made between weak (or weathered/altered) and strong competent rock, and lowly (PE < 40%), moderately (40% < PE < 60%) and highly (60% < PE < 80%) fractured rock. Also, the most relevant areas observed for broken-up and compact jets are added.

Finally, Figure 6.64 compares the crack propagation rates between the theoretical values as determined by the Paris' equation (grey area valid for all rock types) and

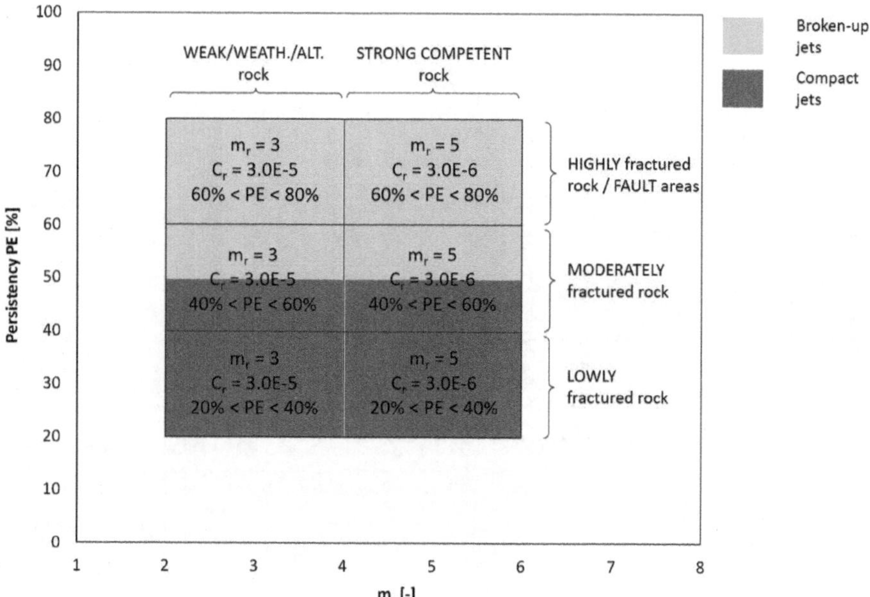

FIGURE 6.63 Summary of best-fit values for practice (m_r, C_r and PE). For weak or weathered/altered rock, and for strong competent rock.

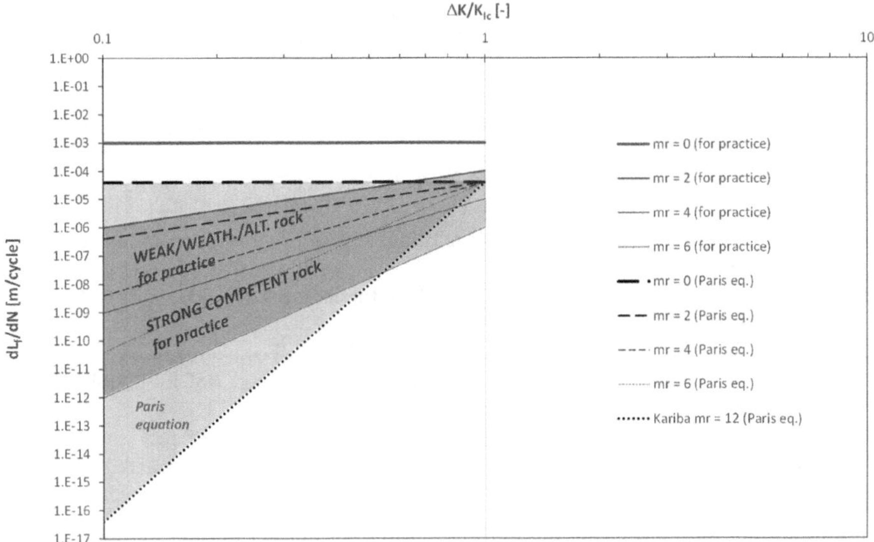

FIGURE 6.64 Summary of crack propagation rates in m/cycle as a function of DK/KIc, for both weak rock and strong competent rock.

the modified best-fit values to be used for practice, based on a series of case studies (i.e. dark grey areas for weak rock and for strong competent rock).

As shown already in Figure 6.59, theoretical and practical values agree fairly well for weak or weathered/altered rock, while corresponding values for strong competent rock are significantly different. In the field, the latter rock types seem to have a fatigue behaviour that is somehow approaching the one for weak rock types.

REFERENCES

Ali, K.H.M. and Karim, O.A., "Simulation of Flow Around Piers", *Journal of Hydraulic Research*, 40, 2, 161–174, 2002.

Ali, K.H.M., Karim, O.A. and Connor, B.A., "Flow Patterns Around Bridge Piers and Offshore Structures", in *Proceedings of ASCE Water Resources Engineering Conference, August 10–15*, San Francisco, pp. 208–213, 1997.

Annandale, G.W., "Erodibility", *Journal of Hydraulic Research*, 33, 471–494, 1995.

Annandale, G.W., *Scour technology: mechanics and engineering practice*, McGraw-Hill Professional, New York, 1 ed., 2006.

Asadollahi, P. and Tonon, F., "Stability of Rock Blocks Subjected to High-Velocity Water Jet Impact", in *44th US Rock Mechanics Symposium and 5th US-Canada Rock Mechanics Symposium*, American Rock Mechanics Association, Salt Lake City, 2010.

Atkinson, B.K., *Fracture mechanics of rock*, Academic Press Inc., London, 1987.

Avila, H. and Pitt, R., "The Calibration and use of CFD Models to Examine Scour from Stormwater Treatment Devices – Hydrodynamic Analysis", in *11th International Conference on Urban Drainage*, Edinburgh, Scotland, UK, pp. 1–10, 2008.

Badas, M.G., Rossi, R. and Garau, M., "May a Standard VOF Numerical Simulation Adequately Complete Spillway Laboratory Measurements in an Operational Context? The Case of Sa Stria Dam", *Water*, 12, 1606, 2020. https://doi.org/10.3390/w12061606

Barla, G., Bonini, M. and Cammarata, G., "Stress and seepage analyses for a gravity dam on a jointed granitic rock mass", in H. Konietzky (ed.), *Numerical Modeling of Discrete Materials in Geotechnical Engineering*, Balkema, Rotterdam, pp. 263–268, 2004.

Barton, N., Bandis, S. and Bakhtar, K., "Strength, Deformation and Conductivity Coupling of Rock Joints", *International Journal of Rock Mechanics and Mining Sciences*, 22, 3, 121–140, 1985.

Beltaos, S., "Oblique Impingement of Plane Turbulent Jets", *Journal of the Hydraulics Division*, 102, 1177–1192, 1976a.

Beltaos, S., "Oblique Impingement of Circular Turbulent Jets", *Journal of Hydraulic Research*, 14, 807–814, 1976b.

Beltaos, S. and Rajaratnam, N., "Plane Turbulent Impinging Jets", *Journal of Hydraulic Research*, IAHR, 11, 1, 29–59, 1973.

Bohrer, J.G., Abt, S.R. and Wittler, R.J., "Predicting Plunge Pool Velocity Decay of Free Falling, Rectangular Jet", *Journal of Hydraulic Engineering*, ASCE, 124, 10, 1043–1048, 1998.

Bollaert, E.F.R., "Transient water pressures in joints and formation of rock scour due to high-velocity jet impact", *PhD Thesis*, LCH-EPFL, Lausanne, Switzerland, 2002.

Bollaert, E.F.R., "The Influence of Joint Aeration on Dynamic Uplift of Concrete Slabs of Plunge Pool Linings", in *Proceedings of the 30th IAHR Congress*, Thessaloniki, Greece, 2003.

Bollaert, E.F.R., "A Comprehensive Model to Evaluate Scour Formation in Plunge Pools", *International Journal of Hydropower & Dams*, 2004, 1, 94–101, 1998.

Bollaert, E.F.R., "Rock scour at hydraulic structures: a practical engineering approach", *Geo-Strata*, 2010.

Bollaert, E.F.R., "Wall Jet Rock Scour in Plunge Pools: A Quasi-3D Prediction Model", *International Journal on Hydropower & Dams,* 19, 4, 70–76, 2012.

Bollaert, E.F.R., "One-Degree-of-Freedom Rigid Body Dynamics of Rock Blocks in Turbulent Flows", in *Proceedings of the 2013 IAHR Congress*, Tsinghua University Press, Beijing, 2013.

Bollaert, E.F.R., "The Rocsc@r Cloud: An Innovative Digital Platform to Compute and Record Rock Scour", *International Journal on Hydropower & Dams*, 28, 5, 60–70, 2021.

Bollaert, E., Pirotton, M. and Schleiss, A., "Multiphase Transient Flow and Pressures in Rock Joints Due to High Velocity Jet Impact: An Experimental and Numerical Approach", in *Proceedings of the 3rd International Symposium on Environmental Hydraulics*, Arizona State University, Tempe, 2001.

Bollaert, E.F.R. and Schleiss, A.J., "Physically Based Model for Evaluation of Rock Scour Due to High-Velocity Jet Impact", *Journal of Hydraulic Engineering*, 131, 3, 153–165, 2005. https://doi.org/10.1061/(ASCE)0733-9429(2005)131:3(153)

Bollaert, E.F.R., Federspiel, M. and Schleiss, A.J., "The Influence of Added Mass on Rock Block Uplift in Plunge Pools", in *Proceedings of the 2013 IAHR Congress*, Tsinghua University Press, Beijing, 2013.

Boroomand, M.R., Salehi Neyshabouri, S.A.A. and Aghajanloo, K., "Numerical Simulation of Sediment Transport and Scouring by an Offset Jet", *Canadian Journal of Civil Engineering,* 34, 10, 1267–1275, 2007.

Bosman, A., "High dam scour hole geometry prediction for fully developed jets plunging into shallow pools on bedrock", *PhD Dissertation*, Civil Engineering Faculty, Stellenbosch University, Cape Town, South Africa, 2021.

Castillo, L.G., "Metodología experimental y numérica para el diseño de los cuencos de disipación de energía. Aplicación al vertido en presas bóveda. ", *PhD Thesis*. Departamento de Ingeniería Hidráulica, Marítima y Medio Ambiental, Universidad Politécnica de Cataluña, Spain. [in Spanish], 1989.

Castillo, L.G. and Carrillo, J.M., "*Scour estimation of the Paute-Cardenillo Dam*", in *Proceedings of the International Perspectives on Water Resources & the Environment*, Quito, Ecuador, 2014.

Castillo, L.G., Carrillo, J.M. and Blázquez, A., "Plunge Pool Mean Dynamic Pressures: A Temporal Analysis in Nappe Flow Case", *Journal of Hydraulic Research*, 53, 101–118, 2015.

Castillo, L.G. and Carrillo, J.M., "Scour, Velocities and Pressures Evaluations Produced by Spillway and Outlets of Dam", *Water*, 8, 68, 2016. https://doi.org/10.3390/w8030068

Castillo, L.G. and Carrillo, J.M., "Comparison of Methods to Estimate the Scour Downstream of a Ski Jump", *International Journal of Multiphase Flow*, 92, 171–180, 2017.

Castillo, L.G. and Carrillo, J.M., "Technical University of Cartagena Research. Study Cases: Paute Cardenillo and Toachi Dams", in *International Workshop on Overflowing Erosion of Dams and Dikes Concrete Dams Overtopping Erosion*, Aussois, France, 2017.

Chanel, P.G. and Doering, J.C., "An Evaluation of Computational Fluid Dynamics for Spillway Modelling", in *16th Australasian Fluid Mechanics Conference*, Gold Coast, Australia, pp. 1201–1206, 2007.

Chesterton, O.J., Ucuncu, M. and Borman, D., "CFD Modelling for Dams and Reservoirs – Best Practice Workflows, Specification and Review", *Dams and Reservoirs*, 29, 4, 148–157, 2019.

Constantinescu, S.G., Ettema, R. and Koken, M., "An Eddy-Resolving Technique to Predict the Unsteady Horseshoe Vortex and Wake of a Cylindrical Pier", in *Proceedings of Second International Conference on Scour and Erosion*, Stallion Press, Singapore, vol 2, pp. 202–211, 14–17 November 2004.

Cundall, P.A., "A Computer Model for Simulating Progressive Large Scale Movements in Blocky Rock Systems", in *Proceeding of Symposium Rock Fracture (ISRM)*, Nancy, vol 1, pp. 2–8, 1971.

Cundall, P.A., "UDEC – A generalized distinct element program for modelling jointed rock", *Report PCAR-1-80*, European Research Office, US Army, pp. 80, 1980.

Dasgupta, B., Basu, D., Das, K. and Green, R., "Development of Computational Methodology to Assess Erosion Damage in Dam Spillways", in *USSD 31st Conference*, San Diego, CA, April 11–15, 2011.

Dasgupta, B., Das, K., Basu, D., Green, R. and McGinnis, R., "Computational Approach to Predict Rock Erosion in Unlined Spillways", in *24th ICOLD Congress*, Q. 94, Kyoto, Japan, 2012.

Damjanac, B. and Fairhurst, C., "Ecoulement tridimensionnel d'eau sous pressions dans les milieux fracturés", in H. Delage and D. Gennaro (eds), *La Sécurité des Grands Ouvrages*, Presses Ponts et Chaussées, Paris, pp. 5–16, 2000.

Deere, D.U. and Miller, D.W., "The Rock Quality Designation (RQD) Index in Practice, Classification Systems for Engineering Purposes", in *ASTM STP*, American Society for Testing and Materials, Philadelphia, PA, pp. 91–101, 1967. https://doi.org/10.1520/STP48465S

Dey, D. and Eldho, T.I., "Effect of Spacing of Two Offset Jets on Scouring Phenomena", *Journal of Hydraulic Research*, 47, 1, 82–89, 2009. https://doi.org/10.3826/jhr.2009.3044

Eberhardt, E. Stead, D. and Coggan, J.S., "Numerical analysis of initiation and progressive failure in natural rock slopes – The 1991 Randa rockslide", *International Journal of Rock Mechanics and Mining Sciences*, 41:1, 69–87, 2004.

Epely-Chauvin, G., De Cesare, G. and Schwindt, S., "Numerical Modelling of Plunge Pool Scour Evolution in Non-Cohesive Sediments", *Engineering Applications of Computational Fluid Mechanics*, 8, 4, 447–487, 2014. https://doi.org/10.1080/19942060.2014.11083301

Ervine, D.A., Falvey, H.R. and Withers, W., "Pressure Fluctuations on Plunge Pool Floors", *Journal of Hydraulic Research*, IAHR, 35, 2, 1997.

Ewing, D.J.F., "Allowing for Free Air in Waterhammer Analysis", in *Proceedings of the 3rd International Conference on Pressure Surges*, Canterbury, England, 1980.

Federspiel, M.P.E.A., "Response of an embedded block impacted by high-velocity jets", *PhD Thesis*, LCH-EPFL, 2011.

Gardner, M. and Sitar, N., "Modeling of Dynamic Rock-Fluid Interaction Using Coupled 3-D Discrete Element and Lattice Boltzmann Methods", *Rock Mechanics and Rock Engineering*, 52, 12, 2019. https://doi.org/10.1007/s00603-019-01857-x

Gardner, M., "Toward a Complete Kinematic Description of Hydraulic Plucking of Fractured Rock", *Journal of Hydraulic Engineering*, 149, 7, 2023. https://orcid.org/0000-0002-5238-630.

George, M.F., Ph.D. Dissertation. University of California, Berkeley, 2015.

George, M.F., Christiansen, C., Rickel, A., Israel, B. and Annandale, G.W., "Erodibility Evaluation of an Unlined Rock Spillway: Comparison Between the Erodibility Index Method and a New Method Based on Block Theory", in *Proceedings of the 4th International Seminar on Dam Protections against Overtopping*, Madrid (Spain), 2022. https://doi.org/10.26077/56a7-33b7

Gimenes, E. and Fernandez, G., "Hydromechanical Analysis of Flow Behavior in Concrete Gravity Dam Foundations", *Canadian Geotechnical Journal*, 43:3, 244–259, 2006.

Goodman, R.E., "Block Theory and its Application", *Géotechnique*, 45, 3, 383–423.

Goodman, R.E., Taylor, R.L. and Brekke, T.L., "A model for the Mechanics of Jointed Rock", *Journal of Soil Mechanics and Foundations Division ASCE*, 94:SM3, 637–659, 1968.

Hart, R.D., Cundall, P.A. and Lemos, J.V., "Formulation of a Three-Dimensional Distinct Element Model – Part II: Mechanical Calculations for Motion and Interaction of A System Composed of Many Polyhedral Blocks", *International Journal of Rock Mechanics and Mining Sciences*, 25:3, 117–125, 1988.

Hartung, F. and Hausler, E., "Scours, stilling basins and downstream protection under free overfall jets at dams", in *Proceedings of the 11th Congress on Large Dams,* Madrid, pp. 39–56, 1973.

Hirt, C.W. and Nichols, B.D., "Volume of Fluid (VOF) Method for the Dynamics of Free Boundaries", *Journal of Computational Physics*, 39, 201–225, 1981. https://doi.org/10.1016/0021-9991(81)90145-5

Hoek, E. and Brown, E.T., "Empirical Strength Criterion for Rock Masses", *J. Geotechnical Engineering Division. ASCE,* 106:GT9, 1013–1035, 1980.

Hoek, E., "https://www.rocscience.com/hoek/corner/Practical_Rock_Engineering.pdf," *Practical rock engineering*, 2006.

ISRM (International Society for Rock Mechanics), "Suggested Methods for the Quantitative Description of Discontinuities in Rock Masses", *International Journal of Rock Mechanics and Mining Sciences & Geomechanics Abstracts*, 15, 319–368, 1978.

Jing, L. and Stephansson, O., *Fundamentals of discrete element methods for rock engineering – theory and application*, Elsevier, Amsterdam, pp. 545, 2007.

Kamanbedast, A.A. and Aghamajidi, R., "Ski Jump Length of the Spillway Using FLOW3DMathematical Model (Case Study: Gotvand Dam)", *Technical Journal of Engineering and Applied Sciences*, 3, 23, 3399–34040, 2013.

Karim, OA. and Ali, K.H.M., "Prediction of flow patterns in local scour holes caused by turbulent water jets", *Journal of Hydraulic Research*, 38, 4, 279–287, 2000. https://doi.org/10.1080/00221680009498327

Kazerani, T. and Zhao, J., "Micromechanical Parameters in Bonded Particle Method for Modelling of Brittle Material Failure", *International Journal for Numerical and Analytical Methods in Geomechanics*, 34:18, 1877–1895, 2010.

Keaton, J., Mishra, S.K. and Clopper, P.E., "NCHRP Report 717 : Scour at Bridge Foundations on Rock", *Transportation Research Board*, Washington, DC, 2012.

Khosronejad, A., Kang, S., Borazjani, I. and Sotiropoulos, F., "Curvilinear Immersed Boundary Method for Simulating Coupled Flow and Bed Morphodynamic Interactions Due to Sediment Transport Phenomena", *Advances in Water Resources*, 34, 7, 829–843, 2011.

Kieffer, D.S. and Goodman, R.E., "Assessing Scour Potential of Unlined Rock Spillways with the Block Scour Spectrum", *Geomechanics and Tunnelling*, 5, 5, 527–536, 2012. https://doi.org/10.1002/geot.201200039

Kirsten, H.A.D., "A Classification System for Excavation in Natural Materials", *The Civil Engineer in South Africa*, 24, 7, 292–308, 1982.

Lamas, L.N. and Sousa, L.R., "The use of a hydromechanical numerical model to understand the behaviour of pressure tunnels and shafts", in L.R. Sousa and N.F. Grossman (eds), *Proceeding of the EUROCK93,* Balkema, Lisboa, pp. 961–968, 1993.

Lemos, J.V., "Discrete element analysis of dam foundations", in V.M. Sharma, K.R. Saxena and R.D. Woods (eds), *Distinct Element Modelling in Geomechanics*, Balkema, Rotterdam, pp. 89–115, 1999.

Lemos, J.V., "Block Modelling of Rock Masses – Concepts and Application to Dam Foundations", *European Journal of Environmental and Civil Engineering*, 12, 7–8, 915–949, 2008.

Lemos, L.V., "Discontinuum Modelling in Rock engineering", *Soils and Rocks*, São Paulo, 36, 2, 137–156, May-August, 2013.

Lesleighter, E.J., Bollaert, E.F.R., McPherson, B. and Scriven, D.C., "Spillway Rock Scour Analysis Composite of Physical & Numerical Modelling", in *Proceedings of the 6th International Symposium on Hydraulic Structures*, Portland, OR, USA, 2016.

Li, A. and Liu, P., "Mechanism of Rock-Bed Scour Due to Impinging Jet", *Journal of Hydraulic Research,* 48, 1, 14–22, 2010. https://doi.org/10.1080/00221680903565879

Liu, P.Q., Dong, J.R. and Yu, C., "Experimental Investigation of Fluctuating Uplift on Rock Blocks at the Bottom of the Scour Pool Downstream of Three-Gorges Spillway", *Journal of Hydraulic Research,* IAHR, 36, 1, 55–68, 1998.

Maleki, S. and Fiorotto, V., "Blocks Stability in Plunge Pools under Turbulent Rectangular Jets", *Journal of Hydraulic Engineering*, 145, 4, 147–157, 2019. https://doi.org/10.1061/(ASCE)HY.1943-7900.0001573

Marson, C., Fiorotto, V. and Caroni, E. "Lining Design in Spillway Stilling Basins", in *Proceeding of the 5th International Conference on Dam Engineering*, Lisbon, Portugal, pp. 299–306, 2007.

Martin, C.S. and Padmanabhan, M., "Pressure Pulse Propagation in Two-Component Slug Flow", *Journal of Fluids Engineering*, 101, 44–52, 1979.

Mas Ivars, D., Pierce, M.E., Darcel, C., Reyes-Montes, J., Potyondy, D.O., Paul Young, R. and Cundall, P.A., "The Synthetic Rock Mass Approach for Jointed Rock Mass Modelling", *International Journal of Rock Mechanics and Mining Sciences*, 48, 2, 219–244, 2011.

Mendoza-Cabrales, C., "Computation of Flow Past a Pier Mounted on a Flat Plate", in *Proceedings of ASCE Water Resources Engineering Conference, July 25–30,* San Francisco, pp. 889–904, 1993.

Menter, F.R., "Zonal Two Equation k-turbulence Models for Aerodynamic Flows", in *AIAA 24th Fluid Dynamic Conference, 6–9 July*, Orlando, Florida, 1993.

Menter, F.R., "Two-Equation Eddy-Viscosity Turbulence Models for Engineering Applications", *AIAA Journal*, 32, 8, 1598–1605, 1994. https://doi.org/10.2514/3.12149

Morgans, R.C., Dally, B.B., Nathan, G.J., Lanspeary, P.V. and Fletcher, D.F., "Application of the revised Wilcox (1998) k–ω turbulence model to a jet in co-flow", in *Second International Conference on CFD in the Minerals and Process Industries*, CSIRO, Melbourne, Australia, pp. 479–484, 6–8 December, 1999.

Müller, G., Wolters, G. and Cooker, M.J., "Characteristics of Pressure Pulses Propagating Through Water-Filled Cracks", *Coastal Engineering*, 49, 1, 83–98, 2003. https://doi.org/10.1016/S0378-3839(03)00048-6

Munjiza, A., *The combined Finite-Discrete element method*, John Wiley, Chichester, pp. 352, 2004.

Otto, B., "Scour potential of highly stressed sheet-jointed rocks under obliquely impinging plane jets", *PhD thesis*, James Cook University of North Queensland, Townsville, 1989.

Palmström A., "The volumetric joint count – A useful and simple measure of the degree of rock mass jointing", *IAEG Congress*, New Delhi, pp. 221–228, 1982.

Pan, Y.-W., Li, K.-W. and Liao, J.-L., "Mechanics and Response of a Surface Rock Block Subjected to Pressure Fluctuations: A Plucking Model and its Application", *Engineering Geology*, 171, 1–10, 2014.

Paris, P.C., Gomez, M.P., Anderson, W.E., "A Rational Analytic Theory of Fatigue", *The Trend in Engineering*, 13, 9–14, 1961.

Paris, P.C. and Sih, G.C., "Stress analysis of cracks, Fracture toughness Testing and its Applications", *ASTM STP 381*, Philadelphia, pp. 63–77, 1965.

Pells, S., "Erosion of rock in spillways", *PhD Dissertation*, School of Civil and Environmental Engineering Faculty of Engineering, University of New South Wales, 2016.

Pells, P.J., Bieniawski, Z.T., Hencher, S.R. and Pells, S.E., "Rock Quality Designation (RQD): Time to Rest in Peace", *Canadian Geotechnical Journal*, 54, 6, 825–834, 2017. https://doi.org/10.1139/cgj-2016-0012

Pierce, M., Mas Ivars, D., Cundall, P. and Potyondy, D., "A synthetic rock mass model for jointed rock", in E. Eberhardt (ed.), *Proceedings of the 1st Canada-US Rock Mechanics Symposium*, Taylor & Francis, Vancouver, London, 2007.

Potyondy, D., "The bonded-particle model as a tool for rock mechanics research and application: current trends and future direction", in *Proceeding of the 7th Asian Rock Mechanics Symposium*, Seoul, Korea, pp. 73–105, 2012.

Potyondy, D. and Cundall, P.A., "A Bonded-Particle Model for Rock", *International Journal of Rock Mechanics and Mining Sciences*, 41, 8, 1329–1364, 2004.

Reich, F., "Umlenkung eines Freien Flüssigkeitsstrahles an einer ebenen Platte", *Zeitschrift Verein deutscher Ingenieure*, 71, 8, 261–264, 1927.

Reinius, E., "Rock Erosion", *International Journal on Water Power and Dam Construction*, 38, 6, 43–48, 1986.

Reyes-Montes, J.M, Pettitt, W.S. and Young, R.P., "Enhanced spatial resolution of caving-induced microseismicity", in N. Schunnesson (ed.), *Proceedings of the 5th International Conference and Exhibition on Mass Mining (MassMin 2008),* Lule, Univ Tech Press, Lulea, Sweden, pp. 961–70, 2008.

Richardson, J.E. and Panchang, V.G., "Three-Dimensional Simulation of Scour-Inducing Flow at Bridge Piers, *Journal of Hydraulic Engineering*, ASCE, 124, 5, 530–540, Rooseboom, 1998.

Salaheldin, T.M., Irman, J. and Chaudhry, M.H., "Numerical Modelling of Three-Dimensional Flow Field Around Circular Piers", *Journal of Hydraulic Engineering*, ASCE, 130, 2, 91–100, 2004.

Savage, B.M. and Johnson, M.C., "Flow Over Ogee Spillway: Physical and Numerical Model Case Study", *Journal of Hydraulic Engineering*, 127, 8, 640–649, 2001. https://doi.org/10.1061/(ASCE)0733-9429(2001)127:8(640)

Serafim, J.L., "Influence of interstitial water on the behaviour of rock masses", in K.G. Stagg and O.C. Zienkiewicz, (eds), *Rock Mechanics in Engineering Practice*, John Wiley, Chichester, pp. 55–97, 1968.

Shi, G.-H. and Goodman, R.E., "Discontinuous deformation analysis – A new method for computing stress, strain and sliding of block systems", in P.A. Cundall, A.M. Sterling and R.L. Starfield (eds), *Key Questions in Rock Mechanics: Proceedings of the 29th U.S. Symposium*, Balkema, Rotterdam, pp. 381–393, 1988.

Schlichting, H. and Gersten, K., *Boundary layer theory*, Springer Verlag, New York, 2000.

Starfield, A.M. and Cundall, P.A., "Towards a Methodology for Rock Mechanics Modelling", *International Journal of Rock Mechanics and Mining Sciences*, 25:3, 93–106, 1988.

Swanson, P.L., "Subcritical Crack Growth and Other Time-and Environment-Dependent Behavior in Crustal Rocks", *Journal of Geophysical Research*, 89, B6, 4137–4152, 1984.

USBR, "Uplift and Crack Flow Resulting from High Velocity Discharges Over Open Offset Joints", *Report DSO-07-07*, US Bureau of Reclamation, Denver, CO, USA, 2007.

Versteeg, H.K. and Malalasekera, W., *An introduction to computational fluid dynamics*, 2nd Edition. Pearson Education Limited, England, 2007.

Vonkeman, J.K., "Coupled fully three-dimensional hydro-morphodynamic modelling of bridge pier scour in an alluvial bed", *PhD Dissertation*, Faculty of Civil engineering, Stellenbosch University, Cape Town, South Africa, 2019.

Wahl, T., "Computing the Trajectory of Free Jets", *Journal of Hydraulic Engineering*, 134, 2, 256–260, 2008. https://doi.org/10.1061/(ASCE)0733-9429(2008)134:2(256)

Wahl, T. and Heiner, B.J., "Hydraulic Jacking in Concrete Spillway Chutes", *The Journal of Dam Safety*, ASDSO, 20, 2, 2023.

Wibowo, J., Villanueva, E., Temple, D., and Yule, D., "Earth and Rock Surface Spillway Erosion Risk Assessment", in *U.S. Symposium on Rock Mechanics*, Paper No. 05-813. American Rock Mechanics Association, Alexandria, VA, 2005.

Wittke, W., *Rock mechanics – theory and applications with case histories*, Springer-Verlag, Berlin, pp. 1076, 1990.

Wylie, E.B. and Streeter, V.L., *Fluid transients*, McGraw-Hill Int., Book Company, New York, US, 1978.

Xiong, W., Cai, C.S., Kong, B. and Kong, X., "CFD Simulations and Analyses for Bridge Scour Development Using a Dynamic Mesh Updating Technique", *Journal of Computing in Civil Engineering,* 30, 1, 1–11, 2014.

7 Digital Scour Applications

INTRODUCTION

The objective of this chapter is to present engineering applications of the different digital scour models discussed in Chapter 6. The applications run on a digital software platform within the rocsc@r digital environment. As explained in Chapter 4, this platform implements multiple semi-empirical and physics-based scour computational methods in two dimensions, by using 2D vertical profiles. Hereafter, scour holes observed downstream of different types of hydraulic structures have been numerically reproduced using this platform and its digitalized methods.

The computations are performed in a deterministic manner, i.e. based on well-defined parametric settings, or in a probabilistic manner, i.e. stochastically defining parametric values, following a cumulative probability distribution.

Moreover, a thorough parametric sensitivity study and analysis points out the most pertinent parametric settings to be applied for practice.

Furthermore, scour computations based on a sequential fluid-solid coupling using 2D and 3D CFD modelling are presented and discussed. These make use of a dedicated API (Application Programming Interface) digital connection between the FLOW-3D CFD software and the rocsc@r software. The advantages of such a coupling are multiple:

- more detailed 2D and full 3D assessment of flow hydraulics and rock scour,
- digital scour modelling for any type of turbulent flow that can be modelled by CFD,
- detailed geometrical situations, such as local flow perturbations, complex pool shapes, etc. can be accounted for,
- scour mitigation measures may be easily implemented and tested on their influence on scour formation.

Examples of CFD fluid-solid coupling are provided for 2D and 3D cases, in dam engineering (plunge pool and channel scour) and river engineering (bridge pier scour).

DEFINITION OF SCOUR DEPTH

Different terminologies of the scour depth exist in literature and it is necessary here to precisely define the terms to be used and which scour depths are numerically reproduced.

One of the most widely used terms is the *ultimate* or *equilibrium* scour depth, defined as an end-condition of deepest scour that may occur at a hydraulic structure for given design flood conditions. The concept of ultimate depth is largely criticized, however, because it is not compatible with the generally accepted fact that scour never ceases during the lifetime of a structure. Effectively, progressive fracturing of rock and/or ball-milling of blocks, splitting up detached blocks into smaller pieces

DOI: 10.1201/9781003319610-7

that can be ejected, theoretically never ends as long as flood events will continue to pass over the spillway of the structure.

As a matter of fact, the following terms and definitions seem more adapted to the sequence of flood events responsible for progressive scour formation at hydraulic structures:

- *Flood event scour depth*: This is defined as the scour depth that is observed following a distinct, time-limited flood event on a spillway structure. This depth may be predicted by one or more of the following computational methods:
    - *fracturing scour potential*: scour potential during the flood event as determined by brittle and fatigue fracturing of rock joints
    - *block uplift scour potential*: scour potential for a distinct flood event as determined by vertical uplift of rock blocks
    - *block detachment scour potential*: scour potential for a distinct flood event as determined by peeling-off (or detachment) of rock blocks by wall jet flows
    - *combined scour potential*: scour potential for a distinct flood event as determined by combination of the aforementioned break-up methods, eventually by including downstream mounding and ball-milling effects.
- *Expected lifetime scour depth*: This is defined as the scour depth that may occur during the lifetime of the structure, by a succession of distinct flood events of a given time duration and intensity. This depth may be predicted by combining the aforementioned computational methods, but needs sound determination of the succession of flood events. It is challenging to compute, and is very rarely observable in situ. Some structures, however, exhibit scour formation since a long time, accumulating the effects of many flood events. One example is Kariba Dam in Zambia-Zimbabwe, where design flood conditions have been passing through the spillway gates for several thousands of hours since 1962, resulting in excessively deep scour formation that would continue today if no mitigation measures were implemented at the site. Another example is Chancy-Pougny Dam on the Swiss-French border, exhibiting scour formation since 1924, and still continuing today.

In the following, except for Kariba Dam, Choranche Dam and Chancy-Pougny Dam for which a sequence of events has been considered, the computed scour mainly concerns one distinct flood event that occurred on the studied hydraulic structure, and thus mainly flood event scour depths are discussed. When only considering rock mass break-up by block detachment, i.e. by neglecting progressive fracturing of the rock, the flood event scour depth may be considered as the equilibrium scour depth for that particular event.

DETERMINISTIC-BASED SCOUR COMPUTATIONS

This subchapter presents deterministic-based computations of scour downstream of hydraulic structures. It deals with 2D jet flows, such as ski-jump spillways, free overflow spillways, low-angle pressurized outflows and so on. First, two distinct case studies are discussed in detail. Second, the computational methods are calibrated by applying them to several worldwide cases for which the scour formation has been soundly recorded for distinct flood events. Next, a parametric sensitivity study on these cases provides a range of plausible values to be adopted for practice.

The Stellenbosch Laboratory Experiments

Large-scale laboratory experiments of scour hole development in broken-up rock, generated by rectangular-shaped oblique jets impinging in an artificial plunge pool, have been performed at the hydraulics laboratory of the Civil Engineering Department at Stellenbosch University, South Africa (Bosman and Basson 2020; Bosman 2021). Some of these experiments have been used as a case study for the different computational methods that apply to fully broken-up rock (see Chapter 6), i.e.:

- DI (2005) method based on C_I values (Bollaert 2002; Bollaert and Schleiss 2005)
- MDI (2021) method based on C_p' values (Bollaert 2021)
- DP (2021) method based on Ω values (Maleki and Fiorotto 2019)
- QSI (2012) method based on quasi-steady forces (Bollaert 2012)
- MQSI (2021) method based on quasi-steady and pulsating forces (Bollaert 2021)
- EIM (1995) method based on Kirsten erodibility index (Annandale 1995)

Experimental Set-Up (Bosman and Basson 2020)

The physical model is illustrated in section and plan view in Figure 7.1 and consists of a zero-sloped 0.25-m-wide rectangular canal at issuance, replicating an uncontrolled dam spillway. The canal could be adjusted to three different fixed heights above the movable rock bed: 3, 4 and 5 m (prototype heights for 1:20 model scale: 60, 80 and 100 m respectively). The tailwater depth at the plunge pool was adjustable between 0.5 and 1 m (prototype: 10 and 20 m respectively). The plunging jets generated unit discharges between 0.39 and 0.89 $m^3/s/m$ (prototype: 35–80 $m^3/s/m$).

The broken-up rock was modelled by tightly hand-packed concrete paver blocks (cobblestones), generating a uniform three-dimensional open-ended horizontal and vertical rock joint network. The two rock sizes tested were rectangular concrete pavers with x, y, z dimensions of $0.1 \times 0.1 \times 0.05$ m and $0.1 \times 0.1 \times 0.075$ m (prototype: $2 \times 2 \times 1$ and $2 \times 2 \times 1.5$ m respectively). The paver block densities used were 2,355.4 kg/m^3 (for 1-m-high blocks) and 2,388.1 kg/m^3 (for 1.5-m-high blocks) respectively.

A total of 31 experimental tests were carried out by the authors. From these, three tests have been chosen to be reproduced by the computational models of the software (Figure 7.2). During each of the experimental tests, the scour of the rock bed by the free-falling rectangular jet was monitored by Bosman and Basson (2020) until scour in the plunge pool stopped and equilibrium conditions were reached (after 6–8 h of testing), corresponding to a test case with downstream mounding. The bed profile of the equilibrium scour hole was then surveyed.

The same test was then repeated, but instead of leaving the scoured blocks in the pool, the blocks deposited downstream were continuously removed until equilibrium conditions were reached, i.e. after approximately 16–20 h, corresponding to a test

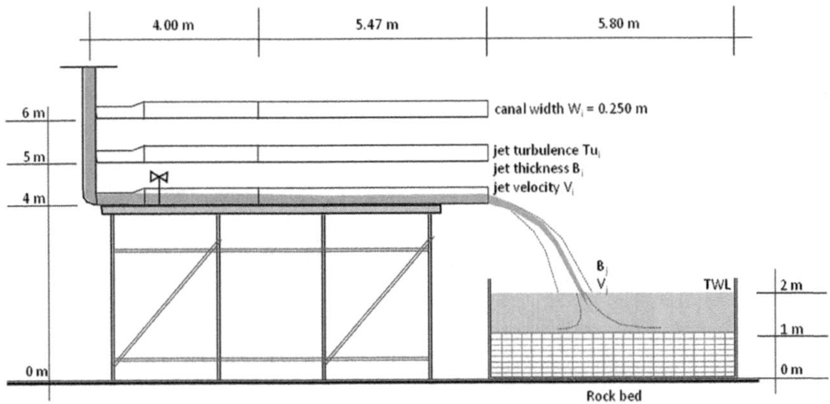

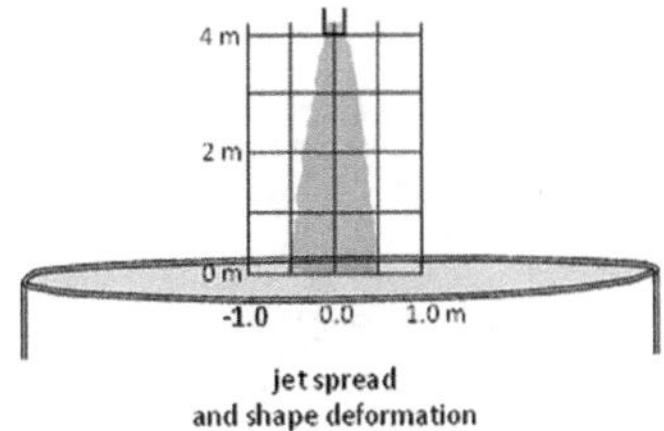

FIGURE 7.1 Experimental set-up of plunging jets impinging on broken-up rock bed. Adapted from Bosman 2021.

case without downstream mounding. The bed profile of the so-formed equilibrium bed profile was surveyed again.

The topography of the bed profiles was surveyed using a 3D laser scanner at a high resolution of 10,000 pixels/360°.

Experimental Parameters

The main parameters of the test runs that have been reproduced by the numerical model are summarized in Table 7.1, for both 1/20 scaled values and corresponding prototype values. Using prototype values in the remaining of this chapter, the unitary discharge equals 80 m^2/s, for a hydraulic head of 98.2 m and a tailwater depth of 20 m. The block heights are 1.0 and 1.5 m, for side lengths of 2 m by 2 m. Test 9A of the authors is reproduced twice, once with mounding and once without mounding downstream of the eroded blocks.

The jet thickness at impact in the pool B_j as given by the authors refers to the total thickness of the jet, i.e. the sum of the contracted core thickness and the outer jet spread due to turbulence. The jets generated by the Stellenbosch test installation have the following main characteristics:

- square-shaped at issuance from the spillway,
- strong deformation during fall, into a flat and wide shape upon impact in the pool,

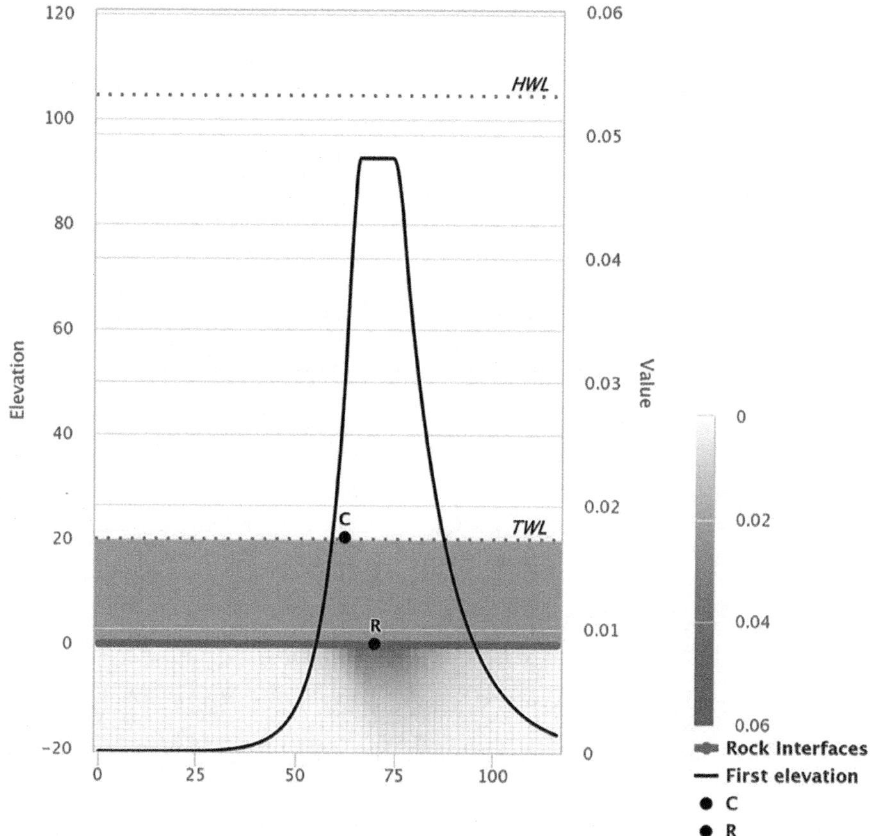

FIGURE 7.2 Numerically computed fluctuating dynamic pressure coefficient.

- initial thickness may be approached by the equivalent diameter of a circular jet, while gravitational contraction during fall more corresponds to the one for nappe flow/rectangular-shaped jets, because of the strong deformation,
- trajectories as reported by Bosman (2021) are generally longer than the theoretical ones based on ballistic equations and air drag, jet angles at impact were lower than theoretically expected,
- based on Bosman (2021), the jets generated by the installation are considered as highly broken up upon impact, with a jet break-up ratio >1.85,
- jet diffusion through the plunge pool has not been recorded, but experimental and numerical studies performed by Calitz (2015) on the same installation, although for a different range of unitary discharges, showed that the velocity decay for fully broken-up rectangular-shaped jets follows an exponential decay law that is rather typical for circular-shaped jets.

Tests 9A_a and 14A (Table 7.1) are performed without downstream mounding. Test 9A_b has been performed with downstream mounding. Test 9A_a is described in detail hereafter, for the other tests performed, only the results are given.

TABLE 7.1
Parameters of Numerically Reproduced Experiments

1/20 Model Scale

Test		Issuance				Pool Impact					Rock Bed		
–	Spec. disc.	Head	Flow depth	Flow width	Flow vel.	TWL	Jet thick	Jet width	Jet vel.	Imp. angle	Block size	Joint angle	Mound.
N°	q	H	B_i	W_i	V_i	y	B_j	W_j	V_j	α_j	z_b	α_b	–
–	m²/s	m	m	m	m/s	m	m	m	m/s	°	m	°	–
9A_a	0.89	4.91	0.31	0.25	2.7	1.0	0.87	0.68	9.5	60	0.050	0	No
9A_b	0.89	4.91	0.31	0.25	2.7	1.0	0.87	0.68	9.5	60	0.050	0	Yes
14A	0.89	4.91	0.34	0.25	2.5	1.0	0.77	1.02	9.5	62	0.075	0	No

Prototype Values

Test		Issuance				Pool Impact					Rock Bed		
–	Spec. disc.	Head	Flow depth	Flow width	Flow vel.	TWL	Jet thick	Jet widt	Jet vel.	Imp. angle	Block size	Joint angle	Mound.
N°	q	H	B_i	W_i	V_i	y	B_j	W_j	V_j	α_j	z_b	α_b	–
–	m²/s	m	m	m	m/s	m	m	m	m/s	°	m	°	–
9A_a	80	98.2	6.2	5	12.1	20	17.4	13.6	42	60	1.0	0	No
9A_b	80	98.2	6.2	5	11.2	20	17.4	13.6	42	60	1.0	0	Yes
14A	80	98.2	6.8	5	11.2	20	15.4	20.4	42	62	1.5	0	No

Source: Adapted from Bosman (2021).

Test 9A_a: $q = 80\ m^2/s$, $H = 100\,m$, $TWL = 20\,m$, $z_b = 1.0\,m$, without mounding

Test 9A_a corresponds to the maximum specific discharge, the maximum hydraulic head and the maximum tailwater level used. The block height is 1.0 m. No mounding was allowed during the test.

Jet Trajectory and Break-up

As the overfall jets of the installation have been generated by means of a pressurized flow that enters a zero-sloped canal before issuance, it might be that the pressure profile at the brink significantly differs from a static head profile. For the computed jet trajectories to match with the laboratory-observed ones, an additional pressure head equal to the flow depth has been added to the computations.

Based on the Froude number at issuance $Fr_i = 1.54$ and a turbulence intensity at issuance of $Tu_i \sim 0.08$, the jet break-up length equals (based on Castillo 2007) $L_b = 45\,m$. Hence, the ratio of jet trajectory length L_j over jet break-up length L_b equals 2.37 and the jet is highly broken up at impact in the pool.

Jet Contraction and Deformation

The jet contracts because of gravitational acceleration. Based on the available photographs and the measurements performed by the authors, the jet width at impact corresponds to 2–3 times the jet width at issuance. Hence, the initially (quasi) square-shaped jet deforms into a much wider, rectangular-shaped jet upon impact. As such, the jet thickness at impact has been computed based on the computed flow velocity at impact V_j, using the formula valid for rectangular jets:

$$B_j = B_i \cdot \frac{V_i}{V_j} = 6.2\ \text{m} \cdot \frac{12.0\ \text{m/s}}{41.7\ \text{m/s}} = 1.79\ \text{m} \tag{7.1}$$

This value is used to determine the development and progressive diffusion of the jet through the plunge pool water depth. The jet angle at impact is computed as 68.8°. This is certainly somewhat different than the 60° as stated by the authors but does correspond to the observed jet trajectory as provided by the same authors.

Jet Diffusion through the Pool

The jet diffusion through the pool is governed by an inner diffusion angle of ~6°–7° and an outer diffusion angle of ~14°. The theoretical length of the core of the jet is ~4.0 B_j. The jet develops linearly through the pool from its pool impact (point C in Figure 7.2) down to the rock bed until the stagnation point of impact (point R in Figure 7.2) that is situated at $x \sim 70\,m$.

The corresponding fluctuating dynamic pressures computed at the rock bed are presented in Figure 7.2. According to the 2D plunging jet model described in Chapter 6 (Tables 6.3 and 6.6), a best fit was obtained using the following parametric settings for lateral pressure decay ratios: $n_2 = n_4 = 3$, $n_1 = 0.20$ towards upstream of the stagnation point, and $n_3 = 0.10$ towards downstream.

The jets are a priori rectangular upon impact in the pool. Nevertheless, based on Bohrer et al. (1998) and experimental and practical evidence, for broken-up rectangular jets, the velocity decay according to circular jets should be used. This was confirmed based on Calitz (2016), who made detailed physical and numerical studies on the same installation (albeit for lower specific discharges and at a smaller scale); both his laboratory tests and 3D numerical modelling of broken-up jets showed axial velocity decays that match the ones for circular jets.

As such, in the following, axial velocity decays valid for circular jets (and also for rectangular broken-up jets) have been applied.

Scour Results

Sound calibration of the methods generated good to excellent agreement between measurements and computations (Figure 7.3). Not only the area of deepest scour formation is correctly modelled, but also the slopes of the scour hole towards upstream and downstream are adequately reproduced by the different computational methods on the digital platform.

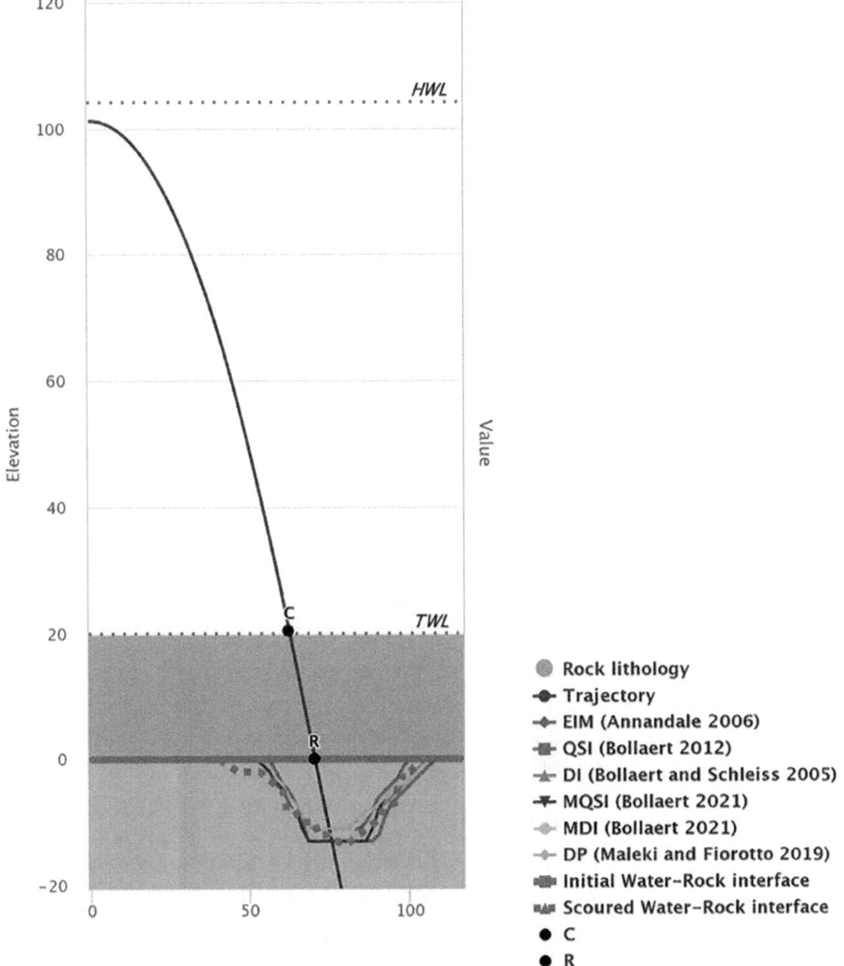

FIGURE 7.3 Results of digital platform numerical reproduction of test 9A_a.

For the DI method, the critical net uplift height ratio n_b is calibrated at 0.10, i.e. within the range of usual values based on experience from practice. For the MDI method, however, a lower value of 0.02 is found. This is probably related to the very high degree of break-up of the jet upon impact, generating very low pressure fluctuations on the blocks (see Figure 7.2).

For the QSI and MQSI methods, a best-fit net uplift pressure coefficient of 0.40 was obtained, matching with theoretical values for significantly protruding blocks, and in agreement with the highly irregular interface of the rock bed during scour formation as well as with experiences made with the MQSI method on other available case studies.

For the DP method, the n-value best fit was found at ~7–8, close to the max. range of values proposed by Maleki and Fiorotto (2019). Finally, by applying the EIM method, best fit was obtained for a mass strength number M_s of ~20, corresponding to a UCS strength of the cobblestones of ~22.5 MPa (i.e. which is a plausible value for standard pavers), and a theoretical ASP (Applicable Stream Power) value of 1.0.

Test 14A: q = 80 m²/s, H = 100 m, TWL = 20 m, z_b = 1.5 m, without mounding

The second numerical reproduction corresponds to the maximum specific discharge, the maximum hydraulic head and the maximum tailwater level used. The block height is now 1.5 m. The influence of an increased block height is thus checked for by the numerical models. No mounding was allowed during the test.

Scour Results

Figure 7.4 shows that good agreement is observed between measurements and computations for all methods. Moreover, for the DP, DI, MDI, QSI and MQSI methods, the best fit was obtained for the same parametric settings than for test 9A.

For the EIM method, however, the best-fit ASP value is now about 0.50 instead of 1.0 as was found for the 1-m-high rock blocks. The EIM is not able to correctly account for the increased rock block height in the analysis, because this parameter is (quasi) not part of the index. The only parameter that slightly changes with increasing rock block height is the ground structure number J_s that increases from 1.09 to about 1.12.

Test 9A_b: q = 80 m²/s, H = 100 m, TWL = 20 m, z_b = 1.0 m, with mounding

The third test that has been performed numerically is a repetition of test 9A but with mounding of blocks during the test. It corresponds to the maximum specific discharge, the maximum hydraulic head and the maximum tailwater level used. The block height is 1.0 m. The numerical methods have been recalibrated to this different physical situation (with mounding).

Scour Results

As could be expected, the parameters to obtain a best fit are somewhat different from the ones obtained for test 9A without mounding. For the DI method, the critical net uplift height of 0.50 instead of 0.10 without mounding is still within the range of usual values based on experience from practice. For the MDI method, the critical value is 0.05 instead of 0.02. For the QSI methods, a best-fit net uplift pressure

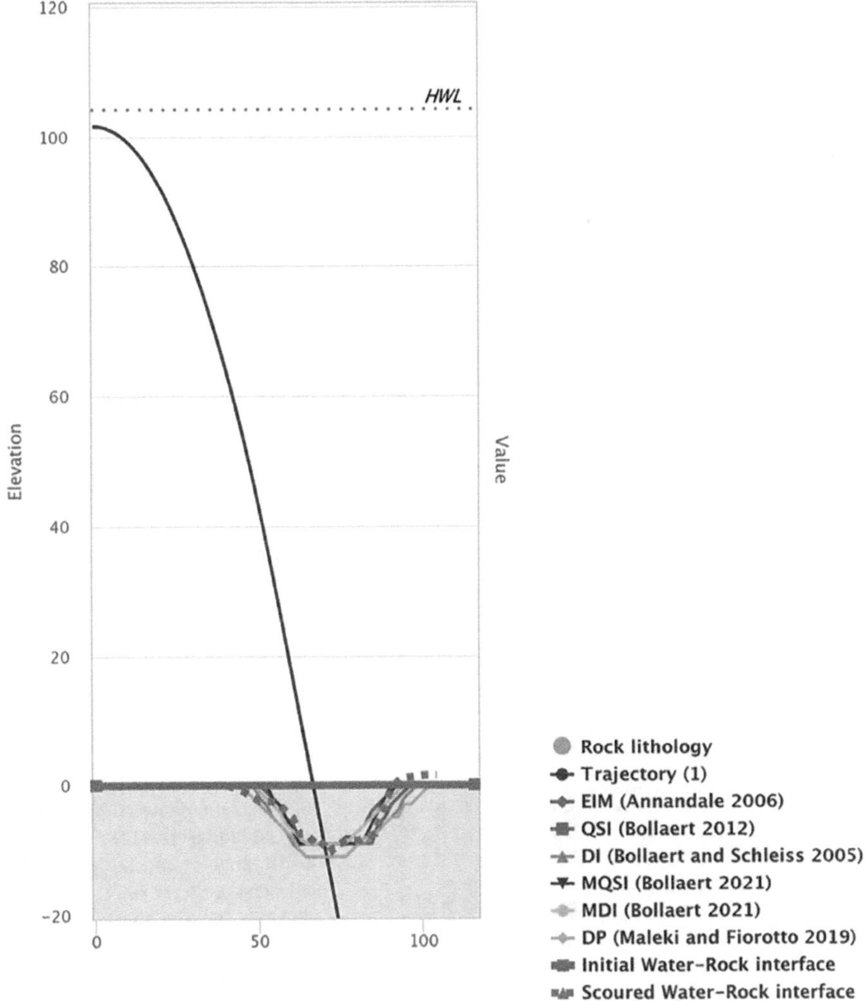

FIGURE 7.4 Results of digital platform numerical reproduction of test 14A.

coefficient on the blocks of 0.27 instead of 0.40 was obtained. The best-fit n-value of the DP method now equals 5 instead of 7–8. The erodibility index method has been calibrated by adapting the ASP (Applicable Stream Power) coefficient. The best-fit value was found at ASP = 0.60 (Figure 7.5).

Wivenhoe Dam (Australia)

Introduction

In 2011, Wivenhoe Dam, located on the Brisbane River, Australia, experienced its flood of record. This flood led a path of destruction through the suburbs of the city of Brisbane, and raised concern for the dam owners when they discovered a huge pile of eroded boulders deposited downstream of the spillway as the tailwater subsided.

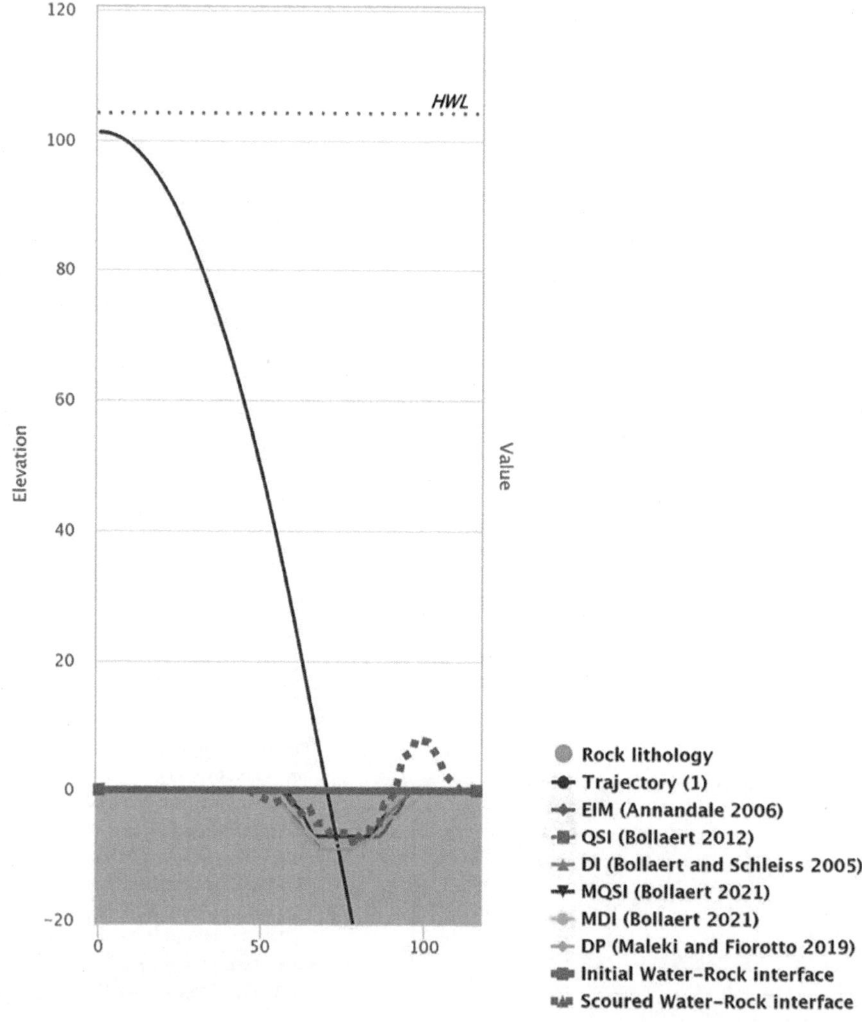

FIGURE 7.5 Results of digital platform numerical reproduction of test 9A with mounding.

The largest of these boulders was around the size of a bus, and was estimated to weigh around 1,200 t. The head differential during this event rarely exceeded 30 m, and the velocities exiting the flip bucket were no more than 25 m/s. With hydraulic characteristics that are moderate when compared to many large dams, it was difficult to imagine how such a discharge could lead to such a dramatic outcome (Lesleighter et al. 2013).

A noteworthy aspect of the review of Wivenhoe spillway rock scour was the amount of detail available on the structure. The spillway was designed using a physical hydraulic model study in the late 1970s, and the prototype has experienced a

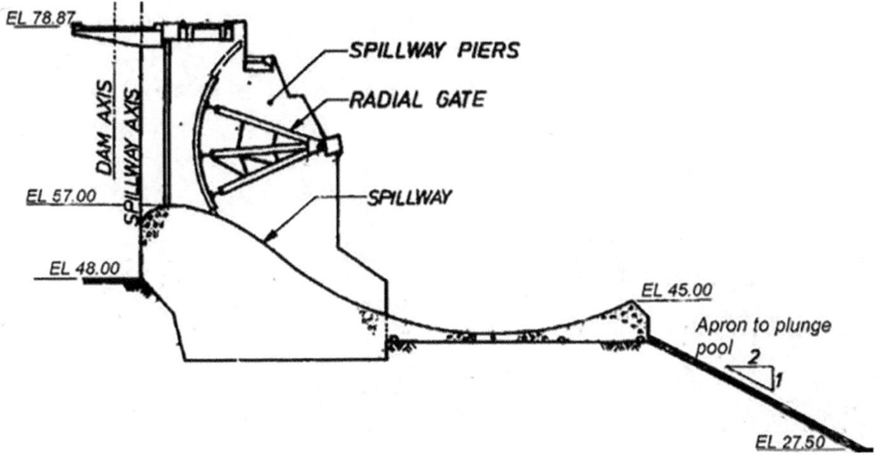

FIGURE 7.6 Longitudinal profile of Wivenhoe Dam upstream of Brisbane, Australia (Lesleighter et al. 2013).

number of discharges, most of which had been followed by a bathymetric survey. The information available allowed for detailed comparisons to be made of the physical model performance versus actual behaviour.

Dam and Plunge Pool

Wivenhoe Dam was constructed in the early 1980s, on the Brisbane River, Australia. The earth-fill dam incorporates a gated ski-jump spillway on the left abutment, which discharges into a pre-excavated unlined plunge pool and outlet channel. The spillway has five radial gates each 12 m wide and 16 m high. Figure 7.6 shows the profile of the spillway and concrete apron below the bucket (Lesleighter et al. 2013).

The spillway and the plunge pool were studied in a physical model during the design phase, and this led to the selection of the plunge pool geometry. On the basis of the model studies, using a mobile material, but recognizing the expected ability of the sandstone to resist scour, pre-excavation of the plunge pool was performed, comprising benches stepping down in increments of 3 m, to the lowest invert level of EL 17.0 m Australian Height Datum (AHD).

2011 Flood Event

Major floods occurred at the site in January 2011, with post-analysis of the inflows indicating that the flood was less frequent than the one in the year 2000 Annual Exceedance Probability (AEP). While the spillway had discharged on a number of occasions previously, the January discharges were prolonged and at larger magnitudes leading to dramatic erosion of rock from the plunge pool. The discharge from Wivenhoe spillway during the floods peaked at 7,600 m^3/s (i.e. q ~ 110 $m^3/s/m$), which was 50% more than the design flood for the plunge pool. In the days following the peak discharge, releases on the order of 3,500 m^3/s prevailed for a period of 4 days. Figure 7.7 shows an aerial view of the spillway during the flood, and Figure 7.8 shows the inflow-outflow hydrographs during this time (Lesleighter et al. 2013).

FIGURE 7.7 Photo of dam and spillway during the 2011 flood event (Lesleighter et al. 2013).

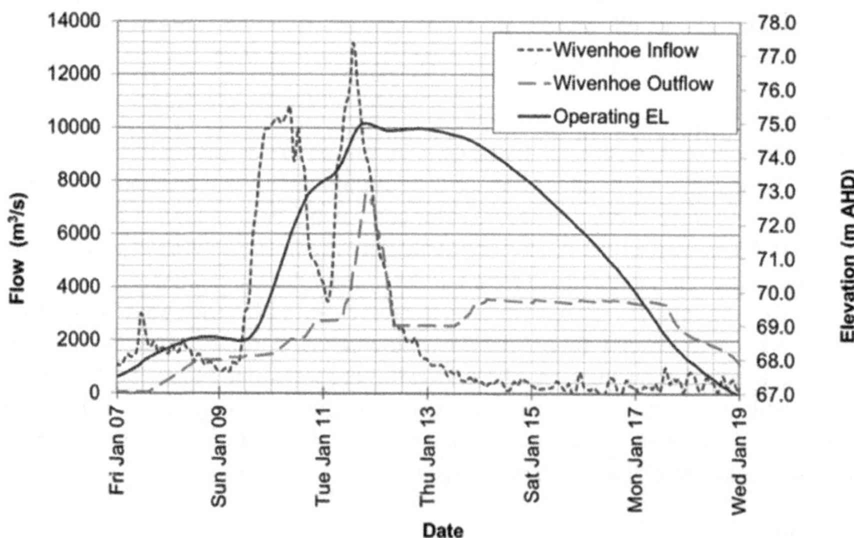

FIGURE 7.8 The 2011 flood event hydrograph (Lesleighter et al. 2013).

Plunge Pool Scour

Due to the high discharges occurring in tributaries downstream of the dam, the tailwater remained elevated for several days, and it was only when the tailwater subsided 4 days after passing the peak discharge that the top of an enormous rock mound that had developed in the spillway channel was revealed. The pile of rocks was approximately 10 m high, and nearly the full width of the channel, as seen in Figure 7.9.

Boulders of up to 15 × 10 × 3 m, weighing over 1,000 t, were observed in the pile of eroded rocks in the spillway channel. The boulders appeared to have separated on

FIGURE 7.9 Rock mound in tailwater channel following the 2011 flood event (Bollaert et al. 2014).

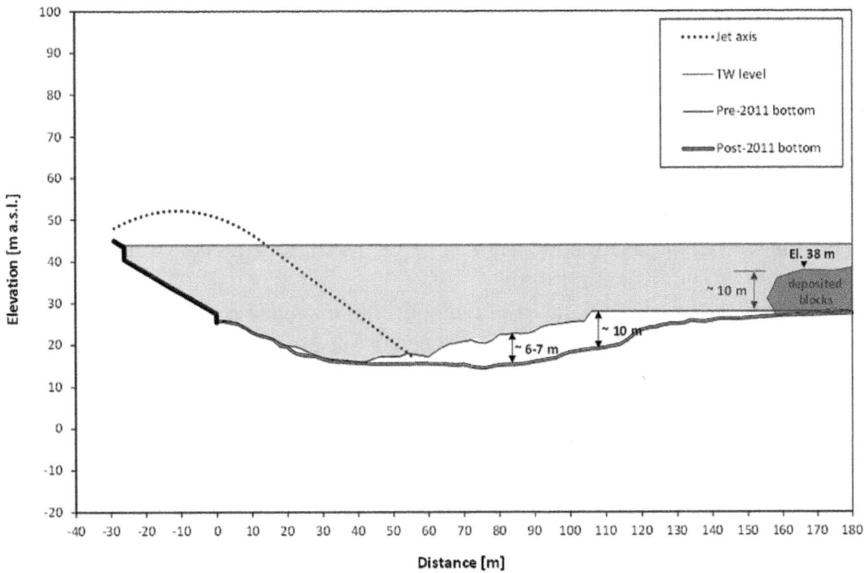

FIGURE 7.10 Longitudinal section of scoured plunge pool.

the near-horizontal bedding planes. The scoured material had block heights between 1 and 3 m, for block side lengths from 4 to 12 m. The ratio of block height-to-side length is thus 1:4.

Bathymetric surveys of the plunge pool were undertaken in 2000 following the 1999 flood, and also in January 2011 after the flood, allowing for assessment of the progression of scour over time. It can be seen that the majority of scour which occurred during the January 2011 flood, as indicated in the longitudinal section in Figure 7.10, removed material from the downstream extent of the plunge pool to extend its length by more than 40 m and its base down to 2 m below design (Lesleighter et al. 2013).

The photograph in Figure 7.9 illustrates the extent of the rock mound in the tailwater channel following the flooding. Many of the monoliths in these pictures indicate bedding-plane spacings in the range of 1–3 m (Bollaert et al. 2014).

Formerly, the tailwater channel was levelled at EL 28, with a left side channel at EL 26. Figure 7.10 shows the longitudinal section, following the January flood, with details of the scoured profile.

Geology

The plunge pool has been excavated in a sandstone with a density of 2,100 kg/m^3. Typically, as evidenced both in exposures in the excavation and in the monoliths which have been lifted out of the plunge pool, the sandstone body has numerous horizontal bedding planes at a spacing of 1–3 m. Photographs show the typical sizes of the rock that has been lifted out, with a large amount of smaller rock which is most likely due to breakdown of larger sizes and the tumbling action in the extreme turbulence in the plunge pool. As well as the horizontal bedding planes which constitute the main defect system in the rock body, there is evidence of up to three joint sets, all of which are sub-vertical and tight. The rock quality and strength is characterized by a UCS strength of 25–30 MPa, and an RQD which is essentially 100%.

Numerical Computations of Observed Scour

The scour observed during the 2011 flood has been numerically reproduced by the digital platform. The peak discharge of ~7,600 m^3/s, observed during 6 h, and the 24 h duration discharge of 5,500–6,000 m^3/s, have been used for the computations. Rectangular-shaped jets are considered, with an initial turbulence intensity of 4%. The issuance velocity is around 24 m/s, for a jet initial thickness of about 4.5 m and a lateral jet width of 68 m at impact.

Based on the bedding planes and the observations made following the event, the representative block size and shape have been defined as a block with a height of 1 m and a side length of 4 m. The different computational methods have been calibrated based on the observed scour formation by using such a block size and shape.

For the CFM method, the calibration parameter is the initial degree of fracturing PE (i.e. persistence of the joints in %), by assuming single-edge joints based on the presence of bedding planes. For the DP method, the calibration parameter is the n-value. For the QSI and MQSI methods, the calibration parameter is the net uplift coefficient C_{up}. Furthermore, the MDI method has been calibrated using a critical uplift displacement n_b, and a time duration of pressure pulsations that is governed by the resonance frequency of the joints ($2L/c$). Also, the EIM has used the following parameters to define the index: a mass strength number M_s of 23 MPa (i.e. C_r·UCS, with coefficient of relative density $C_r = 2{,}100/2{,}700 = 0.77$ and UCS ~30 MPa), a block size number $K_b = 29.94$ based on RQD = 100% and $J_n = 3.34$, a discontinuity shear strength number $K_d = 2.67$ based on $J_r = 2.0$ and $J_a = 0.75$, and finally a relative ground structure number $J_s = 1.05–1.14$ (depending on the block aspect ratio).

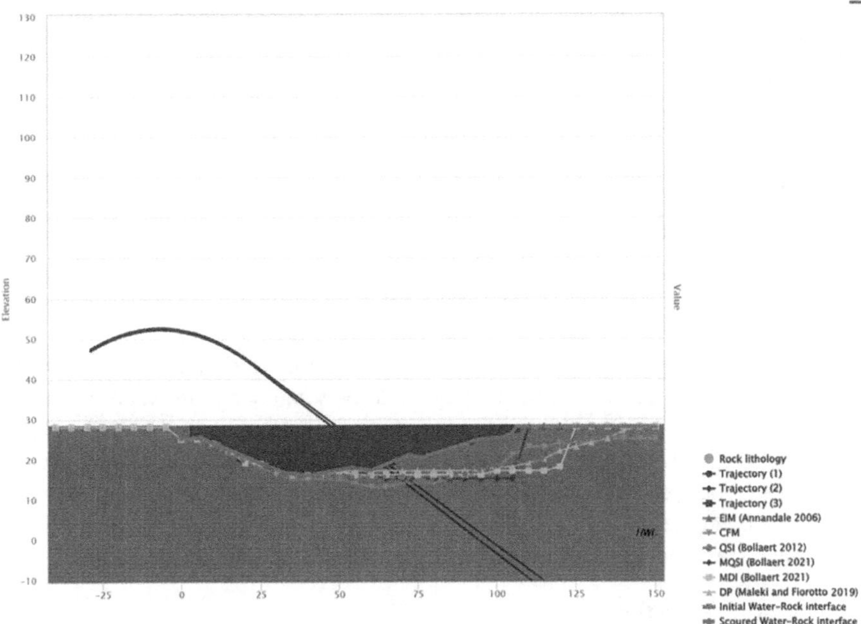

FIGURE 7.11 Comparison of in-situ observed and digital platform computed scour at Wivenhoe Dam following the 2011 flood.

The results for blocks of 1 m by 4 m are discussed hereafter and summarized in Figure 7.11. The CFM method has been calibrated by assuming that the sandstone is weak and erodible rock, with a UCS strength of ~25–30 MPa. The best-fit fatigue coefficient $m_r = 3.0$ and the fatigue exponent $C_r = 3.10^{-5}$. An initial persistence of fracturing of $PE = 40\%$ was found most adequate, corresponding to lowly to moderately fractured rock.

The low-angle impacting jet at Wivenhoe Dam generates a high-velocity wall jet along the rock interface. Hence, break-up methods expressing detachment and downstream displacement of blocks (QSI and MQSI methods) are particularly relevant. For a compact jet upon impingement, a net uplift pressure coefficient of $C_{up} = 0.11$ has been found adequate, which is significantly lower than the value of 0.40 calibrated for the Stellenbosch tests. For the MDI method, the critical net uplift coefficient n_b has a best fit of 0.15, which is in the usual range of values but higher than the 0.02 value found for the Stellenbosch tests. For the DP method, the n-factor equals 2, while a value of 7–8 was calibrated for the Stellenbosch tests. Finally, the EIM shows an ASP (Applicable Stream Power) coefficient of 0.20 (0.50–1.0 for Stellenbosch tests).

The computational models thus are capable to reproduce the observed scour formation, nevertheless the parametric settings are not always in agreement with the ones found for the Stellenbosch tests. This is not so surprising given the completely different scales and flow situations of both cases.

In the following, a significant number of in-situ observed scour holes have been systematically recomputed by the digital platform, allowing to perform a parametric sensitivity analysis and to point out the most relevant range of parametric values to be used for practice.

Calibration of Computational Models

Database of Scour Cases

The different computational methods of the digital platform have been calibrated by numerical reproduction of in-situ observed scour for distinct flood events at multiple dam spillways.

Table 7.2 illustrates the flow characteristics of the flood events for which significant scour was observed. Distinction is made between scour in weak or altered/weathered rock and scour in strong competent rock to conform with the theory explained in Chapter 6. Some dam spillways are mentioned more than once, because of multiple scour profiles, or different flow and/or geomechanical assumptions used during the computations.

Table 7.3 presents the rock mass characteristics of the same flood events.

Computational Methods

The following scour computational methods have been analysed in detail:

- *CFM method (fracture mechanics)*: This method generates scour potential with time based on fatigue fracturing, and provides an average value as well as lower and upper boundaries of scour, depending on class estimates of initial degree of fracturing (persistency PE in %) following the theory outlined in Chapter 6.
- *MQSI method*: This method provides flood-related equilibrium scour potential for the given hydraulic and geomechanics conditions, based on flow velocities along the water-rock interface.
- *MDI method*: This method provides flood-related equilibrium scour potential for the given hydraulic and geomechanics conditions, based on dynamic pressure fluctuations (RMS values) acting along the water-rock interface.
- *DP method*: This method provides flood-related equilibrium scour potential for the given hydraulic and geomechanics conditions, based on net uplift forces acting along the water-rock interface.
- *EIM method*: This method provides flood-related equilibrium scour potential for the given hydraulic and geomechanics conditions, based on applicable stream power acting along the water-rock interface.

It is essential to point out that the CFM method does not provide a flood-related final scour depth like the other methods, but a scour progression depending on the intensity and time duration of overflow discharges. Within this regard, different situations can be distinguished:

TABLE 7.2
Database of Scour Cases for Parametric Sensitivity Study (Flow Parameters)

Rock Classif.	N°	Case	Flow							
			Period	Type	Jet Core	Init. Turb.	Air Drag	Break-up	Angle	Air Conc.
			–	–	K	Tu_i	K_{air}	–	α	α_{air}
			–	–	–	%	–	–	°	%
WEAK rock or ALTERED / WEATHERED HARD rock	1	STELLENB. (SA)	2021	plung. jet	4	0.08	1.0	2.37	62.0	40
	2	STELLENB. (SA)	2021	plung. jet	4	0.08	1.0	2.38	62.0	40
	3	PARADISE (AUS)	2013	stilling basin	–	–	–	–	–	–
	4	BOONDOOMA DYKES (AUS)	2011	channel flow	–	–	–	–	–	–
	5	WIVENHOE (AUS)	2011	ski-jump	4	0.04	1.0	0.65	30.4	40
	6	CHANCY-POUGNY (CH)	1924–2015	ski-jump	4	0.04	1.0	0.17	22.3	40
	7	CHORANCHE (F)_LB	1963–2003	plung. jet	4	0.02	0.9	1.44	75.1	40
	8	CHORANCHE (F)_RB	1963–2003	plung. jet	4	0.02	0.9	1.44	75.1	40
	9	CHUCÁS 0.25 m BLOCKS (CRI)	2017	ski-jump	4	0.04	0.9	1.41	52.1	40
	10	CHUCÁS 0.50 m BLOCKS (CRI)	2017	ski-jump	4	0.04	0.9	1.41	52.1	40
	11	WAC BENNETT 19 MPA (CAN)	1972	ski-jump	4	0.04	0.8	4.17	53.3	40
	12	WAC BENNETT 69 MPA (CAN)	1972	ski-jump	4	0.04	0.8	4.17	53.3	40
	13	KARAKAYA (TUR)	1986–1988	ski-jump	4	0.04	0.9	2.44	60.7	40

(Continued)

TABLE 7.2 (*Continued*)
Database of Scour Cases for Parametric Sensitivity Study (Flow Parameters)

			Flow							
			Period	Type	Jet Core	Init. Turb.	Air Drag	Break-up	Angle	Air Conc.
			–	–	K	Tu_i	K_{air}	–	α	α_{air}
Rock Classif.	N°	Case	–	–	–	%	–	–	°	%
STRONG COMPETENT rock	14	LAOUZAS RB (F)	1971–1982	plung. jet	4	0.01	1	1.11	74.5	40
	15	JAGUARA (BRA)	1976–1990	ski-jump	4	0.04	0.65	1.32	48.7	40
	16	KARIBA 1976 (ZAM/ZIM)	1972–1976	jet adj.	4	0.04	0.9	0.59	75.0	40
	17	KARIBA 1981 (ZAM/ZIM)	1972–1981	jet adj.	4	0.04	0.9	0.59	75.0	40
	18	PICOTE (POR)	1958–1964	ski-jump	4	0.04	0.9	1.97	47.0	40

TABLE 7.3
Database of Scour Cases for Parametric Sensitivity Study (Rock Mass Parameters)

			Rock Mass									
			Lithol.	**Dens.**	**UCS**	**RQD**	**Er. ind.**	**Joint Length**	**Joint Persist.**	**Joint Shape**	**Block Shape**
			–	ρ	–	–	**EI**	**Lf**	***PE***	–	***z/x***
Rock Classif.	**N°**	**Case**	–	**kg/m³**	**MPa**	**%**	–	**m**	**%**	–	**m/m**
WEAK rock or ALTERED / WEATHERED HARD rock	1	STELLENB. (SA)	cobbl	2350	23	0	799	1.00	100	2D	1/2
	2	STELLENB. (SA)	cobbl	2350	23	0	821	1.00	100	2D	1.5/2
	3	PARADISE (AUS)	meta.	2000	70	20	–	0.30	50	2D	0.30/0.30
	4	BOONDOOMA DYKES (AUS)	rhyo./pegm.	2500	72	20	–	0.33	70	2D	0.33/0.33
	5	WIVENHOE (AUS)	sand	2100	23	–	1930	1.0	40	2D	1/4
	6	CHANCY-POUGNY (CH)	marl-sand.	2600	5	–	179	0.25	40	2D	0.25/1
	7	CHORANCHE (F)_LB	clay-lime.	2650	50	90	60	0.30	50	2D	0.3/1
	8	CHORANCHE (F)_RB	clay-lime.	2650	50	90	60	0.30	70	2D	0.3/1
	9	CHUCÁS 0.25 m BLOCKS (CRI)	and. lava	2700	20	40	80	0.25	60	2D	0.25/1
	10	CHUCÁS 0.50 m BLOCKS (CRI)	and. lava	2700	20	40	80	0.50	60	2D	0.5/0.5
	11	WAC BENNETT 19 MPA (CAN)	shale/sand.	2750	19	70–95	1115	0.50	70	2D	0.5/1

(Continued)

TABLE 7.3 (*Continued*)
Database of Scour Cases for Parametric Sensitivity Study (Rock Mass Parameters)

Rock Classif.	N°	Case	Rock Mass								
			Lithol.	Dens.	UCS	RQD	Er. ind.	Joint Length	Joint Persist.	Joint Shape	Block Shape
			–	ρ	–	–	EI	Lf	*PE*	–	*z/x*
			–	kg/m³	MPa	%	–	m	%	–	m/m
	12	WAC BENNETT 69 MPA (CAN)	shale/sand.	2750	69	70–95	4050	0.50	70	2D	0.5/1
	13	KARAKAYA (TUR)	fault. gneiss	2650	23.5	50	352	0.50	50	2D	0.5/1
STRONG COMPETENT rock	14	LAOUZAS RB (F)	migmatite	2700	156	50	245	0.30	80	2D	0.3/1
	15	JAGUARA (BRA)	quartzite	2600	150	-	501	0.40	60	2D	0.2/1
	16	KARIBA 1976 (ZAM/ZIM)	gneiss	3000	120	75	3234	0.75	50	2D	0.75/1
	17	KARIBA 1981 (ZAM/ZIM)	gneiss	3000	120	75	3234	0.75	50	2D	0.75/1
	18	PICOTE (POR)	porph.gran.	2600	77	-	7700	0.50	70	2D	0.5/1

- In-Situ Rock Mass Is Not Completely Broken Up (Fractured)
  - Scour by progressive fracturing (CFM) is less than equilibrium scour by block uplift/detachment (MDI/MQSI/DP): scour calibration is governed by joint fracturing (brittle and fatigue).
  - Scour by progressive fracturing (CFM) is equal to or deeper than equilibrium scour by block uplift/detachment (MDI/MQSI/DP): scour calibration is governed by block uplift/detachment. Blocks can still fracture but not be uplifted or detached anymore.
- *In-situ rock mass is completely broken up (fractured)*: scour calibration is governed by the equilibrium scour depth as computed based on rock block uplift or detachment (MDI/MQSI/DP).

In other words, when a significant difference exists between in-situ observed scour and flood-related equilibrium scour potential based on block uplift/detachment, the fracturing process is ruling (and retarding) the scour process. For such cases, calibration of parameters used by the MDI, MQSI and DP methods will substantially differ from the values that generate an estimate for the flood-related equilibrium scour depth.

On the other hand, when quasi no difference is observed between in-situ observed scour and flood-related equilibrium scour potential based on block uplift/detachment, the fracturing process becomes irrelevant. Calibration of parameters used by the MDI, MQSI and DP methods will be very close to the values that generate an estimate for the flood-related equilibrium scour potential.

Scour Analysis for Weak/Weathered/Altered Rock

Figure 7.12 compares the in-situ observed scour depth with the digital platform computed scour depth, for weak and/or weathered/altered rock. For the CFM method, a bandwidth of scour depths is provided, which conform to the theoretical analysis outlined in Chapter 6.

The corresponding parametric values are illustrated in Table 7.4 for CFM (with average, lower and upper boundaries) and in Table 7.5 for MDI/MQSI/DP/EIM methods. The latter methods propose parametric settings to calibrate in-situ observed scour and parametric settings to estimate the equilibrium scour depth.

For the MDI method, this equilibrium scour depth is computed for $n_b = 0.10$. For the MQSI method, the equilibrium scour depth is computed for $C_{up} = 0.40$. Finally, for the DP method, the equilibrium scour depth is computed for $n = 6.0$.

Scour Analysis for Strong Competent Rock

Figure 7.13 compares the in-situ observed scour depth with the digital platform computed scour depth for strong competent rock. For the CFM method, a bandwidth of scour depths is provided, which conform to the theoretical analysis outlined in Chapter 6.

The corresponding parametric values are illustrated in Table 7.4 for CFM (with average, lower and upper boundaries) and in Table 7.5 for MDI/MQSI/DP/EIM methods. The latter methods propose parametric settings to calibrate in-situ observed scour and parametric settings to estimate the equilibrium scour depth.

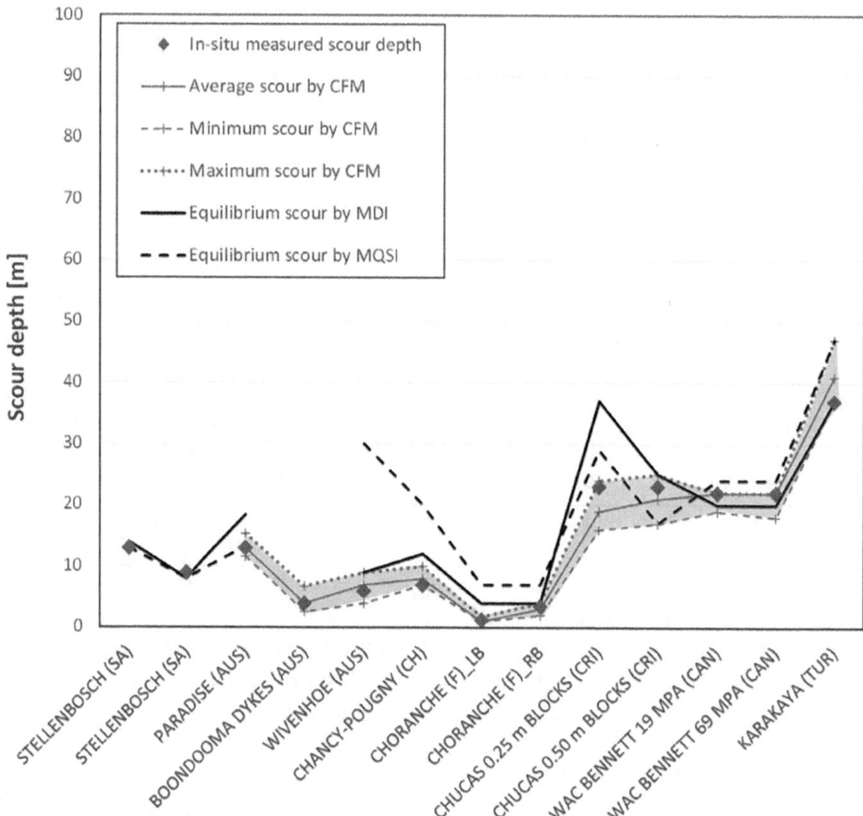

FIGURE 7.12 Digital platform computed scour versus in-situ observed scour for weak and/or weathered/altered rock.

Recommendations for Practice

In the following, the essential parameters that have been calibrated during the digital platform scour computations are synthetized and transformed into recommendations for practice.

Error Estimate

An error estimate is provided for all computational methods by comparing the in-situ observed scour with the digital platform computed scour. This is done separately for weak or weathered/altered rocks in Figure 7.14 and in Table 7.6, and for strong competent rocks in Figure 7.15 and also in Table 7.6.

CFM Method

For the average scour computed by the model, the error estimate provided in Figures 7.14 and 7.15 indicates an error bandwidth of +15%/–20% of the measured scour. For the minimum scour computations, the estimate is 0%–83% lower than the measured scour, and for the maximum scour computed, the estimate is 0%–79% higher than the measured scour.

TABLE 7.4
Calibration of CFM Parameters for Weak/Weathered/Altered Rock (N° 1–13) and for Strong Competent Rock (N° 14–18)

		In-situ	CFM Calibration				CFM min. Scour					CFM Avg. Scour					CFM max. scour				
		Scour	PE	mr	Cr	Sc.	PE	mr	Cr	Sc.	Dif.	PE	mr	Cr	Sc.	Dif.	PE	mr	Cr	Sc.	Dif.
N°	Case	m	%	–	–	m	%	–	–	m	%	%	–	–	m	%	%	–	–	m	%
1	STELLENB.	13.0	–	–	–	–	–	–	–	–	–	–	–	–	–	–	–	–	–	–	–
2	STELLENB.	9.0	–	–	–	–	–	–	–	–	–	–	–	–	–	–	–	–	–	–	–
3	PARADISE	13.0	50	3.0	3e-5	13.0	40	3.0	3e-5	11.6	–10.8	50	3.0	3e-5	13.0	0.0	60	3.0	3e-5	15.3	17.7
4	BOOND.	4.0	70	3.0	3e-5	4.0	60	3.0	3e-5	2.6	–35.0	70	3.0	3e-5	4.0	0.0	80	3.0	3e-5	6.8	68.8
5	WIVENHOE	6.0	40	2.9	3.5e-5	6.0	30	3.0	3e-5	4.0	–33.3	40	3.0	3e-5	7.0	16.7	50	3.0	3e-5	9.0	50.0
6	CHANCY-P.	7.0	40	3.0	3e-5	7.0	40	3.0	3e-5	7.0	0.0	50	3.0	3e-5	8.0	14.3	60	3.0	3e-5	10.0	42.9
7	CHOR. LB	1.3	50	2.9	3.5e-5	1.3	40	3.0	3e-5	1.0	–20.0	50	3.0	3e-5	1.0	–20.0	60	3.0	3e-5	1.8	40.0
8	CHOR. RB	3.5	70	2.7	4.5e-5	3.5	60	3.0	3e-5	2.0	–42.9	70	3.0	3e-5	3.0	–14.3	80	3.0	3e-5	4.0	14.3
9	CHUCÁS 0.25 m block	23.0	60	3.1	2.9e-5	23.0	40	3.0	3e-5	16.0	–30.4	50	3.0	3e-5	19.0	–17.4	60	3.0	3e-5	24.0	4.3
10	CHUCÁS 0.50 m block	23.0	60	3.3	2e-5	23.0	40	3.0	3e-5	17.0	–26.1	50	3.0	3e-5	21.0	–8.7	60	3.0	3e-5	25.0	8.7
11	BENNETT 19 MPa	22.0	70	3.0	3e-5	22.0	60	3.0	3e-5	19.0	–13.6	70	3.0	3e-5	22.0	0.0	80	3.0	3e-5	22.0	0.0
12	BENNETT 69 MPa	22.0	80	3.0	3e-5	22.0	60	3.0	3e-5	19.0	–18.2	70	3.0	3e-5	22.0	0.0	80	3.0	3e-5	22.0	0.0
13	KARAKAYA	37.0	50	3.0	3.0e-5	37.0	50	3.0	3e-5	37.0	0.0	60	3.0	3e-5	41.0	10.8	70	3.0	3e-5	47.0	27.0

(Continued)

TABLE 7.4 (*Continued*)
Calibration of CFM Parameters for Weak/Weathered/Altered Rock (N° 1–13) and for Strong Competent Rock (N° 14–18)

N°	Case	In-situ Scour	CFM Calibration				CFM min. Scour					CFM Avg. Scour					CFM max. scour				
			PE	mr	Cr	Sc.	PE	mr	Cr	Sc.	Dif.	PE	mr	Cr	Sc.	Dif.	PE	mr	Cr	Sc.	Dif.
		m	%	–	–	m	%	–	–	m	%	%	–	–	m	%	%	–	–	m	%
14	LAOUZAS RB	2.2	80	5.5	1.5e-6	2.2	60	5.0	3e-6	1.20	−45.5	70	5.0	3e-6	1.8	−18.2	80	5.0	3e-6	2.6	18.2
15	JAGUARA	27.0	60	4.8	3.5e-6	27.0	60	5.0	3e-6	25.0	−7.4	70	5.0	3e-6	31.0	14.8	80	5.0	3e-6	37.0	37.0
16	KARIBA 1976	64.0	50	4.9	3.0e-6	64.0	40	5.0	3e-6	58.0	−9.4	50	5.0	3e-6	65.0	1.6	60	5.0	3e-6	80.0	25.0
17	KARIBA 1981	68.0	50	4.9	3.0e-6	68.0	40	5.0	3e-6	58.0	−14.7	50	5.0	3e-6	68.0	0.0	60	5.0	3e-6	82.0	20.6
18	PICOTE	19.0	70	5.5	1.5e-6	19.0	60	5.0	3e-6	13.0	−31.6	70	5.0	3e-6	21.0	10.5	80	5.0	3e-6	34.0	78.9

TABLE 7.5
Calibration of MDI/MQSI/DP/EIM Parameters for Weak/Weathered/Altered Rock (N° 1–13) and for Strong Competent Rock (N° 14–18)

		In-situ	MDI		MDI (Equilibrium)			MQSI		MQSI (Equilibrium)			DP		EIM	
		scour	n_b	sc.	n_b	sc.	Diff.	$C_{UP,QSI}$	sc.	$C_{UP,QSI}$	sc.	Diff.	n	sc.	ASP	sc.
N°	Case	m	–	*m*	–	*m*	%	–	*m*	–	*m*	%	–	*m*	–	*m*
1	STELLENB.	13.0	0.02	13.0	0.10	14.0	7.7	0.40	13.0	0.40	13.0	0.0	8.0	13.0	1.0	13.0
2	STELLENB.	9.0	0.02	9.0	0.10	8.0	−11.1	0.42	9.0	0.40	8.00	−11.1	8.0	9.0	0.5	9.0
3	PARADISE	13.0	0.59	13.0	0.10	18.4	41.5	0.39	13.0	0.40	13.2	1.5	–	–	–	–
4	BOOND.	4.0	–	–	–	–	–	–	–	–	–	–	–	–	–	–
5	WIVENHOE	6.0	0.15	6.0	0.10	9.0	50.0	0.11	6.0	0.40	30.0	400	2.0	6.0	0.2	6.0
6	CHANCY-P.	7.0	0.30	7.0	0.10	12.0	71.4	0.18	7.0	0.40	20.0	186	2.2	7.0	0.4	7.0
7	CHOR. LB	1.25	0.50	1.25	0.10	4.0	220	0.10	1.25	0.40	7.0	460	1.5	1.25	0.1	1.3
8	CHOR. RB	3.5	0.12	3.5	0.10	4.0	14.3	0.14	3.5	0.40	7.0	100	4.0	3.5	0.4	3.5
9	CHUCÁS 0.25 m block	23.0	0.55	23.0	0.10	37.0	60.9	0.30	23.0	0.40	29.0	26.1	4.5	23.0	0.4	23.0
10	CHUCÁS 0.50 m block	23.0	0.12	23.0	0.10	25.0	8.7	0.55	23.0	0.40	17.0	−26.1	14.0	23.0	0.4	23.0
11	BENNETT 19 MPa	22.0	0.08	22.0	0.10	20.0	−9.1	0.25	22.0	0.40	24.0	9.1	9.0	22.0	1.0	22.0

(Continued)

TABLE 7.5 (*Continued*)
Calibration of MDI/MQSI/DP/EIM Parameters for Weak/Weathered/Altered Rock (N° 1–13) and for Strong Competent Rock (N° 14–18)

		In-situ	MDI		MDI (Equilibrium)			MQSI		MQSI (Equilibrium)			DP		EIM	
		scour	n_b	sc.	n_b	sc.	Diff.	$C_{UP,QSI}$	sc.	$C_{UP,QSI}$	sc.	Diff.	n	sc.	ASP	sc.
N°	Case	m	–	*m*	–	*m*	%	–	*m*	–	*m*	%	–	*m*	–	*m*
12	BENNETT 69 MPa	22.0	0.08	22.0	0.10	20.0	–9.1	0.25	22.0	0.40	24.0	9.1	9.0	22.0	2.0	22.0
13	KARAKAYA	37.0	0.10	37.0	0.10	37.0	0.0	0.20	37.0	0.40	47.0	27.0	9.0	37.0	0.6	37.0
14	LAOUZAS RB	2.2	1.00	2.2	0.10	5.0	127.3	0.25	2.2	0.40	3.6	63.6	0.75	2.2	0.3	2.2
15	JAGUARA	27.0	2.50	27.0	0.10	55.0	103.7	0.09	27.0	0.40	55.0	104	1.5	27.0	0.3	27.0
16	KARIBA 1976	64.0	0.20	64.0	0.10	78.0	21.9	0.25	64.0	0.40	80.0	25.0	11.0	64.0	0.4	64.0
17	KARIBA 1981	68.0	0.15	68.0	0.10	78.0	14.7	0.30	68.0	0.40	80.0	17.6	14.0	68.0	0.5	68.0
18	PICOTE	19.0	0.25	19.0	0.10	38.0	100.0	0.35	19.0	0.40	22.0	15.8	7.0	19.0	2.0	19.0

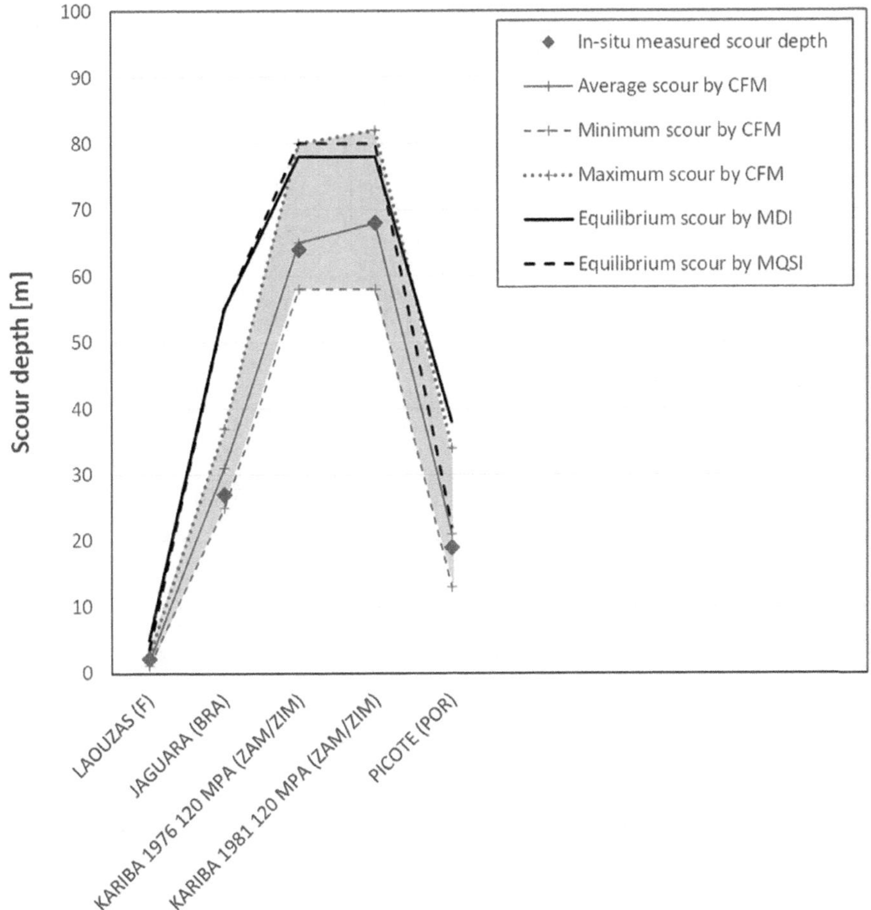

FIGURE 7.13 Digital platform computed scour versus in-situ observed scour for strong competent rock.

The initial degree of fracturing or persistency PE of the joints is illustrated in Figure 7.16. For in-situ case studies, the values are between 40% and 80%. The Stellenbosch laboratory experiments were performed with cobblestones, and thus have a value of 100%.

The m_r–PE and m_r–C_r relationships observed during the calibration process are presented in Figures 7.17 and 7.18, distinguishing between weak and/or weathered/altered rocks, and strong competent rocks. The calibrated m_r values are close to the ones recommended for these types of rock (respectively 3 and 5).

MDI Method

The essential parameter for calibration of the MDI method is the net block uplift height ratio n_b, which stands for the vertical uplift height needed for the block to

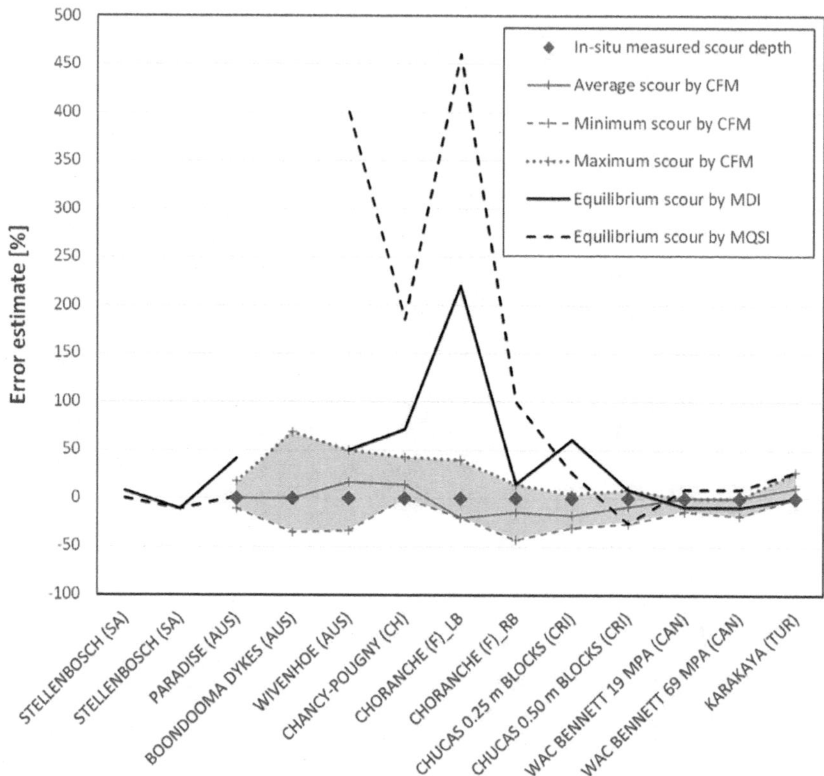

FIGURE 7.14 Error estimate between in-situ scour and digital platform computed scour for weak and/or weathered/altered rock.

be uplifted, divided by the total block height. The lower this value, the easier it becomes to uplift the block and, thus, the deeper becomes the equilibrium scour depth prescribed by the method.

Figure 7.19 illustrates the calibrated n_b values for the different case studies. The lower range of values, to be used for practice, is clearly situated around 0.10. This can happen in the following situations:

- the rock mass is completely or quasi-completely fractured, and the fracturing phase is not or very few relevant to the process, such as for example for Stellenbosch tests or Choranche Dam right bank scour formation,
- the formed scour is so deep that the equilibrium scour is quasi reached, and the fracturing process takes significant time to progress further, such as for example at Kariba Dam.

In other words, the higher the calibrated n_b value, the more the in-situ observed scour differs from equilibrium scour depth based on block uplift. This may indicate

TABLE 7.6
Error Estimate of Digital Computed Scour for Weak/Weathered/Altered Rock (N° 1–13) and for Strong Competent Rock (N° 14–18)

			CFM min.		CFM avg.		CFM max.		MDI eq.		MQSI eq.	
		In-situ scour	sc.	Diff.	sc.	Diff.	sc.	Diff.	sc.	Diff.	sc.	Diff.
N°	Case	m	m	%	m	%	m	%	m	%	m	%
1	STELLENB.	13.0	–	–	–	–	–	–	14.0	7.7	13.0	0.0
2	STELLENB.	9.0	–	–	–	–	–	–	8.0	−11.1	8.00	−11.1
3	PARADISE	13.0	11.6	−10.8	13.0	0.0	15.3	17.7	18.4	41.5	13.2	1.5
4	BOOND.	4.0	2.6	−35.0	4.0	0.0	6.8	68.8				
5	WIVENHOE	6.0	4.0	−33.3	7.0	16.7	9.0	50.0	9.0	50.0	30.0	400
6	CHANCY-P.	7.0	7.0	0.0	8.0	14.3	10.0	42.9	12.0	71.4	20.0	186
7	CHOR. LB	1.25	1.0	−20.0	1.0	−20.0	1.75	40.0	4.0	220.0	7.0	460
8	CHOR. RB	3.5	2.0	−42.9	3.0	−14.3	4.0	14.3	4.0	14.3	7.0	100
9	CHUCÁS 0.25 m block	23.0	16.0	−30.4	19.0	−17.4	24.0	4.3	37.0	60.9	29.0	26.1
10	CHUCÁS 0.50 m block	23.0	17.0	−26.1	21.0	−8.7	25.0	8.7	25.0	8.7	17.0	−26.1
11	BENNETT 19 MPa	22.0	19.0	−13.6	22.0	0.0	22.0	0.0	20.0	−9.1	24.0	9.1

(Continued)

TABLE 7.6 (*Continued*)

Error Estimate of Digital Computed Scour for Weak/Weathered/Altered Rock (N° 1–13) and for Strong Competent Rock (N° 14–18)

		In-situ scour	CFM min.		CFM avg.		CFM max.		MDI eq.		MQSI eq.	
			sc.	Diff.	sc.	Diff.	sc.	Diff.	sc.	Diff.	sc.	Diff.
N°	Case	m	m	%	m	%	m	%	m	%	m	%
12	BENNETT 69 MPa	22.0	18.0	−18.2	22.0	0.0	22.0	0.0	20.0	−9.1	24.0	9.1
13	KARAKAYA	37.0	37.0	0.0	41.0	10.8	47.0	27.0	37.0	0.0	47.0	27.0
14	LAOUZAS RB	2.2	1.20	−45.5	1.8	−18.2	2.6	18.2	5.0	127.3	3.6	63.6
15	JAGUARA	27.0	25.0	−7.4	31.0	14.8	37.0	37.0	55.0	103.7	55.0	104
16	KARIBA 1976	64.0	58.0	−9.4	65.0	1.6	80.0	25.0	78.0	21.9	80.0	25.0
17	KARIBA 1981	68.0	58.0	−14.7	68.0	0.0	82.0	20.6	78.0	14.7	80.0	17.6
18	PICOTE	19.0	13.0	−31.6	21.0	10.5	34.0	78.9	38.0	100.0	22.0	15.8

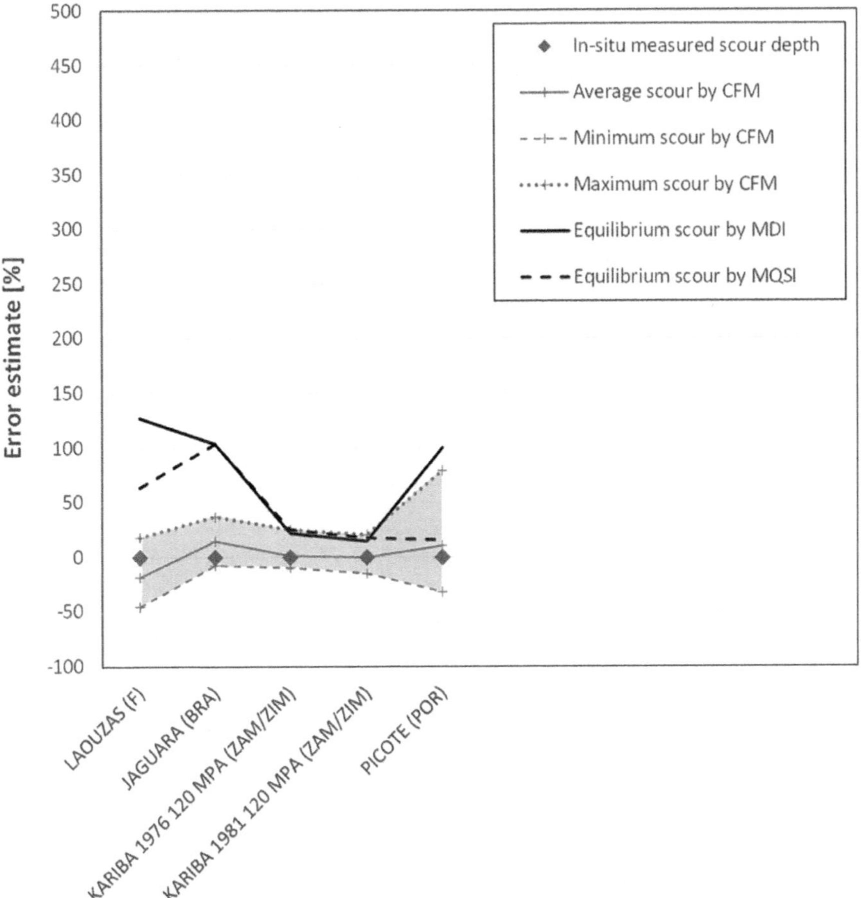

FIGURE 7.15 Error estimate between in-situ scour and digital platform computed scour for strong competent rock.

that the fracturing process is more relevant, and/or that too small block heights have been used during the computations. This can be clearly seen in the results for Chucás Dam, where a n_b value of around 0.10 is obtained for 0.50-m-high blocks, while a n_b value of 0.55 is obtained for 0.25-m-high blocks.

MQSI Method

The essential parameter for calibration of the MQSI method is the net uplift pressure coefficient C_{up} which, when multiplied by the local kinetic energy of the flow oriented along the water-rock interface, provides the local net uplift force for detachment of rock blocks.

Figure 7.20 illustrates the calibrated C_{up} values for the different case studies. The higher range of values, to be used for practice, is situated around 0.40. This can happen in the following situations:

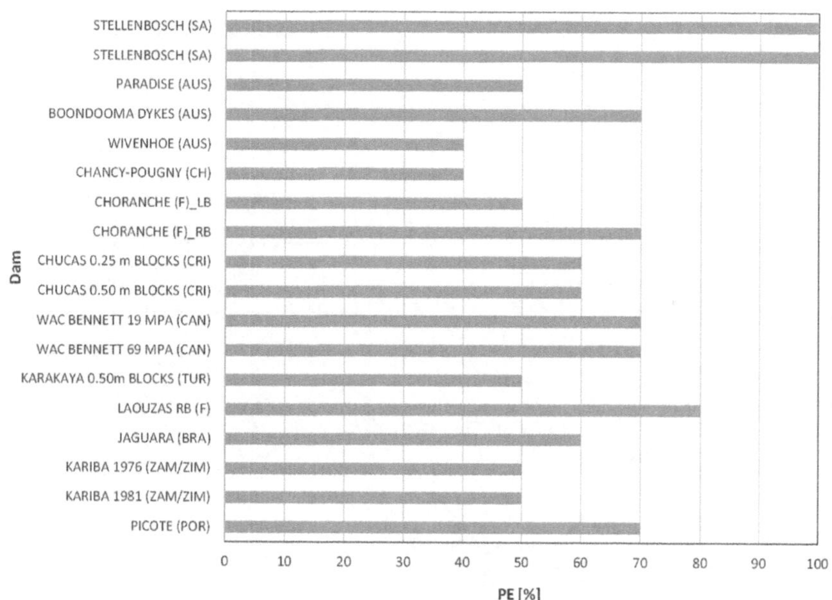

FIGURE 7.16 Persistencies *PE* [%] used for calibration of the CFM method.

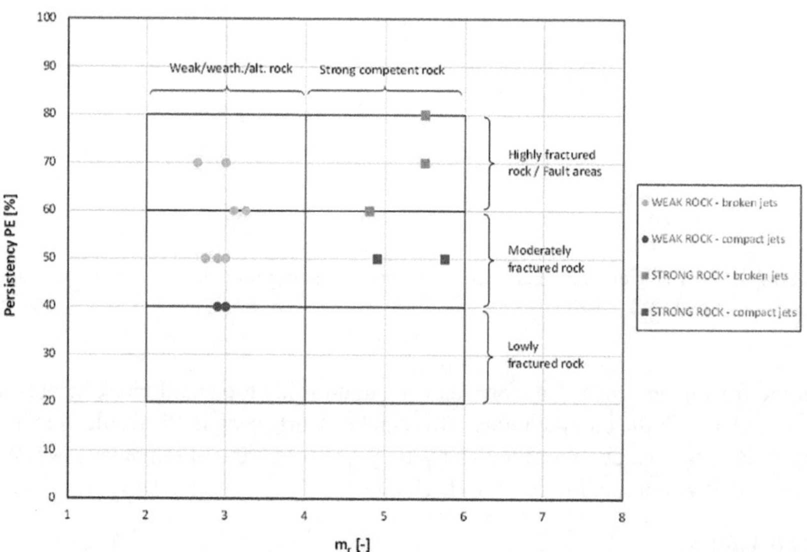

FIGURE 7.17 m_r–PE relationship for calibration of the CFM method.

- the rock mass is completely or quasi-completely fractured, and the fracturing phase is not or very few relevant to the process, such as for example for Stellenbosch tests or Karakaya Dam,
- the formed scour is so deep that the equilibrium scour is quasi reached, and the fracturing process takes significant time to progress further, such as for example at Kariba Dam.

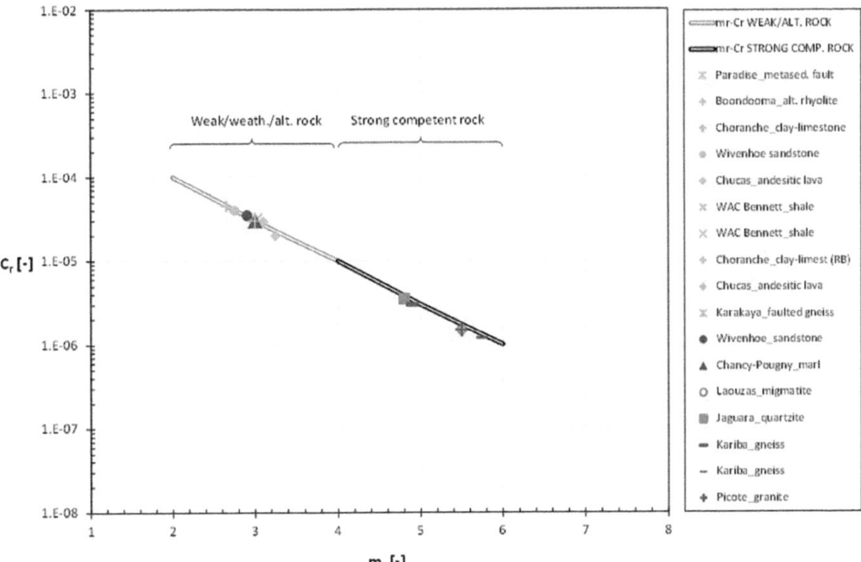

FIGURE 7.18 m_r–C_r relationship for calibration of the CFM method.

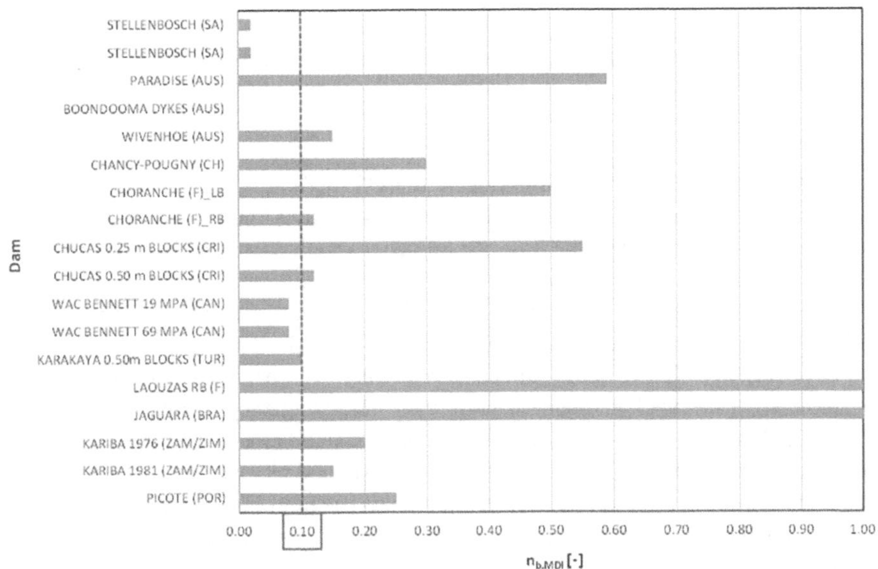

FIGURE 7.19 n_b values used for calibration of the MDI method.

In other words, the lower the calibrated C_{up} value, the more the observed scour differs from equilibrium scour depth based on block detachment by flow velocities. This may indicate that the fracturing process is more relevant, and/or that too small block heights have been used during the computations. This can be clearly seen in the results for Chucás Dam, where a C_{up} value of around 0.55 is obtained

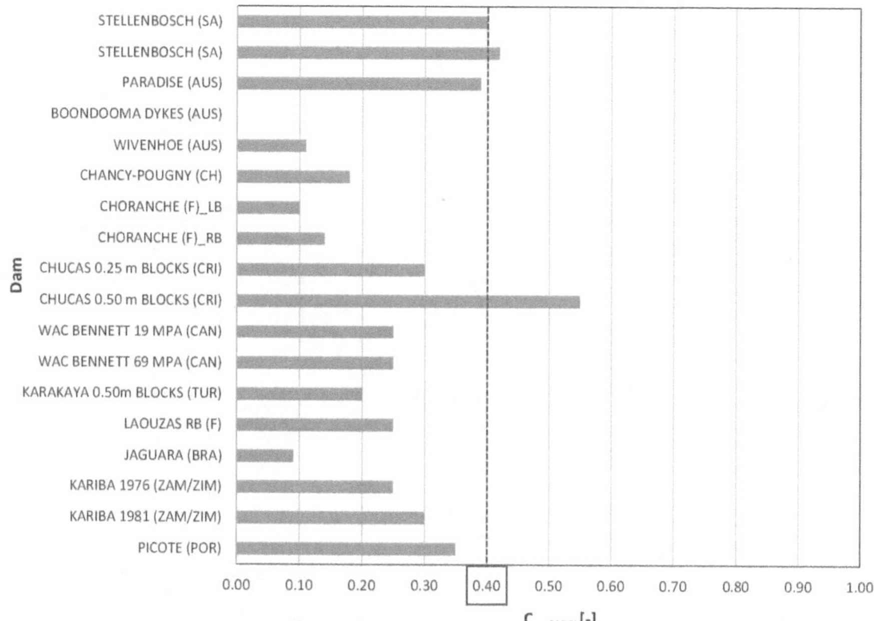

FIGURE 7.20 C_{up} values used for calibration of the MQSI method.

for 0.50-m-high blocks, i.e. out of range, while a C_{up} value of 0.30 is obtained for 0.25-m-high blocks.

DP Method

The essential parameter for calibration of the DP method is the net uplift pressure coefficient n, which stands for the difference between maximum and minimum pressures divided by twice the RMS value. It influences the local net uplift force for uplift of rock blocks.

Figure 7.21 illustrates the calibrated n-values for the different case studies. The values show a rather widespread range between 1 and 14.

However, by relating the n-values to the shape of the block, i.e. the ratio of block height z_b to block side length x_b (or y_b), a clear tendency becomes apparent in Figure 7.22. Values of 14 are obtained for cubical-shaped blocks, i.e. for a shape ratio of 1, but much lower values are observed for flat-shaped blocks. The lower the shape ratio, the lower the n-values. A best-fit equation is given in Figure 7.22.

As such, the calibrated n-values are shape-dependent and may become significantly higher than the values of 6–7 recommended by Maleki and Fiorotto (2019). They express the fluctuating net uplift pressure as $\Omega{\cdot}n{\cdot}C'_p$ times the incoming kinetic energy of the jet, with Ω depending on the non-dimensional length of the block, and its general shape and dimensions. This corresponds to a value of $0.5{\cdot}(C_p^+ - C_p^-)$, which is exactly half of the value used by Fiorotto and Rinaldo (1992) for slab uplift computations.

Also, when comparing with the MDI method, which defines the net uplift pressure coefficient as $4.2{\cdot}C'_p$, typical values of $n=6$–7 and $\Omega=0.35$ would result in 2.1–2.4 times C'_p, i.e. about half of the MDI values. Finally, Liu et al. (1998) considered

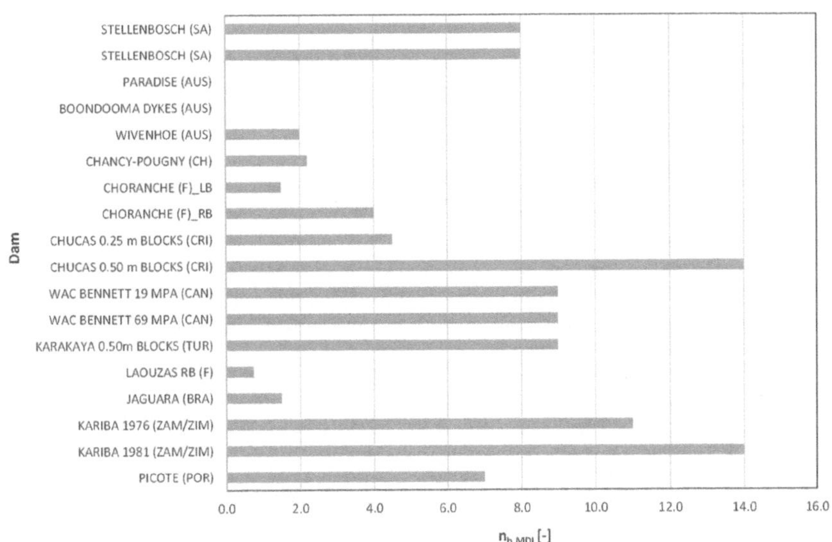

FIGURE 7.21 n-values determined by calibration of the DP method.

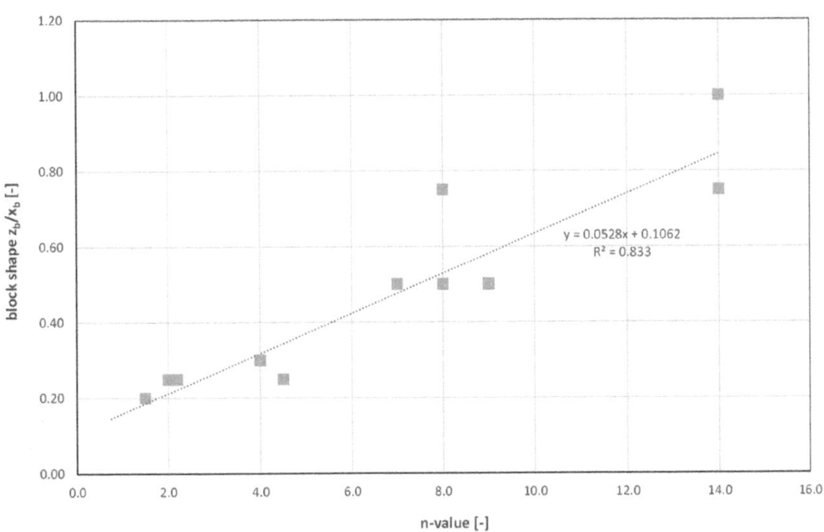

FIGURE 7.22 n-values as a function of rock block shape, obtained by calibration of the DP method.

values of up to $4 \cdot C_p'$ for block uplift. As such, it seems not surprising to find calibrated n-values of up to twice the values recommended by the authors.

EIM Method

There is a priori no calibration parameter for the erodibility index method. Nevertheless, the ASP (Applicable Stream Power) coefficient has been used to this extent. The results are presented in Figure 7.23.

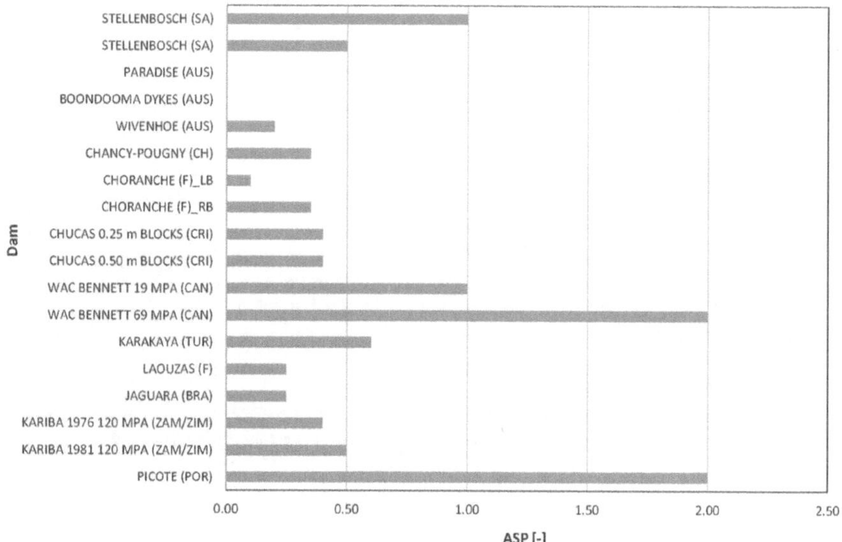

FIGURE 7.23 ASP values obtained by calibration of the EIM method.

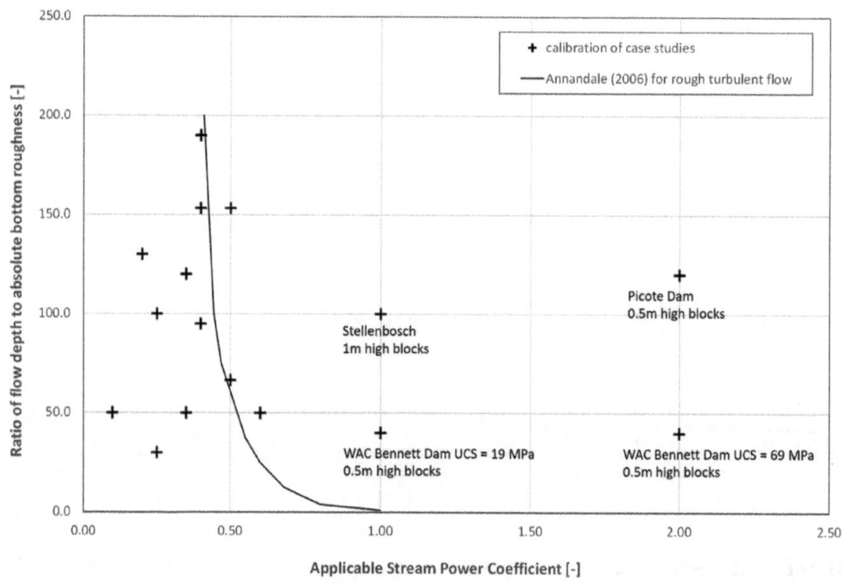

FIGURE 7.24 ASP values as a function of the ratio of flow depth to absolute roughness height.

Based on Annandale (2006), typical values for ASP are 0.40 for rough turbulent flow conditions, provided that the ratio of the flow depth to the absolute bottom roughness (k_s in m) is higher than 50–100. Values much higher may be obtained for lower ratios. The corresponding curve is compared in Figure 7.24 with the calibrated values of the present study.

The absolute bottom roughness height is difficult to determine for bedrock and should incorporate both form drag and frictional bottom roughness. Based on Inoue et al. (2014) and Ferguson et al. (2017), typical bedrock roughness heights in alluvial rivers are 0.05–0.10 m. For scour holes, however, increased roughness is reasonably expected because of the increased angularity of the suddenly exposed blocks and because of potential additional form drag by the shape of the scour hole. In Figure 7.24, a value of 20% of the average rock block height has been chosen as roughness height, i.e. between 0.04 and 0.20 m. It may be observed that, for most cases under study, the calibration results in ASP values well within the range of values as predicted by turbulent flow theory (Annandale 2006).

A few cases, however, need significantly higher stream power values to explain the in-situ observed scour formation. By expressing the ASP values as a function of the degree of break-up of the jets, it may be observed in Figure 7.25 that the highest ASP values in fact correspond to jets that are highly broken up, i.e. for jet break-up ratios >2 (WAC Bennett Dam, Picote Dam, Stellenbosch tests). As the stream power is computed based on the mean and RMS pressure coefficients of the impacting jets (see Chapter 6), the stream power significantly reduces at high ratios of jet break-up.

Some highly broken-up jet cases nevertheless exhibit ASP values within the range of the theoretical curve. For the Stellenbosch tests with a block height of 1.5 m, an ASP coefficient of 0.5 is found. As the erodibility index does (quasi-) not account for the influence of the rock block height, the tests performed for large block heights of 1.5 m exhibit reduced scour and finally result in a lower ASP value during calibration. For Karakaya Dam, with 0.5-m-high blocks, the relatively low ASP value can be attributed to a low UCS value used, resulting in a low erodibility index for the rock mass and thus easily erodible rock.

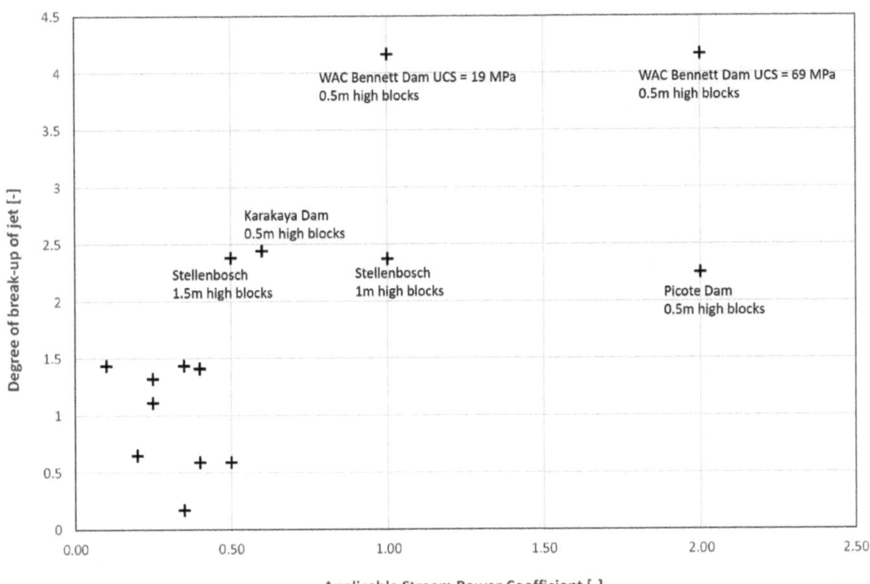

FIGURE 7.25 ASP values as a function of the degree of jet break-up.

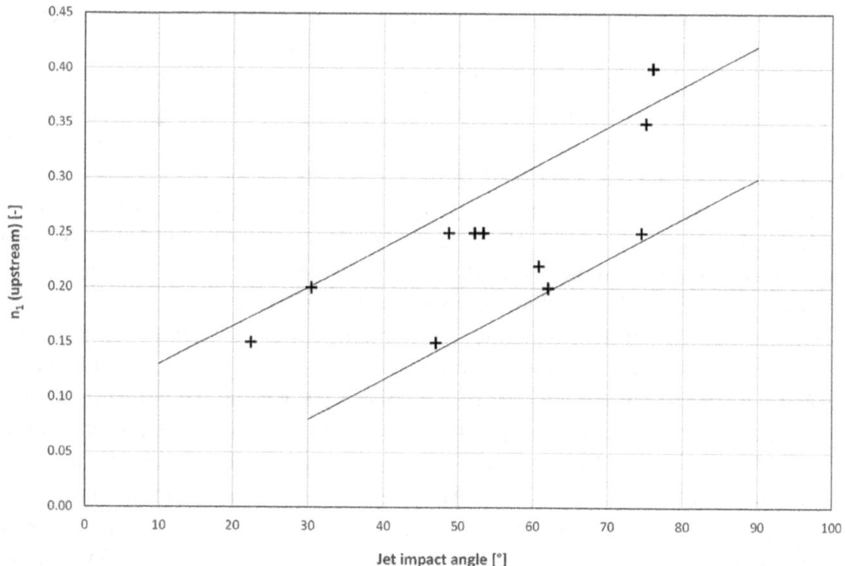

FIGURE 7.26 n_1 values for radial pressure distribution towards upstream, obtained by calibration.

Radial Pressure Distribution Coefficients n_1–n_4

Finally, the radial distribution of dynamic pressures at the water-rock interface, described by the coefficients n_1–n_4 (see Chapter 6), has been soundly calibrated. The results are illustrated in Figure 7.26 for the n_1 coefficient defining the shape of the dynamic pressure curve, and thus the shape of the scour hole, towards upstream. A clear tendency can be outlined, and a bandwidth of data is proposed.

Second, the n_3 values defining the shape of the dynamic pressure curve towards downstream are presented in Figure 7.27. A bandwidth of data is proposed to fit the calibrated values.

Furthermore, for jet impact angles lower than 30°–40°, the n-value is considered independent of the angle instead of tending towards a zero value, based on the assumption that a minimum of flow turbulence and dynamic pressures will subsist towards downstream. For such very low angles, the jet impact will approach wall jet flow or hydraulic jump conditions.

Finally, a clear relationship exists between n_3 (downstream) and n_1 (upstream) values, as shown in Figure 7.28.

For practice, recommended values for n_1 and n_3 are summarized in Table 7.7.

PRELIMINARY SCOUR DIAGNOSTICS

This subchapter presents a series of graphical solutions that allow a preliminary screening of the rock mass scour potential as a function of the quality and degree of fracturing of the rock, the rock block dimensions, the hydraulic head and finally the specific discharge. Second, a direct comparison is made with the available erodibility index scour thresholds.

The scour assessment is preliminary in the sense that a number of simplifying hypotheses have been made. These can be summarized as follows (Figure 7.29):

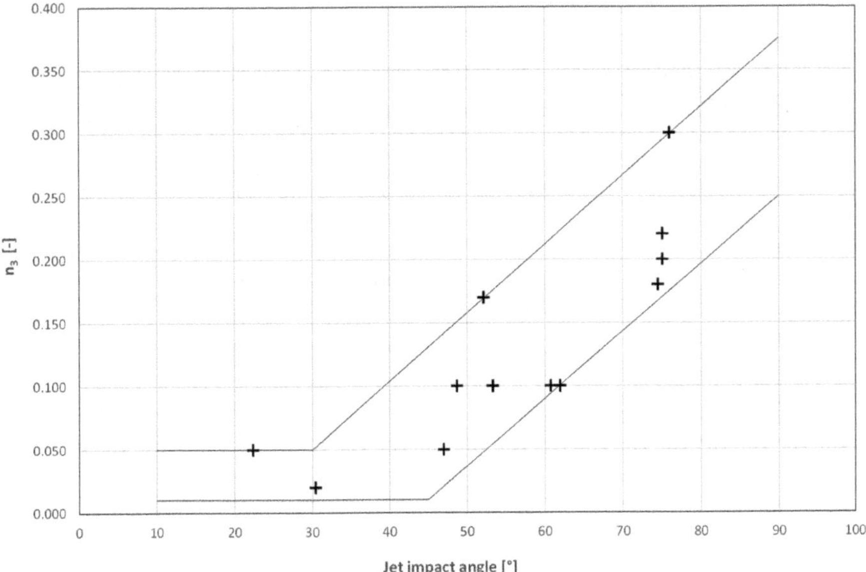

FIGURE 7.27 n_3 values for radial pressure distribution towards downstream, obtained by calibration.

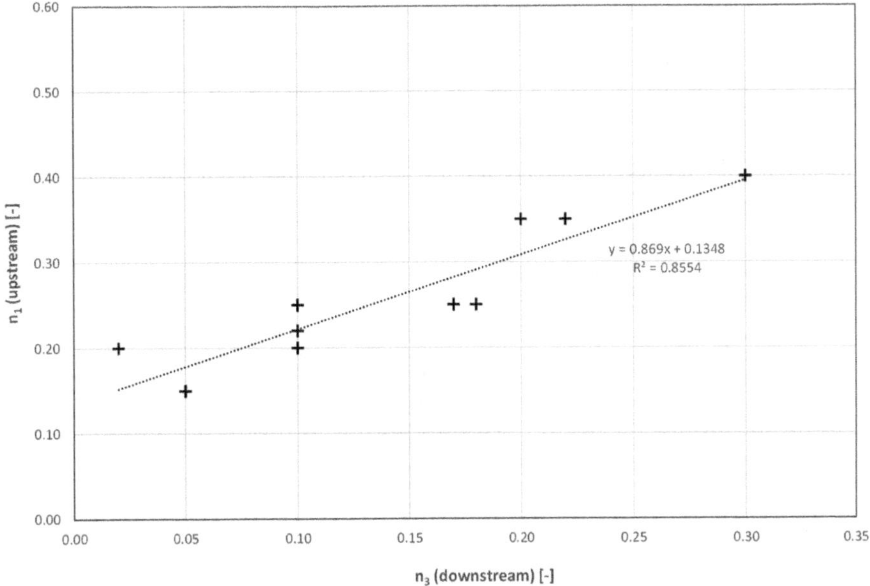

FIGURE 7.28 n_3 (down)–n_1 (up) relationship for radial pressure distribution, obtained by calibration.

TABLE 7.7
Recommended Values for n_3 (Down) and n_1 (Up) Determining the Radial Pressure Distribution as a Function of Jet Impact Angle

Impact Angle	n_1		n_3	
	min	max	min	max
[°]	[–]	[–]	[–]	[–]
10	0.010	0.130	0.010	0.050
30	0.080	0.200	0.010	0.050
45	0.135	0.255	0.010	0.131
60	0.190	0.310	0.090	0.212
75	0.245	0.365	0.170	0.294
90	0.300	0.420	0.250	0.375

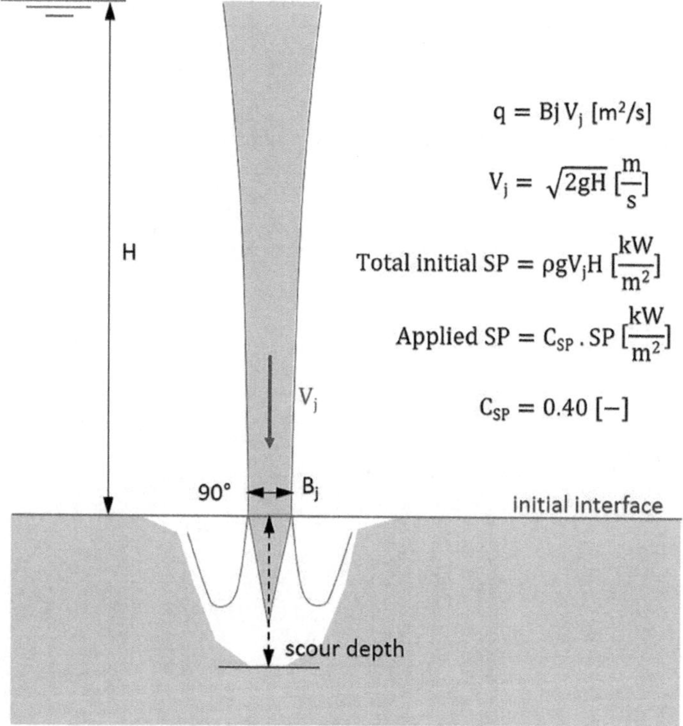

FIGURE 7.29 Schematic of assumptions made for the preliminary scour diagnostics.

1. Jets are vertically impinging (90°) and are of rectangular (2D) shape.
2. Jet impact velocity V_j [m/s] is determined based on the net hydraulic head H [m].
3. No initial tailwater is considered.
4. Jet impact thickness B_j [m] is determined by the ratio of the specific discharge q [m^2/s] to the jet impact velocity V_j [m/s].

5. Total available stream power per unit area at the interface is computed as $\rho g V_j H$ [kW/m^2].
6. Applied stream power per unit area at the interface [kW/m^2] is computed as $C_{SP}=0.40$ times the total available stream power (value based on case studies).
7. Only the deepest point of scour is estimated, i.e. the point of impact of the jet (no 2D scour profiles)

Furthermore, the parametric sensitivity analysis is restricted to the following cases:

1. Only rock joint fracturing (CFM) and dynamic block uplift (MDI) are considered.
2. Rock mass quality for fracturing is weak/altered/weathered rock or strong competent rock, with low (30% persistency) or moderate (50% persistency) degree of fissuring, following the distinction made in Chapter 6. Joints are 2D, open and planar.
3. Scour progression by fracturing is estimated after 1/10/100/1,000 days of constant specific discharge q at a net hydraulic head H.
4. Scour depth by dynamic block uplift (MDI) is estimated for rock blocks that are prismatic, and cubic-shaped (1 m height for 1 m side length) or flat-shaped (0.2 m height for 1 m side length).
5. The net hydraulic head H is 10/50/100/200 m, depending on the type of rock.
6. The specific discharge q is between 50 and 200 m^2/s.
7. The UCS strength of the rock is situated between 2 and 400 MPa.

In case of an initial tailwater depth, its influence on scour may be included by computing the net hydraulic head based on the tailwater level and by simply deducing from the total number of days that number of days that would have been needed to scour the tailwater depth. Also, this tailwater depth should be deduced from the total scour depth. For example, in case of 10 m of scour formation in 10 days and 15 m of scour formation in 100 days, a tailwater depth of 10 m would result in a scour formation of 5 m (15 – 10) in 90 days (100 – 10).

Finally, for lower angled jets, the scour depths should be multiplied by the sinus of the angle of the jet axis with the horizontal upon impact. For example, for 30° and 60° jet angles upon impact, the scour depths should be multiplied by a value of 0.50 respectively 0.87.

Preliminary Diagnostics for Weak and/or Weathered/Altered Rock

These cases are valid for weak rock as defined by one of the following situations:

1. Weak rock, i.e. mainly sedimentary rock and/or rock with UCS strength < 25 MPa (classes R0-R1-R2 according to ISRM (1978)).
2. Strong competent rock that is moderately to strongly weathered and/or altered, typically weathering grades W3 to W5 according to ISRM (1978).
3. Strong competent rock that is significantly tectonically disturbed, due to presences of folding, faulting and/or shearing planes, reducing the original structural integrity of the rock mass. Typically blocky/disturbed, disintegrated or sheared/laminated rock according to the GSI strength index criterion (Hoek 1994).

Two different degrees of fracturing have been considered, based on the database of case studies used in Chapter 6:

1. *Lowly fractured* weak rock with an initial joint persistency of $PE = 30\%$.
2. *Moderately fractured* weak rock with an initial joint persistency of $PE = 50\%$.

For these types of rock, both compact jets and fully broken-up jets are considered, with $H/L_b \sim 0$ respectively $1.35 < H/L_b < 1.60$. Rock block heights are $z_b = 0.20$ m and 1.00 m. The net hydraulic head H is 10 m and 50 m when using joint fracturing, and is variable for dynamic block uplift.

Scour Potential Based on Rock Joint Fracturing (CFM)

Lowly Fractured Weak Rock

Figures 7.30 and 7.31 illustrate the scour potential for compact jets and rock block heights of 1 m (cubic blocks), computed for $H = 50$ m respectively $H = 10$ m.

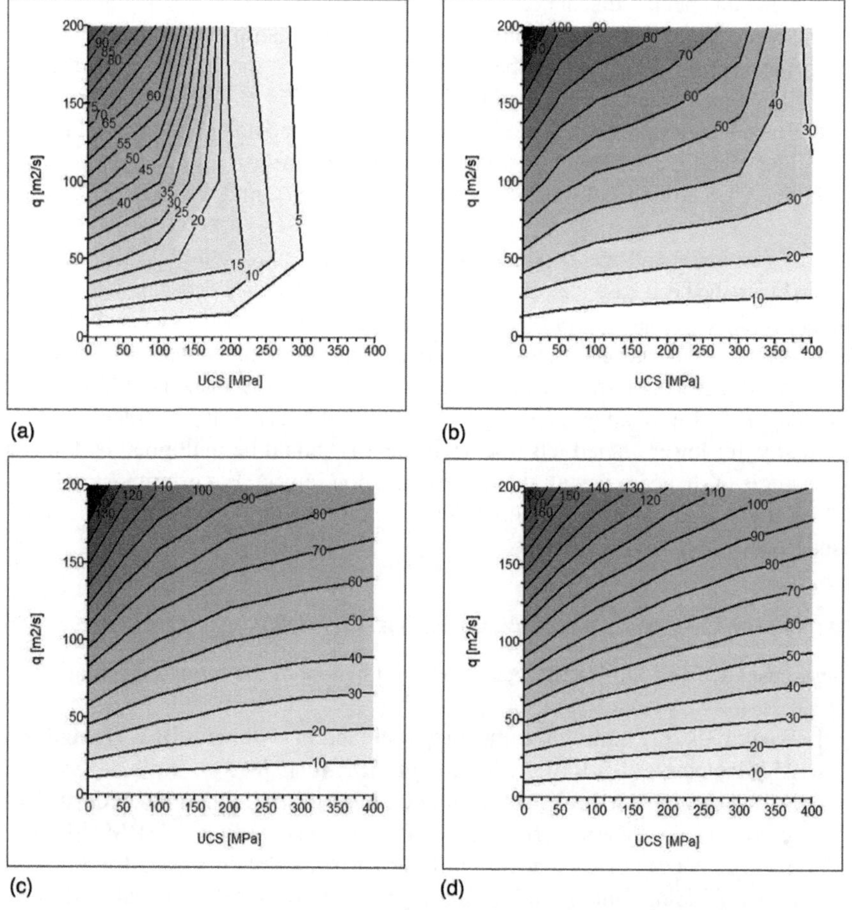

FIGURE 7.30 Scour potential [m] by rock joint fracturing (CFM) for compact jets, $z_b = 1.0$ m and $H = 50$ m: (a) after 1 day, (b) after 10 days, (c) after 100 days, (d) after 1,000 days.

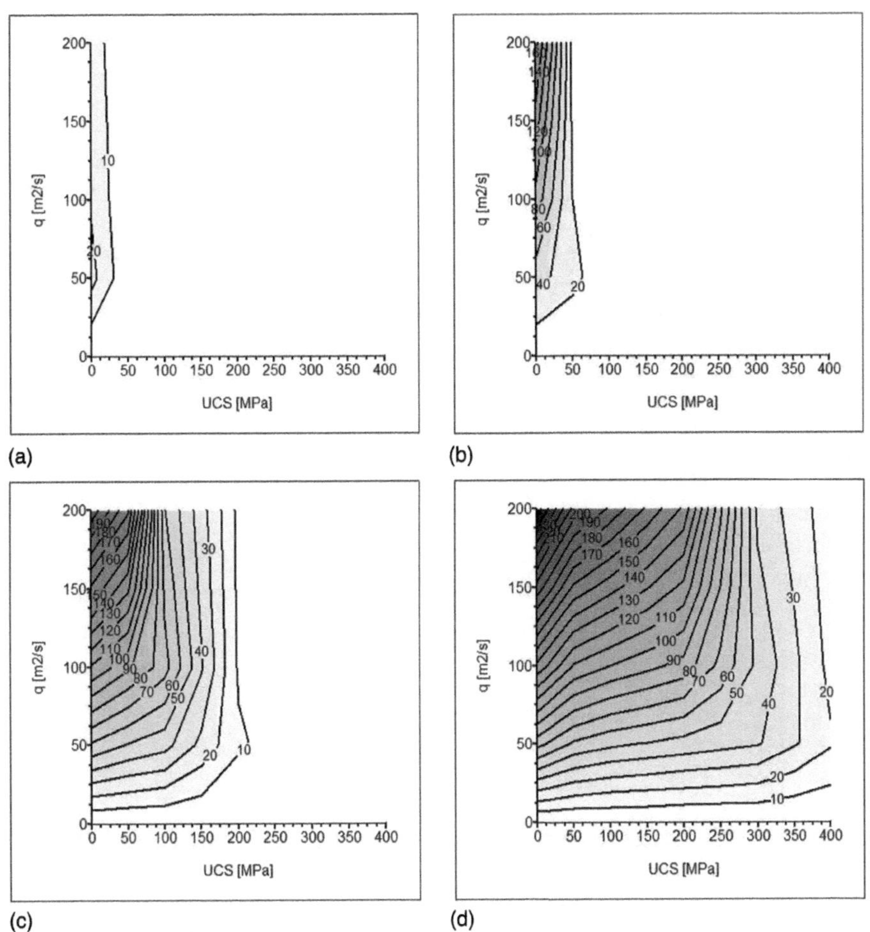

FIGURE 7.31 Scour potential [m] by rock joint fracturing (CFM) for compact jets, $z_b = 1.0$ m and $H = 10$ m: (a) after 1 day, (b) after 10 days, (c) after 100 days, (d) after 1,000 days.

Similarly, Figure 7.32 illustrates the scour potential for fully broken-up jets and rock block heights of 1 m (cubic blocks), computed for $H = 50$ m.

The same computations have been performed for flat-shaped prismatic rock blocks with a height of $z_b = 0.20$ m. Figures 7.33 and 7.34 illustrate the scour potential for compact jets, computed for $H = 50$ m respectively $H = 10$ m.

Similarly, Figure 7.35 illustrates the scour potential for fully broken-up jets and rock block heights of 0.2 m (flat-shaped blocks), computed for $H = 50$ m.

Moderately Fractured Weak Rock

Figures 7.36 and 7.37 illustrate the scour potential for compact jets and rock block heights of 1 m (cubic blocks), computed for $H = 50$ m respectively $H = 10$ m.

Similarly, Figure 7.38 illustrates the scour potential for fully broken-up jets and rock block heights of 1 m (cubic blocks), computed for $H = 50$ m.

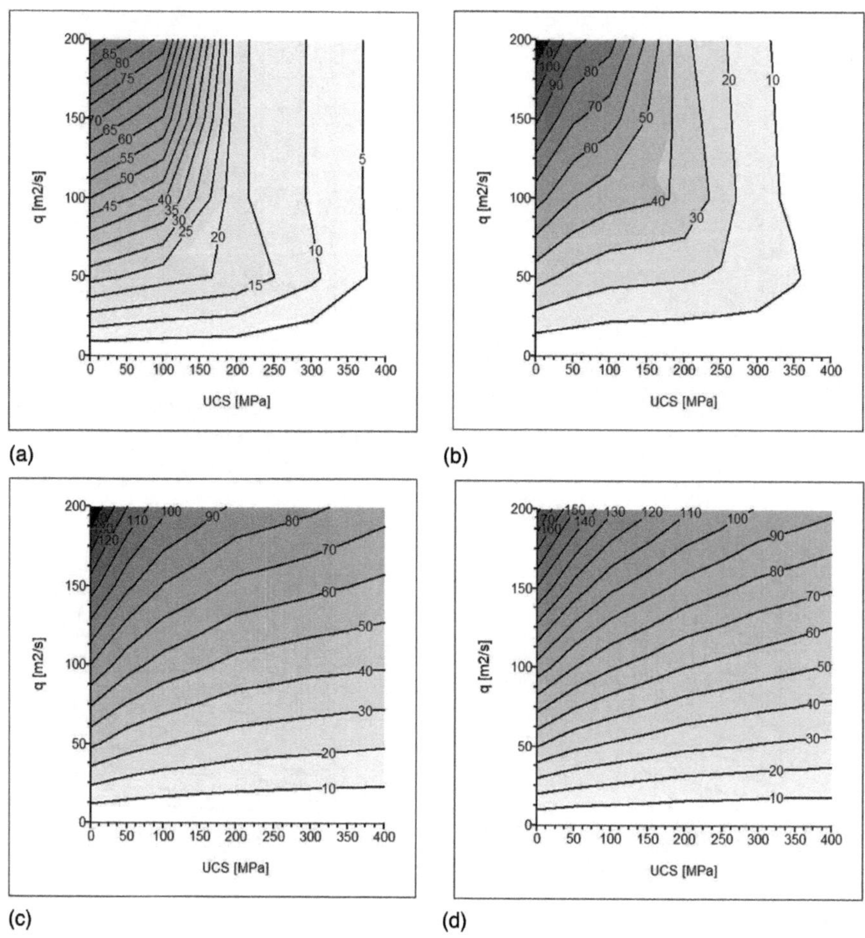

FIGURE 7.32 Scour potential [m] by rock joint fracturing (CFM) for fully broken-up jets, $z_b = 1.0$ m and $H = 50$ m: (a) after 1 day, (b) after 10 days, (c) after 100 days, (d) after 1,000 days.

The same computations have been performed for flat-shaped prismatic rock blocks with a height of $z_b = 0.20$ m. Figures 7.39 and 7.40 illustrate the scour potential for compact jets, computed for $H = 50$ m respectively $H = 10$ m.

Similarly, Figure 7.41 illustrates the scour potential for fully broken-up jets and rock block heights of 0.2 m (flat-shaped blocks), computed for $H = 50$ m.

Scour Potential Based on Dynamic Block Uplift (MDI)

Figure 7.42 presents the scour potential computed for both compact and fully broken-up jets, based on the dynamic block uplift method (MDI). The potential is expressed as a function of the net hydraulic head and for rock block heights of 0.20 m and 1.00 m respectively.

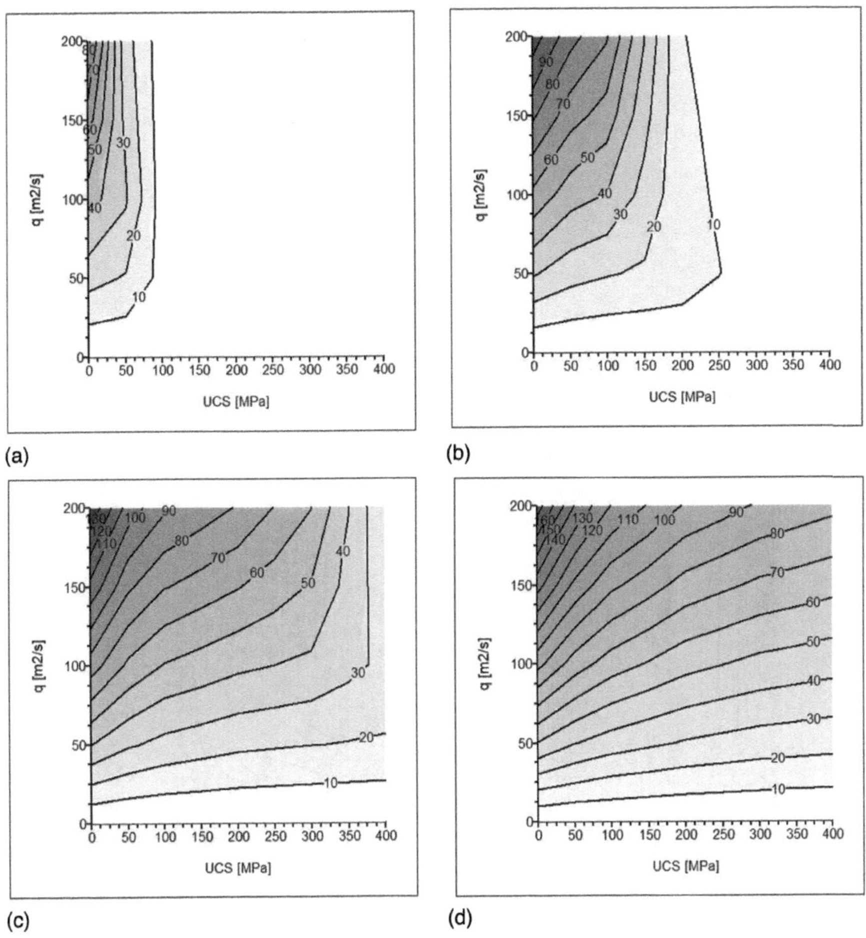

FIGURE 7.33 Scour potential [m] by rock joint fracturing (CFM) for compact jets, $z_b = 0.2\,\text{m}$ and $H = 50\,\text{m}$: (a) after 1 day, (b) after 10 days, (c) after 100 days, (d) after 1,000 days.

Preliminary Diagnostics for Strong Competent Rock

These cases are valid for strong competent rock that may be considered as lowly to not scour vulnerable, such as metamorphic and igneous rock types, with UCS strengths typically >25 MPa (classes R3 to R6 according to ISRM, 1978), and with low to no weathering (W1 to max. W2 according to ISRM, 1978) or tectonic perturbations (i.e. no folding, no faulting, no bedding layers, no shear planes, etc.).

The rock may contain several joint sets and be partially fissured, typically intact/massive or blocky rock with very good to good surface conditions according to GSI strength index criterion (Hoek 2006). Typical examples of strong competent rock are quartzite, granite, dolerite, gneiss.

Similar to weak rock, two degrees of fracturing have been considered, i.e. *lowly* and *moderately fractured* strong competent rock. For these types of rock, compact

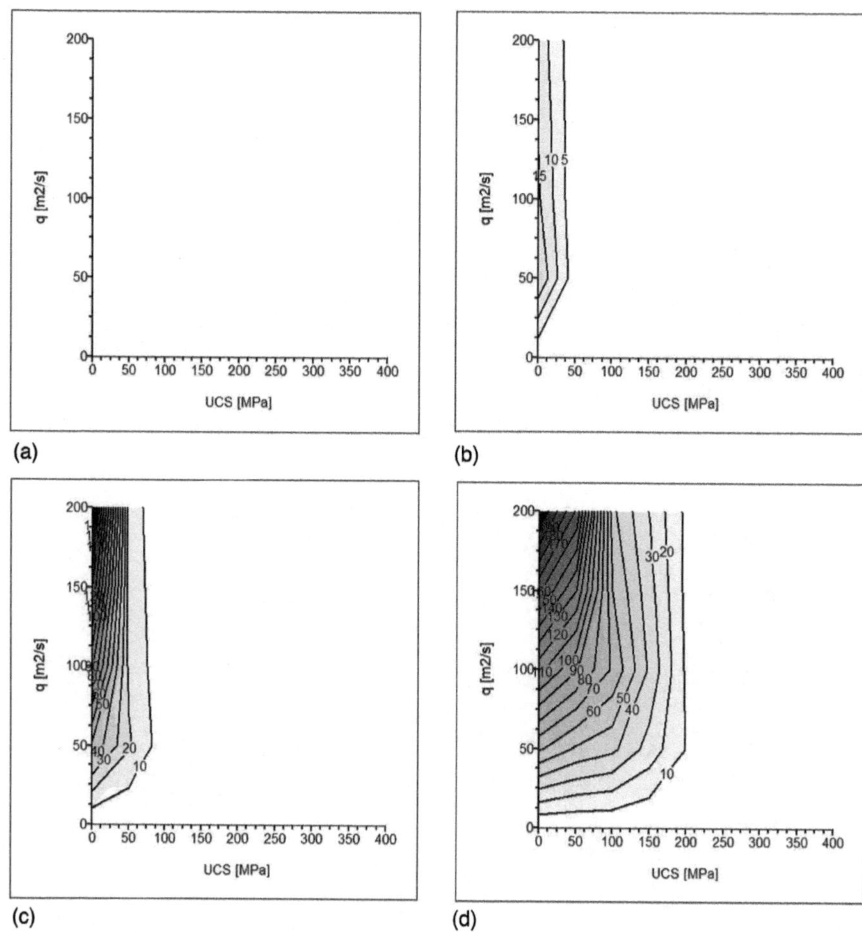

FIGURE 7.34 Scour potential [m] by rock joint fracturing (CFM) for compact jets, $z_b = 0.2$ m and $H = 10$ m: (a) after 1 day, (b) after 10 days, (c) after 100 days, (d) after 1,000 days.

jets are considered, with $H/L_b \sim 0$. Rock block heights are $z_b = 0.20$ m and 1.00 m. The net hydraulic head H is 50 m and 200 m when using joint fracturing as scour criterion (CFM method), and is variable for dynamic block uplift (MDI method).

Scour Potential Based on Rock Joint Fracturing (CFM)

Lowly Fractured Strong Rock

Figures 7.43 and 7.44 illustrate the scour potential for compact jets and rock block heights of 1 m (cubic blocks), computed for $H = 50$ m respectively $H = 200$ m.

The same computations have been performed for flat-shaped prismatic rock blocks with a height of $z_b = 0.20$ m. Figures 7.45 and 7.46 illustrate the scour potential for compact jets, computed for $H = 50$ m respectively $H = 200$ m.

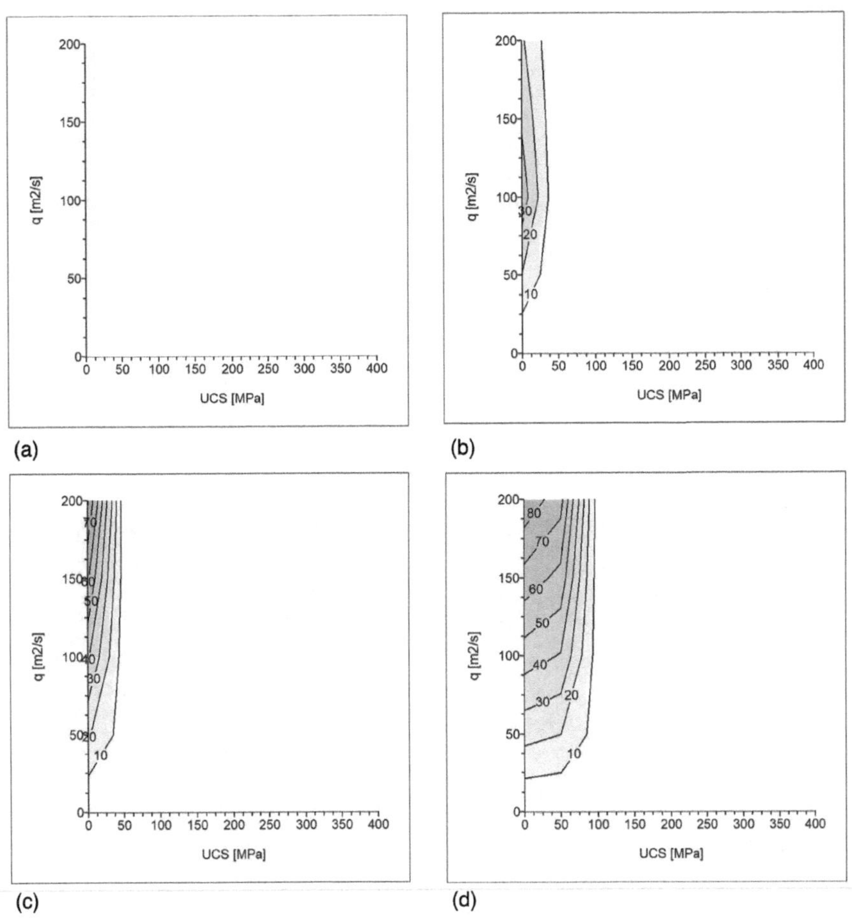

FIGURE 7.35 Scour potential [m] by rock joint fracturing (CFM) for fully broken-up jets, $z_b = 0.2$ m and $H = 50$ m: (a) after 1 day, (b) after 10 days, (c) after 100 days, (d) after 1,000 days.

Moderately Fractured Strong Rock

Figures 7.47 and 7.48 illustrate the scour potential for compact jets and rock block heights of 1 m (cubic blocks), computed for $H = 50$ m respectively $H = 200$ m.

The same computations have been performed for flat-shaped prismatic rock blocks with a height of $z_b = 0.20$ m. Figures 7.49 and 7.50 illustrate the scour potential for compact jets, computed for $H = 50$ m respectively $H = 200$ m.

Scour Potential Based on Dynamic Block Uplift (MDI)

The scour potential computed based on the dynamic block uplift method (MDI) procures results that are similar to the ones obtained for weak rock in Figure 7.42.

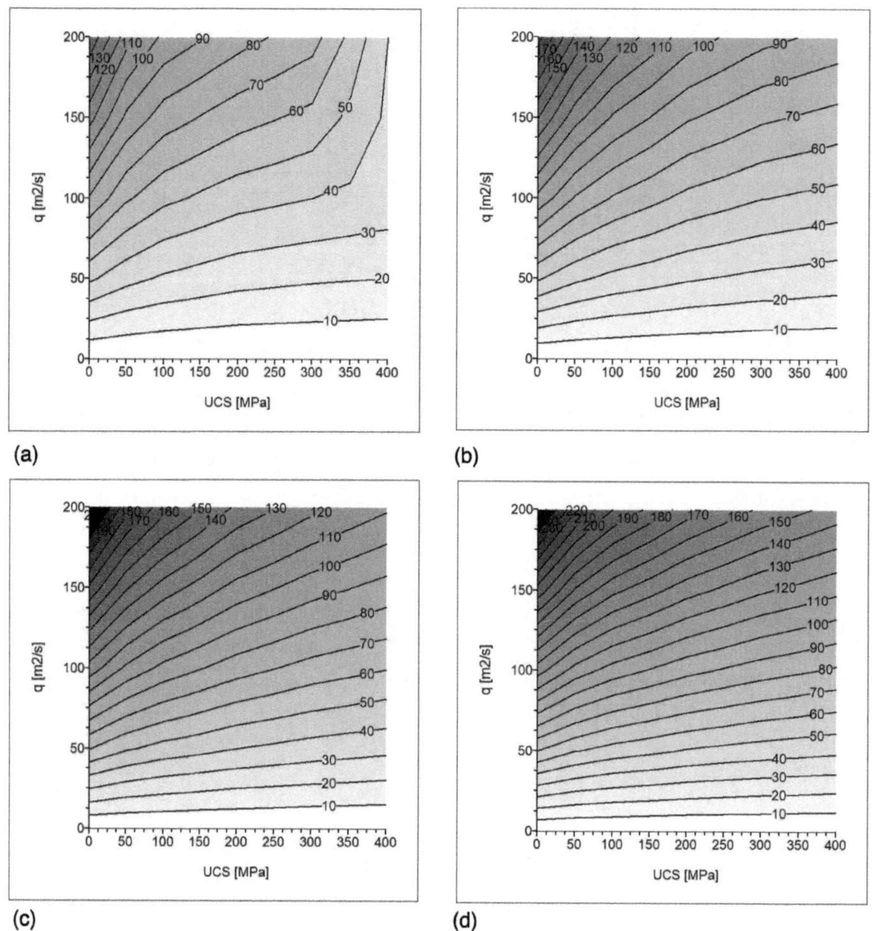

FIGURE 7.36 Scour potential [m] by rock joint fracturing (CFM) for compact jets, $z_b = 1.0$ m and $H = 50$ m: (a) after 1 day, (b) after 10 days, (c) after 100 days, (d) after 1,000 days.

Comparison with Erodibility Index Methods (EIM)

The different Erodibility Index Methods (EIM) make use of scour thresholds or scour classes that graphically allow to check the potential for rock scour, based on the unitary stream power and the Kirsten index for rock resistance. In the following, the application of the CFM and MDI methods (as presented by the graphical solutions for lowly and moderately fractured weak and strong rock) has been implemented into a EIM graph to compare their scour thresholds with the different available EIM scour thresholds/classes.

The following assumptions have been made (see Figure 7.29):

1. Stream power (SP) per unit of area (in kW/m²) is determined by computing γQH per m² of impact area of the plunging jet onto the rock, and by multiplying this value by the applicable stream power coefficient $C_{SP} = 0.40$

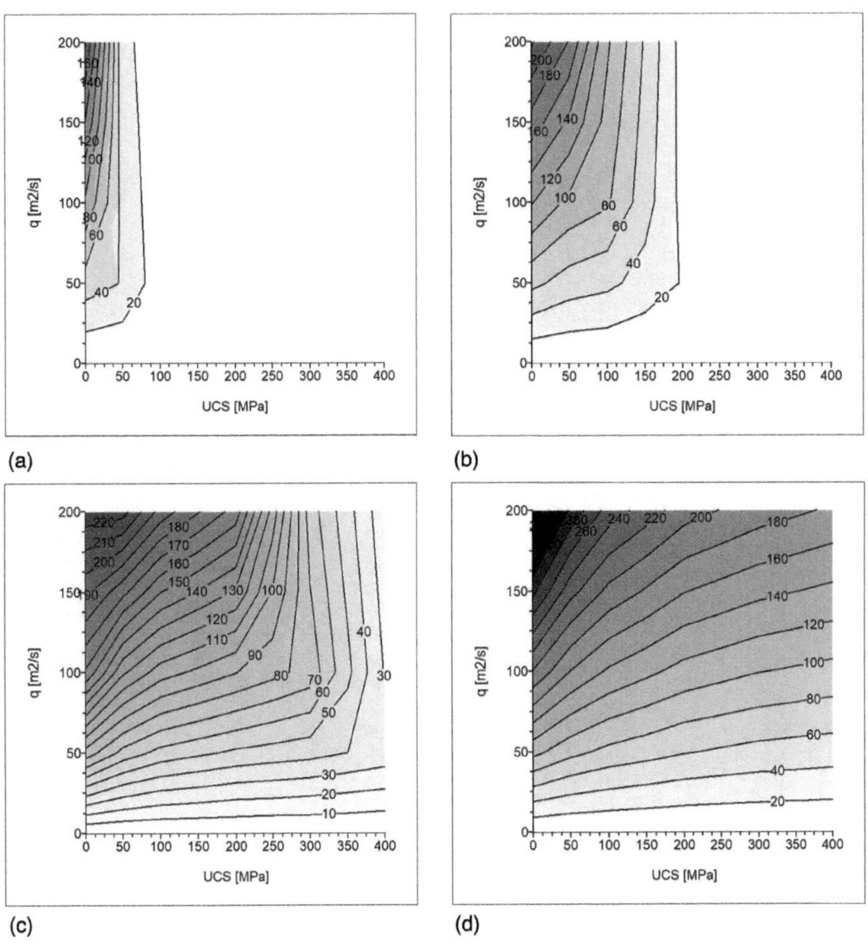

FIGURE 7.37 Scour potential [m] by rock joint fracturing (CFM) for compact jets, $z_b = 1.0\,\text{m}$ and $H = 10\,\text{m}$: (a) after 1 day, (b) after 10 days, (c) after 100 days, (d) after 1,000 days.

(based on the different case studies from the database), i.e. $SP = 0.40 \cdot \gamma \cdot V_j \cdot H$. No initial tailwater level is considered. No head losses are considered during the fall of the jet.

2. The scour curves are valid for vertically impinging jets. For lower angled jets upon impact, the scour depths of the CFM method should be multiplied by the sinus of the angle of the jet axis with the horizontal upon impact. For example, for 30° and 60° jet angles upon impact, the scour depths should be multiplied by a value of 0.50 respectively 0.87.
3. The scour curves are computed for $q = 100\,\text{m}^2/\text{s}$ and for varying net hydraulic heads H. For other specific discharges between 50 and $200\,\text{m}^2/\text{s}$, the results do not vary by more than 5%–10%. For lower specific discharges, however, the presented curves should be adapted.

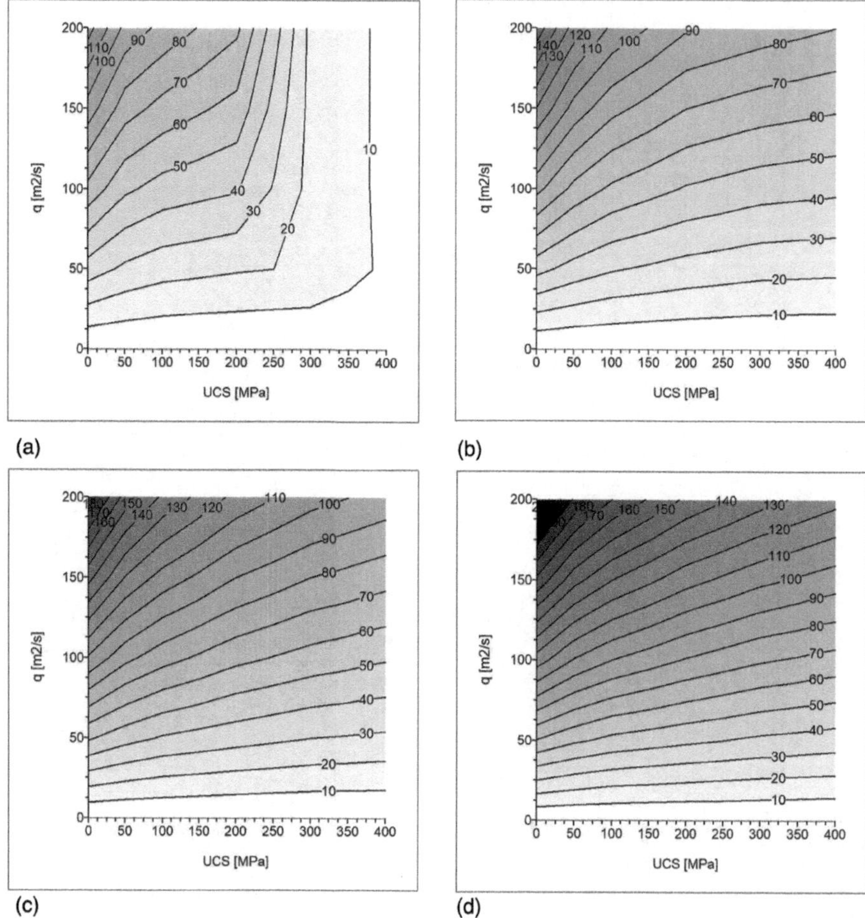

FIGURE 7.38 Scour potential [m] by rock joint fracturing (CFM) for fully broken-up jets, $z_b = 1.0$ m and $H = 50$ m: (a) after 1 day, (b) after 10 days, (c) after 100 days, (d) after 1,000 days.

4. The Kirsten index is computed based on the following values and assumptions:
   a. M_s is based on the UCS strength and a rock density of 2,700 kg/m^3
   b. K_b is based on the RQD value and the $J_n = 3.34$ (i.e. three joint sets plus random)
   c. The RQD value is determined based on the rock block height and the initial degree of fracturing of the rock as follows:

Block height	**Persistency**	**RQD**
z_b [m]	PE [%]	[%]
1.0	30	80
1.0	50	50
0.2	30	80
0.2	50	50

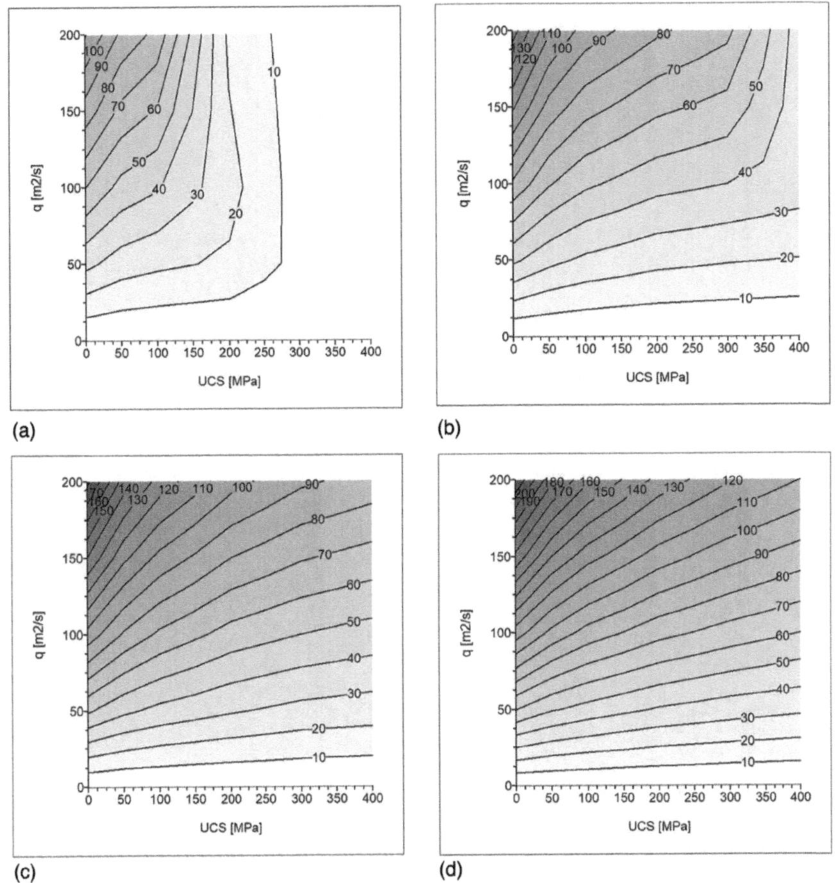

FIGURE 7.39 Scour potential [m] by rock joint fracturing (CFM) for compact jets, $z_b = 0.2$ m and $H = 50$ m: (a) after 1 day, (b) after 10 days, (c) after 100 days, (d) after 1,000 days.

d. The discontinuity bond shear strength number K_d is determined based on a joint roughness number $J_r = 1$, assuming open joints, and a joint alteration number of $J_a = 1.5$, assuming a 1 mm joint containing slightly altered non-cohesive rock mineral as gouge.
e. The relative ground structure number $J_s = 1.14$ for 1-m-high blocks respectively $J_s = 1.26$ for 0.20-m-high blocks, corresponding to a 90° dip angle and a ratio of joint spacing of 1:1 respectively 1:8.

Furthermore, the rocsc@r generated results are compared with the scour thresholds and scour classes as determined by van Schalkwyk (1994), Annandale (1995) and Pells (2016). This is done for the following parametric combinations:

1. Weak rock, lowly fractured ($PE = 30\%$), compact jets, $z_b = 1.0$ m
2. Weak rock, moderately fractured ($PE = 50\%$), compact jets, $z_b = 1.0$ m

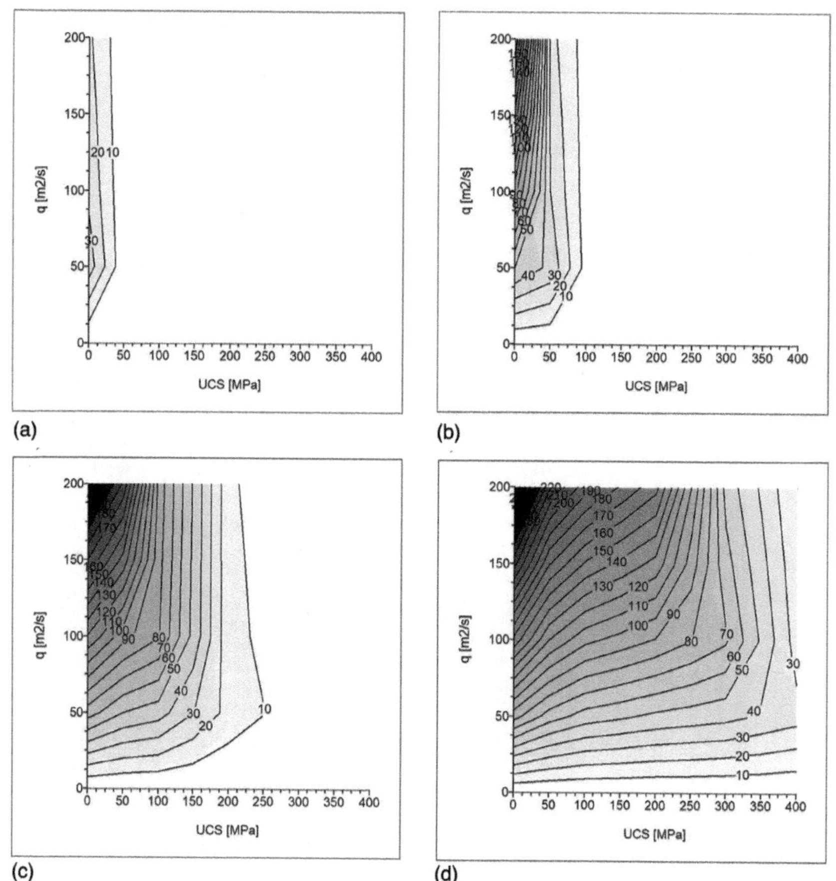

FIGURE 7.40 Scour potential [m] by rock joint fracturing (CFM) for compact jets, $z_b = 0.2$ m and $H = 10$ m: (a) after 1 day, (b) after 10 days, (c) after 100 days, (d) after 1,000 days.

3. Weak rock, moderately fractured ($PE = 50\%$), fully broken jets, $z_b = 1.0$ m
4. Weak rock, moderately fractured ($PE = 50\%$), compact jets, $z_b = 0.2$ m
5. Weak rock, moderately fractured ($PE = 50\%$), fully broken jets, $z_b = 0.2$ m
6. Strong rock, moderately fractured ($PE = 50\%$), compact jets, $z_b = 1.0$ m

Lowly Fractured Weak Rock

Figure 7.51 illustrates the results for compact vertically plunging jets with no initial tailwater. The rock block height is $z_b = 1.0$ m. The general shape of the CFM scour thresholds seems to follow a combination of the initial curves established by van Schalkwyk (1994) and van Schalkwyk et al. (1994b) as well as with the erosion classes as determined by Pells (2016).

The scour threshold as defined by Annandale (1995) represents a linear approximation of the CFM curves for 5–10 m of scour in 100 days of discharge. Also, compared to the other erodibility index thresholds and classes, this scour threshold is situated in an area of moderate to strong erosion. Hence, the Annandale scour

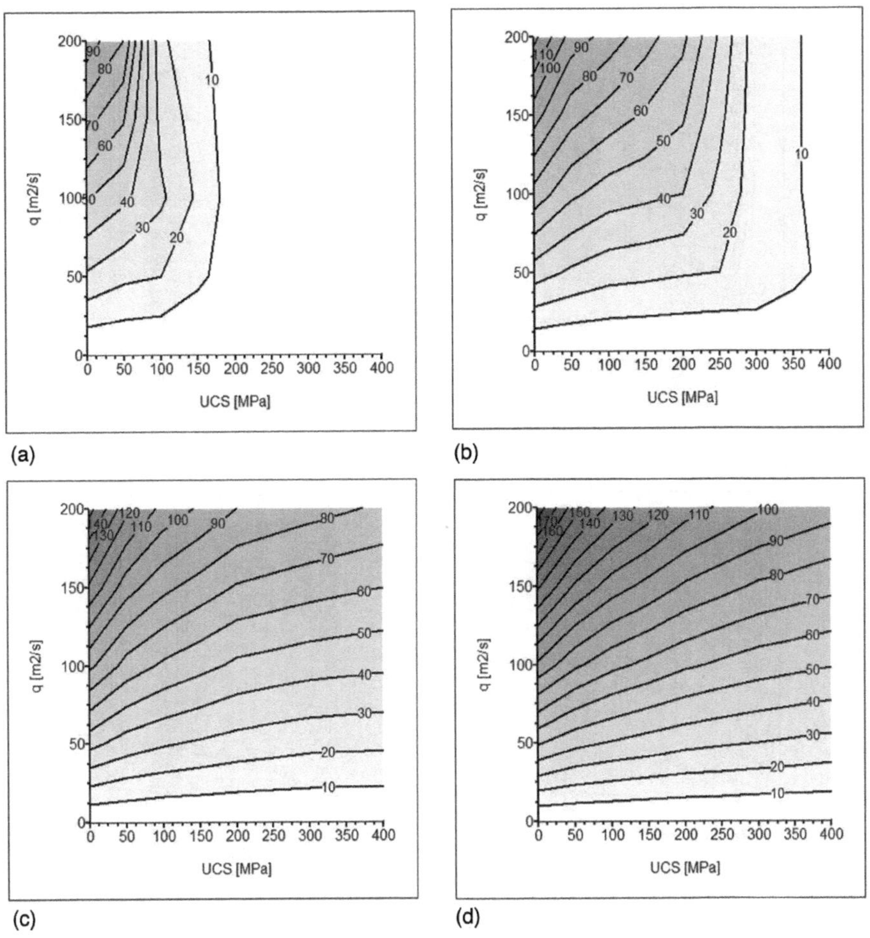

FIGURE 7.41 Scour potential [m] by rock joint fracturing (CFM) for fully broken-up jets, $z_b = 0.2$ m and $H = 50$ m: (a) after 1 day, (b) after 10 days, (c) after 100 days, (d) after 1,000 days.

threshold should be applied with precaution, because significant scour may form at stream power values well below the threshold values. The MDI method scour threshold is situated at very high stream power values of around 1,500 kW/m^2. The 1-m-high rock blocks are difficult to eject from their mass, despite the ease and speed of fracturing of the joints.

Moderately Fractured Weak Rock

Figure 7.52 illustrates the results for compact vertically plunging jets with no initial tailwater. The rock block height is $z_b = 1.0$ m. The CFM scour threshold of 1–10 m of scour in 1,000 days of discharge corresponds quite well to the initial line established by van Schalkwyk (1994).

The curved shape of the computed curves seems to match with the erosion classes as determined by Pells (2016). The scour threshold as defined by Annandale (1995)

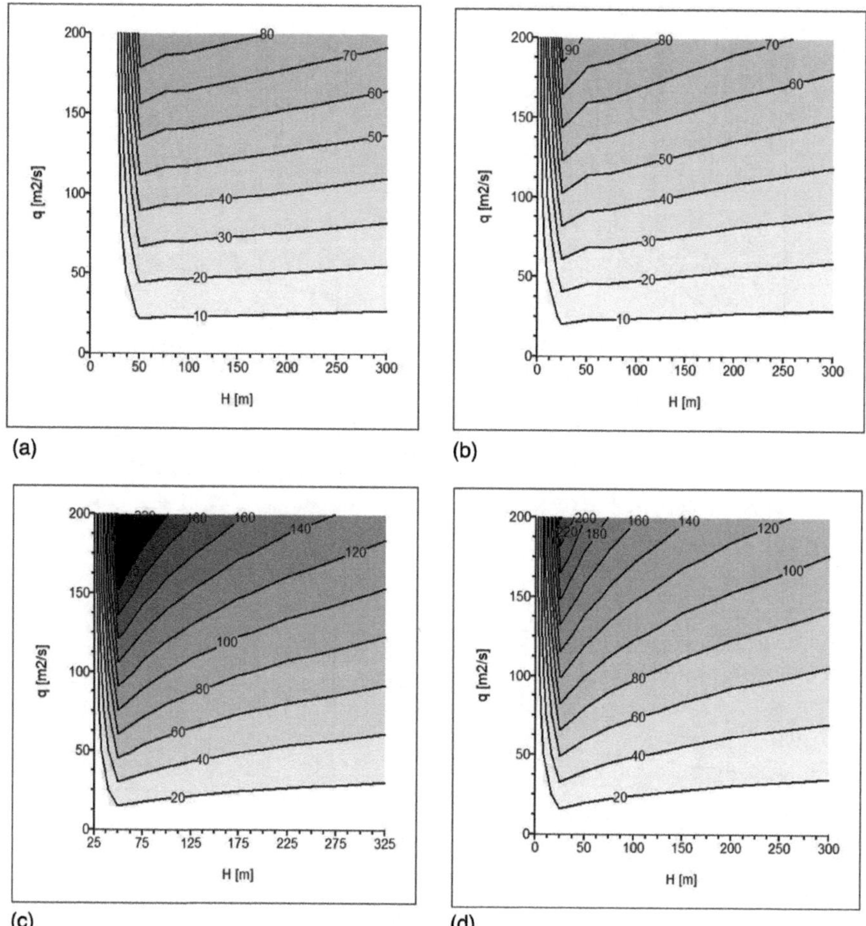

FIGURE 7.42 Scour potential [m] for variable hydraulic heads H and based on dynamic block uplift (MDI): (a) compact jet and $z_b = 1.0$ m, (b) broken-up jet and $z_b = 1.0$ m, (c) compact jet and $z_b = 0.20$ m, (d) broken-up jet and $z_b = 0.20$ m.

represents a linear approximation of the CFM curves for 5–10 m of scour in 10 days of discharge. Hence, the Annandale scour threshold should be applied with a lot of precaution here, because significant scour may rapidly form at stream power values well below the threshold values. The MDI method scour threshold is situated at very high stream power values of around 1,500 kW/m^2. The 1-m-high rock blocks are difficult to eject from their mass, despite the ease and speed of fracturing of the joints.

Second, Figure 7.53 illustrates the results for fully broken vertically plunging jets with no initial tailwater. The rock block height is $z_b = 1.0$ m. The scour threshold as defined by Annandale (1995) represents a linear approximation of the CFM curve for 10 m of scour in 100 days of discharge. Hence, despite the Annandale scour threshold being somewhat more in line with the CFM curves for broken jets, it should still

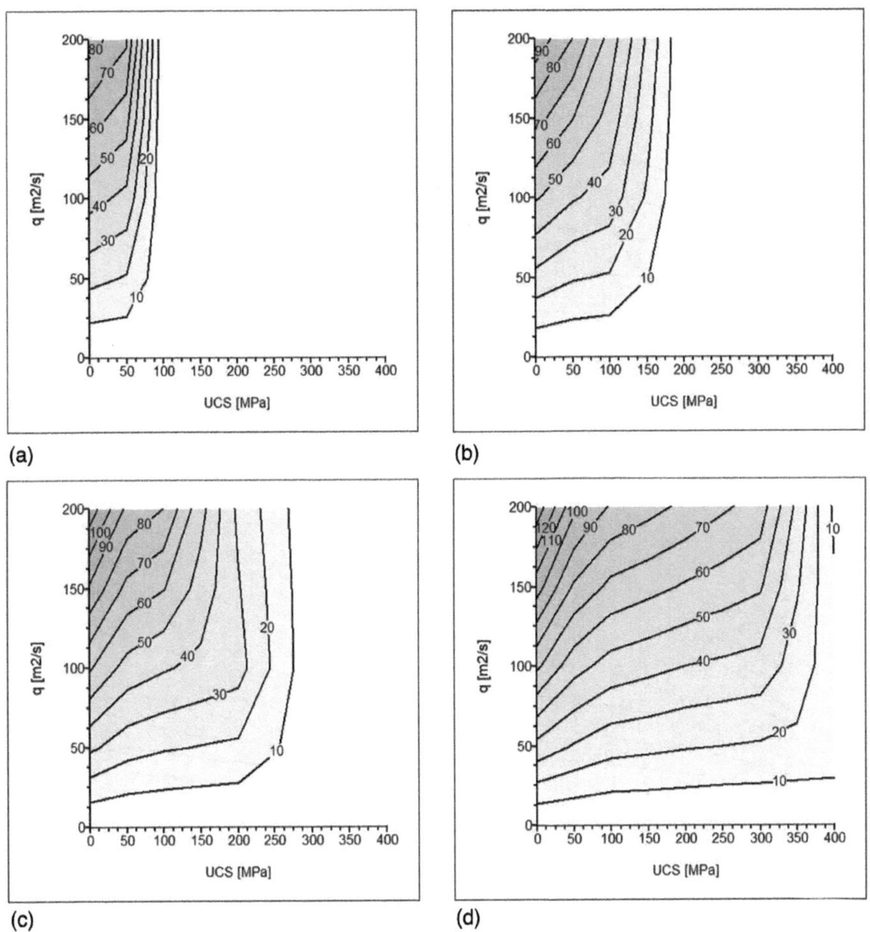

FIGURE 7.43 Scour potential [m] by rock joint fracturing (CFM) for compact jets, $z_b = 1.0$ m and $H = 50$ m: (a) after 1 day, (b) after 10 days, (c) after 100 days, (d) after 1,000 days.

be applied with precaution in case of moderately to strongly fractured weak rock, because significant scour may form on the medium to long term at stream power values well below the threshold values.

The MDI method scour threshold is situated at very high stream power values of around 1,000 kW/m^2. The 1-m-high rock blocks are again difficult to eject from their mass, despite the ease and speed of fracturing of the joints.

The third case of moderately fractured weak rock deals with compact vertically plunging jets for rock block heights of $z_b = 0.20$ m (i.e. flat-shaped blocks, Figure 7.54). The scour threshold as defined by Annandale (1995) represents a linear approximation of the CFM curve for 5–10 m of scour in 100 days of discharge. Hence, it should again be applied with precaution in case of moderately to strongly fractured weak rock, because significant scour may form on the medium to long term at stream power values well below the threshold values.

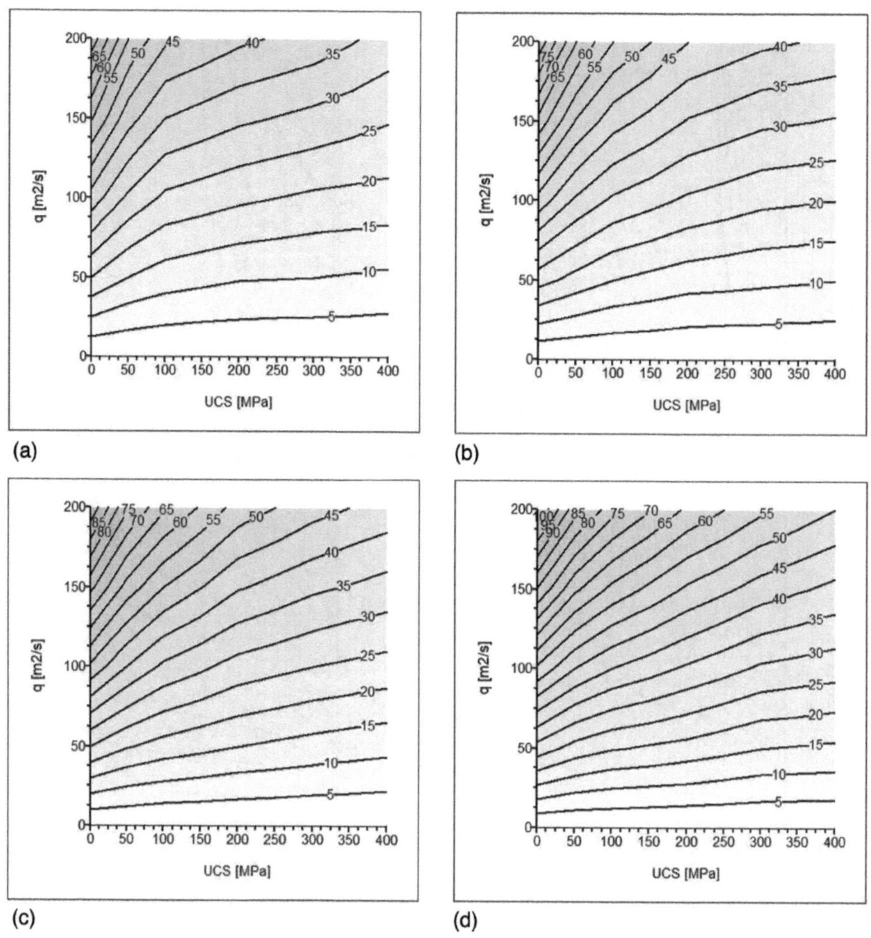

FIGURE 7.44 Scour potential [m] by rock joint fracturing (CFM) for compact jets, $z_b = 1.0$ m and $H = 200$ m: (a) after 1 day, (b) after 10 days, (c) after 100 days, (d) after 1,000 days.

The MDI method scour threshold is situated at very low stream power values of around 20 kW/m^2. The 0.2-m-high rock blocks are easy to eject from their mass.

The last case of moderately fractured weak rock deals with broken vertically plunging jets for rock block heights of $z_b = 0.20$ m (i.e. flat-shaped blocks, Figure 7.55). The scour threshold as defined by Annandale (1995) represents a linear approximation of the CFM curve for 1 m of scour in 100 days of discharge and thus is more or less in line with the results of the CFM method.

The MDI method scour threshold is situated at very low stream power values of around 20 kW/m^2. The 0.2-m-high rock blocks are easy to eject from their mass.

Moderately Fractured Strong Competent Rock

Figure 7.56 illustrates the results for compact vertically plunging jets with no initial tailwater. The rock block height is $z_b = 1.0$ m.

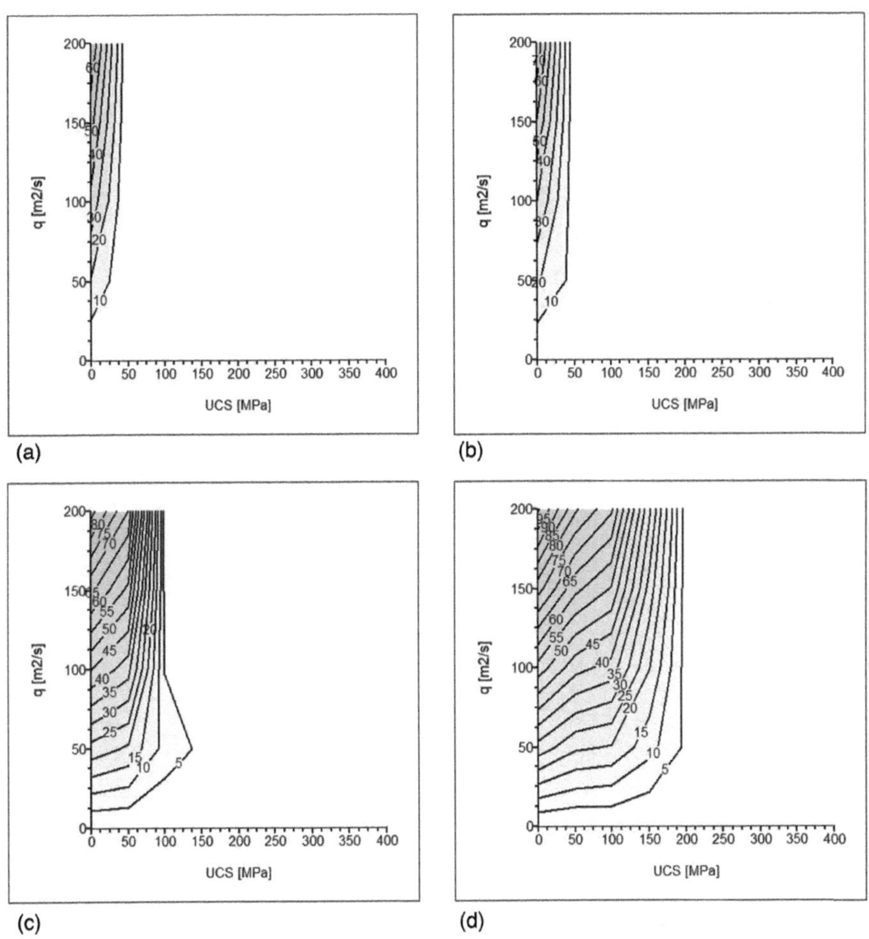

FIGURE 7.45 Scour potential [m] by rock joint fracturing (CFM) for compact jets, $z_b = 0.2$ m and $H = 50$ m: (a) after 1 day, (b) after 10 days, (c) after 100 days, (d) after 1,000 days.

The scour threshold as defined by Annandale (1995) represents a linear approximation of the CFM curves for 1 m of scour in 1,000 days of discharge. Hence, for strong competent rock, the Annandale scour threshold seems to be in accordance with the CFM scour curves. Again, the MDI method scour threshold is situated at very high stream power values of around 1,500 kW/m^2. The 1-m-high rock blocks are difficult to eject from their mass, despite the ease and speed of fracturing of the joints.

PROBABILISTIC-BASED SCOUR COMPUTATIONS

Introduction

Risk-informed analysis and decision-making processes related to safety and stability of hydraulic structures study the probability of occurrence of potential failure

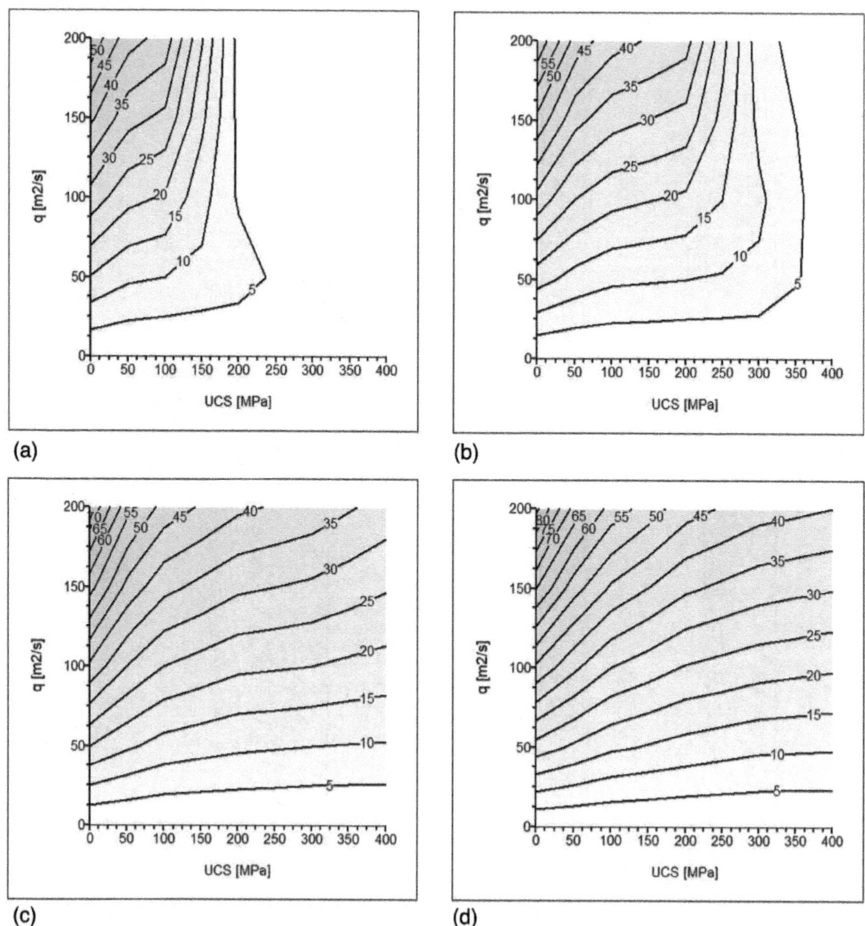

FIGURE 7.46 Scour potential [m] by rock joint fracturing (CFM) for compact jets, $z_b = 0.2$ m and $H = 200$ m: (a) after 1 day, (b) after 10 days, (c) after 100 days, (d) after 1,000 days.

modes (PFMA). The probability of a given scour depth to be reached during the lifetime of the structure, or following a distinct flood event, is of particular importance. The uncertainty inherent to the many input parameters responsible for scour formation asks for a probabilistic-based computational approach, such as the Monte Carlo method.

Monte-Carlo Scour Simulations

The Monte Carlo method is a popular mathematical technique that simulates the range of possible outcomes for an uncertain event. The computations are based on a randomly estimated range of values instead of a fixed set of values for each of the relevant input parameters. The random input parameters may thereby be correlated. Monte Carlo simulations use a marginal probability distribution for any input

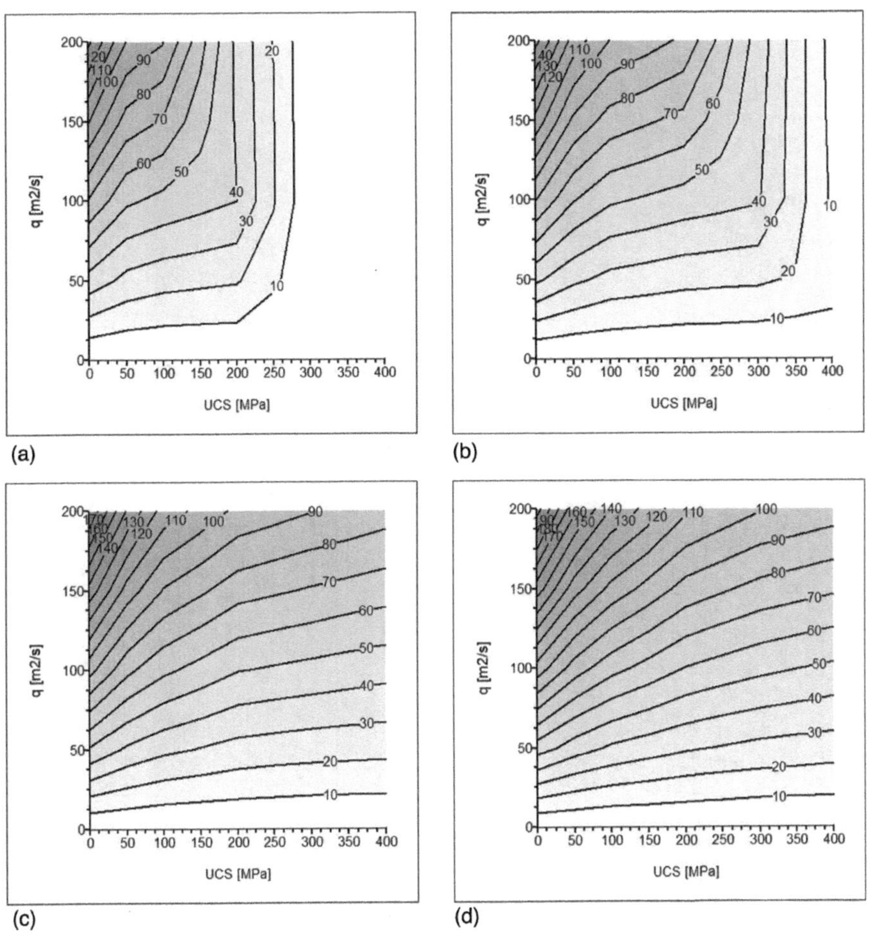

FIGURE 7.47 Scour potential [m] by rock joint fracturing (CFM) for compact jets, $z_b = 1.0$ m and $H = 50$ m: (a) after 1 day, (b) after 10 days, (c) after 100 days, (d) after 1,000 days.

variable that has inherent uncertainty. Then, it recalculates the results many times, using a different set of random numbers within the estimated range at each time. This process generates many probable outcomes, which become more accurate as the number of inputs grows. The different outcomes are generally considered to form a normal distribution, which allows to determine a likelihood of appearance for each of the computed outcomes.

Monte Carlo simulations are simple conceptually but allow to solve problems in complex systems. They are particularly useful for long-term predictions because of their accuracy. Monte Carlo simulations are also a good alternative to machine learning when there isn't enough data to make a machine learning model accurate, which is typically the case in rock scour engineering. Advantages of Monte Carlo simulations include the following:

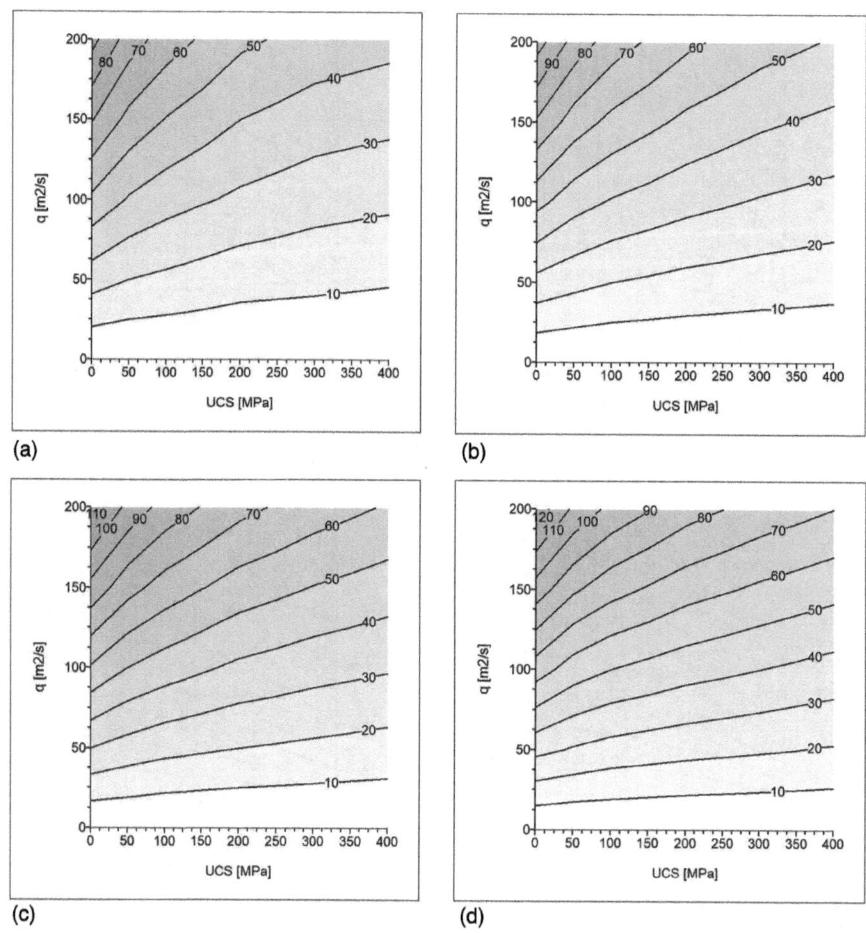

FIGURE 7.48 Scour potential [m] by rock joint fracturing (CFM) for compact jets, $z_b = 1.0$ m and $H = 200$ m: (a) after 1 day, (b) after 10 days, (c) after 100 days, (d) after 1,000 days.

- Improve decision-making. Monte Carlo simulations allow to make decisions with a chosen level of confidence.
- Solve complex problems simply. Monte Carlo simulations show how likely each outcome is based on a vast range of input parameters.
- Visualize the range of possible outcomes and their likelihood of occurring. Monte Carlo simulations make it easy to visualize what a chosen outcome means compared to the family of outcomes.

Probabilistic Parameters

Rock scour involves a large number of input parameters for both the flow environment and the rock mass. In general, most of these parameters contain a certain degree of uncertainty. Some are better known than others, based on experience and/or detailed measurements or test campaigns, and some have more influence on the

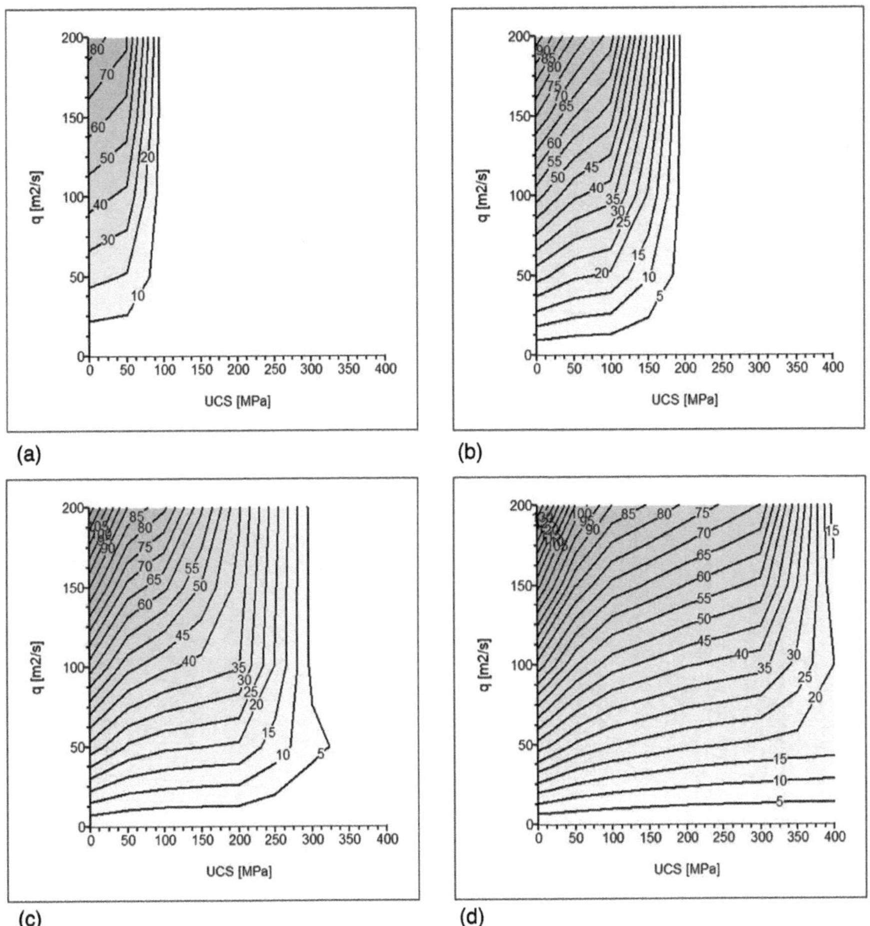

FIGURE 7.49 Scour potential [m] by rock joint fracturing (CFM) for compact jets, $z_b = 0.2$ m and $H = 50$ m: (a) after 1 day, (b) after 10 days, (c) after 100 days, (d) after 1,000 days.

end result than others, based on the physics involved. Generally, the variability of the input parameters is defined based on the available data and/or by simple engineering judgement and analysis.

Furthermore, because of the complexity inherent to rock scour, it is not obvious for the practising engineer to determine the relevance of each parameter. As such, a tool for probabilistic-based scour modelling, allowing to incorporate the uncertainties of these parameters, is of relevance.

Probability Density Functions

Different cumulative density functions may be appropriate for the input parameters, among which the Normal or Gaussian distribution often represents a plausible option. Nevertheless, other functions may be more appropriate depending on the parameters in question.

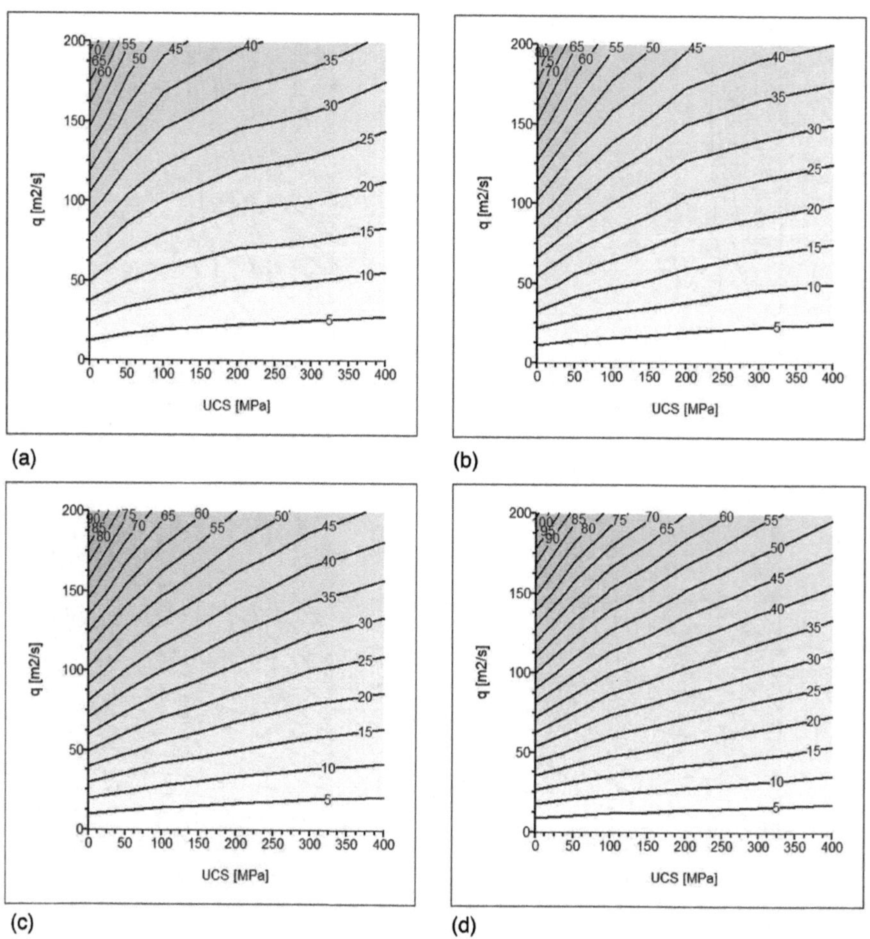

FIGURE 7.50 Scour potential [m] by rock joint fracturing (CFM) for compact jets, $z_b = 0.2$ m and $H = 200$ m: (a) after 1 day, (b) after 10 days, (c) after 100 days, (d) after 1,000 days.

As an example, George (2015) made use of terrestrial LiDAR scan data to define a Beta function for dip angles, dip directions, friction angles and dilation angles, and a Log-Normal function for block protrusions. Závacký et al. (2017) make use of a Log-Normal distribution for the UCS strength of the rock mass, while Yan et al. (2010) rather propose a Normal distribution for the same parameter. Following Debecker et al. (2010), joint friction angles may take different distributions, according to the literature. Hsu and Nelson (2006) assume a Normal distribution for the friction angle of a fracture in shale. Genske and Walz (1991) describe the friction angle of sedimentary rock as Log-Normally distributed.

As such, the most relevant distributions should be studied on a case-by-case basis, based on relevant statistical data from field campaigns.

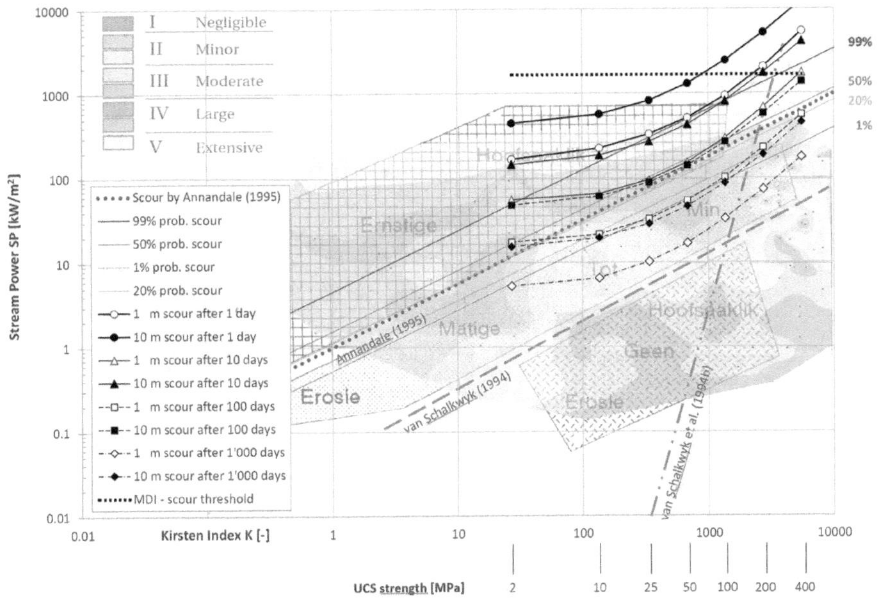

FIGURE 7.51 Scour potential [m] for compact jets, $z_b = 1.0$ m. Comparison with erodibility index scour thresholds from literature (for lowly fractured weak rock).

Correlations between Parameters

Monte Carlo simulations often make use of multiple random parameters that are correlated with each other, requesting to perform a multivariate stochastic analysis making use of a matrix containing the correlations between the parameters. Different mathematical techniques are available to determine the random variables such that their correlation is accounted for in a way that their marginal probability distribution is still respected. Examples are use of a Cholesky decomposition into the product of a lower triangular matrix and its conjugate transpose to determine the covariance matrix R, or defining the eigenvalues and eigenvectors of this matrix R through orthogonal transformation.

Furthermore, in case some of the random variables are non-normally distributed, their correlations are implemented by making use of specific copulas or by a set of Nataf empirical equations. For the latter case, the correlations must be transformed based on a set of semi-empirical formulas derived by Der Kiureghian and Liu (1985) and using the Nataf bivariate distribution model. This set of formulas transforms the correlation coefficient of a pair of non-normal random variables to its equivalent correlation coefficient in a bivariate standard normal space. Through this transformation, multivariate Monte Carlo simulation can thus be performed in a correlated standard normal space.

Correlations between variables may be positive or negative. Implementation of correlations may potentially influence the outcome of the simulations and both increase or decrease the risk of scour formation for a given set of random values.

Figure 7.57 summarizes potential correlations that may reasonably exist between a set of flow and rock mass parameters of relevance to rock scour modelling. The existence of correlations is thereby based on both feedback from literature and engineering judgement.

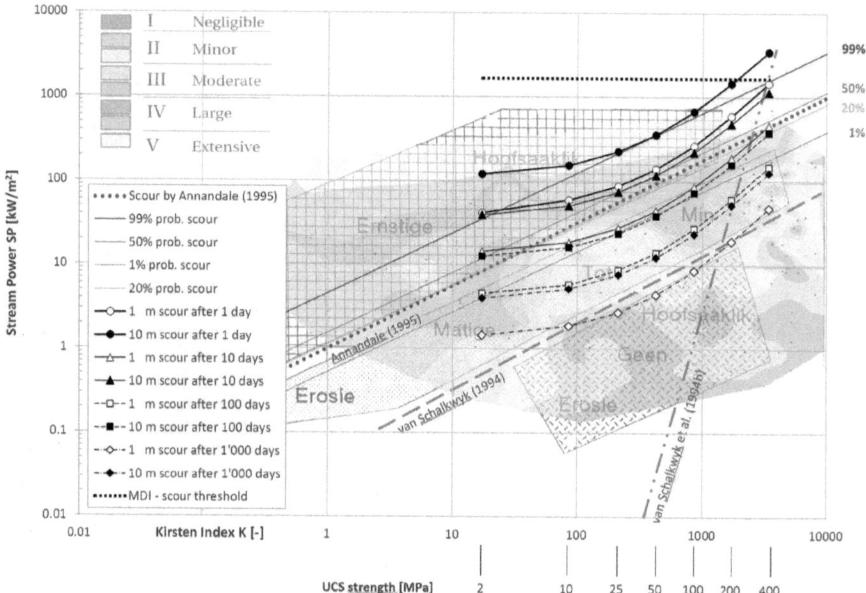

FIGURE 7.52 Scour potential [m] for compact jets, $z_b = 1.0$ m. Comparison with erodibility index scour thresholds from literature (for moderately fractured weak rock).

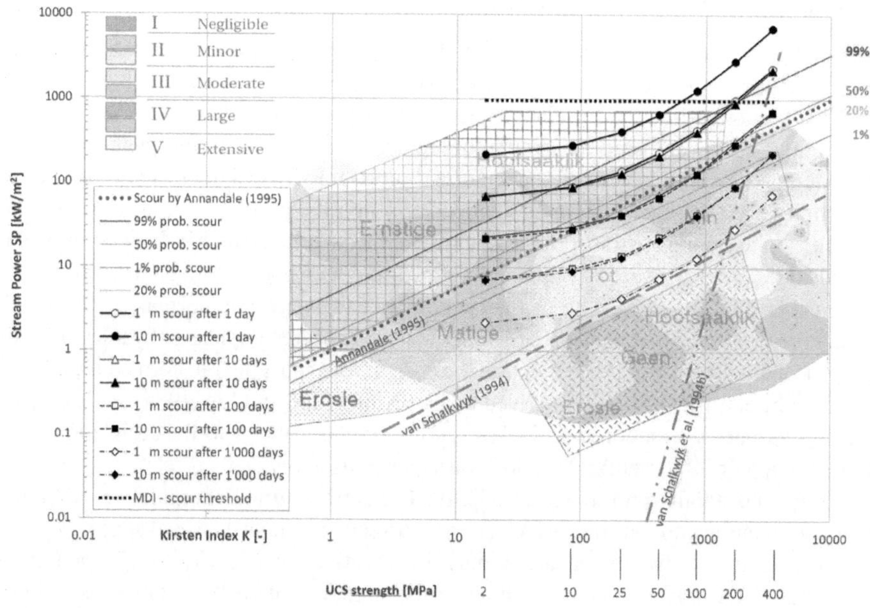

FIGURE 7.53 Scour potential [m] for fully broken jets, $z_b = 1.0$ m. Comparison with erodibility index scour thresholds from literature (for moderately fractured weak rock).

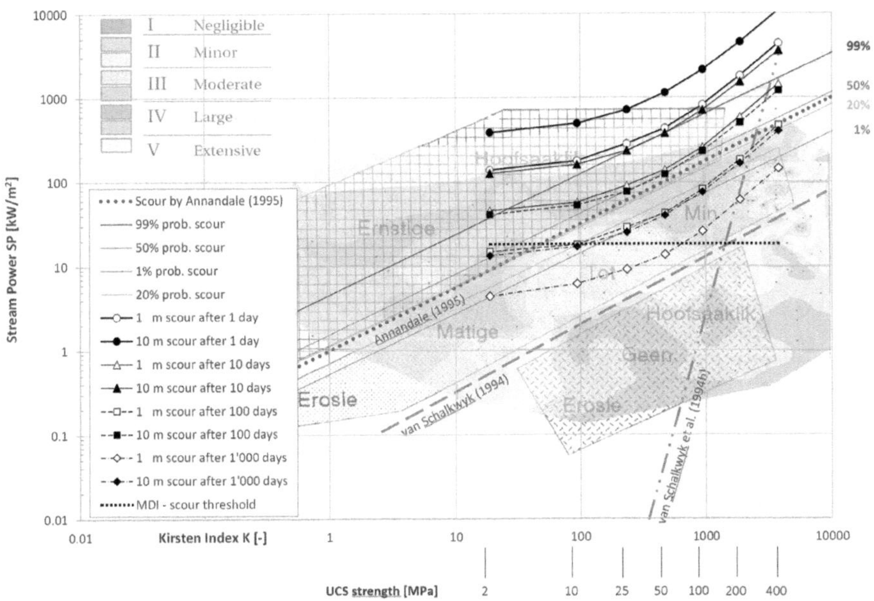

FIGURE 7.54 Scour potential [m] for compact jets, $z_b = 0.2\,\text{m}$. Comparison with erodibility index scour thresholds from literature (for moderately fractured weak rock).

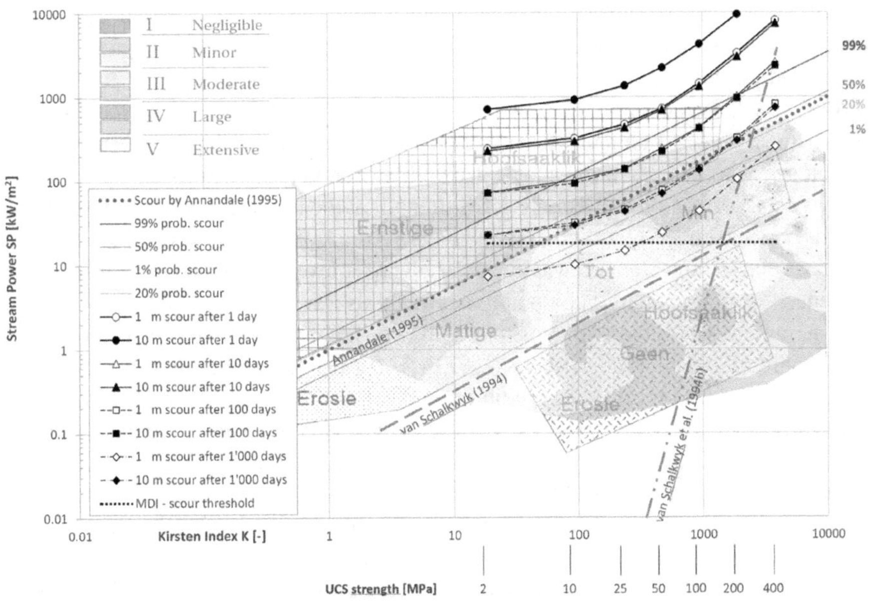

FIGURE 7.55 Scour potential [m] for fully broken jets, $z_b = 0.2\,\text{m}$. Comparison with erodibility index scour thresholds from literature (for moderately fractured weak rock).

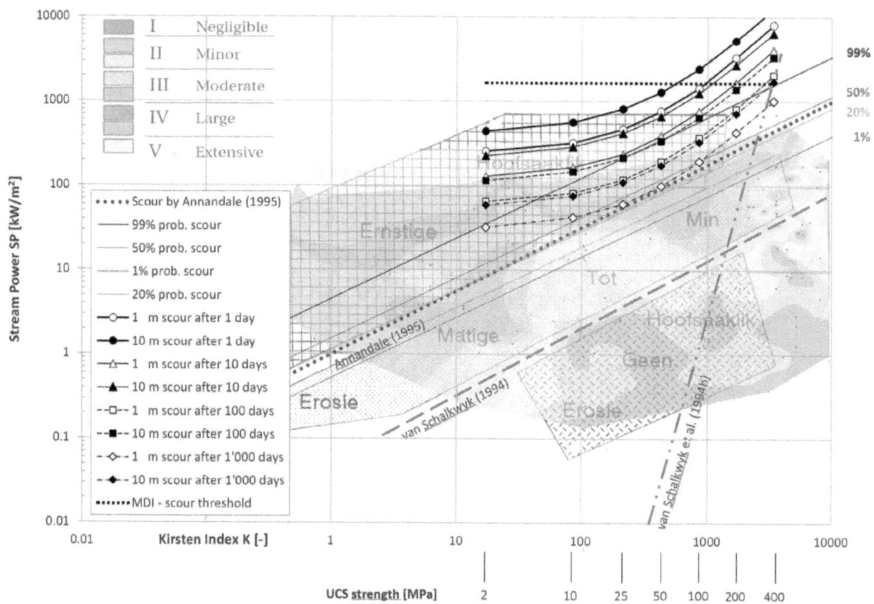

FIGURE 7.56 Scour potential [m] for compact jets, $z_b = 1.0\,\text{m}$. Comparison with erodibility index scour thresholds from literature (for moderately fractured strong competent rock).

The area in dark grey shows that correlations may exist between variables related to geological (field) mapping, such as between dip angles and dip orientations of joint sets, or also between friction angles of different joint sets or between joint friction angles and joint dilation angles and apertures of the same joint set.

Moreover, correlations may exist between the UCS strength and the rock density (Burkhardt et al. 2018), although this is not always the case. As the density is generally not a parameter with a lot of variation, it is generally excluded from the probabilistic simulations. Also, a correlation between the UCS strength of the intact rock mass and its persistency is possible but will most often result in the subdivision of the lithology into different resistance classes of rock, each with their own degree of weathering and persistency.

Hence, for physics-based computational methods (i.e. DP, MDI, MQSI, CFM), the main parameters of interest may generally be considered exempt of correlations. For semi-empirical index-based methods (i.e. EIM), a priori the same holds when using the main parameters composing the erodibility index (i.e. M_s, K_b, K_d, J_s, see Chapter 6).

Nevertheless, correlations may exist between the different geologic parameters determined by field mapping campaigns (George et al. 2014; George 2015), for example between the joint roughness and joint alteration parameters J_r and J_a (including joint aperture influence), between the dip angle and the dip direction, between the joint friction angle and the joint dilation angle, or also between the RQD value and the number of joint sets J_n. As such, when using these base parameters as input, determined for different joint sets separately, correlations may exist and should be accounted for during the simulations.

As a summary, Figure 7.57 illustrates that correlations are estimated non-existent between flow and rock mass parameters, and are estimated very low to quasi-non-existent between the different rock mass parameters used by rock mass break-up methods related to fracturing and block quarrying or plucking, such as the UCS

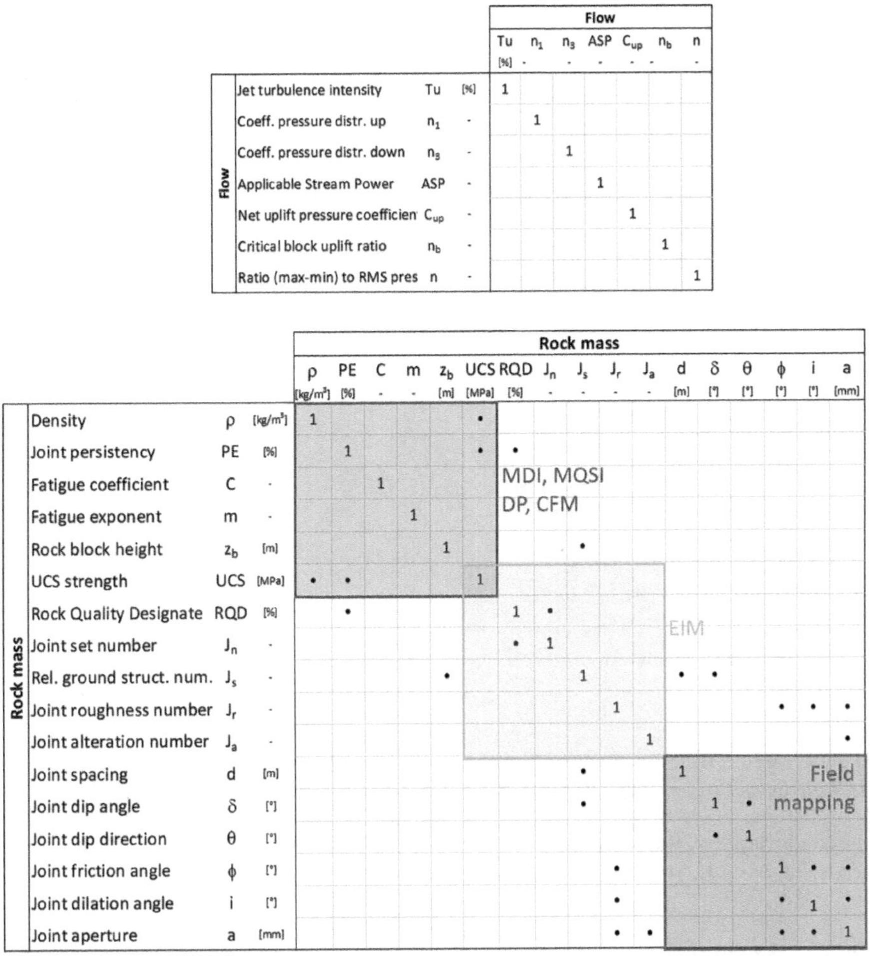

				Flow						
				Tu	n_1	n_3	ASP	C_{up}	n_b	n
				[%]	-	-	-	-	-	-
Flow	Jet turbulence intensity	Tu	[%]	1						
	Coeff. pressure distr. up	n_1	-		1					
	Coeff. pressure distr. down	n_3	-			1				
	Applicable Stream Power	ASP	-				1			
	Net uplift pressure coefficien	C_{up}	-					1		
	Critical block uplift ratio	n_b	-						1	
	Ratio (max-min) to RMS pres	n	-							1

				Rock mass																
				ρ	PE	C	m	z_b	UCS	RQD	J_n	J_s	J_r	J_a	d	δ	θ	ϕ	i	a
				[kg/m³]	[%]	-	-	[m]	[MPa]	[%]	-	-	-	-	[m]	[°]	[°]	[°]	[°]	[mm]
Rock mass	Density	ρ	[kg/m³]	1					•											
	Joint persistency	PE	[%]		1				•	•										
	Fatigue coefficient	C	-			1														
	Fatigue exponent	m	-				1													
	Rock block height	z_b	[m]					1				•								
	UCS strength	UCS	[MPa]	•	•				1											
	Rock Quality Designate	RQD	[%]		•					1	•									
	Joint set number	J_n	-							•	1									
	Rel. ground struct. num.	J_s	-					•				1			•	•				
	Joint roughness number	J_r	-										1					•	•	•
	Joint alteration number	J_a	-											1						•
	Joint spacing	d	[m]									•			1					
	Joint dip angle	δ	[°]									•				1	•			
	Joint dip direction	θ	[°]													•	1			
	Joint friction angle	ϕ	[°]										•					1	•	•
	Joint dilation angle	i	[°]										•					•	1	•
	Joint aperture	a	[mm]										•	•				•	•	1

MDI, MQSI
DP, CFM
EIM
Field mapping

FIGURE 7.57 Correlation estimate between flow parameters and between rock mass parameters (=correlation exists; 1=full correlation). Parameters used by physics-based methods are indicated in the upper dark grey area, erodibility index-based parameters are in the central light grey area.

strength, the joint persistency, the fatigue coefficient and sensitivity, the average block height and the block protrusion.

Furthermore, correlations are considered insignificant between the different turbulent flow parameters, such as the initial turbulence intensity, the flow depth and width, or the upstream and downstream radial pressure distributions.

Correlations based on geological mapping are often related to a certain spatial distribution of the rock mass quality with depth or with distance from the structure. Typical examples are lowly or highly fractured areas alternating with depth in a given rock mass, corresponding to higher respectively lower UCS strengths.

Under such circumstances, for digital rock scour computations, it is more accurate and convenient to subdivide the rock mass model with depth into different lithologies rather than to include the correlation into the probabilistic simulations. Within each

Please choose the parameters to randomized and their configuration with mean and standard deviation.

General

Execution number [-] 100

Number of run of computation to execute

[Rock lithology]

		Enabled	Mean	Stdev
Mass strength (UCS)	[MPa]	☑	20	2
Joint persistency (PE)	[%]	☑	0.60	0.05
Fatigue coefficient (C)	[-]	☑	0.00003	0.000003
Fatigue sensibility (m)	[-]	☑	3	0.3
Block height (zb)	[m]	☐	0	0

Flow

		Enabled	Mean	Stdev
Variant 2: Upstream dynamic pressures n1 (n1)	[-]	☑	0.25	0.02
Variant 2: Downstream dynamic pressures n3 (n3)	[-]	☑	0.17	0.02

FIGURE 7.58 User interface of digital application for probabilistic rock scour analysis, showing the statistical determination of different non-correlated input parameters.

specific lithology, the remaining correlations then generally become non-relevant or non-detectable by the available site data.

Digital Application

A digital application for 2D reliability-based rock scour modelling is presented. It allows to implement the main parameters of interest based on different assumptions regarding the probability density function of each of the parameters, among which are the Gaussian, Beta, Gamma and Log-Normal distributions. Correlations can optionally be accounted for based on the Cholesky decomposition of a correlation matrix provided by the user. In the following, for simplicity and based on Figure 7.57, correlations between input parameters are disregarded. Figure 7.58 illustrates a user interface to statistically define input parameters that are considered relevant to the analysis. These input parameters can be subdivided into three main groups:

- *Rock mass parameters*: parameters related to the geomechanical characteristics of the rock mass, such as the UCS strength, the initial degree of fracturing, the rock block height and so on.
- *Flow parameters*: parameters related to the incoming turbulent flow, such as the initial turbulence intensity, the lateral distribution of dynamic pressures over the water-rock interface and so on.

– *Break-up parameters*: parameters that express the ease of break-up of the rock mass, such as the critical block uplift height, the net uplift pressure coefficient and so on.

Probabilistic Outcome

Second, a number of runs has to be chosen. This number should allow the outcome of the modelling to be correctly represented by a mean value and a standard deviation based on the assumption of a Gaussian density function. The number of runs that is needed to obtain statistically relevant scour values directly depends on the allowable error on the scour depth for a given level of confidence.

Suppose we have a normally distributed random variable x (i.e. scour depth computed at a given location) where the population mean (i.e. in-situ real scour depth) is μ_x and the population variance σ_x^2, and a sample size n is drawn from the population where the sample mean is $\overline{x}$ and the sample variance S_x^2. If this is done a large number of times, one obtains a distribution of $\overline{x}$ and S_x^2 with the following characteristics:

- the expected mean of $\overline{x}$ is μ_x the population mean
- the expected variance of $\overline{x}$ may be expressed in terms of the variance of x, and is given by:

$$\sigma_{\overline{x}}^2 = \frac{\sigma_x^2}{n} \tag{7.2}$$

in which n stands for the number of runs of the sample. The corresponding lower and upper confidence limits and levels for the mean are then written as:

$$(L, U)_{CI} = \overline{x} \pm z_c \cdot \frac{S_x}{\sqrt{n}} \tag{7.3}$$

in which CI=confidence interval and z_c stands for the number of standard deviations from the mean, and the sample mean $\overline{x}$ and standard deviation S_x are used rather than the equivalent real values for the total population, because the latter are not known precisely.

For example, for CI=95%, there's 95% confidence that a sample (i.e. scour result) will be within the (L, U) range of the true mean value (i.e. measured scour depth), provided that a large enough number of runs has been performed such that the sample statistics sufficiently approaches the population statistics.

Furthermore, an error estimate may be made by considering the confidence interval to represent twice the maximum error. The percentage of error can then be written as:

$$\text{error} = \frac{100 z_c S_x}{\overline{x}\sqrt{n}} \tag{7.4}$$

Transforming this equation as a function of "n" allows to determine the number of runs needed to be CI% confident that the computed scour depth will not differ by more than error% from the true scour depth. This equation yields:

$$n = \left[\frac{100 z_c S_x}{\overline{x} \cdot \text{error}}\right]^2 \tag{7.5}$$

For practical rock scour computations, the error can also be expressed in [m] instead of in [%]. Also, the error estimate and related necessary number of runs for a given confidence interval will be different for each location along the water-rock interface.

The ideal situation would be to consider an iterative Monte Carlo simulation that has no upper bound to the number of runs to be performed. As the runs proceed, the scour depth estimates accumulate into a sample of increasing size. As more runs take place the sample approaches the population. The estimates of the sample statistics approach that of the population. Thus, the sample mean and variance may be determined after each new run and used in the above equation to determine how many iterations are needed to achieve a specified maximum percentage error with a specified confidence level. This number of iterations converges quickly and even for sample sizes an order of magnitude lower than the number required, the calculation of that number is quite stable.

Case Study: Stellenbosch Experiments

A first case study is provided for the free overfall jet of the Stellenbosch laboratory experiment (test 9A). For this example, a total of 450 runs of 2D scour computations has been performed during the Monte Carlo simulation. The Erodibility Index Method has been used (EIM) to compute rock scour potential. The following non-correlated normally distributed input parameters have been implemented by a mean and standard deviation to account for their uncertainty (Table 7.8):

Figure 7.59 presents the stochastically defined values for the UCS strength and the jet turbulence intensity. Figure 7.60 illustrates the results for 100 of the 450 computed 2D scour hole curves (dark grey lines), together with the median, maximum and minimum scour depths obtained for each grid point (black lines).

TABLE 7.8
Statistical Values Used for Input Parameters (450 Runs)

Parameter	Mean Value	Standard Deviation
UCS strength [MPa]	20	3
Turbulence intensity [%]	3	0.3
Pressure coefficient $\boldsymbol{n}_1$ [–]	0.20	0.02
Pressure coefficient $\boldsymbol{n}_3$ [–]	0.10	0.02

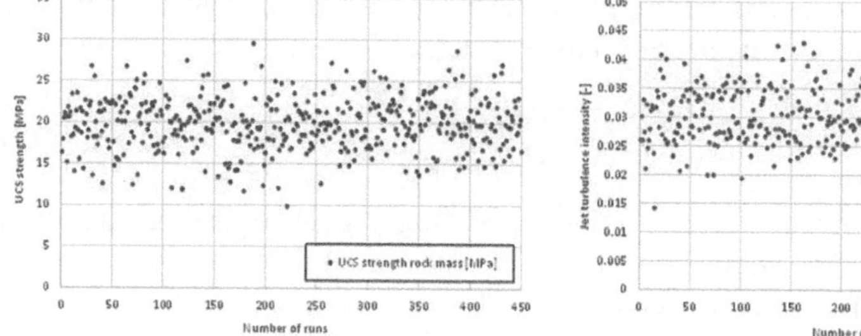

FIGURE 7.59 Computed values for UCS strength and turbulence intensity (450 runs).

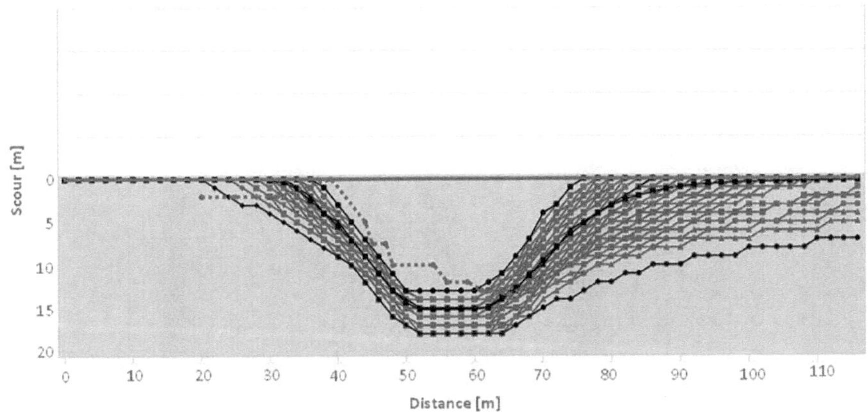

FIGURE 7.60 Computed 2D scour holes using probabilistic-based EIM (450 runs).

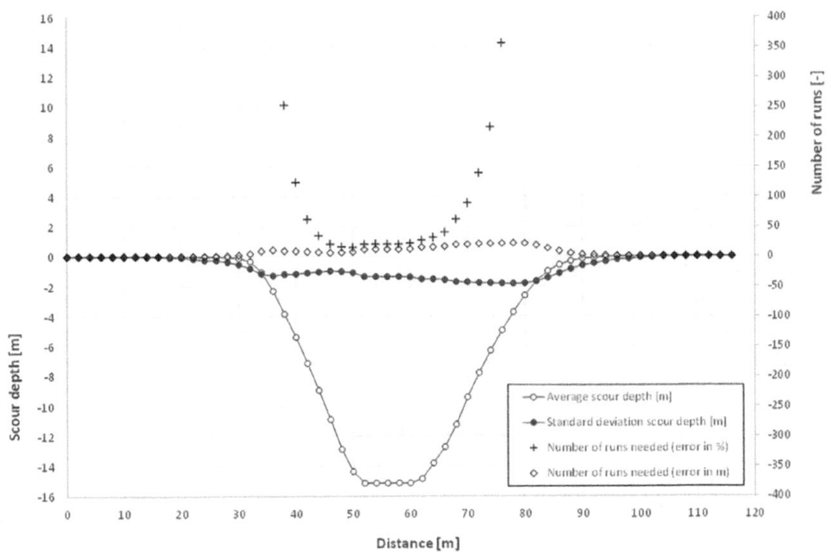

FIGURE 7.61 Mean and standard deviation of computed scour depths – Number of runs needed for a 99% confidence interval and 5% error estimate, using probabilistic-based EIM (450 runs).

For a confidence level chosen at 99% and a maximum allowable error estimate of 5% of the measured scour depth, or a maximum allowable error estimate of 1 m deviation from the measured scour depth, the necessary number of runs is presented in Figure 7.61 (grey dots and grey plus signs).

This number depends on the measured scour depth and thus is different at each computational node of the numerical grid. For the deepest part of the scour hole, the number of runs needed is around 20–30. For the sides of the scour hole, however, the number of runs needed depends on the way the error is computed and may increase up to a few hundreds.

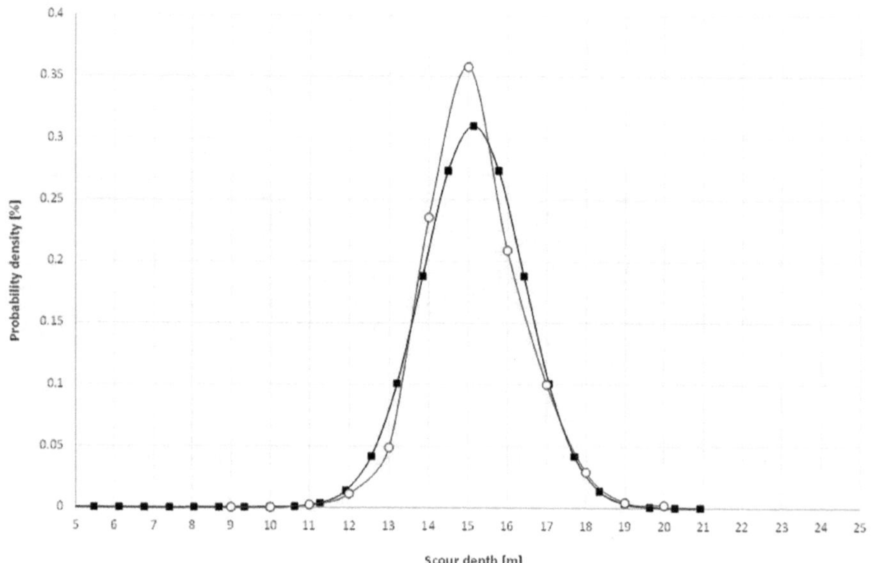

FIGURE 7.62 Probability density function of computed scour depths at deepest scour – comparison with Gaussian distribution, using probabilistic-based EIM (450 runs).

Furthermore, the standard deviation is presented at each computational node of the grid and mounts up to 1–2 m depending on the location, for a mean scour depth between 0 and max. 15 m.

Finally, Figure 7.62 illustrates the obtained probability density function of the scour curves at the deepest point of scour (grey curve with circular dots), and compares with the Gaussian distribution (black curve). Good agreement is observed.

Case Study: Chucás Dam 2017 Flood Event

Site and Project Features

The Chucás hydroelectric project, owned by Enel Costa Rica, was developed as a BOT project (Capuozzo and Jiménez 2017). It is located about 40 km west of San José, the country's capital. The project, in operation since November 2016, dams the Tárcoles river with a 63-m-high dam. A surface penstock, about 400 m long and 6.5 m in diameter, leads to a powerhouse where two Francis units generate a total output of 50 MW (Figure 7.63).

For flood control, the dam is equipped with four radial gates, 15 m by 12.4 m, capable of discharging 5,400 m^3/s under design conditions (Capuozzo and Jimenez 2017). This large discharge is conveyed through a ski-jump spillway chute discharging into the rock bed downstream of the dam.

Scour formation in this unprotected rock mass was expected and a pre-excavation was planned but never executed. The corresponding plunge pool formation, in this

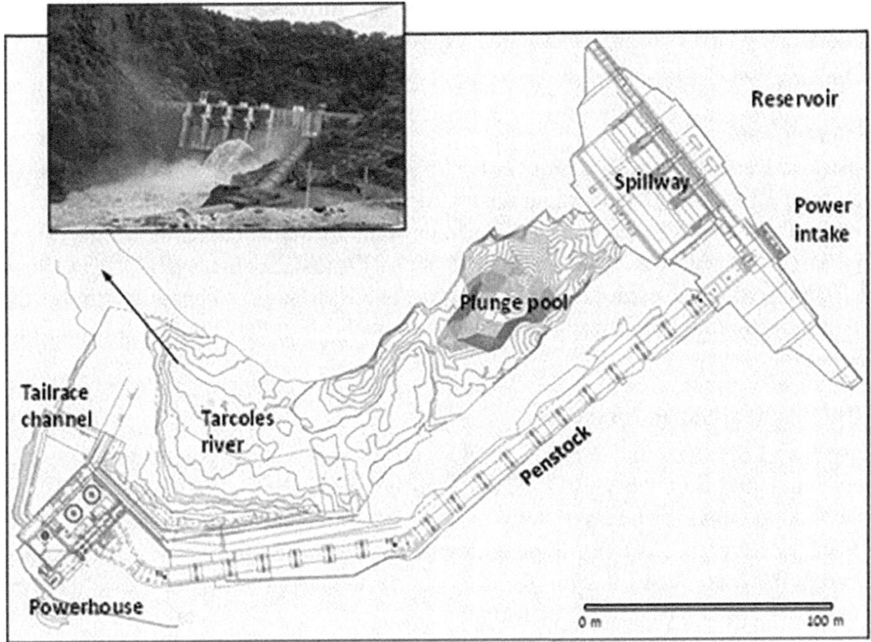

FIGURE 7.63 General layout of Chucás HEP. Courtesy of M. O. Jiménez.

particular case, may endanger the exposed penstock, because of the risk of lateral expansion of the scour hole, which might affect this structure. In addition, there is the risk that backwards erosion of the plunge pool may affect the dam foundations.

Geology

Based on Capuozzo and Jiménez (2017), the in-situ rock mass consists of volcanic rocks such as andesitic lava, ignimbrite pyroclastic flows and occasionally thin layers of tuff and ash deposits. Good outcrops are observed on both banks of the river; however, the bedrock cannot be observed as a result of the constant flow of the river. The rock mass consists of aphanitic (fine grain) and aphanitic porphyritic (coarse grain) andesitic lava flow. The aphanitic mass is hard and highly fissured, and has good weathering resistance. The aphanitic porphyritic contains abundant calcite giving a brecciated textured appearance (called the "coarse grain"); it is soft to moderately hard, moderately fissured andesite, susceptible to weathering. The rock mass has been affected by a hydrothermal process, which has made it very susceptible to accelerated weathering when exposed to the environmental conditions of the area (rain and sun).

Local site investigation with boreholes indicates a variable rock quality designation (RQD); however, at the left abutment the rock is more fissured. The RQD percentage at the left abutment ranges from 0% to 76% with an overall average of 40%, which classifies the rock mass as poor quality.

The joint spacing produces flat-shaped rock blocks of only 0.04–0.5 m side length. These results would indicate that a plunge pool formation is very likely to occur relatively rapidly. As the smaller range of values correspond to gravel and cobbles generated by ball-milling, a block side length of 0.5 m has been used, together with a height-to-side length ratio of 1:2.

Design Flood

Based on Capuozzo and Jiménez (2017), the flood discharges were calculated based on about 50 years of data from an upstream hydrometric station, located a few kilometres upstream of the project site. Based on standard statistical methods, the design discharge (1:1,000 years) was estimated at 5,400 m^3/s and the safety check flood (1:10,000 years) was estimated at 8,100 m^3/s. The hydrological characteristics of the basin indicate that floods normally have a relatively short duration, in the order of a few hours.

2017 Tropical Storm Incident

Based on Capuozzo and Jiménez (2017), the tropical storm Nate hit Costa Rica between 4 and 6 October 2017. The storm remained almost stationary for 2 days, reaching precipitation levels of between 200 and 400 mm over the whole basin of the Chucás powerplant. As a result, peak flows of more than 3,000 m^3/s were discharged through the radial gates, lasting for about 24 h. According to the records, this flood has a return period of more than 100 years.

Although the powerplant suffered relatively little damage, and was back in operation after 2 days, it was clear that important scour downstream of the dam had taken place because of the natural formation of the plunge pool. Bathymetric measurements showed that a plunge pool had been formed, with a maximum depth of 25 m, larger than the 20 m initially assumed based on hydraulic model tests. Figure 2.6 in Chapter 2 shows the jet trajectory and the scour damage observed following the 2017 flood event. Fortunately, the observed damage and scour did not pose threats to the existing structures. It was estimated that about 15,000 m^3 of material was removed and deposited downstream, as shown in Figure 7.64 (right-hand side).

FIGURE 7.64 Jet trajectory and rock deposits downstream of the plunge pool of Chucás HEP. Courtesy of M. O. Jiménez.

Numerical Reproduction of Observed Scour Formation

The scour formed during tropical storm Nate has been numerically reproduced. The maximum observed discharge of 3,000 m^3/s has been used for the scour computations. The computations are based on probabilistic-based modelling of progressive fracturing of rock joints (CFM method).

A total of 100 runs has been performed by the Monte Carlo simulation. The input parameters that have been stochastically defined at each run are the UCS strength, the initial joint persistency PE, the fatigue exponent m_r and fatigue coefficient C_r, and finally the dynamic pressure coefficients towards upstream and downstream (n_1 and n_3).

Figure 7.65 presents the results. The in-situ observed scour hole is presented by the dark grey dotted curve, while the light grey curves represent the different runs performed. The black curves correspond to the minimum, average and maximum computed values of scour formation.

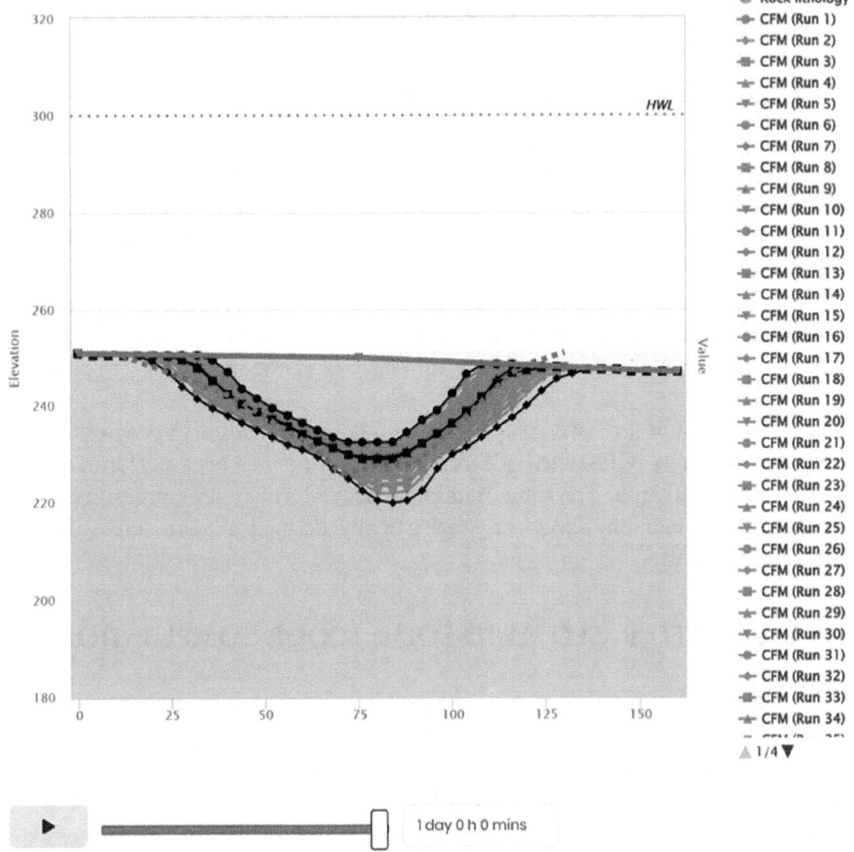

FIGURE 7.65 Comparison of in-situ observed scour and digital computed probabilistic scour of the 2017 flood event at Chucás Dam.

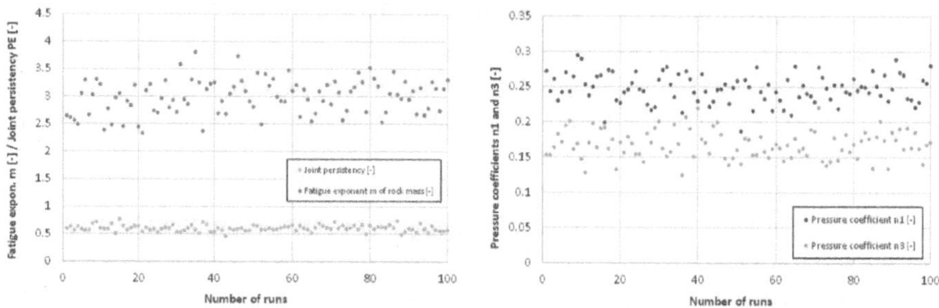

FIGURE 7.66 Computed values for fatigue parameters and pressure distributions (100 runs).

This method allows to obtain probabilistic scour as a function of time period of discharge. By considering a total time period of 24 h, it may be observed how the computed scour is on average in good agreement with the observed scour for this particular event.

Figure 7.66 illustrates the values stochastically defined for the fatigue parameters and for the pressure distribution coefficients over the water-rock interface.

Case Study: Bed-Parallel Turbulent Flow with Head-Cut Migration

This example deals with a fictitious case of a bed-parallel high-velocity turbulent flow, as is typically encountered in river channels or downstream of bottom outlets. The flow velocity is 22.3 m/s, for a 6 m flow depth. The water-rock interface is protected by a concrete lining up to a distance of $X=35$ m, followed by unlined bedrock further downstream.

The example points out the potential influence of head-cut migration of the scour hole towards upstream, by progressive undermining and destruction of the concrete lining. Computations with and without influence of scour regression are compared in Figure 7.67.

The computations are based on probabilistic-based modelling of progressive fracturing of rock joints (CFM method). A total of 20 runs has been performed by the Monte Carlo simulation. The time duration of scour formation on prototype is 12 h. The input parameter that has been stochastically defined at each run is the UCS strength of the bedrock.

ADVANCED CFD-BASED FLUID-SOLID SCOUR COMPUTATIONS

Introduction

The scour computations in the rocsc@r digital environment make use of simplified fluid-solid coupling by predefined 2D matrices that describe the hydrodynamic parameters. These matrices are strictly only valid for flat water-rock interfaces and are not updatable during the computations. Moreover, they are currently only available for plunging jet flows, restricting the number of flow situations that can be handled by the coupling.

A more accurate fluid-solid coupling asks for an automatic update of the hydrodynamics of the turbulent flow as a function of scour progression. This may be obtained by coupling the scour model with a CFD model during the computations.

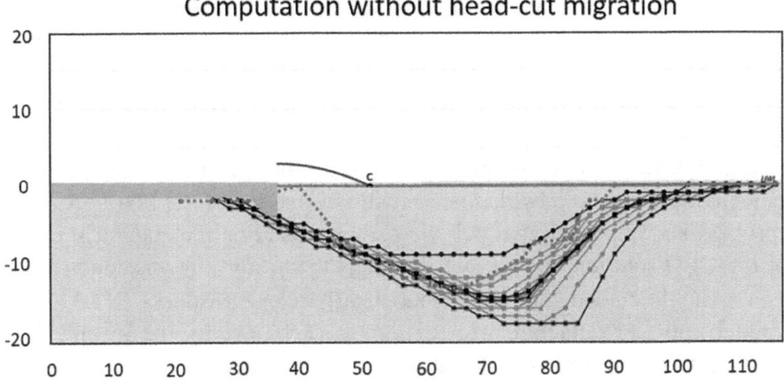

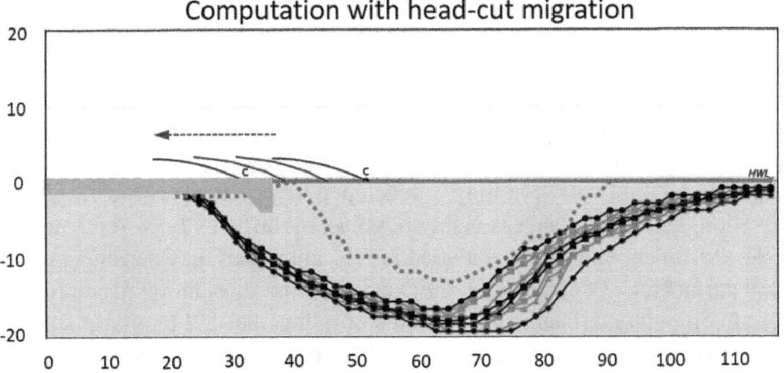

FIGURE 7.67 Influence of head-cut migration on probabilistic-based scour computations in unlined rock downstream of a concrete lining.

In the following, such an advanced fluid-solid coupling is presented between the rocsc@r digital environment and the FLOW-3D® CFD software. As discussed in Chapter 4, the connection is performed by means of the RemoteSc@r API (Application Programming Interface) and applies the following methodology:

- CFD hydraulic simulation using steady-state boundary conditions, during a short time frame, necessary to reach statistically meaningful values of hydraulic parameters at the water-rock interface (average values, RMS values)
- recording and transfer of hydraulic parameters to rocsc@r by use of a generic exchange file (csv format)
- scour computation in rocsc@r based on these hydraulic parameters, until a user-defined scour depth (layer height) or time duration of scour is reached (i.e. geomechanical time step)
- transfer of scoured water-rock interface to CFD by use of STL file
- update of CFD hydraulic simulation based on the previous result (restart)
- cyclic repetition until no further scour occurs anymore between the iterations

To model fracturing of the rock mass into distinct blocks, the fluid-solid coupling is governed by the geomechanical time scale. This time scale typically results in periods of quasi steady-state turbulent flow conditions during one or several hours on prototype, separated by sudden non-negligible modifications of the water-rock interface once a block is fractured and expelled.

The update of the CFD-computed hydraulics is thus sequential and governed by the geomechanical timestep, which is typically of one to several hours. As the fracturing process is computed separately and solely based on the statistical characteristics of the CFD solution, the CFD only needs to run for a few seconds to tens of seconds, i.e. the time needed to obtain statistically relevant values.

To model uplift or detachment of distinct rock blocks by the instant dynamic action of the flow, the geomechanical action mathematically takes place (quasi-) instantaneously (i.e. zero time step), for a water-rock interface that is updated on a layer-per-layer basis. For layer heights equivalent to block heights, a hydraulic update is performed at each block movement. For layer heights that are larger than the block height, however, the hydraulic update needs several layers of blocks to be removed.

In any of these cases, the hydraulic time step of the computations is governed by the time needed to determine statistically relevant values for the hydrodynamic parameters, by assuming quasi steady-state flow conditions.

The presented fluid-solid coupling is developed for practical engineering purposes and thus should run as fast as reasonably possible, by sufficiently correctly modelling global flow turbulence such that statistically relevant values may be generated for the hydraulic parameters. Especially the CFD may be time-consuming depending on the number of computational cells and the time durations needed to obtain statistically relevant values. Hereafter, these essential CFD aspects have been checked, and computed results are discussed, for both 2D and 3D cases.

2D Fluid-Solid Coupling

A first series of applications deals with 2D CFD-based fluid-solid coupling. To minimize computational times, the CFD computations are performed in 2D, i.e. by using a lateral flow width (Y-axis) of only one cell, and the rock scour computations are performed along a 2D vertical profile in the X–Z plane.

Plunge Pool Scour by Free Overfall Jets

The free overfall jet of the Stellenbosch laboratory experiment (i.e. test 9A_a) described earlier on in this chapter has been numerically reproduced. This test corresponds to the maximum specific discharge, the maximum hydraulic head and the maximum tailwater level used. The block height is 1.0 m. No mounding was allowed during the test.

CFD Modelling Parameters

Modelling parameters are presented in Table 7.9. Both a LES (run 1, MDI method) and a RNG k–ε (run 2, EIM method) turbulence model have been used. For computational methods based on RMS values of dynamic pressure fluctuations, such as EIM, MDI or CFM methods, a mesh size of 0.5 m has been applied. This was found to be

TABLE 7.9
CFD Modelling Parameters Applied during 2D Fluid-Solid Coupling of Stellenbosch Laboratory Experiments

Parameter	Run 1 (MDI method)	Run 2 (EIM method)
Turbulence model	LES model	RNG k–ε model
Volume of fluid advection	Automatic (Split or Unsplit Lagrangian)	Automatic (Split or Unsplit Lagrangian)
Momentum advection	First-order approximation	Second-order mon. approx.
Time step control	Stability and convergence	Stability and convergence
Time duration per steady-state run	10 s	20 s
Mesh size	0.5 m (MDI)/1.0 m (MQSI)	0.5 m
Pressure solver	Implicit (GMRES)	Implicit (GMRES)
Air entrainment model	Yes, with flow bulking and buoyancy	Yes, with flow bulking and buoyancy
Air entrainment coefficient	0.50	0.50
Roughness height dam	0.01 m	0.01 m
Roughness height rock mass	1.0 m	1.0 m
Layer height iterations	1.0 m	1.0 m
Upstream approach channel	No	Yes

the largest mesh size that still adequately estimates RMS fluctuations. For computational methods based on average flow velocities at the water-rock interface, a larger mesh size of 1 m could generally be used.

Air entrainment was accounted for by the CFD model based on a dispersed multi-phase model, where the two phases are assessed by a one-mixture constitutive equation depending on the turbulent kinetic energy, gravity, surface tension and an air entrainment rate coefficient to be defined by the user. The entrainment model is fully coupled with the hydrodynamic solver and is complemented by a bubble transport model, allowing fluid bulking effects (variable density) and buoyancy (drift-flux model).

As the turbulent kinetic energy was found very low for the LES model, and no turbulence was artificially injected along the upstream boundary, air entrainment revealed to be quasi-insignificant using the LES model. For the RNG k–ε model, air entrainment rate used was in accordance with standard parametric settings of the CFD environment and found significant.

The roughness height of the water-rock interface was set at 1.0 m. Values of rock roughness heights for practice would be typically between 0.10 and 1.0 m, depending on the joint sets and in-situ rock block heights. A no-slip boundary condition was applied at the water-rock interface. The average wall velocity computed along the water-rock interface is governed by a logarithmic law-of-the-wall and assuming hydraulically rough turbulent flow.

TABLE 7.10
Hydraulic Boundary Conditions Applied during 2D Fluid-Solid Coupling of Stellenbosch Laboratory Experiments

Jet velocity at issuance	12.04 m/s
Jet thickness at issuance	4.0 m
Flow depth at issuance	20.0 m

Hydraulic Boundary Conditions

The initial and boundary conditions are summarized in Table 7.10 and have been taken from the laboratory model tests at Stellenbosch. Nevertheless, as the jet is only modelled in 2D, there is no lateral spread. As such, the initial jet thickness at issuance from the dam has been adapted from 6 to 4 m, such that the correct jet thickness is obtained at impact in the downstream plunge pool for the LES model, i.e. 1.75 m. For the RNG k–ε model, the corresponding jet thickness was slightly higher at 2.5 m. Furthermore, the upstream canal used during the laboratory model tests has not been reproduced during run 1, but was reproduced during run 2.

Jet Trajectory and Flow Velocities

Figure 7.68 presents the computed jet trajectory and flow velocities for both CFD runs. The jet impacts the plunge pool at $X = 53$ m, which is slightly more upstream than the value of $X = 62$ m as observed during the laboratory runs. When using 2D jet diffusion theory, for the computed jet trajectories to match with the observed ones, an additional pressure head equal to the flow depth had been added into the trajectory computations of rocsc@r. This is not the case in the current 2D CFD modelling.

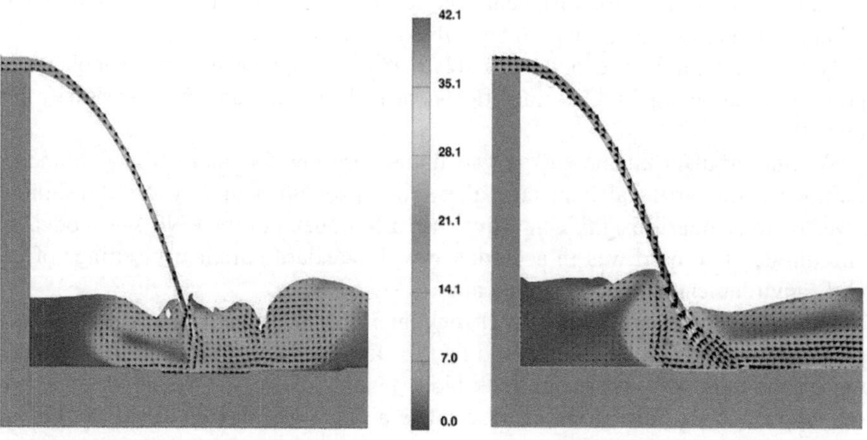

FIGURE 7.68 Jet trajectory and flow velocities computed by 2D LES model (left-hand graph) and 2D RNG k–ε model (right-hand graph).

The computed average flow velocity at impact is 42.1 m/s for run 1 and 40.5 m/s for run 2, i.e. close to the value of 42.4 m/s observed during the laboratory runs (Bosman 2021).

Jet Break-Up, Contraction and Deformation

As the CFD computations are only performed in two dimensions, the jet is not able to deform laterally. Furthermore, no initial turbulent kinetic energy is present. Hence, the jet contracts during its fall but does not change its squared-shape initial shape into a much wider rectangular-shaped jet upon impact as was observed during the laboratory tests. Also, the CFD generated jet is not broken up at impact, while the laboratory jets were described as highly broken-up (degree of break-up >1.85) by Bosman (2021).

Jet Diffusion through the Pool

The CFD computed jet develops through the pool depth as a rectangular jet, and not as a circular jet like it was found during the laboratory model tests by Calitz (2015), or like it was modelled in rocsc@r using the 2D flow matrices. This means that the axial velocity decay will be lower in the CFD, and thus the corresponding jet velocities along the rock interface will be higher. In other words, the CFD computed jets are too compact and overestimate the flow velocities at the water-rock interface, by lack of aeration, turbulence and related jet break-up and velocity decay.

CFD Fluid-Solid Scour Results

Figure 7.69 illustrates several iterations of scour results obtained during the two 2D CFD runs. The results have been shifted towards downstream to match with the laboratory jets and scour hole formation. The laboratory-observed scour formation is presented by the thick dotted grey line.

For both runs, fair agreement is observed between the laboratory measurements and the computed scour. Not only the area of deepest scour formation is well modelled, but also the slopes of the scour hole towards upstream and downstream are adequately reproduced. Run 1 used the MDI method, simulated 10 s per CFD run, and converged in 35 iterations and 1 h 45 min on a standard INTEL i5 processor. Run 2 used the EIM method, simulated 20 s per CFD run, and converged in 30 iterations and 4 h 05 min on an AMD Ryzen 9 processor.

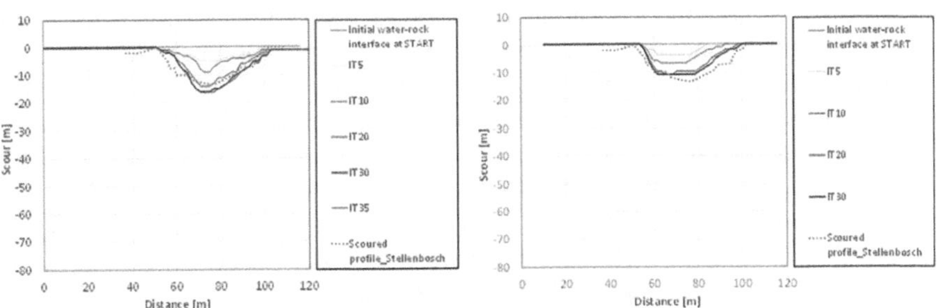

FIGURE 7.69 Scour computed using 2D fluid-solid coupling (left side: MDI method using LES model, right side: EIM method using RNG model).

For run 1, the critical net uplift height ratio n_b equalled 0.20, i.e. within the range of usual values based on experience from practice, but nevertheless higher than the value of 0.02 that was calibrated by using the internal 2D flow matrices.

For run 2, the applicable stream power coefficient equalled 0.10, i.e. a quite low value but nevertheless already observed on other case studies using the 2D jet diffusion theory.

Comparison with 2D Jet Theory

For the two CFD runs of Table 7.9, the RMS pressure fluctuations generated in the CFD are different from the values based on 2D flow matrices. This is logic because the former values are computed by CFD and for a compact jet, while the latter values are based on diffusion theory for a highly broken-up jet (i.e. which results in very low values).

A comparison is made in Figure 7.70. The CFD pressures follow a globally decreasing trend during the iterations. The average pressure coefficients start at about 0.35 and remain above 0.10–0.15. They are higher than the corresponding values based on jet diffusion theory. The RMS values start at 0.14 and a lower bound of 0.04 seems to be maintained by the CFD.

Furthermore, locally high values may be generated by the CFD at sudden changes of the bottom geometry. All in all, while average CFD pressures are somewhat high, the RMS values are in good agreement with the theoretical values valid for compact 2D jets (Figure 7.70).

In a next step, a general comparison has been made between the CFD solution and the internal 2D flow matrix solution based on 2D jet diffusion theory for a flat bottom. For this comparison, an additional CFD run has been performed with a perfectly flat bottom and for a tailwater depth (TWL) of 20 m. The LES turbulence model was used, with a first-order solution for the momentum advection, and a uniform mesh size of 0.5 m.

Different solutions have been generated by rocsc@r based on 2D jet diffusion theory: one generating a widely dispersed pressure curve as is generally observed on prototype scour holes (i.e. n_2 and n_4 values of 3), and a second and third one generating more concentrated pressure curves that may occur in deep and shallow scour holes such as Kariba Dam (i.e. n_2 and n_4 values of 1), for different initial turbulence intensities (TI). Furthermore, the computed pressures have been compared with laboratory measurements made by Wei et al. (2020) for similar obliquely impinging aerated jets.

First, the average dynamic pressures generated by the impacting jet at the water-rock interface are presented in Figure 7.71 (upper part). The CFD pressures agree fairly well with the concentrated pressure curves of 2D jet diffusion theory and with the laboratory measurements, although CFD remains slightly higher.

Second, the fluctuating part of the dynamic pressures generated by the impacting jet (root-mean-square or RMS values) at the water-rock interface is also presented and compared in Figure 7.71 (lower part). The CFD values agree well with the laboratory measurements by Wei et al. (2020) at both the point of jet impact and laterally outside of this area. The lateral decay of the CFD computed RMS values at first follows a very strong slope, in agreement with theoretical values for a concentrated dynamic pressure curve. Next, this lateral decay strongly diminishes and the RMS values remain almost constant with increasing lateral distance from the impact point.

This behaviour is in agreement with theoretical values for a strongly dispersed dynamic pressure curve, i.e. for which RMS values remain significant even at large distances.

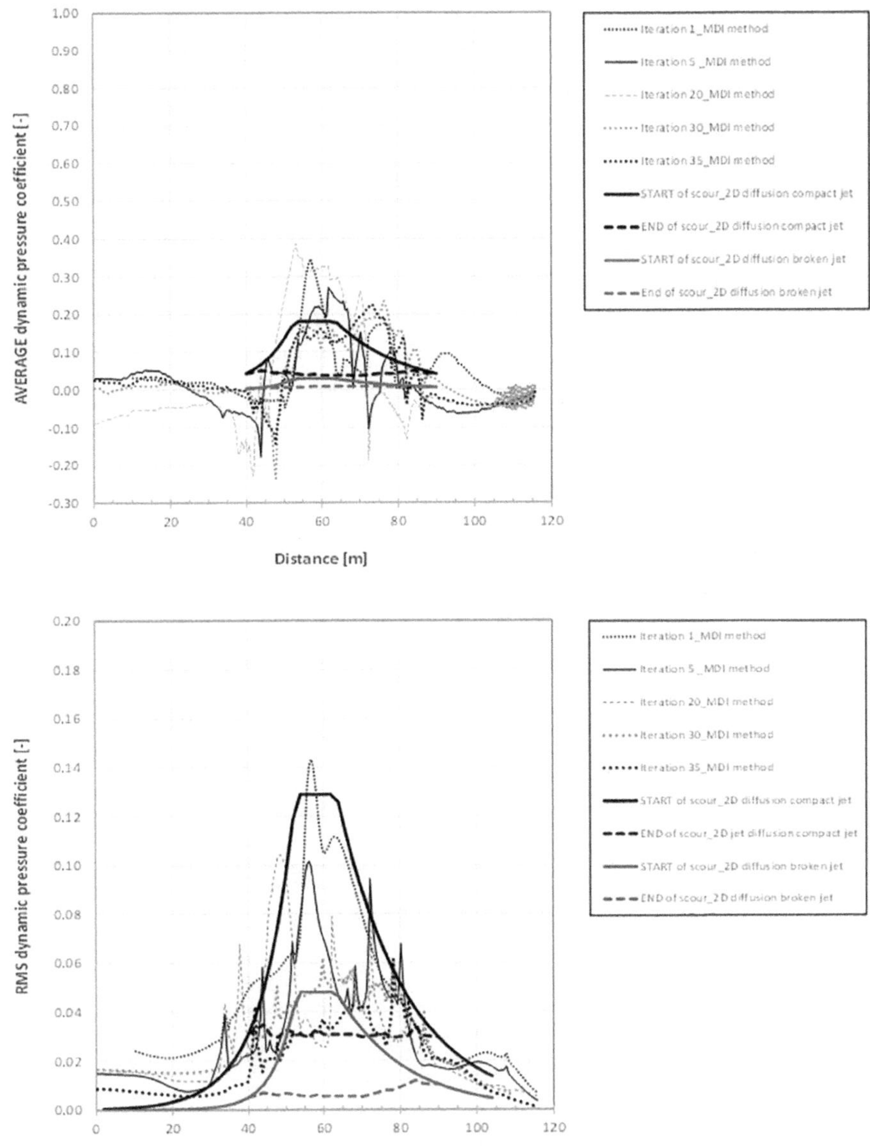

FIGURE 7.70 Evolution of average dynamic pressures (upper part) and RMS dynamic pressure fluctuations (lower part) during CFD computations. Comparison with values based on 2D jet diffusion theory, for compact and broken-up jets.

Next, Figure 7.72 compares the time-averaged wall jet velocities that develop along the water-rock interface following jet impingement. In general, the CFD computed curve has a shape similar to the theoretical curve for broken-up jets and a jet core extension coefficient (i.e. K) of 4 times the jet diameter at impact. However, in

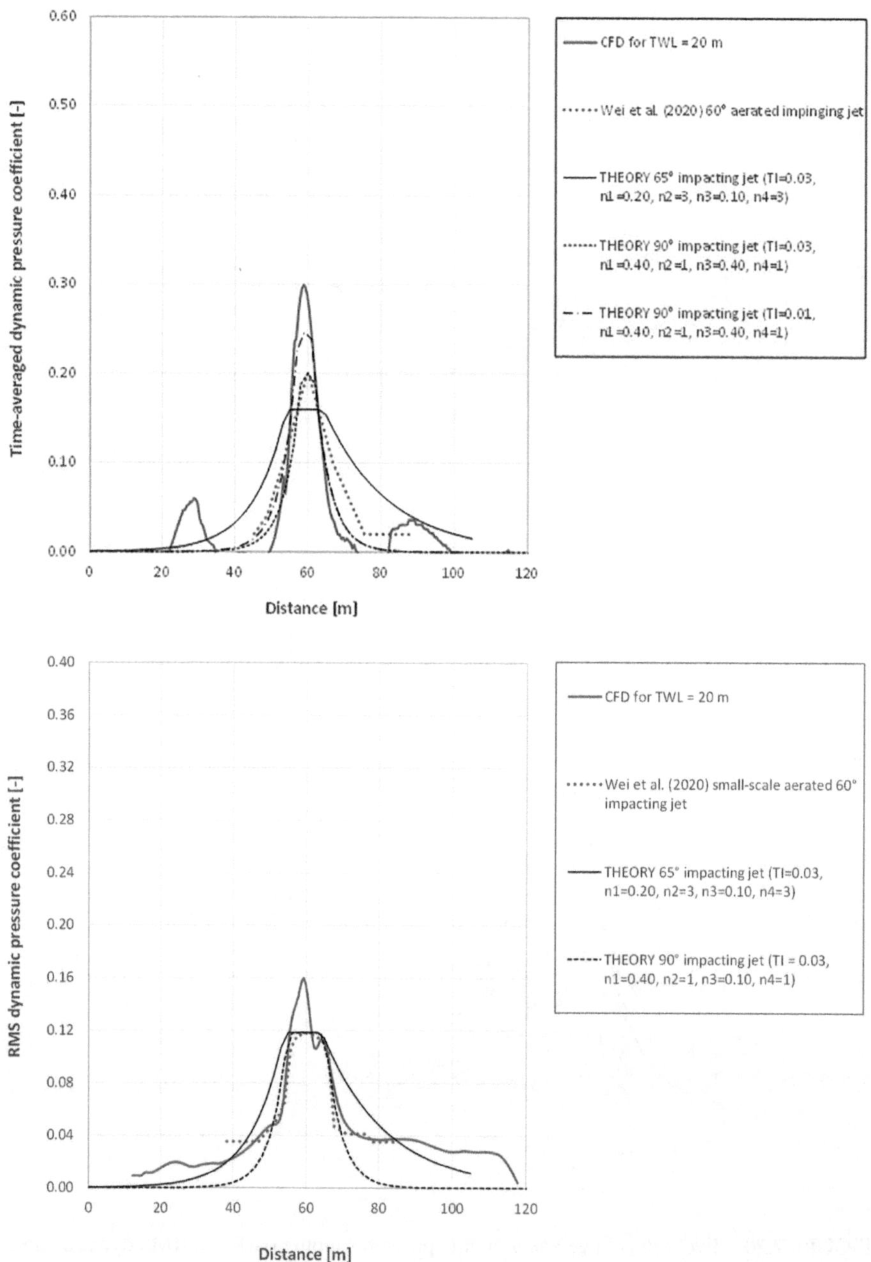

FIGURE 7.71 Comparison of time-averaged pressures (up) and RMS pressure fluctuations (down) based on 2D jet diffusion theory, laboratory experiments, and 2D LES model (TWL = 20 m).

the area of jet impingement, the CFD computed flow velocities are somewhat higher and are situated in between values valid for broken-up and compact jets.

Furthermore, the CFD computed velocities show a dip at the point of jet impact where significant stagnation pressure builds up. This is logic because this area is not

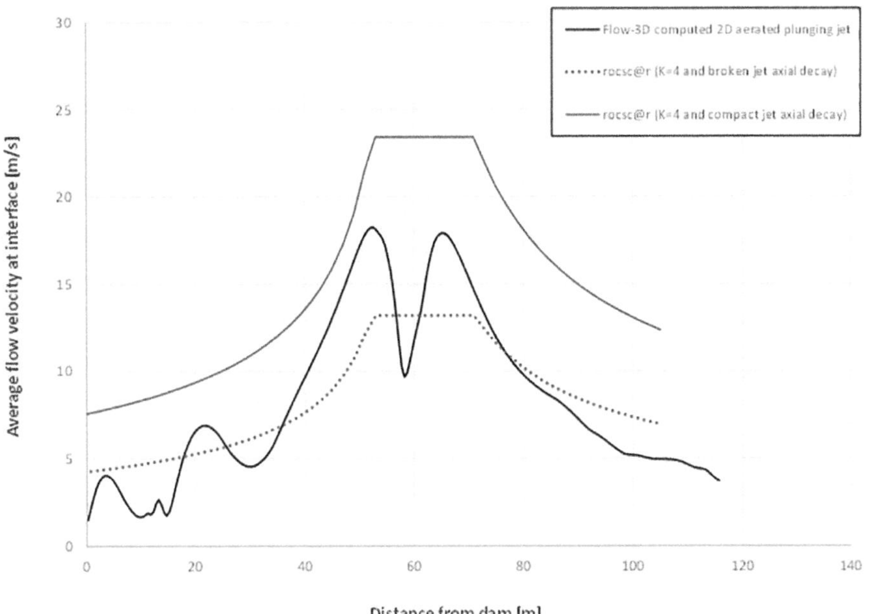

FIGURE 7.72 Comparison of flow velocities at water-rock interface based on 2D jet diffusion theory and computed by 2D LES modelling (TWL = 20 m).

part of the wall jets, that generally start at a lateral distance of 1–2 jet diameters. The theoretical flow velocities do not show this dip because it is automatically removed from the analytical equations for convenience of scour computations.

As a summary, a detailed comparison between CFD values and measured or theoretical values for 2D diffusing jets explains why the current LES turbulence model generates a scour hole that is in good agreement with the laboratory-measured one, or the one computed based on 2D jet diffusion theory in rocsc@r.

Finally, the influence of the tailwater level on the dynamic pressures and flow velocities has been studied for tailwater levels of 20, 40 and 60 m (Figure 7.73). CFD computed flow velocities show a decrease with increasing tailwater. A similar trend is observed for the flow velocities based on 2D jet diffusion theory (for rectangular jets).

As explained, the CFD generates jets that are more compact than observed in the laboratory and predicted by 2D jet diffusion theory. As such, Figure 7.69 has shown that the CFD computed flow velocities, for a usual jet core extension coefficient K of 4 times the jet diameter at impact, are situated somewhere in between values for compact and broken-up jets.

Figure 7.74 illustrates that, for broken-up jets and a jet core extension coefficient K of 6 times the jet diameter at impact, fair agreement is obtained between 2D jet diffusion theory and the CFD results.

Influence of Turbulence Parameters

Finally, the influence of turbulence parameters such as the turbulence model, the mesh size and the momentum advection resolution has been investigated. Figure 7.75 compares RMS values for LES and RNG k–ε turbulence models, for mesh sizes

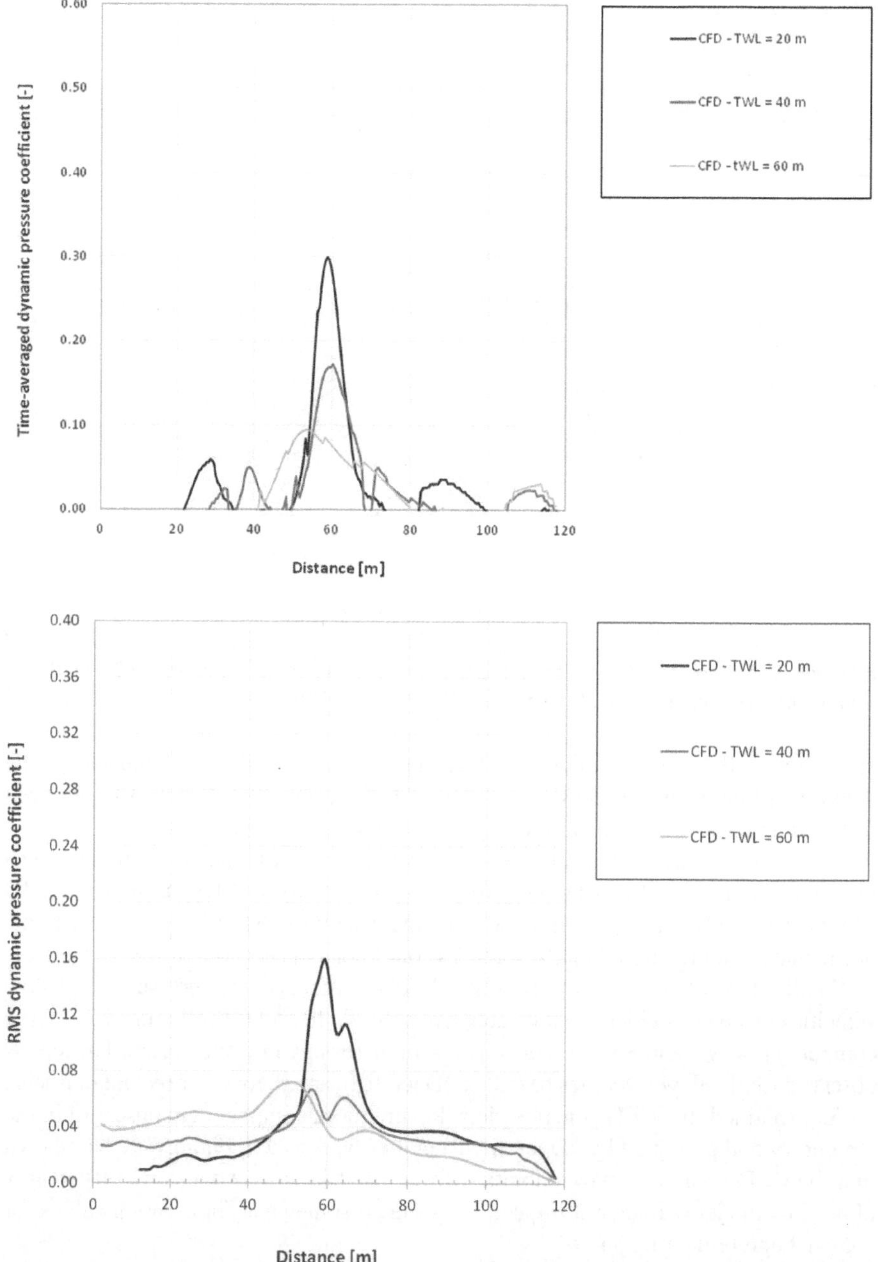

FIGURE 7.73 Comparison of time-averaged and RMS dynamic pressures at water-rock interface based on 2D jet diffusion theory and computed by 2D LES modelling (TWL = 20-40-60 m).

of 0.5 m and 1.0 m, and first-order, second-order monotonicity preserving, and second-order momentum advection equations.

The LES turbulence model generates a large range of RMS values, depending on the mesh size and the momentum advection order. For a 1 m grid size, only very

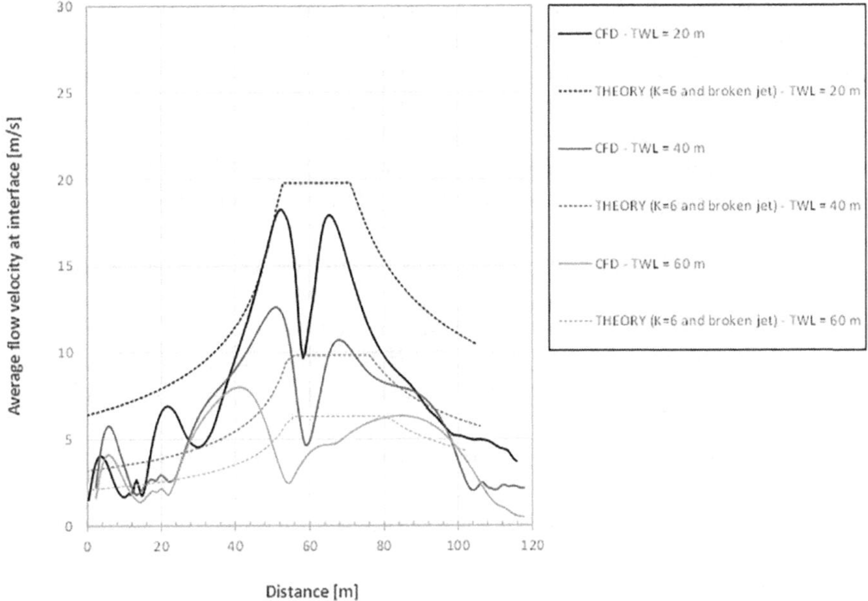

FIGURE 7.74 Influence of tailwater level on flow velocities at water-rock interface based on 2D jet diffusion theory and computed by 2D LES modelling.

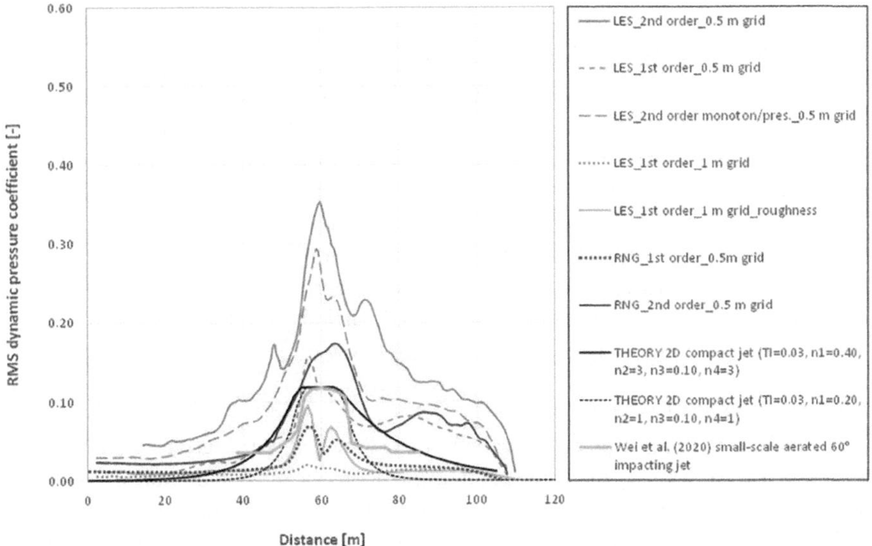

FIGURE 7.75 Influence of turbulence parameters on RMS values computed by CFD.

low RMS values are being generated and this mesh size seems unsuited to correctly reproduce pressure fluctuations. The only way to model significant RMS values with such a grid is by introducing flow perturbating aspects such as an increased bottom roughness or increased air entrainment.

Second, significant RMS values are generated for a mesh size of 0.5 m. The values are in agreement with 2D jet diffusion theory for the first-order momentum advection scheme, and are significantly higher in case of second-order momentum advection schemes.

The RNG k–ε turbulence model is a RANS-based model and thus theoretically not able to generate turbulent pressure fluctuations. Nevertheless, the following aspects inherent to real-life jets were found present also during the CFD computations and seem to generate a certain unsteady behaviour of the turbulent flows under study:

- small (natural) variability in inflow conditions, generating small jet instabilities
- small variability in inflow conditions because of the shape and the roughness of the spillway structure, which may generate unstable, undulating or even wavy conditions
- air entrainment and air transport, generating jet instabilities (especially for broken-up jets)
- highly variable and suddenly changing geometry of the water-rock interface.

Figure 7.75 illustrates that the RNG k–ε model generates the RMS values fairly well for a second-order momentum advection scheme, while the first-order scheme generates somewhat lower values.

As such, it is obvious that a correct choice of the turbulence parameters for CFD-based fluid-solid coupling is of utmost importance to obtain plausible results. Before any coupling, preliminary CFD runs should allow determining the most plausible set of parameters.

Channel Scour by Horizontal Turbulent Flow

The second application considers scour of an unlined bedrock of a 2D river channel downstream of a gated hydraulic structure. The case involves a high-velocity turbulent flow of depth $h_1 = 2$ m entering the computational domain parallel to the erodible channel bottom, at an initial velocity of 40 m/s and for a downstream tailwater depth set at 20 m. The bedrock is considered to consist of rock blocks similar to the first coupling test case, i.e. prismatic blocks with a height of 1 m and a side length of 2 m, and completely fractured. The inflow Froude number equals 9.0, and the theoretical sequent depth is 24.5 m. The flow is considered partially developed, i.e. with a developing bottom boundary layer and an ideal-fluid flow region above this layer.

Comparison with Wall Jet/Hydraulic Jump Theory

A comparison has been made with available theory and experiments on similar flow situations, i.e. free or submerged hydraulic jump conditions, or submerged wall jet conditions. The CFD computations have been made using LES (first-order momentum advection) and RNG k–ε (second-order momentum advection) turbulence models, for both a smooth water-rock interface ($k_s = 0.01$ m), typical for concrete-lined stilling basins, and a very rough water-rock interface ($k_s = 1$ m), more representative for eroding rock. Figure 7.76 compares the 2D CFD computed flow velocities at the water-rock interface with 3D CFD modelling performed by Valero et al. (2016) and with a range of experimentally determined velocity decay laws for both hydraulic jump flows and wall jet flows. Good agreement is observed with the 3D CFD, and the 2D velocity decay is also in good agreement with the submerged wall jet decay, especially for the LES results.

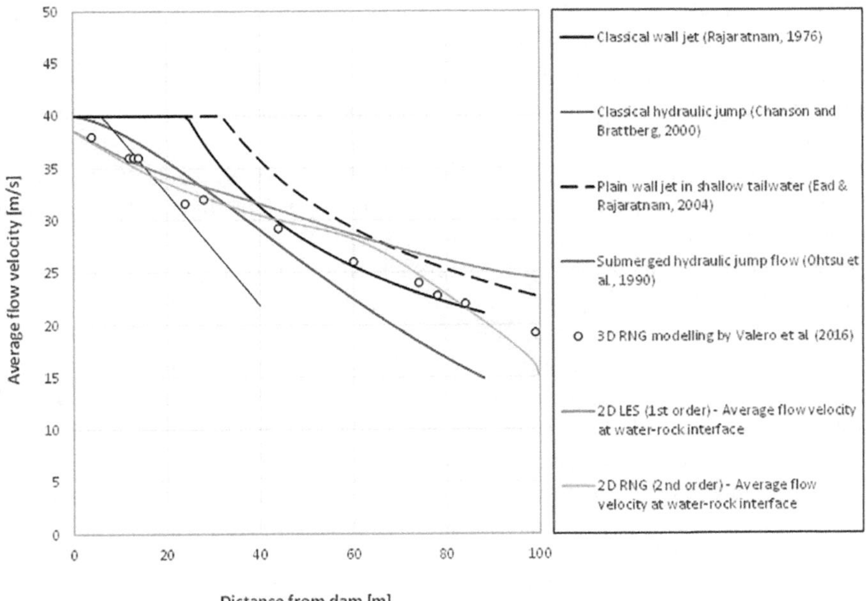

FIGURE 7.76 Average flow velocities at smooth water-rock interface: comparison of 2D CFD computations with experimental decay laws and with 3D CFD by Valero et al. (2016).

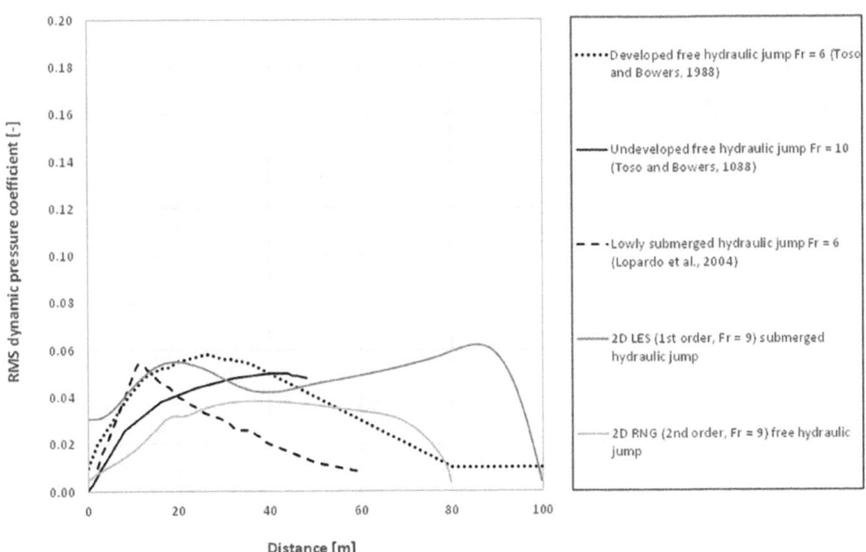

FIGURE 7.77 RMS pressure fluctuations at smooth water-rock interface: comparison of 2D CFD computations with experimentally determined laws.

Second, Figure 7.77 compares the 2D CFD computed RMS pressure fluctuations at the water-rock interface with a range of experimentally determined laws for hydraulic jump flows.

The LES model is in good agreement with experimental data up to a relative distance of $X/h_1 = 20$, but deviates further downstream. The absence of RMS decay is due to the downstream boundary influence during the modelling, which maintains a lowly submerged jump or wall jet, while the tailwater level is strictly speaking lower than the sequent depth (Figure 7.78 upper left-hand figure).

The RNG k–ε model has slightly lower values but for a shape that is in much better agreement with the experiments. Figure 7.78 (upper right-hand figure) illustrates that a free hydraulic jump forms during the modelling.

Next, 2D CFD computations have been repeated but for a very rough (rocky) water-rock interface. The corresponding flows are illustrated in Figure 7.78 (lower graphs).

Figure 7.79 again compares the 2D CFD computed flow velocities at the water-rock interface with 3D CFD modelling performed by Valero et al. (2016) and with a range of experimentally determined velocity decay laws for both hydraulic jump flows and wall jet flows. The increased roughness strongly enhances bottom boundary layer development and interface velocity decay with downstream distance.

Also, Figure 7.80 compares the 2D CFD computed RMS pressure fluctuations at the water-rock interface with a range of experimentally determined laws for hydraulic jump flows. For the LES model, RMS values strongly increase. For the RNG k–ε model, however, a decrease in RMS values is observed. More stable jump conditions

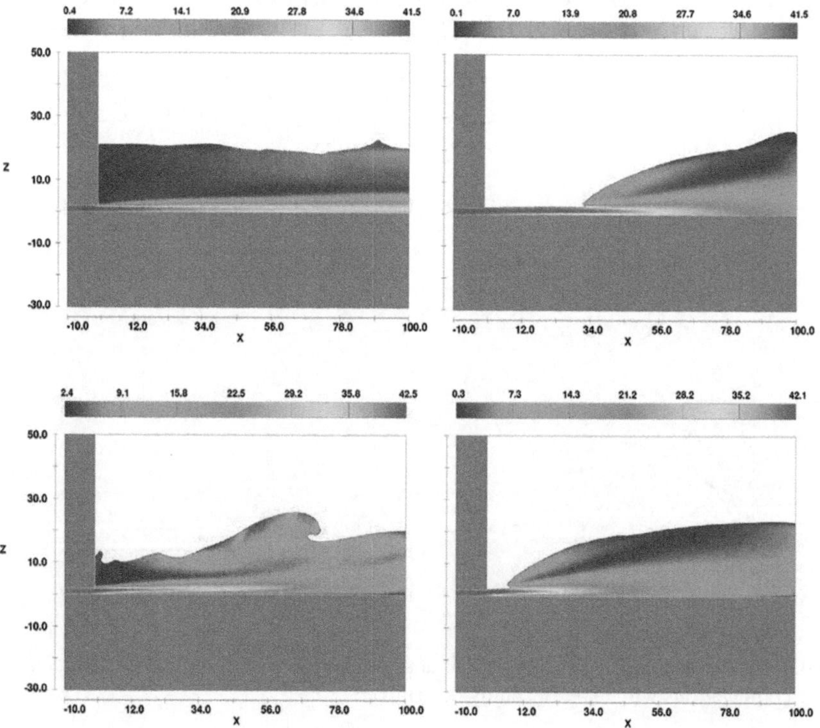

FIGURE 7.78 Average flow velocities at water-rock interface: comparison of 2D CFD computations using LES model (left) and RNG k–ε model (right), for both smooth (up) and rough (down) water-rock interfaces.

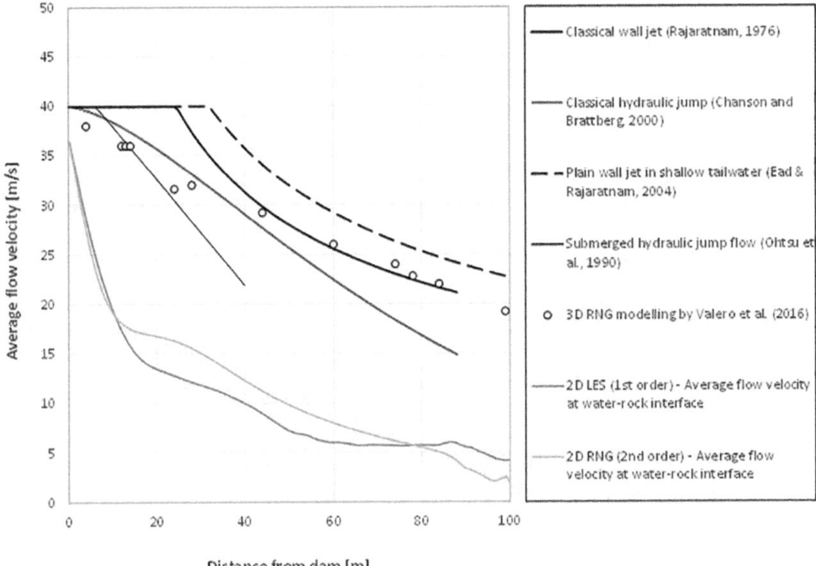

FIGURE 7.79 Average flow velocities at rough water-rock interface: comparison of 2D CFD computations with experimental decay laws and with 3D CFD by Valero et al. (2016).

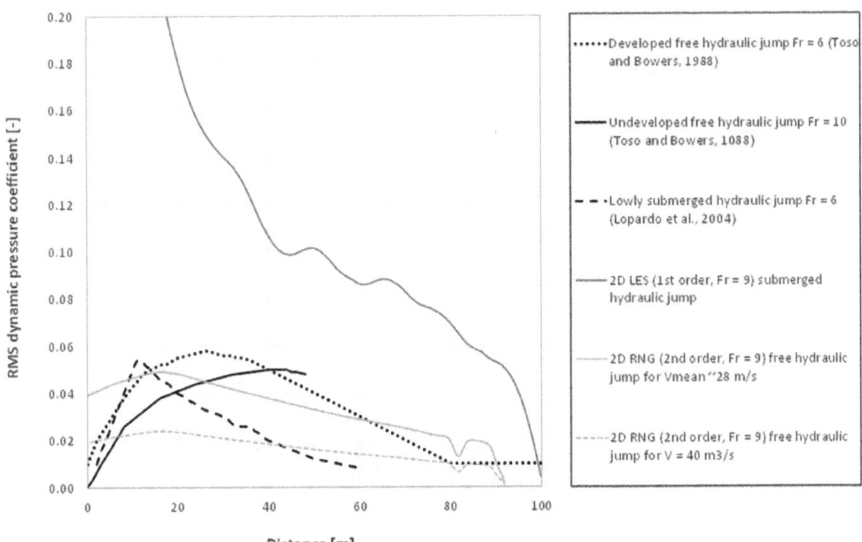

FIGURE 7.80 RMS pressure fluctuations at rough water-rock interface: comparison of 2D CFD computations with experimentally determined laws.

are computed because of the increased roughness, together with a detachment of the flow turbulent shear layer from the bottom.

As such, the average jet inflow velocity becomes significantly lower than the initial velocity of 40 m/s. By adopting such a lower initial inflow velocity (~28 m/s), computed RMS values are in agreement with values for a smooth bottom.

CFD Fluid-Solid Modelling Parameters

Modelling parameters for the fluid-solid coupled scour computations are presented in Table 7.11. The LES turbulence model has been used, with a first-order momentum advection scheme. A mesh size of 1.0 m has been applied for the MQSI method (velocity-based), and of 0.5 m for the MDI method (RMS pressure-based). Air entrainment was accounted for. The roughness height of the water-rock interface was set at 1.0 m. A no-slip boundary condition was applied at the water-rock interface.

CFD Fluid-Solid Scour Results

The 2D hydrodynamic solution computed by the CFD at the end of the scour process is visualized in Figure 7.81, showing the average flow velocities [m/s] for the MQSI

TABLE 7.11
CFD Modelling Parameters Applied during 2D Fluid-Solid Coupling of Horizontal Wall Jet Flows

Turbulence model	LES (Large Eddy Simulation)
Volume of fluid advection	Automatic (Split or Unsplit Lagrangian)
Momentum advection	First-order approximation
Time step control	Stability and convergence
Time duration per steady-state run	10 s (for MQSI)/20 s (for MDI)
Mesh size	1.0 m (for MQSI)/0.5 m (for MDI)
Pressure solver	Implicit (GMRES)
Air entrainment model	Yes, with flow bulking and buoyancy
Air entrainment coefficient	0.50
Roughness height dam	0.01 m
Roughness height rock mass	1.0 m
Layer height for fluid-solid iterations	2.0 m (for MQSI)/1.0 m (for MDI)

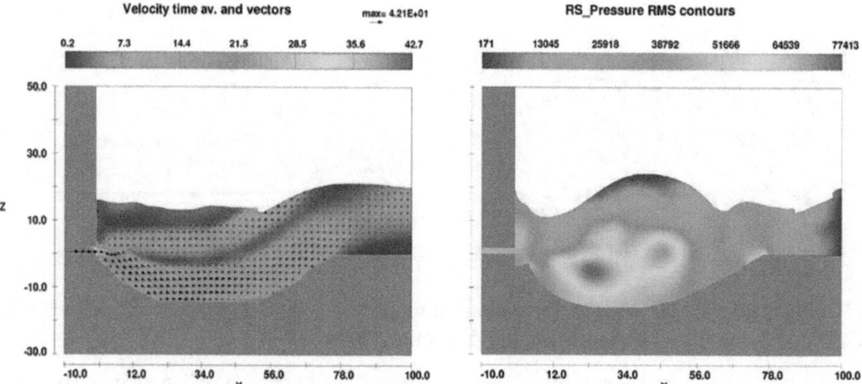

FIGURE 7.81 2D CFD-based fluid-solid coupled scour potential by horizontal turbulent channel flow, based on MQSI method (left-hand side) and MDI method (right-hand side).

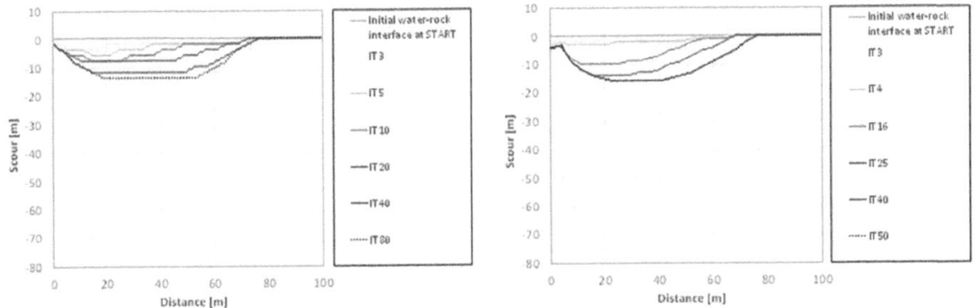

FIGURE 7.82 CFD-based coupled iterations of scour by horizontal turbulent channel flow, based on the MQSI method (left graph) and the MDI method (right graph).

method (left hand) and the RMS pressure fluctuations [N/m^2] for the MDI method (right hand).

For the MQSI computational method, iterative vertical layers of 2 m have been used, and a net uplift coefficient of $C_{up,MQSI} = 0.15$. For the MDI computational method, iterative vertical layers of 1 m have been used, for a critical net uplift displacement factor of 0.15.

The numerical grid used by the rocsc@r software is 4 m in the *X*-direction and 1 m in the *Z*-direction.

Figure 7.82 illustrates successive iterations of scour computed by the CFD fluid-solid coupling for both the MQSI method (left graph) and the MDI method (right graph). A scour depth of maximum 14–16 m is observed over a large area downstream of the gates. During progressive scour formation, the jet becomes directed towards the bed, and a hole is formed that appears to be in accordance with scour holes observed for granular material and submerged wall jets (Dey and Sarkar 2006; Rajaratnam 1981).

The corresponding evolution of the average flow velocities and RMS pressure fluctuations may be observed in Figures 7.83 and 7.84 respectively.

For the MQSI method, the stabilizing force (i.e. submerged weight without frictional forces) equals 13,150 N/m^2. For the applied net uplift coefficient of 0.15, the critical flow velocity equals 13.1 m/s, which seems coherent with the CFD results in Figure 7.83.

For the MDI method, the stabilizing force is again the submerged weight (no frictional forces). During block acceleration, added mass is also accounted for to counteract the block uplift. For the applied critical net uplift displacement factor of 0.15 (i.e. necessary uplift height divided by block height), the critical RMS value equals 0.05, which again seems fairly coherent with the CFD results in Figure 7.84.

Plunge Pool Scour by Ski-Jump Jets

The next application considers CFD fluid-solid coupling for scour downstream of a ski-jump jet. The previously discussed case of the 2011 flood event at Wivenhoe Dam in Australia has been reused here.

CFD Modelling Parameters

Modelling parameters are presented in Table 7.12. The LES turbulence model has been used in two dimensions, together with the MQSI method for break-up by peeling-off of

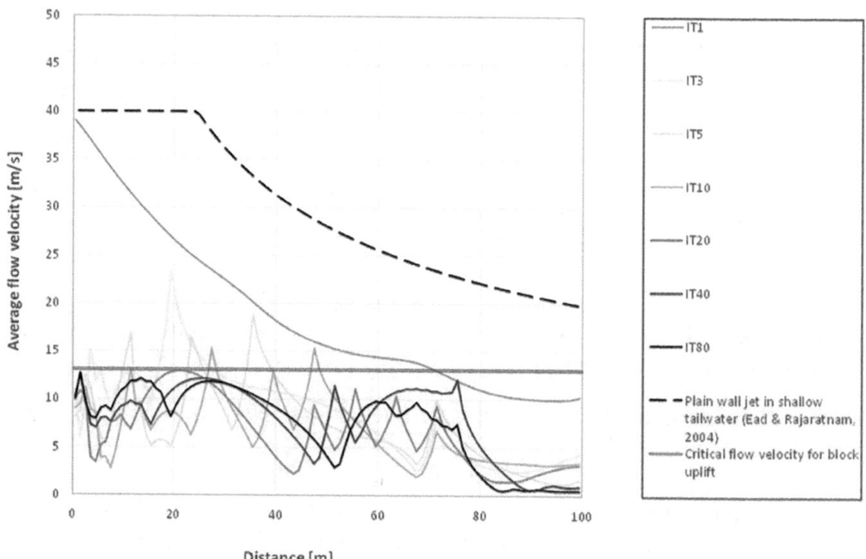

FIGURE 7.83 Average flow velocities at rough water-rock interface during CFD-based fluid-solid coupled scour computations.

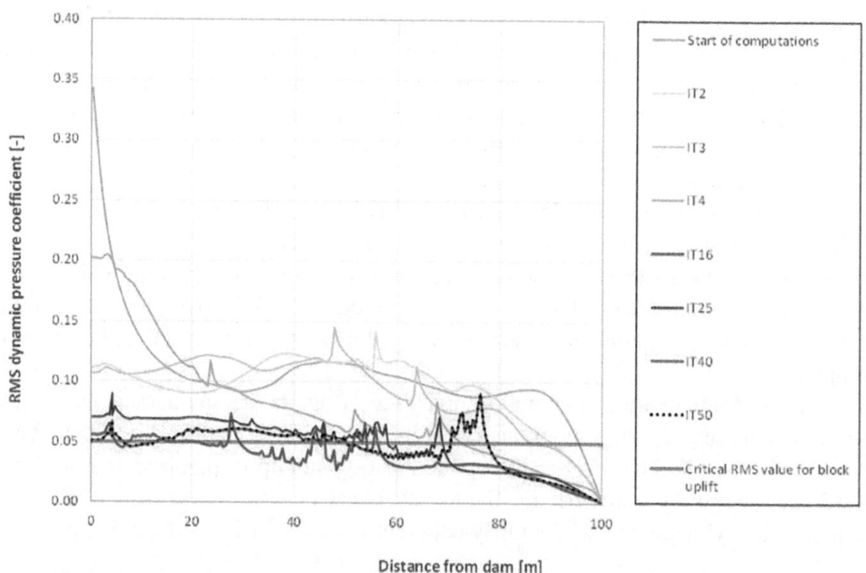

FIGURE 7.84 RMS pressure fluctuations at rough water-rock interface during CFD-based fluid-solid coupled scour computations.

rock blocks by turbulent flows quasi-parallel to the water-rock interface, which is applicable to the case of Wivenhoe Dam. As such, a large mesh size of 1 m was used during the computations, for a layer height of scour formation in rocsc@r of 2 m.

Air entrainment was not really accounted for by the CFD model because the entrainment method in the CFD is triggered by the turbulent kinetic energy in

TABLE 7.12
CFD Modelling Parameters Applied during 2D Fluid-Solid Coupling of Wivenhoe Dam 2011 Flood Event

Parameter	Run 1 (MQSI method)
Turbulence model	LES model
Volume of fluid advection	Automatic (Split or Unsplit Lagrangian)
Momentum advection	First-order approximation
Time step control	Stability and convergence
Time duration per steady-state run	5 s
Mesh size	1.0 m
Pressure solver	Implicit (GMRES)
Air entrainment model	Yes, with flow bulking and buoyancy
Air entrainment coefficient	0.50
Roughness height dam	0.01 m
Roughness height rock mass	1.0 m
Layer height iterations	2.0 m

the flow. By using an LES turbulence model for a large grid size, and without any upstream injected turbulent kinetic energy, the air entrainment revealed to be very low to quasi non-existent.

The roughness height of the water-rock interface was set at 1.0 m. A no-slip boundary condition was applied at the water-rock interface. The average wall velocity computed along the water-rock interface is governed by a logarithmic law-of-the-wall and assuming hydraulically rough turbulent flow.

Jet Trajectory and Flow Velocities

Figure 7.85 presents the computed jet trajectory and flow velocities for a max. discharge of 7,600 m^3/s. The jet impacts the plunge pool at X = 50–55 m. The computed average flow velocity at impact is ~30 m/s. The jet is deflected at the water-rock interface and generates a significant wall jet towards downstream. This wall jet may generate rock scour by peeling-off of protruding rock blocks along the downstream face of the pre-excavated plunge pool. This has been numerically modelled.

Jet Break-Up, Contraction and Deformation

As the CFD computations are only performed in two dimensions, the jet is not able to deform laterally. Furthermore, no initial turbulent kinetic energy is present and the flow acceleration during fall is relatively low. Hence, the jet does not change its initial shape and does not really contract during its fall. Also, the CFD generated jet is not broken up at impact.

Jet Diffusion through the Pool

The compact CFD computed jet develops through the pool depth as a rectangular jet, generating a significant wall jet flow along the water-rock interface of 20–25 m/s.

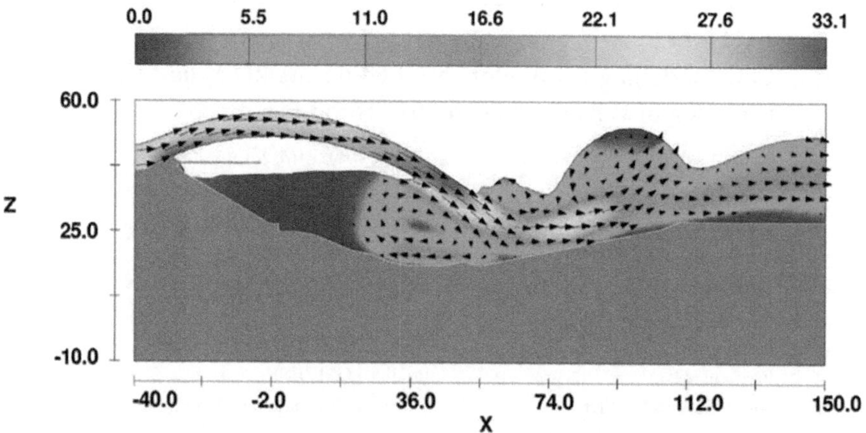

FIGURE 7.85 Jet trajectory and flow velocities computed by 2D LES model.

CFD Fluid-Solid Scour Results

Figure 7.86 illustrates the iterative results obtained by the coupled scour computations. A best-fit net uplift pressure coefficient of $C_{up}=0.17$ was found. The computation converged in 78 iterations and took 58 min on a standard i5 processor.

3D Fluid-Solid Coupling

A second series of applications deals with full 3D CFD fluid-solid coupling. This increases the computational times in comparison with 2D applications, but has the benefit to procure a physically more plausible modelling of real-life flow turbulence. In rocsc@r, a series of 2D vertical profiles are defined along the *Y*-axis, in order to generate a full 3D rock mass.

Scour by Free Overfall Jets

Like for the 2D fluid-solid coupling, the free overfall jet of the Stellenbosch laboratory experiment (test 9A_a) has been numerically reproduced. This test corresponds to the maximum specific discharge, the maximum hydraulic head, and the maximum tailwater level used. The block height is 1.0 m. No mounding was allowed during the test. Figure 7.87 illustrates a general overview of the Stellenbosch test installation, together with a plan and perspective view of measured scour formation, as well as a photo of the scour formation after the test.

CFD Modelling Parameters

Modelling parameters are presented in Table 7.13. Different runs have been made, involving k–ε, RNG k–ε or k–Ω turbulence models. Like for the 2D tests, for computational methods based on RMS values of dynamic pressure fluctuations, such as MDI or CFM methods, a mesh size of 0.5 m has been used. For computational methods based on average flow velocities at the water-rock interface, a mesh size of 0.5 or 1 m has been used. One of the advantages of using the k–ε, RNG k–ε or k–Ω

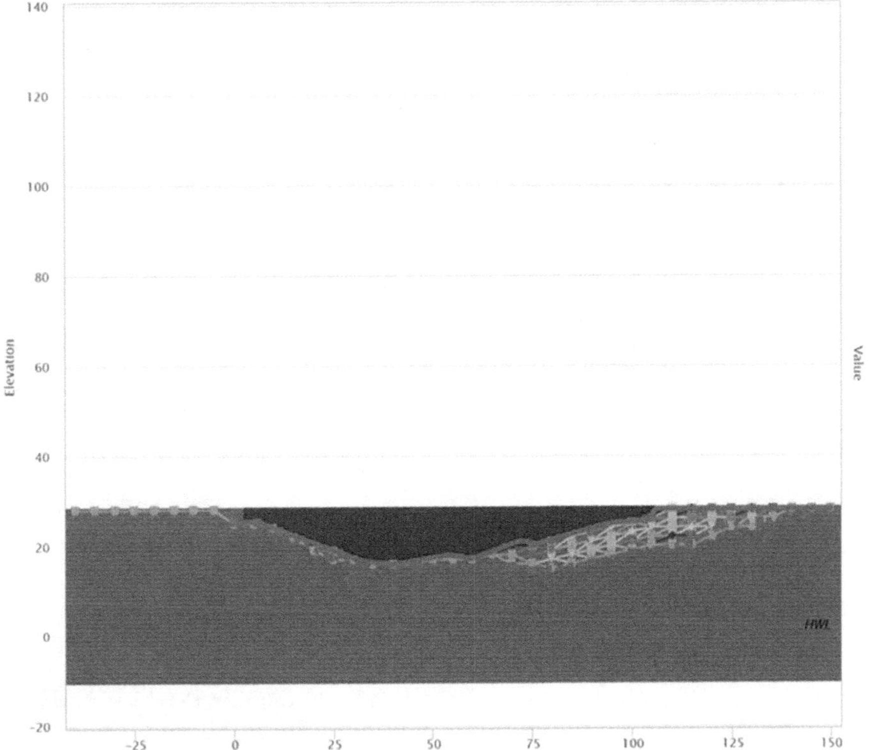

FIGURE 7.86 Plunge pool scour iterations computed by fluid-solid coupling.

turbulence models is the relative ease of generating air entrainment as compared to the LES turbulence model.

The modelling generated too compact jets with insignificant degrees of break-up compared to the laboratory jets. However, while the 2D modelling allowed reproducing adequate scour holes, the 3D modelling resulted in too deep and localized scour holes. Very high air entrainment, or very high bottom roughness of the upstream channel flow, had to be introduced to numerically reproduce flow instabilities that allow approaching the behaviour of fully broken-up jets as observed during the laboratory tests. This revealed to procure CFD scour holes in much better agreement with the laboratory-observed scour holes.

Hence, in the following, two 3D runs are compared: run 1 with air entrainment as prescribed by the CFD manual and resulting in (too) compact jets, and run 2 with very high air entrainment, out of the manual prescriptions, but resulting in an unstable, broken-up jet. The main parameters of both runs are compared in Table 7.13. Iterative vertical (Z-)layers of 1 m have been used. The numerical grid used by the rocsc@r software is 2 m in the *X*- and *Y*-directions. Run 1 converged in 25 iterations and took 20 h on an AMD Ryzen 9 5900X 12-core processor with 64 Gb of installed RAM, while run 2 also converged in 25 iterations but took 35 h on the same 12-core processor.

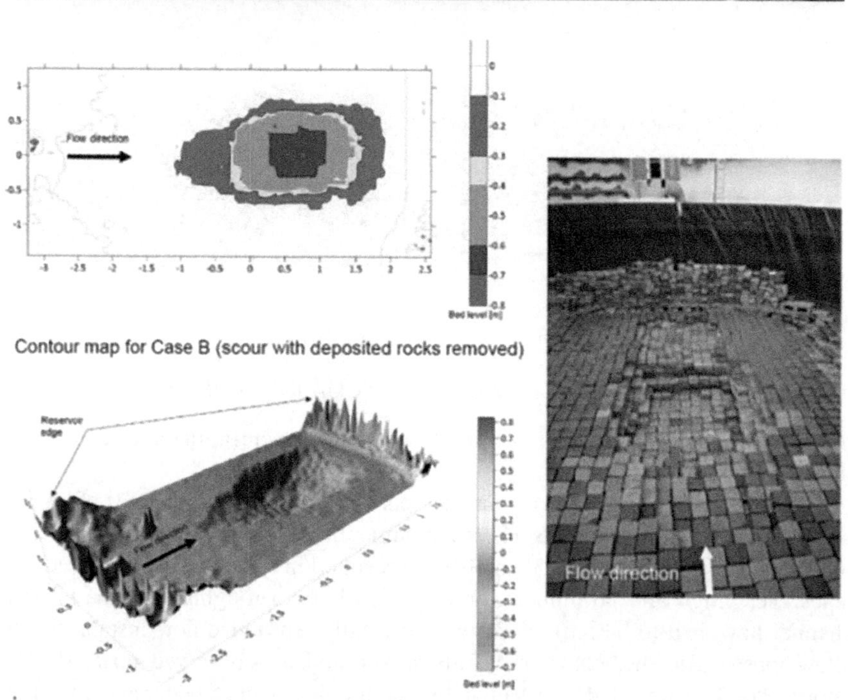

FIGURE 7.87 General overview and plan and perspective view of measured scour at Stellenbosch laboratory installation. Courtesy of Dr A. Bosman.

Hydraulic Boundary Conditions

The initial and boundary conditions have been taken from the laboratory model tests at Stellenbosch. The jet is modelled in 3D and exhibits lateral spread. As illustrated in Figure 7.88, this lateral spread remains low for the compact jets of run 1, while the lateral spread generated by the broken-up jets of run 2 is much closer to the laboratory-observed jets.

TABLE 7.13
CFD Modelling Parameters Applied during 3D Fluid-Solid Coupling of Stellenbosch Laboratory Experiments

Parameter	Run 1 (compact jet)	Run 2 (broken jet)
Turbulence model	k–ω model	k–ε model
Volume of fluid advection	Automatic (Split or Unsplit Lagrangian)	Automatic (Split or Unsplit Lagrangian)
Momentum advection	Second-order monotonicity	Second-order monotonicity
Time step control	Stability and convergence	Stability and convergence
Time duration per steady-state run	5 s	5 s
Mesh size	0.5 m	0.5 m
Computational method	MDI	MDI
Critical net uplift height ratio n_b	0.30	0.40
Pressure solver	Implicit (GMRES)	Implicit (GMRES)
Air entrainment model	Yes, with flow bulking and buoyancy (drift flux)	Yes, with flow bulking and buoyancy (drift flux)
Air entrainment coefficient	0.50	5.00
Roughness height upstream canal	0.01 m	0.01 m
Roughness height rock mass	0.01 m	0.01 m
Layer height for fluid-solid iterations	1.0 m	1.0 m

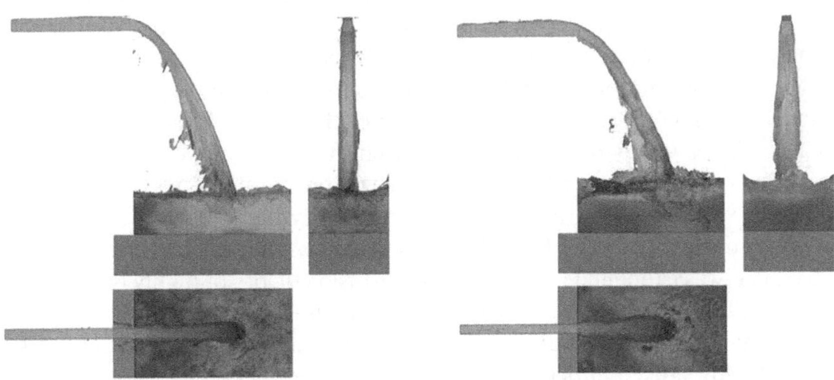

FIGURE 7.88 3D CFD modelling of Stellenbosch laboratory experiments for compact jets (left-hand side, run 1) and broken-up jets (right-hand side, run 2).

Jet Trajectory and Flow Velocities

The jet impacts the plunge pool at $X = 61$ m, which is very close to the laboratory-observed distance of $X = 62$ m. The computed average flow velocity at impact is 42.9 m/s, i.e. close to the value of 42.4 m/s measured during the experiments (Bosman 2021).

Jet Break-Up, Contraction and Deformation

The laboratory jets were described as highly broken-up (degree of break-up >1.85) by Bosman (2021), showing significant lateral spread upon impact, together with a highly curved flat-shaped footprint. The 3D modelling allows the jet to deform laterally. For CFD run 2, the jet contracts during its fall and changes its almost squared-shape initial shape into a much wider, flat and curved shape upon impact as was observed during the laboratory tests. For CFD run 1, the numerical jet is not broken up at impact, but still deforms its initial shape into a curved shape upon impact.

Jet Diffusion through the Pool

For compact 3D CFD jets (run 1), the computed jet develops through the pool depth as a compact rectangular jet, and not as a broken-up jet like observed during the laboratory experiments. Hence, the jet velocities computed along the water-rock interface are high, as illustrated in Figure 7.89 (upper graph). The CFD computed jet is too compact and overestimates the flow velocities at the water-rock interface, by lack of aeration, turbulence and related break-up.

Second, for broken-up 3D CFD jets (run 2), the computed jet develops through the pool depth as a broken-up jet. The corresponding jet velocities along the water-rock interface are significantly lower than for run 1, as illustrated in Figure 7.89 (lower graph).

CFD Fluid-Solid Scour Results

Figure 7.90 illustrates the scour results obtained by 3D CFD fluid-solid iterations by use of the dynamic block uplift method (MDI), for both compact (left-hand graph, run 1) and broken-up (right-hand graph, run 2) turbulent jets. The laboratory-observed scour formation is presented by the thick dotted grey line.

The best fit compared with the laboratory measurements is obtained for broken-up jets. Not only the area of deepest scour formation is well modelled, but also the slopes of the scour hole towards upstream and downstream seem adequately reproduced.

Scour by compact jets is clearly more concentrated around the point of jet impact. Also, scour by compact jets is slightly deeper than by broken-up jets, but this is because of the slightly different critical net uplift height ratio n_b applied during the modelling (i.e. 0.30 for compact jets and 0.40 for broken-up jets).

The n_b values for 3D CFD modelling are slightly higher than the value of 0.20 calibrated during the 2D CFD modelling, and are significantly higher than the value of 0.02 that was calibrated based on 2D jet diffusion theory of highly broken-up jets.

In fact, the parametric sensitivity study for real-life scour cases showed that the critical value based on 2D diffusion theory is around 0.10, but could be smaller in case of highly broken-up jets. Nevertheless, n_b values between 0.10 and 0.50 were observed frequently in the different case studies tested.

The apparently rather high n_b values obtained during the 3D CFD coupling result from different RMS pressure fluctuations as compared with the RMS values based on 2D jet diffusion theory.

First, for compact jets, the upper graph of Figure 7.91 shows the evolution of the RMS pressure fluctuations through the centre profile ($y=0$). Fair agreement is observed with RMS values based on 2D jet diffusion theory, and the last few iterations respect the critical RMS coefficient for a n_b value of 0.30, i.e. 0.07. The corresponding scour formation, however, is slightly too deep as illustrated in Figure 7.90 (left graph).

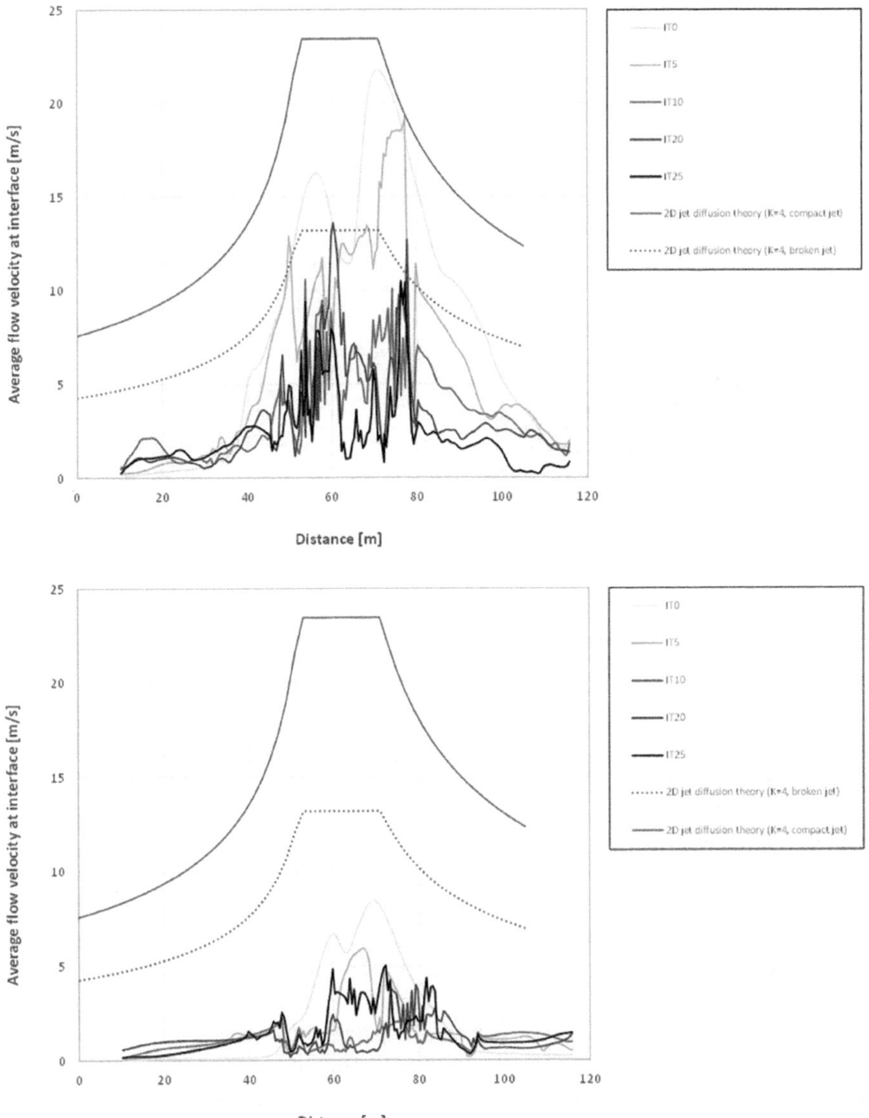

FIGURE 7.89 Average flow velocity at water-rock interface by 3D CFD modelling of Stellenbosch laboratory experiments for compact jets (upper graph, run 1) and broken-up jets (lower graph, run 2).

Second, for broken-up jets, the lower graph of Figure 7.91 shows the evolution of the RMS pressure fluctuations through the centre profile ($y=0$). The values are significantly higher than the equivalent values based on 2D jet diffusion theory. From iteration 22 and further on, the critical RMS coefficient for a n_b value of 0.40 is respected, i.e. 0.09. The corresponding scour formation agrees with the laboratory-measured scour as illustrated in Figure 7.90 (right graph).

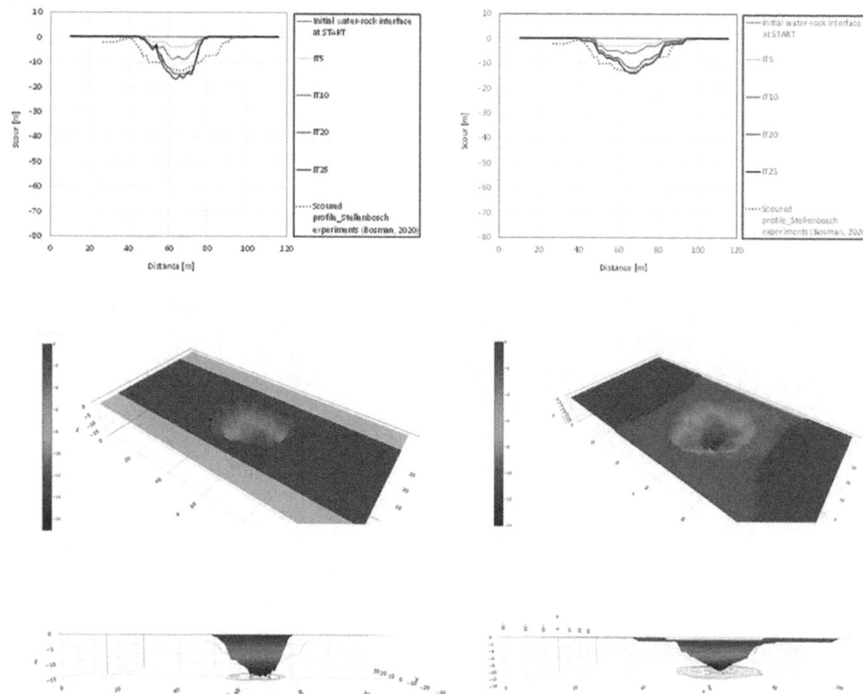

FIGURE 7.90 Scour computed using 3D fluid-solid coupling (MDI method) for compact jet (left-hand side, run 1) and broken-up jet (right-hand side, run 2).

Bridge Pier Scour

3D fluid-solid coupled scour computations have been applied to the virtual case of bedrock scour in a river channel containing three bridge piers. The aim of this case is not to outline the quality of the flow turbulence and vortices computed around the bridge piers, but to illustrate fluid-solid coupled scour computations in a different flow environment.

Figure 7.92 illustrates the rectangular-shaped channel of 40 m width, containing three rectangular bridge piers 4 m wide and 10 m long. Flow conditions are characterized by a 5 m flow depth along the downstream boundary and uniformly distributed inflow with an average flow velocity of 5 m/s. The channel slope equals ~ zero.

The bedrock consists of completely fractured bedding layers of sedimentary rock with characteristic blocks of 0.20 m height and 1 m side length in both *X*- and *Y*-directions, with a UCS strength of 20 MPa, a density of 2,360 kg/m^3 and planar tight joints. The RQD is 40% and the joint network may be described by three joint sets plus random joints. The joints are smooth and planar, with apertures of 1–5 mm and a joint alteration coefficient of 6. The bottom has an equivalent roughness height of $k_s = 0.10$ m.

CFD Modelling Parameters

Modelling parameters are presented in Table 7.14. Different runs have been made with the RNG k–ε turbulence model. A coarse mesh size of 1 m has been used.

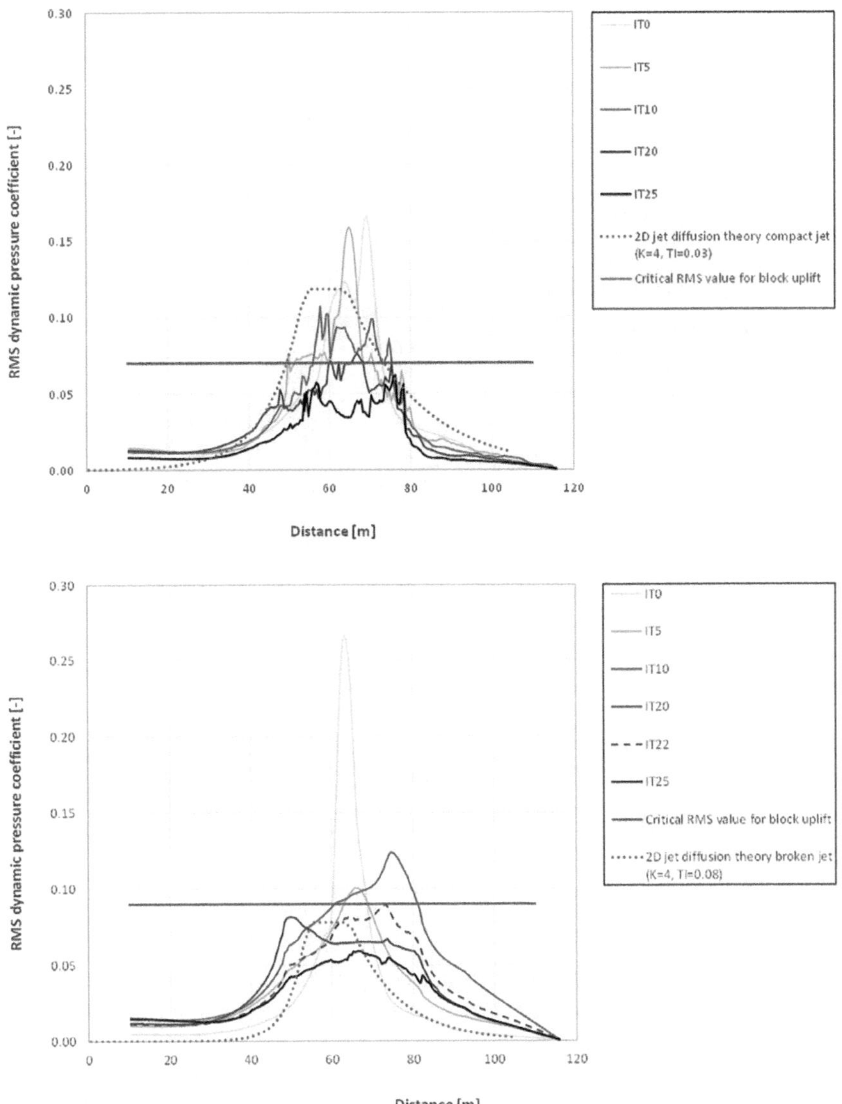

FIGURE 7.91 RMS pressure fluctuations at water-rock interface by 3D CFD modelling of Stellenbosch laboratory experiments for compact jets (upper graph, run 1) and broken-up jets (lower graph, run 2).

In the following, two of the available 3D runs are compared: run 1 based on the MQSI method (i.e. velocity-based), with a net uplift pressure coefficient of 0.20, and run 2 based on the EIM method, in which the stream power is determined as the product of bottom shear stress τ and bottom flow velocity V. The applicable stream power coefficient is set at 1.0. The erodibility index $EI = 35.2$, resulting in a critical stream power $SP_c = 14.45$ kW/m^2.

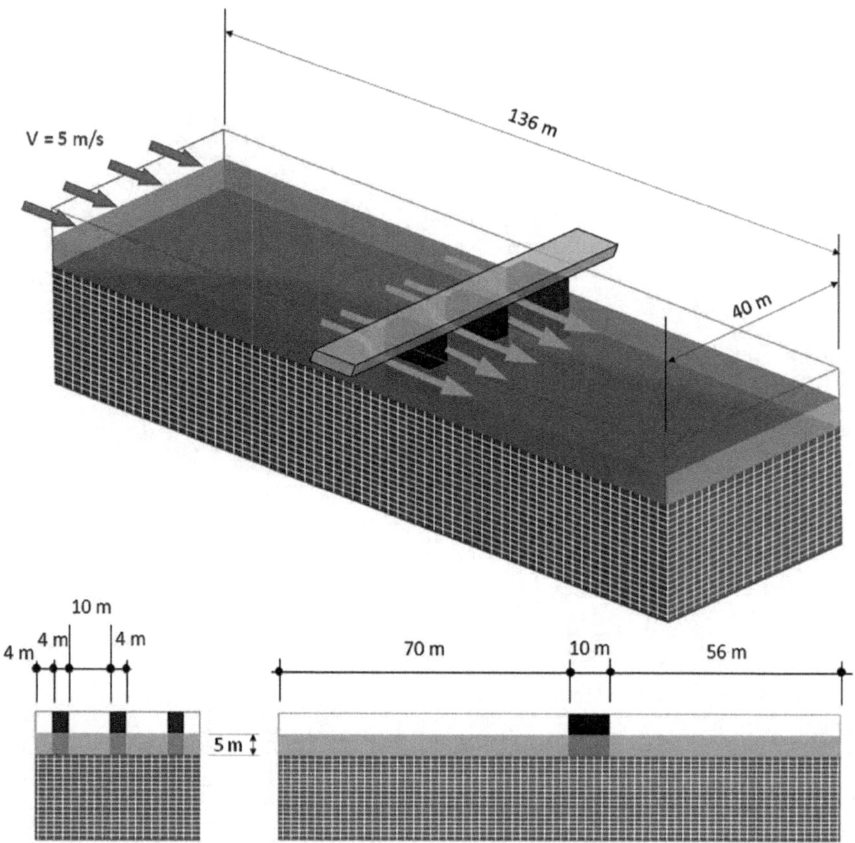

FIGURE 7.92 Hydraulic and geometric parameters of a 3D fluid-solid coupling case study involving bridge pier scour in a river channel.

The main parameters of both runs are compared in Table 7.14. Iterative vertical (Z-)layers of 1 m have been used. The numerical grid used by the rocsc@r software is 2 m in the *X*- and *Y*-directions.

Run 1 converged in 25 iterations and took 2 h and 10 min on an AMD Ryzen 9 5900X 12-core processor with 64 Gb of installed RAM, while run 2 was for 20 iterations that took 1 h and 10 min on the same processor.

CFD Fluid-Solid Coupling Results

The flow conditions in the modelled river reach are presented in Figure 7.93 by the flow velocities at the start of the scour computations (left-hand side) and at the end of the scour computations (right-hand side). The plan view shows the flow velocities at the flow surface.

The flow contraction and acceleration generated by the bridge piers is extending significantly downstream of the piers. Flow deceleration and increased water surface are observed upstream of the piers at start of the modelling. These phenomena disappear following bedrock scour formation.

TABLE 7.14
CFD Modelling Parameters Applied during 3D Fluid-Solid Coupling of Bridge Pier Scour in River Channel

Parameter	Run 1 (MQSI method)	Run 2 (EIM method)
Turbulence model	RNG model	RNG model
Volume of fluid advection	Automatic (Split or Unsplit Lagrangian)	Automatic (Split or Unsplit Lagrangian)
Momentum advection	Second-order monotonicity	Second-order monotonicity
Time step control	Stability and convergence	Stability and convergence
Time duration per iteration	3 s	3 s
Mesh size	1.0 m	1.0 m
Computational method	MQSI	EIM
Net uplift pressure coefficient	0.20	–
Mass strength number M_s	–	17.48
Block size number K_b	–	11.98
Disc. shear strength number K_d	–	0.16
Relative joint structure number J_s	–	1.05
Stream power equation used	–	$\tau \cdot V$ (shear stress × velocity)
Applicable stream power coeff.	–	1.0
Pressure solver	Implicit (GMRES)	Implicit (GMRES)
Air entrainment model	Yes, with flow bulking and buoyancy (drift flux)	Yes, with flow bulking and buoyancy (drift flux)
Air entrainment coefficient	0.50	0.50
Roughness height bedrock (k_s)	0.10 m	0.10 m
Layer height for fluid-solid iterations	1.0 m	1.0 m

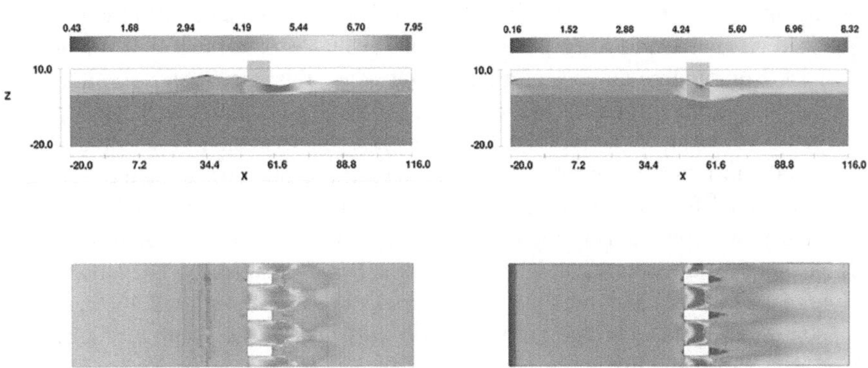

FIGURE 7.93 Flow velocities in the river reach for run 1: (a) at start of computation (left graphs); (b) at end of computation (right graphs).

Second, Figure 7.94 illustrates the bedrock scour computed for both run 1 (MQSI method, left-hand side) and run 2 (EIM method, right-hand side).

Run 1 makes use of the MQSI method modelling rock block detachment using time-averaged flow velocities at the water-rock interface to compute net lift forces on

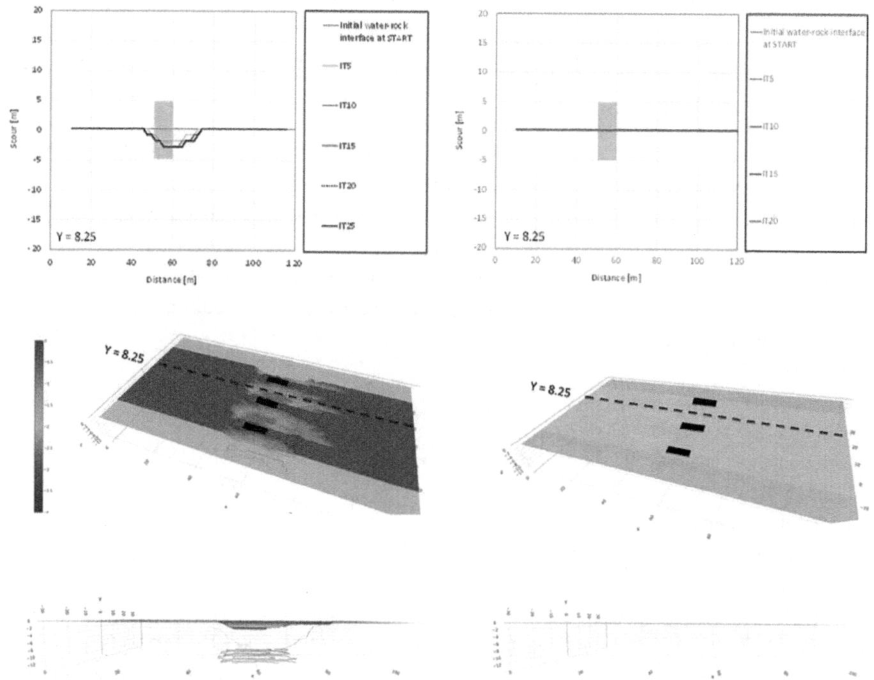

FIGURE 7.94 Bridge pier scour computed using 3D fluid-solid coupling using MQSI method (left-hand side, run 1) and EIM method (right-hand side, run 2).

rock blocks. Net uplift pressure coefficients of 0.20 and 0.30 have been applied during the runs (i.e. values to be multiplied by $\rho V^2/2$ to obtain the net lift force).

Based on analytical and empirical evidence (Pells 2016; Reinius 1986), as well as feedback from a large number of plunging jet cases, this corresponds to an average value for computing lift forces on rock blocks protruding into the flow. Scour depths of 3–4 m (C_p = 0.20) and 5–6 m (C_p = 0.30) are observed in between the bridge piers, starting slightly upstream of the piers and extending significantly downstream of the piers.

For a net uplift coefficient of 0.20, Figure 7.95 illustrates the computed flow velocities and flow depths during the different fluid-solid iterations. Scour forms at the bridge pier for flow velocities higher than the critical flow velocity. The latter equals 5.1 m/s for a 0.20-m-thick rock block and a rock mass density of ρ = 2,360 kg/m^3.

Flow depths at the bridge piers increase with increasing scour formation and decreasing flow velocities.

Run 2 makes use of the EIM method, based on the erodibility index of the bedrock and the river channel stream power, whereby the applicable stream power is computed as the product of the bottom shear stress τ and the bottom flow velocity V. Figure 7.94 (right-hand graphs) shows that no scour formation is predicted by the fluid-solid computations.

The computed shear stress, flow velocity and stream power are presented in Figure 7.96 along a longitudinal section (Y = 8.25). Stream power values are 0.10–0.15 kW/m^2 upstream of the bridge and mount up to 3 kW/m^2 near the downstream end of

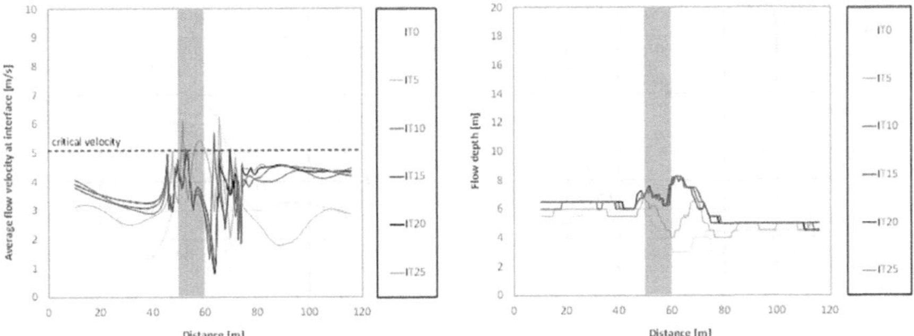

FIGURE 7.95 Bridge pier scour computed using 3D fluid-solid coupling using MQSI method ($C_p = 0.20$): evolution of flow velocities (left-hand graph) and flow depths (right-hand graph) in between the piers (for $Y = 8.25$).

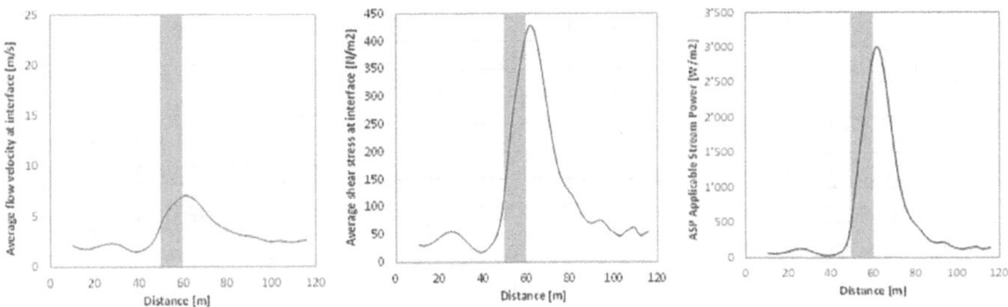

FIGURE 7.96 Bottom shear stress, bottom flow velocity and stream power computed by 3D CFD in between bridge piers in a river channel.

the bridge. Hence, for an erodibility index of the bedrock of 35.18, the critical stream power equals 14.45 kW/m^2. The stream power computed in between the bridge piers is thus not sufficient to start scour formation of the bedrock.

Comparison with Available Computational Methods

The 3D CFD computed scour results are compared with other CFD runs, performed for different parametric settings, and with different analytical methods available to practitioners for rock scour computations near bridge piers in rivers, i.e.:

- *Erodibility Index method by Annandale and Smith (2001)*: Based on flume tests performed by Johnson and Jones (1992), critical shear stress values were recorded in bridge pier scour holes using fine-grained bottom material, showing that stream power decreases with increasing scour formation, until the critical stream power is reached. Smith (1994) and Smith et al. (1999) deduced from this an exponential relationship expressing the ratio of stream power inside the scour hole SP to stream power in the upstream river channel SP_a as a function of non-dimensionless scour depth z_{sc}/B, written as follows:

$$\frac{SP}{SP_a} = 8.42 \cdot e^{\left(-0.712 \cdot \frac{z_{sc}}{B}\right)} \tag{7.6}$$

in which B stands for the pier width perpendicular to the flow direction. The applicable unit stream power in the upstream channel is thereby computed as follows (George et al. 2022):

$$SP_a = \frac{7.853}{1{,}000} \cdot \left(\frac{f}{8}\right)^{3/2} \cdot \rho \cdot V^3 \tag{7.7}$$

in which the friction factor of the channel bottom is written as:

$$f = \left(\frac{1}{2 \cdot \log\left(12 \cdot \frac{R}{k}\right)}\right)^2 \tag{7.8}$$

where the hydraulic radius R may be taken equal to the flow depth h for large river channels compared to the flow depth, and k is the equivalent roughness coefficient according to Darcy-Weisbach.

- *Comprehensive Scour Model by Bollaert (2010) and published in Keaton et al. (2012)*: Bollaert (2010) computed 1D transient pressure pulses and related uplift forces and impulsions on distinct rock blocks at bridge piers, allowing to express the critical flow velocity and the non-dimensional maximum scour depth at the bridge pier as a function of several parameters, such as the height of the block, the protrusion of the block into the flow, the angle of flow attack and the slope of the river channel. The results are available as a series of curves and spreadsheets in the appendices of the NCHRP-717 research report of the FHWA.
- *HEC-18 Pier Scour Equation for non-cohesive material (Arneson et al. 2012)*: Bridge pier scour equation generally used and recommended in the HEC-18 engineering circular of the FHWA, valid for granular (non-cohesive) soils:

$$\frac{z_{sc}}{B} = 2.0 \cdot K_1 \cdot K_2 \cdot K_3 \cdot \left(\frac{h}{B}\right)^{0.35} \cdot Fr^{0.43} \tag{7.9}$$

where B stands for the width of the bridge pier perpendicular to the flow direction, h is the flow depth directly upstream of the pier, Fr is the Froude number directly upstream of the pier, K_1 is a correction factor for pier nose shape, K_2 is a correction factor for angle of attack of flow and K_3 is a correction factor for bed condition.

Table 7.15 compares the critical flow parameters following the different computational methods. For Erodibility Index-based methods, the critical stream power at the bedrock equals 14.45 kW/m². For Quasi-Steady Impulsion-based methods, the critical flow velocity at the bedrock equals 5.1 m/s respectively 4.2 m/s for net uplift pressure coefficients of 0.20 respectively 0.30. For computations based on 1D transient flow modelling involving turbulent wall pressures and their amplifications in rock joints (CSM), the critical flow velocity at the bedrock is less than 3.6 m/s, the latter

TABLE 7.15
Critical Flow Parameters Following Different Computational Methods for Bridge Pier Scour in River Channel

		MQSI			
EIM		$C_{p,up} = 0.20$	$C_{p,up} = 0.30$	$C_{p,up} = 0.40$	**CSM**
Erod. Index	**Critical *SP***	**Critical Vel.**	**Critical Vel.**	**Critical Vel.**	**Critical Vel.**
EI	**SPc**	**Vc**	**Vc**	**Vc**	**Vc**[a]
Bedrock	At bridge	At bridge	At bridge	At bridge	At bridge
–	kW/m^2	m/s	m/s	m/s	m/s
35.2	14.45	5.1	4.2	3.6	<3.6

[a] For block height = 25 cm.

value being taken from curves valid for a block height of 0.25 m, no block protrusion and small channel slopes.

The rock mass characteristics, and thus also the critical flow parameters, are assumed constant with depth.

The scour results obtained by applying these different methods are summarized in Table 7.16. First, the EIM method by Annandale and Smith (2001) predicts no scour formation. The upstream stream power is thereby computed based on the bottom shear stress and the flow velocity, resulting in a max. value at the bridge pier of 2.13 kW/m^2, which is significantly lower than the critical value of 14.45 kW/m^2.

Second, the 3D CFD modelling using the EIM approach (i.e. run 2) also predicts no scour formation. Similarly to the first approach, the stream power is thereby computed based on the bottom shear stress and the flow velocity near the bedrock, and reaches a max. value of 3 kW/m^2, again significantly lower than the critical value.

Next, comparison is made with 3D CFD modelling using the EIM approach in which the stream power is computed based on the mean and RMS values of dynamic pressures at the water-rock interface, together with the local flow velocity near the bedrock. This approach is generally used for plunging jets, but may be considered adequate also for any highly turbulent flows involving pressure fluctuations, such as the horseshoe vortex generated by a bridge pier. The so computed stream power is one order of a magnitude higher than the one based on shear stress, and a max. scour depth of 6 m is computed at the bridge pier. The max. computed stream power in the scour hole may locally increase with increasing scour formation, because of the additional flow turbulence generated by the shape evolution of the scour hole. Nevertheless, when the scour hole shape approaches its equilibrium shape, the max. stream power becomes less than the critical one.

The 3D CFD modelling based on flow velocities at the water-rock interface (MQSI), generating quasi-steady net lift forces on protruding rock blocks, has been performed for net uplift pressure coefficients of 0.20 (i.e. run 1) respectively 0.30. Compared to feedback from plunging jet cases, this corresponds to average values. Corresponding max. scour depths are 3–4 m respectively 5–6 m, values in line with

TABLE 7.16
Scour Depths at Bridge Pier Founded on Rock, Computed Following Different Computational Methods

Scour Depth		EIM (Annandale and Smith 2001)				EIM (3D CFD, Fluid-Solid Coupling)						MQSI (3D CFD, Fluid-Solid Coupling)				CSM (Bollaert 2010)	HEC-18 (2012)
		$SP = f(\tau, V)$				$SP = f(\tau, V)$			$SP = f(C_p, C'_p, V)$			$C_{p,up} = 0.20$		$C_{p,up} = 0.30$			
		Applicable SP			Sc.	Applicable SP		Sc.	Applicable SP		Sc.	Vel.	Sc.	Vel.	Sc.	Sc.	Sc.
z_{sc}	z_{sc}/B	SP_a	SP/SP_a	SP	?	SP_a	SP	?	SP_a	SP	?	V	?	V	?	?	?
abs.	Rel.	Up	Ratio	At bridge		Up	At bridge		Up	At bridge		At br.		At br.			
m	m/m	kW/m²	–	kW/m²	–	kW/m²	kW/m²	–	kW/m²	kW/m²	–	m/s	–	m/s	–	–	–
0	0.00	0.25	8.42	2.13	NO	0.1–0.15	3.00	NO	0–2.65	17.60	YES	6.6	YES	6.6	YES	YES	YES
1	0.25	0.25	7.05	1.78	NO	–	–	NO	0–2.65	39.50	YES	6.1	YES	5.9	YES	YES	YES
2	0.50	0.25	5.90	1.49	NO	–	–	NO	0–2.65	38.80	YES	5.7	YES	5.3	YES	YES	YES
3	0.75	0.25	4.94	1.25	NO	–	–	NO	0–2.65	46.00	YES	5.1	YES	4.6	YES	YES	YES
4	1.00	0.25	4.13	1.04	NO	–	–	NO	0–2.65	32.50	YES	–	NO	4.3	YES	YES	YES
5	1.25	0.25	3.46	0.87	NO	–	–	NO	0–2.65	29.50	YES	–	NO	4.25	YES	YES	YES
6	1.50	0.25	2.89	0.73	NO	–	–	NO	0–2.65	15.00	YES	–	NO	4.2	NO	YES	YES
7	1.75	0.25	2.42	0.61	NO	–	–	NO	–	–	NO	–	NO	–	NO	YES	YES
8	2.00	0.25	2.03	0.51	NO	–	–	NO	–	–	NO	–	–	–	NO	YES	YES
9	2.25	0.25	1.70	0.43	NO	–	–	NO	–	–	NO	–	–	–	NO	YES	YES
10	2.50	0.25	1.42	0.36	NO	–	–	NO	–	–	NO	–	–	–	NO	YES	NO

the scour potential of 6 m computed by the EIM using dynamic pressure fluctuations to compute the stream power.

By using net uplift forces and impulsions computed by 1D transient pressure amplifications in rock joints (CSM, Bollaert 2010), significant scour potential is derived from the available design curves (even without accounting for block protrusion).

Finally, scour potential estimated based on the HEC-18 pier scour equation for non-cohesive soils is 8–9 m and should a priori represent a safe-side scour assumption for rock scour.

Synthesis

The present case study of rock scour potential generated by a bridge pier in a river channel generates results going from no scour potential to very large scour potential, depending on the applied computational method. This outcome appears disappointing at first sight to a practitioner. However, a more profound analysis of the applicability and main parameters and basic assumptions of each of the methods provides useful additional insight.

First, all the methods that predict no scour potential, analytical or CFD, are based on the Erodibility Index Method with a stream power computed by using the bottom shear stress. The concept of using the bottom shear stress as a proxy for scour triggering, however, is questionable for cases where the flow environment may be qualified by locally intense turbulence generating dynamic pressure fluctuations that dissipate the energy, such as plunging jets, hydraulic jumps, submerged wall jets and so on. It seems a priori very plausible to range the 3D flow turbulence generated by a bridge pier, i.e. horseshoe vortex and wake, into this category of flows also. As such, shear stress-based methods may not be suited to describe these flows.

This seems to be confirmed by the results obtained based on the Erodibility Index Method making use of time-averaged and fluctuating dynamic pressures to compute the stream power. The computed scour potential is in agreement with the potential computed by the MQSI method based on quasi-steady lift forces on blocks that protrude into the flow and deviate the flow velocity at the water-rock interface.

The CSM method based on 1D transient pressure pulses and their amplification in rock joints seems to provide large scour predictions, especially when the effect of turbulent pressure fluctuations and their amplifications in joints is combined with quasi-steady lift forces by block protrusion. When discarding the latter (i.e. using design curves for $C_p = 0$), the critical velocity of the design curves is in close agreement with the critical velocity of the MQSI method for a net uplift coefficient of 0.40–0.50 (i.e. higher range of values). Second, the design curves expressing the scour potential as a function of the velocity in the upstream channel seem difficult to apply quantitatively, and only inform about the probability of scour onset.

Moreover, uncertainties in flow and rock mass input parameters may significantly affect the scour results, and the lack of well-documented case studies makes parameter calibration of the available rock scour computational methods (EIM, CFM, MDI, MQSI, DP) a difficult task to perform. When making use of available parametric calibrations performed in the field of turbulent flows at dams, however, the different physics-based rock break-up methods used by the 3D CFD fluid-solid coupling (MQSI, EIM) seem to provide coherent scour results.

As such, the following guidelines for rock scour computations at bridge piers are formulated for the practitioner in this field:

- Perform site investigations to determine the necessary flow and rock mass parameters with sufficient accuracy.
- Make a first-hand assessment by using the EIM by Annandale and Smith (2001), the CSM method by Bollaert (2010) and the HEC-18 equation for non-cohesive soils (Arneson et al. 2012), keeping in mind that the latter two methods rather provide safe-side scour estimates, and that the former method may potentially significantly underestimate scour potential.
- Perform a parametric sensitivity analysis.
- In case scour potential exists and may affect bridge stability, perform a more detailed scour analysis by 3D CFD modelling of the flow parameters and related fluid-solid coupling with the rock mass, using different rock mass break-up methods.
- In the absence of parametric calibration, use parameter values as recommended for turbulent flows in plunge pools and hydraulic jumps.

REFERENCES

Annandale, G.W., "Erodibility", *Journal of Hydraulic Research*, 33, 471–494, 1995.

Annandale, G.W., *Scour technology: mechanics and engineering practice*, McGraw-Hill Professional, New York, 1 ed., 2006.

Annandale, G.W. and Smith, S., "Calculation of Bridge Pier Scour Using the Erodibility Index Method", *Report No. CDOT-DTD-R-2000-9*, Colorado DOT, Denver, US, 2001.

Arneson, L.A., Zevenbergen, L.W., Lagasse, P.F. and Clopper, P.E., "Evaluating scour at bridges - 5th edition", *HEC-18 circular, Publication No. FHWA-HIF-12-003, Federal Highway Administration,* US, 2012.

Bohrer, J.G., Abt, S.R. and Wittler, R.J., "Predicting Plunge Pool Velocity Decay of Free Falling, Rectangular Jet", *Journal of Hydraulic Engineering*, ASCE, 124, 10, 1043–1048, 1998.

Bollaert, E.F.R., "Transient water pressures in joints and formation of rock scour due to high-velocity jet impact", *PhD Thesis*, LCH-EPFL, Lausanne, Switzerland, 2002.

Bollaert, E.F.R., "Rock scour at hydraulic structures: a practical engineering approach", *Geo-Strata*, 2010.

Bollaert, E.F.R., "Wall Jet Rock Scour in Plunge Pools: A Quasi-3D Prediction Model", *International Journal on Hydropower & Dams*, 131, 4, 153–165, 2012.

Bollaert, E.F.R., "The Rocsc@r Cloud: An Innovative Digital Platform to Compute and Record Rock Scour", *International Journal on Hydropower & Dams*, 28, 5, 60–70, 2021.

Bollaert, E.F.R. and Schleiss, A.J., "Physically Based Model for Evaluation of Rock Scour due to High-Velocity Jet Impact", *Journal of Hydraulic Engineering*, 131, 3, 153–165, 2005. https://doi.org/10.1061/(ASCE)0733-9429(2005)131:3(153)

Bollaert, E.F.R., Stratford, C.E., and Lesleighter, E.J., "Numerical modelling of rock scour, Case study of Wivenhoe Dam", Int. Conf. on Scour and Erosion, Perth, Australia, 2014.

Bosman, A., "High dam scour hole geometry prediction for fully developed jets plunging into shallow pools on bedrock", *PhD Dissertation*, Civil Engineering Faculty, Stellenbosch University, Cape Town, South Africa, 2021.

Bosman, A. and Basson, G., "Physical Model Study of Bedrock Scour Downstream of Dams Due to Spillway Plunging Jets", *Journal of the South African Institution of Civil Engineering*, 62, 3, 36–52, 2020.

Burkhardt, M., Kim, E. and Nelson, P., "EMI Database Analysis Focusing on Relationship between Density and Mechanical Properties of Sedimentary Rocks", *Geomechanics and Engineering*, 14, 5, 491–498, 2018, https://doi.org/10.12989/gae.2018.14.5.491

Calitz, J.A., "Investigation of air concentration and pressures of a stepped spillway equipped with a crest pier", *Master's Thesis*, Stellenbosch, South Africa, Stellenbosch University, 2015.

Capuozzo, L. and Jiménez, O., "Model and Prototype Studies for the Chucàs Hydro Scheme Costa Rica", *International Journal of Hydropower and Dams*, 24, 6, 52–56, 2017.

Castillo, LG., "Pressures characterization of undeveloped and developed jets in shallow and deep pool", Congress-International Association of Hydraulic Engineering and Research, 32(2), (p. 645). Venice, Italy, 2007.

Debecker, B., Tavallali, A. and Vervoort, A., "Probability Distribution of Rock Properties: Effect on the Rock Behaviour", in *6th International Symposium on Ground Support in Mining and Civil Engineering Construction,* KU Leuven, Research Unit Mining, Belgium, 2010.

Der Kiureghian, A. and Liu, P.L., "Structural Reliability under Incomplete Probability Information", *Journal of Engineering Mechanics Engineering Mechanics*, 112, 1, 85–104, 1985.

Dey, S. and Sarkar, A., "Scour Downstream of an Apron Due to Submerged Horizontal Jets", *Journal of Hydraulic Engineering*, 132, 3, https://doi.org/10.1061/(ASCE)0733-9429(2006)132:3(246), 2006.

Ferguson, R.I., Sharma, B.P., Hardy, R.J., Hodge, R.A. and Warburton, J., "Flow resistance and hydraulic geometry in contrasting reaches of a bedrock channel", *Water Resources Research*, 53, 3, pp. 2278-2293, https://doi.org/10.1002/2016WR020233, 2017.

Fiorotto, V. and Rinaldo, A., "Fluctuating Uplift and Lining Design in Spillway Stilling Basins", *Journal of Hydraulic Engineering*, (1992)118:4(578), 1992.

Genske, D. and Walz, B., "Probabilistic Assessment of the Stability of Rock Slopes", *Structural Safety*, 9, 179–195, 1991.

George, M.F., *PhD Dissertation*. University of California, Berkeley, 2015.

George, M.F., Sitar, N. and Der Kiureghian, A., "System Reliability Analysis of Rock Scour," in *Proceeding of the 34th United States Society on Dams (USSD) Conference*, San Francisco, CA, USA, pp. 478–494, 2014.

George, M.F., Christiansen, C., Rickel, A., Israel, B. and Annandale, G.W., "Erodibility Evaluation of an Unlined Rock Spillway: Comparison Between the Erodibility Index Method and a New Method Based on Block Theory", in *Proceedings of the 4th International Seminar on Dam Protections against Overtopping*, Madrid (Spain), 2022. https://doi.org/10.26077/56a7-33b7

Hoek, E., "Strength of rock and rock masses", *ISRM News Journal*, 2(2), 4–16, 1994

Hoek, E., *Practical rock engineering*, 2006. Available at: https://www.rocscience.com/assets/resources/learning/hoek/Practical-Rock-Engineering-Full-Text.pdf

Hsu, S.C. and Nelson, P., "Material Spatial Variability and Slope Stability for Weak Rock Masses", *Journal of Geotechnical and Geoenvironmental Engineering*, 132, 2, 183–193, 2006.

Inoue, T., Izumi, N., Shimizu, Y. and Parker, G., "Interaction among alluvial cover, bed roughness, and incision rate in purely bedrock and alluvial-bedrock channel", *Journal of Geophysical Research*, 119, 10, pp. 2123–2146. https://doi.org/10.1002/2014JF003133, 2014.

ISRM (International Society for Rock Mechanics), "Suggested Methods for the Quantitative Description of Discontinuities in Rock Masses", *International Journal of Rock Mechanics and Mining Sciences & Geomechanics Abstracts*, 15, 319–368, 1978.

Johnson, P. A. and Jones, J. S., "Shear Stress at the Base of Bridge Pier", Transportation Research Record 1350, pp. 14–18, Transportation Research Board National Research Council, 1992.

Keaton, J.R., Mishra, S.K. and Clopper, P.E., "NCHRP Report 717: Scour at Bridge Foundations on Rock", *Transportation Research Board*, Washington, DC, 2012.

Lesleighter, E., Stratford, C. and Bollaert, E., "Plunge Pool Rock Scour Experiences and Analysis Techniques", IAHR Congress, Chengdu, China, 2013.

Liu, P., Dong, J. and Yu, C., "Fluctuating uplift on rock blocks at the bottom of a scour pool by overfall jets", *Science China Technological Sciences*, 41, 130–139. https://doi.org/10.1007/BF02919675, 1998.

Maleki, S. and Fiorotto, V., "Blocks Stability in Plunge Pools under Turbulent Rectangular Jets", *Journal of Hydraulic Engineering*, 145, 4, 2019. https://doi.org/10.1061/(ASCE)HY.1943-7900.0001573

Pells, S., "Erosion of rock in spillways", PhD Dissertation, School of Civil and Environmental Engineering Faculty of Engineering, University of New South Wales, 2016.

Rajaratnam, N., "Erosion by Plane Turbulent Jets", *Journal of Hydraulic Research*, 19: 4, 339 — 358, https://doi.org/10.1080/00221688109499508, 1981

Reinius, E., "Rock Erosion", *International Journal on Water Power and Dam Construction*, 38, 6, 43–48, 1986.

Smith, S.P., "Preliminary Procedure to Predict Bridge Scour in Bedrock", Colorado 561 Department of Transportation, 1994.

Smith, S.P., Annandale, G.W., Johnson, P.A., Jones, J.S. and Umbrell, E.R., "Pier Scour in Resistant Material: Current Research on Erosive Power", in *Proceedings of Managing Water: Coping with Scarcity and Abundance, 27th Congress of the International Association of Hydraulic Research*, San Francisco. CA, pp. 160–165, 1999.

Valero, D., Bung, D., Crookston, B., and Matos, J., "Numerical investigation of USBR type III stilling basin performance downstream of smooth and stepped spillways", in B. Crookston and B. Tullis (Eds.), *Hydraulic Structures and Water System Management. 6th IAHR International Symposium on Hydraulic Structures*, Portland, OR, 27–30 June (pp. 652–663), 2016. doi:10.15142/T340628160853.

van Schalkwijk, A., "Minutes - Erosion of Rock in Unlined Spillways", ICOLD, Q.71 R.37, 1056–1062, 1994.

van Schalkwijk, A., Jordaan, J. and Dooge, N., "Erosion of Rock in Unlined Spillways," *International Commission on Large Dams*, Paris. Q.71–E.37, 555–571, 1994b.

Wei, W., Xu, W., Deng, J. and Liu, B. "Experimental Study of Impact Pressures on Deep Plunge Pool Floors Generated by Submerged Inclined Jets with Controlled Aeration", *Journal of Hydraulic Engineering*, Vol. 146, Issue 4, https://doi.org/10.1061/(ASCE)HY.1943-7900.0001704, 2020.

Yan, C., Ding, D., Tang, Y. and Bi, Z., "Probability Distribution of Strength Parameters and Deformation Parameters of Surrounding Rock and Reliability Analysis", *Geotechnical Engineering*, 31, S2, 349–353, 2010.

Závacký, M., Štefaňák, J., Horák, V. and Miþa, L., "Statistical Estimate of Uniaxial Compressive Strength of Rock Based on Shore Hardness", *Procedia Engineering*, 191, 248–255, 2017. https://doi.org/10.1016/j.proeng.2017.05.178

Index

For Product Safety Concerns and Information please contact our EU representative GPSR@taylorandfrancis.com Taylor & Francis Verlag GmbH, Kaufingerstraße 24, 80331 München, Germany

Batch number: 10397790

Printed by Printforce, the Netherlands